Principles of PLANT VIROLOGY

Genome, Pathogenicity, Virus Ecology

Principles of
PLANT VIROLOGY
Genome, Pathogenicity, Virus Ecology

S. ASTIER
Institut National de la Recherche Agronomique (INRA)
Versailles, France

J. ALBOUY
Institut National de la Recherche Agronomique (INRA)
Versailles, France

Y. MAURY
Institut National de la Recherche Agronomique (INRA)
Versailles, France

C. ROBAGLIA
Université de la Méditerranée, Aix-Marseille II, Faculté des Sciences de Luminy
Marseille, France

H. LECOQ
Institut National de la Recherche Agronomique (INRA)
Avignon, France

Science Publishers
Enfield (NH) Jersey Plymouth

SCIENCE PUBLISHERS
An imprint of Edenbridge Ltd., British Isles.
Post Office Box 699
Enfield, New Hampshire 03748
United States of America

Website: *http://www.scipub.net*

ISBN 978-1-57808-316-9 (HC)
978-1-57808-503-3 (PB)

Library of Congress Cataloging-in-Publication Data

[Principes de virologie végétale. English]
Principles of plant virology/S. Astier ... [et al.].
p. cm.
Includes bibliographical references and index.
ISBN 978-1-57808-316-9
1. Virus diseases of plants. 2. Plant viruses. 3. Plant viruses--Control. I. Astier, S.

SB736.P65 2006
632'.8--dc22

2006042180

Published with arrangement with INRA, Paris

Translation of: **Principes de virologie végétale; génome, pouvoir pathogène, écologie des virus.**
French edition: © INRA, Paris, 2001
ISBN 2-7380-0937-9

Revised and updated by S. Astier, C. Robaglia and H. Lecoq for the English Edition in 2006

Corresponding Author: **Hervé Lecoq**, INRA, Station de Pathologie Végétale, Domaine Saint Maurice, BP94, 84140 Montfavet, France.
E-mail: Herve.Lecoq@avignon.inra.fr

Ouvrage publié avec le concours du Ministère Français de la Culture—Centre National du Livre. (This book has been published with the help of French Ministère de la Culture—Centre National du Livre)

Published by Science Publishers, Enfield, NH, USA
An imprint of Edenbridge Ltd.
Printed in India

Acknowledgment

The authors are particularly grateful to Isabelle JUPIN (Institut Jacques Monod, Universités Paris VI and VII) for her exhaustive analysis of the French version of this book and for her valuable propositions to improve its content.

We acknowledge all the colleagues who were involved in the French edition and particularly Gilbert MOLIN (INRA, Versailles) for the drawing of the virus genetic maps included in Chapter 15.

We wish to thank Cécile DESBIEZ (INRA, Avignon), Ali ELJACK (University of Gezira, Wad Medani, Sudan), Youssif FADLALLA (University of Gezira, Wad Medani, Sudan), Hervé HUET (Bio-Oz, Ashkelon, Israel), Benny RACCAH (ARO, Bet Dagan, Israel) and Mark TEPFER (ICGEB, Ca'Tron, Italy) for their valuable help in reviewing parts of the English version of the manuscript.

This book is dedicated to all the researchers who contributed to the outstanding development of plant virology and to those who will continue this exciting experience in the future.

Contents

THE VIRUS IN THE AGRO-ENVIRONMENT

Introduction

First steps in virology

The first virus discovered, at the end of the 19th century, was *Tobacco mosaic virus* (TMV), through the complementary contributions of three scientists. Adolf Mayer published in 1886 a detailed description of a highly contagious disease of tobacco crops in the Netherlands and proposed to name it "tobacco mosaic". He suggested that the causal agent could be bacteria that he was not able to see under the microscope or to cultivate *in vitro*. Dimitri Ivanovsky carried out a decisive experiment in 1886 in Russia showing that "the mosaic disease retains its infectious qualities even after filtration through Chamberland filtration candles". As bacteria are retained by such filters, a new concept of filterable pathogens was born.

Martinus Beijerinck in the Netherlands carried on the research work initiated by Mayer. In a paper published in 1898, he named the agent of the tobacco disease "virus", and he confirmed it to be filterable ("a soluble living germ") and definitively different from bacteria. He also showed that the "virus" multiplies in young tobacco leaves and suggested a possible migration through the phloem.

Human infections today known as viral diseases were identified a very long time ago, as shown by records of poliomyelitis and smallpox in ancient Egypt. In China, by the 17th century, variolation was developed to limit smallpox outbreaks. In the 19th century, vaccination against smallpox was extensively used. In his studies on rabies, Louis Pasteur named the agent "a virus", a term describing a poison or infectious matter. But the nature of the infectious agent was not understood as completely different from bacteria. Following Beijerinck's results on tobacco mosaic, filterable agents were recognized as responsible for foot and mouth disease of cattle by E. Loeffler and P. Frosch in 1898, for yellow fever by W. Reed and co-workers in 1901, and for poliomyelitis by K. Landsteiner and E. Popper in 1909.

Numerous viral diseases have been described since 1900, the viral nature of the agent being assessed by its filterability. In plants, attention was first focused on the description of symptoms, host ranges and indicator plants, and transmission by mechanical means or by insects. Studies on the structure and composition of viral particles started in 1935 when W. Stanley published in *Nature* a paper showing that particles of TMV could be obtained as a proteinaceous crystal that was still infectious. Later, Bawden et al. (1936) showed that the TMV particle was, more precisely, a nucleoprotein containing a pentose nucleic acid and Kausche et al. (1939), using the electron microscope, could visualize for the first time the morphology of a virus particle, the rod-shaped TMV.

It is now known that viruses multiply in the widest possible variety of organisms: in archaeobacteria, bacteria (bacteriophages), algae, fungi, plants, invertebrates, and vertebrates. Viruses are responsible for more than half of the infectious diseases that affect human beings (including influenza, smallpox, poliomyelitis, and AIDS) and some cancers. They are also responsible for many diseases that reduce yields of cultivated plants. The study of viruses developed considerably during the 20th century, particularly the 1990s.

What are the specific characters of viruses, their relationships with their hosts, and the mechanisms by which they contaminate new cells and new hosts? How can viral diseases be diagnosed and controlled? How can viruses be classified and their evolution described? These are the questions that we have addressed in this book, discussing our present knowledge of the field of plant viruses (limited here essentially to viruses in Angiosperms) and proposing matter for further study.

Readers of this book are invited to extend their investigations by consulting the original works listed in the bibliographic references. It should be kept in mind that any scientific description is bound to evolve, and that every result opens the door to new questions that, in turn, will push back the boundaries of the unknown.

■ The viral particle: a simple geometric architecture

In relation to cells (which measure about 10 µm for an animal cell, 100 µm for a plant cell), viral particles are smaller by several orders of magnitude. They are measured in nanometres (10^{-9} m). Viruses are invisible under a light microscope, and their observation under an electron microscope shows particles of a wide variety of sizes and shapes (Chapter 1, Fig. 1.1). They are surrounded by a protein capsid, which is an assemblage of elements organized in regular geometric forms (helix or icosahedron), and sometimes also a lipid membrane (Chapter 1).

■ The viral genome: a relatively small amount of information

The analysis of elements constituting a virus shows a very simple basic structure: a nucleic acid (DNA or RNA) surrounded by a protein shell (the capsid), which is sometimes covered by an envelope. The quantity of genetic information carried by the virus, estimated by the number of genes, is much smaller than that of the smallest of cells. Plant viruses generally code for 4 to 12 proteins and the more complex animal viruses up to 250 proteins. Beyond this number, we enter into the domain of cellular organisms and there is an increase in the quantity of information that corresponds mainly to the set of some 500 genes required for protein synthesis. This leap indicates the difference between acellular elements, including viruses, and cellular organisms (phytoplasma, 625 genes; *Escherichia coli*, 2800; yeast, 7000; *Drosophila*, 13,600; *Arabidopsis thaliana*, 25,500; man, about 30,000).

To express these proteins, the virus depends entirely on the host system of protein synthesis. To invade this system and exploit it is its primary need (Chapter 2).

The support of information: DNA or RNA

While cellular organisms have double-stranded DNA genomes, the viral genome is made up of DNA or RNA. Viruses are the only taxonomic group in which the genomes present such a diversity of nature (DNA or RNA) and structure (single or double strand, monopartite or multipartite). The viral genera are divided very unequally into six groups defined on the basis of the nature of the genome (Fig. 1). Plant viruses are present in five of these six groups:

— Single-stranded positive RNA: most plant viruses are found in this group. The genomic RNA is "positive" (messenger).
— Single-stranded negative RNA: the strand complementary to the parental RNA is the messenger.

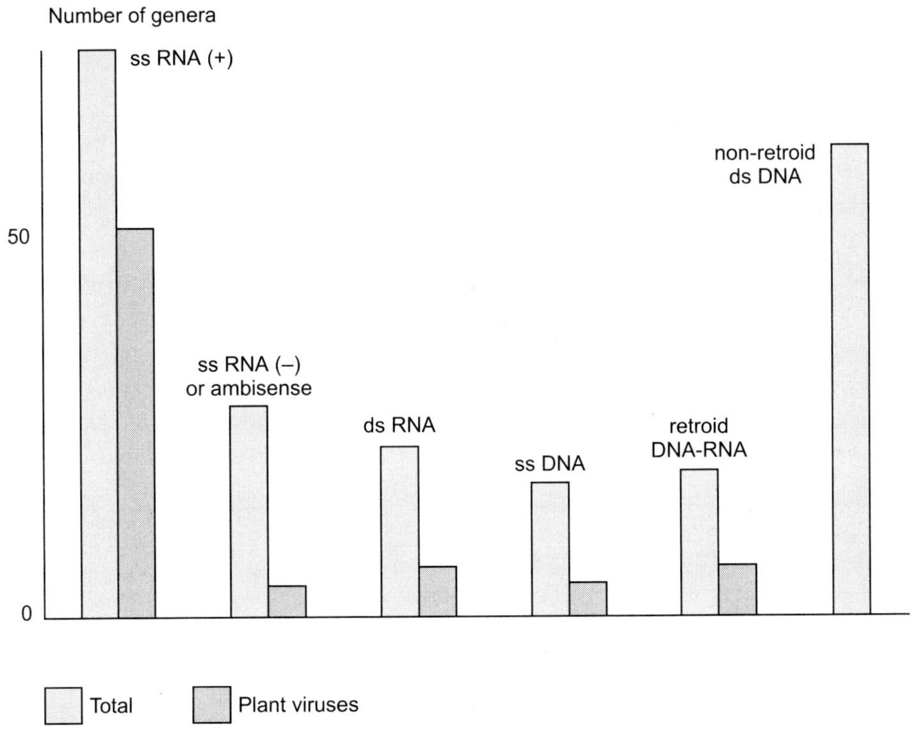

Figure 1 Genera of viruses

— Double-stranded RNA: the genome is made up of several segments of double-stranded RNA.
— Single-stranded DNA: the genome is made up of one or several molecules of circular single-stranded DNA.
— Double-stranded retroid DNA: the genome is made up of double-stranded circular DNA and the replication includes a reverse transcription step.

Viruses are intracellular parasites

The viral particle has no metabolism; it is inert. To become active and multiply, the virus needs to penetrate a cell and mobilize the metabolic resources for its own replication. All the energy it needs is provided by the cellular metabolites. The physicochemical properties of the viral particle and the characteristics of its relationship with a living cell together define a virus as an "integrated biological system" (Van Regenmortel and Fauquet, 2000).

When the virus has successfully entered the cell, it releases its genetic information. If the cell provides the requisite machinery, a cycle begins during which the viral proteins and copies of the genome are produced. At the end of the cycle, many daughter particles are assembled from the pool of components: nucleic acid and capsid, as well as other viral proteins in some cases.

There are various stages in the viral cycle, and they often overlap:

— *Penetration* into the cell, *decapsidation.*
— *Translation.* The viral message is translated by cellular ribosomes, which synthesize the structural and non-structural viral proteins (Chapter 2). These proteins are involved in one or several functions (replication, movement, transmission, protection, etc.).
— *Replication.* This is the process that ends in the multiplication of the incoming nucleic acid into several copies; this process involves the formation of complementary chains. One or several viral proteins are necessary for the replication. They participate directly or indirectly in the replication (Chapter 3). Because of the simplicity of their structure and their specific enzymes, which escape certain regulations by the host, viruses can multiply much faster than the cells that harbour them. They use the metabolism and resources of the cells and rarely cause cell death. Their presence is often manifested in external symptoms such as mosaic, yellowing, discolouring, rolling, or stunting as well as internal symptoms such as inclusions (Chapter 9).
— *Encapsidation.* The components, nucleic acid and proteins, are assembled spontaneously to form new particles. A viral particle cannot grow or divide.

From the first infected cells, the virus invades the entire plant by using intercellular communications that carry out physiological exchanges between cells (Chapter 4). This invasion does not occur without a reaction from the plant, which in each cell produces mechanisms to degrade the viral messenger (Chapter 5).

Sometimes, the plant detects the infection at an early stage and manages to circumvent it (Chapter 6). The plants sensitive to a virus make up the host range of that virus.

Transmission of viruses

In nature, most plant viruses are selected continuously for transmission by vectors. Viruses are strictly obligatory parasites that, in order to multiply, must enter living cells. To perpetuate themselves, viruses contaminate other individuals derived from the infected plant (internal pathway, vertical transmission) or infect new individuals of the same species and other plant species (external pathway, horizontal transmission), most often through the intermediary of a vector (Chapter 8).

Diagnosis of viral diseases

The diagnosis of viral diseases is based on increasingly elaborate techniques. Viruses are difficult to diagnose and identify because of their diversity and number. The symptomology alone is insufficient: current methods are based on infectious, immunological, and genomic properties (Chapter 9).

Integrated control

Integrated control of viruses requires a thorough epidemiological understanding. In the absence of the capacity to fight viral infections directly, control methods are mainly prophylactic: selection of healthy plants, control of vectors, search for resistance genes (Chapters 10 and 11). These means often allow for effective control of the viruses responsible for the most serious epidemics.

A better understanding of the mechanisms of viral infection and plant defences, along with the development of techniques of cellular and molecular biology, could lead to new methods of control. Through genetic engineering, it is now possible to introduce into a plant genome nucleotide sequences that make it resistant or tolerant to viral infection (Chapter 12).

Evolution of viruses in the living world

Like all genomes, viral nucleic acids vary over time. Mutation, recombination, and reassortment occur with a particularly high frequency in viral genomes. This variability, far from being a handicap, is in fact essential to the efficiency and adaptability of a virus. The sequencing of viral genomes allows us to form phylogenetic groups resulting from a modular evolution. The origin of viruses raises problems, but we can propose the hypothesis of sequences of cellular origin that, over time and on various occasions, were organized into parasites of the cell (Chapter 13).

How to define a virus

A virus is an infectious and potentially pathogenic parasite. The viral particle is a nucleoproteic structure having a single type of nucleic acid, either DNA or RNA (Lwoff, 1957). Its nucleic acid contains up to 12 genes in plant viruses and codes for proteins that, in a living cell, fulfil the functions required for its survival, replication and spread.

Replication function: The nucleic acid codes for at least one protein indispensable to its replication. This character is essential. It differentiates a virus from a plasmid, a transposon, or a viroid.

Other functions are necessary or not, depending on the virus, notably the following:

— protection in the cell and outside: the capsid;
— movement from one cell to another and into the plant;
— transmission from one plant to another by a vector.

Van Regenmortel and Fauquet (2000) make the important distinction between the entity called virus and the viral particle or virion as follows: "A virion has intrinsic physicochemical and structural properties that suffice to characterize it exhaustively. A virus, on the other hand, has in addition relational properties that exist only by virtue of its relation with other entities such as the host or vector. These relational and emergent properties are revealed only when a virus infects a cell and is integrated into its metabolic activities during the replication cycle".

Classification of viruses

The classification of viruses has gradually become organized into species, genera, and families, defined by a set of common characters. Replication strategies play a dominant role in this classification. Plant viruses are classified into 81 genera at present (Chapter 14). Each of these genera is described in the profiles at the end of this book.

The classification is always a work in progress. It is continually modified by the International Committee of Taxonomy of Viruses on the basis of new information. This new information is acquired by two complementary means: one is the increasingly precise description of viral particles and the inventory of viruses and the other is the deciphering of the genetic message using techniques of molecular biology. The present and the future of virology essentially involve the relationships between the virus and its host, i.e., the way in which a small element, formed of a few genes enveloped in proteins, can exploit a cell in all its complexity and then gradually can invade its entire agro-environment.

The vocabulary of virology

This book attempts to propose a description of the story of plant viruses at various levels of their existence: the cell, the plant, the population, and the smaller or larger ecological niche. Virology borrows a number of terms from cellular and molecular biology. Some of these terms are defined in the glossary.

The names of viruses pose a particular problem. While cellular organisms have a binomial terminology in Latin, the names of viruses generally arise from the symptoms caused in a host. The English names as well as their abbreviations are internationally recognized. To remedy the sometimes cryptic effect of acronyms, the abbreviation, full name, and genus of the virus cited are indicated on the first page of each chapter. According to rule, the species, genus, family, and order are written in italics with an initial capital. This arrangement involves mainly viruses considered as taxonomic entities. It is not essential for names of species and genera of viruses considered as concrete objects (Pringle, 1999b).

The Virus, the Cell and the Plant

1

CHAPTER

Viral Structures

The morphological description of viral particles began in 1935. Until then, viruses were defined as infectious agents small enough to cross filters that prevented the passage of bacteria. Stanley (1935) applied methods of protein fractionation to the sap of diseased plants and purified particles of tobacco mosaic virus in the form of liquid crystals. The fine paracrystalline needles obtained, even when highly diluted, proved capable of reproducing the disease, and Stanley concluded that they were an "autocatalytic protein" that needed living cells in order to multiply. However, very shortly afterwards, Bawden et al. (1936) discovered that they were not a pure protein, but a nucleoprotein containing RNA. Simultaneously, a first bacterial virus was described as a nucleoprotein containing DNA. The study and description of these infectious molecules capable of organizing the multiplication of their nucleic acid and then transferring it from one cell to another then began.

Architecture of the virion

The viral particle, also called *virion*, is composed of a nucleic acid (DNA or RNA) and a proteic shell called a capsid; depending on the virus, it may also have an outer coat derived from membranes of the host, enzymes, and zinc or calcium ions in small quantity. The capsid surrounds the nucleic acid; it is formed of subunits. Each subunit is a polypeptide chain of around 150 to 400 amino acids, coded by the viral nucleic acid; it is one of the viral proteins in which the sequence is most variable between species, strains, and isolates. On the outside of the cell, the capsid protects the nucleic acid. Inside the cell, it plays a particularly important role in various

Viruses cited

AMV (*Alfalfa mosaic virus, Alfamovirus*), BMV (*Brome mosaic virus, Bromoviridae*), CaMV (*Cauliflower mosaic virus, Caulimovirus*), CCMV (*Cowpea chlorotic mottle virus, Bromovirus*), CPMV (*Cowpea mosaic virus, Comovirus*), PCV (*Peanut clump virus, Pecluvirus*), SBMV (*Southern bean mosaic virus, Sobemovirus*), TBSV (*Tomato bushy stunt virus, Tombusvirus*), TCV (*Turnip crinkle virus, Carmovirus*), TMV (*Tobacco mosaic virus, Tobamovirus*), TNV (*Tobacco necrosis virus, Necrovirus*), TVMV (*Tobacco vein mottling virus, Potyviridae*), TYMV (*Turnip yellow mosaic virus, Tymovirus*).

processes, depending on the virus: e.g., movement in the plant, transmission by vectors (Pacot-Hiriart et al., 1997).

The size of viruses was at first estimated by their sedimentation coefficient in ultracentrifugation. From 1939 onward, their morphology was revealed by observations under electron microscope showing the rod form of the TMV (*Tobacco mosaic virus* or *Tobamovirus*) and the tadpole shape of a bacteriophage. The electron microscope could be used to give a general image of the virion, either by examining ultrathin sections or staining suspended particles with a negative stain. The images obtained under the best conditions could be subjected to methods of quantitative analysis and a three-dimensional image reconstruction. The resolution was 3 to 4 nm in negative staining and 1.5 to 2 nm in electron cryomicroscopy. This resolution makes it possible to distinguish some details of structures, but it remains insufficient for observing molecular interactions. X-ray crystallographic diffraction pattern analysis provides a much higher resolution and allows the description of the internal structure and relationships between the nucleic acid and the protein. The first complete structure was obtained in 1978 with crystals of TBSV (*Tomato bushy stunt virus, Tombusvirus*), then of poliovirus. From the proteins for which the structure is known by crystallographic methods and for which the sequence is also known, predictions can be made on the structure of homologous proteins.

■ Viral capsids

In plant viruses, viral capsids are generally composed of copies of a single protein

Since 1954, excellent X-ray diffraction diagrams of the TMV particle have shown the helical structure of the particle (Franklin, 1955). From these observations as well as other experimental results, Crick and Watson (1956) formulated a general hypothesis about the structure of the simplest viruses: given that the virion is made of an infectious viral RNA surrounded by proteins, the outer structure can be economically achieved only by a large number of identical elements arranged symmetrically and surrounding the nucleic acid. This hypothesis was largely confirmed; the viral genome, because of its compactness, carries the information for a small number of proteins including the capsidial subunits. Plant viruses generally have a single capsidial protein, sometimes two (*Comovirus*) or three (*Sequiviridae, Phytoreovirus*). Animal viruses and bacteriophages frequently have two or three different capsidial proteins. The capsid is constructed by the spatial repetition of many subunits connected by non-covalent bonds; for these bonds to be similar throughout the structure, a regular symmetrical arrangement of subunits is necessary. Two types of arrangements are found in most viruses:

- Viruses with helical symmetry are seen under electron microscope in the form of rods or flexuous filaments.
- Viruses with icosahedral symmetry most often have an isometric form.

Some methods to study viral particles and their components

Ultracentrifugation can be used to migrate and sediment viral particles by means of very high centrifugal forces (up to 100,000 g). In analytic centrifugation, the centrifugal forces and the forces of friction end up at an equilibrium and the particle thus migrates at a speed that allows us to define a sedimentation constant characteristic of the particle, expressed in Svedberg units (S). A density gradient ultracentrifugation makes it possible to classify constituents of a plant extract and separate the viral particles that concentrate at the zone corresponding to their density. Differential ultracentrifugation allows the separation of particles from contaminants by several cycles of centrifugation at low speed and high speed.

The electron microscope uses an electron beam that is propagated in vacuum. It can be used to observe particles only outside their aqueous medium. This is the most direct means of observing the size and morphology of particles as well as their relation to cellular structures. However, since proteins and nucleic acids are not opaque to electrons, other means must be found to visualize them. Metallic shading, followed by negative staining by uranyl acetate or phosphotungstic acid opaque to electrons, has made it possible to increase the contrast and to see not only the forms but also certain details of particle structure. Ultrathin sections can be used to observe the virus in relation with cellular structures under electron microscope after fixation or after freezing (cryo-electromicroscopy).

X-ray diffraction by viral crystals gives the most precise information about the three-dimensional structure of the particle. If X-rays pass through a crystal structure, they are diffracted in an orderly pattern. The atomic arrangement can be deduced from angles of scattering pattern. This type of analysis can be applied only to simple viruses that can be obtained in the crystallized state.

Electrophoresis subjects the molecules (proteins or nucleic acids) to an electric field. The most commonly used method for proteins consists in dissociation into monomers, then migration in polyacrylamide gels of controlled porosity, in a buffer containing an anionic detergent (SDS) that uniformly charges the polypeptides. The electric field then causes them to migrate as an inverse function of their molecular weight because of the molecular sieve effect of the gel. Under controlled conditions, this method can be used to determine with some precision the molecular weights in comparison with those of known proteins. It can be followed by a transfer on to a support for the purpose of serological detection. The *sequence* of a protein can be established by microsequencing short fragments or by deduction from the gene sequence. Predictions can then be made about its structure.

Nucleic acids migrate in agarose or polyacrylamide gels that are poorly reticulated and their molecular weight is determined by simultaneous migration of combinations of nucleic acids of known molecular weight. Some special precautions must be taken to protect the RNAs from the action of omnipresent ribonucleases. Once the nucleic acids are purified, they can be cleaved by enzymes, hybridized, sequenced, and so on.

Some plant viruses are surrounded by a lipoprotein coat derived from cell membranes; their internal structure, called the nucleocapsid, resembles that of viruses with helical symmetry.

Isometric particles

1. *Polerovirus* CABYV (*Cucurbit aphid-borne yellows virus, Luteoviridae*)
2. *Cucumovirus* CMV (*Cucumber mosaic virus, Bromoviridae*)
3. *Tombusvirus* PetAMV (*Petunia asteroid mosaic virus, Tombusviridae*)
4. *Tymovirus* MRMV (*Melon rugose mosaic virus*)

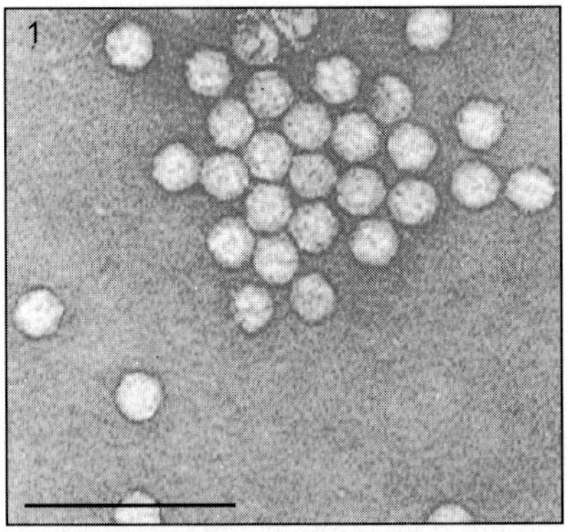

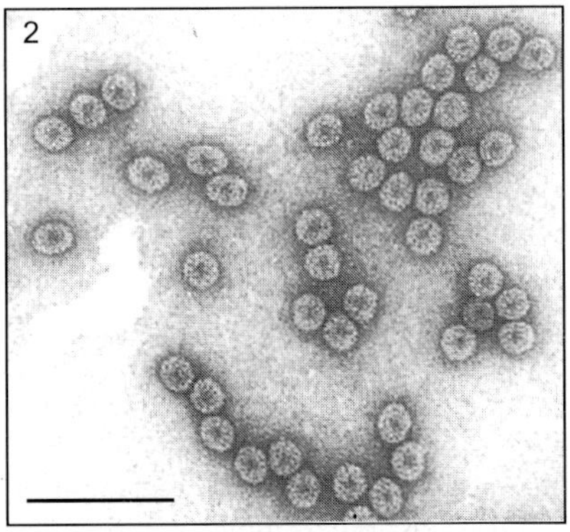

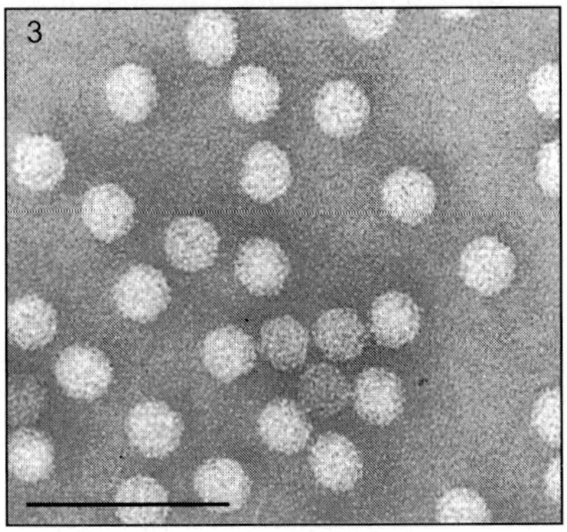

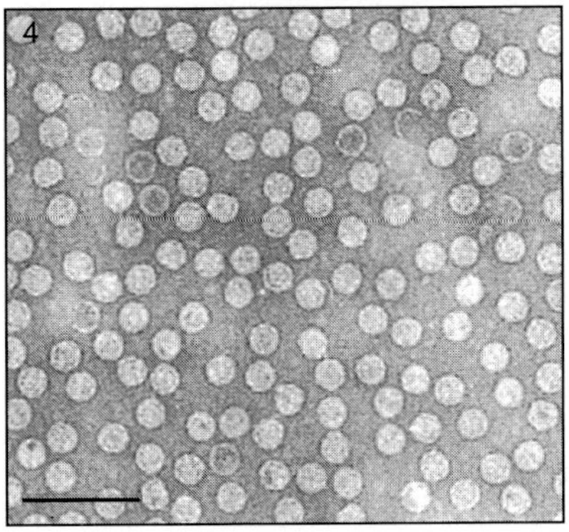

Geminate particles

5. *Mastrevirus* CpCDV (*Chickpea chlorotic dwarf virus, Geminiviridae*)

Enveloped particles

6. *Tospovirus* TSWV (*Tomato spotted wilt virus, Bunyaviridae*)

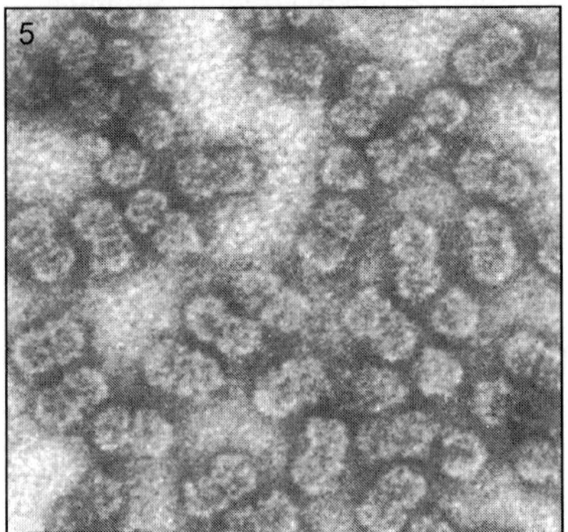

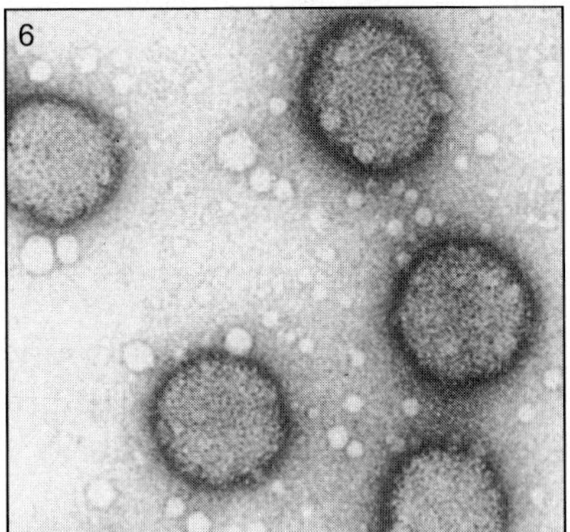

Bacilliform particles

7. *Alfamovirus* AMV (*Alfalfa mosaic virus, Bromoviridae*)
8. *Badnavirus* CSSV (*Cacao swollen shoot virus, Caulimoviridae*)

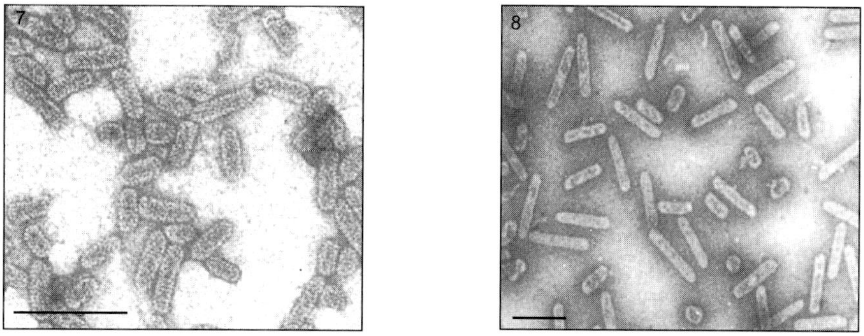

Particles with helical symmetry

9. *Tobamovirus* TMV (*Tobacco mosaic virus*)
10. *Potyvirus* SMV (*Soybean mosaic virus, Potyviridae*)
11. *Potyvirus* ZYMV (*Zucchini yellow mosaic virus, Potyviridae*)
12. *Allexivirus* GarMbFV (*Garlic mite-borne filamentous virus*)

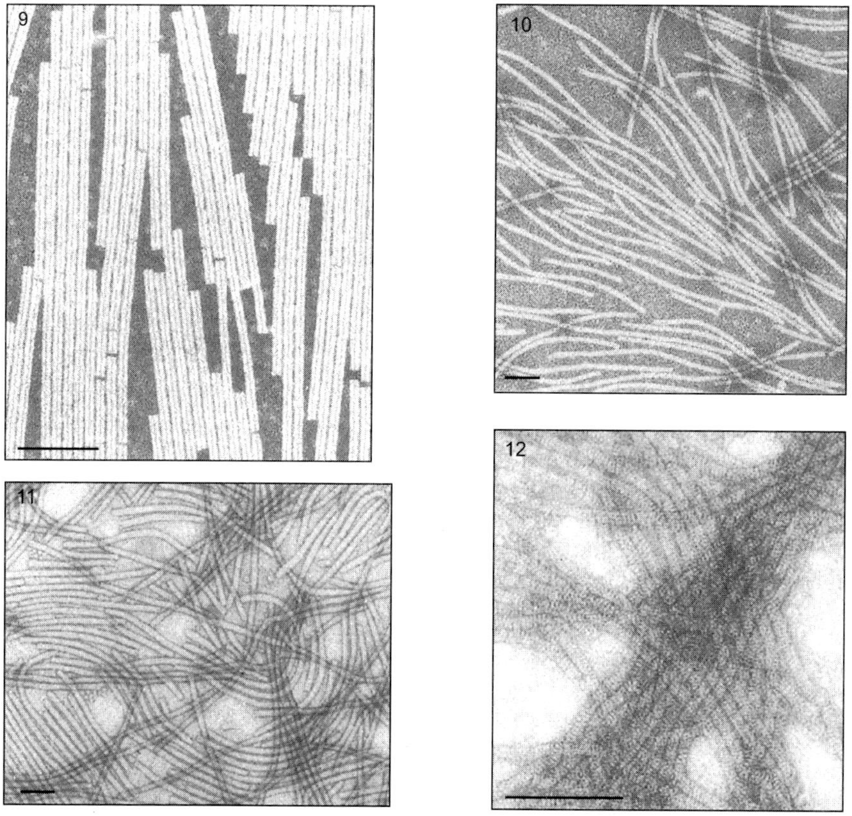

Figure 1.1 Purified viral particles, negatively stained and observed under electron microscope (bar = 100 nm).

The morphology and size of the virion are significant characters for taxonomy

The morphology and size of the virion are conserved within genera and families (Chapter 14). Therefore, an observation under the electron microscope often allows the proposal of a primary classification (Fig. 1.1). The size of viral particles is extremely variable. Compared with the largest animal viruses, plant viruses have relatively small sizes, of 17-20 to 60-80 nm for isometric viruses, 18 × 300 to 12 × 2200 nm for viruses with helical symmetry (Fig. 1.2).

■ TMV and viruses with helical symmetry

The TMV particle has a highly stable structure; it is easy to purify in large quantities, so it can quickly be studied using different methods. When highly concentrated suspensions of the virus are placed in capillary tubes, slow movements of the liquid then suffice to orient the rod-shaped particles. Particles aggregate in bundles parallel to the axis of the tube and constitute a gel that has the properties of a quasi-crystal by regular hexagonal arrangement in the plane perpendicular to the axis. Such samples have been used to obtain an X-ray crystallographic diffraction pattern, with a precision of 0.36 nm, then of 0.29 nm (Namba et al., 1989). By fixing mercury atoms on the single cysteine residue of the subunit, one can count and measure the subunits on one turn of the helix. Information about the structure of the virus is deduced from the diffraction pattern by complex mathematical calculations with successive approximations to reduce the margin of error and highlight the significant details. These crystallographic methods have made it possible to describe the intact particle (RNA and protein) at an atomic level.

These studies, along with the observations of Franklin (1955) and Watson (1954), demonstrated that the TMV capsid is not formed of stacked-up discs but made up of subunits arranged in a helix, characterized by its radius and pitch. Each turn contains 16 and 1/3 subunits and the pitch of the helix is 2.3 nm. The virion measures around 300 nm in length, so it has 300 × 16.3/2.3 or around 2100 identical subunits. The particle, with a diameter of 17 to 18 nm, has at its centre a canal of diameter 4 nm filled with liquid, and a single RNA molecule is imbricated with the proteins at a distance of around 4 nm from the axis (Fig. 1.3). All the subunits are in equivalent positions by translation and rotation around the axis of the helix.

The structure of the capsidial subunit and its relationship with viral RNA are known precisely

When the sequence of a polypeptide chain is established, the stable domains of the structure resulting from its folding can be predicted: α-helices generated by the rotation of the polypeptide chain on itself, β-sheets in which the chain folds on itself, and turns and loops that connect the sheets and helices into a globular mass

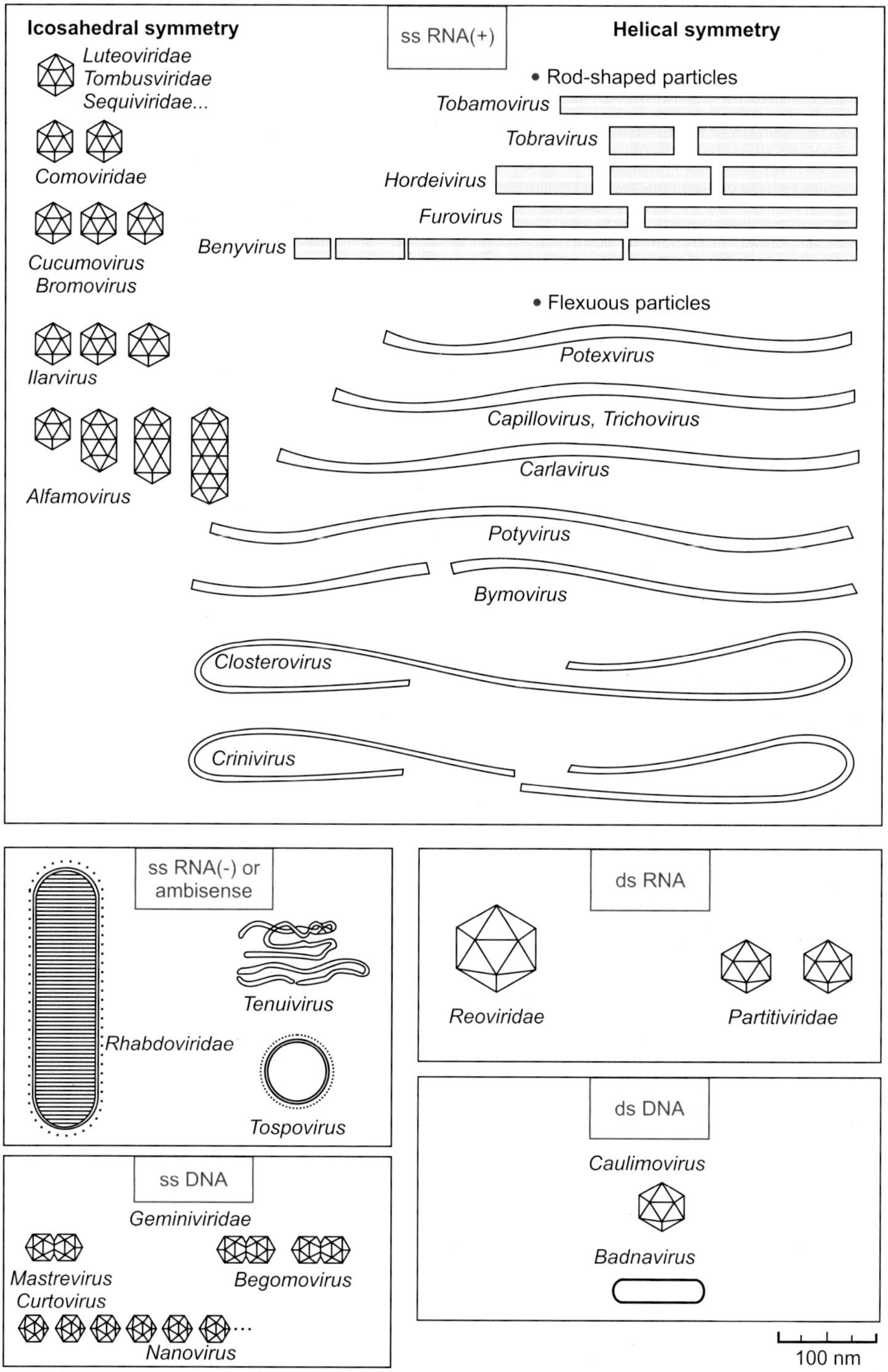

Figure 1.2 Table of plant viruses classified according to morphology. All the viral genera are not mentioned here (see Chapter 14). Some dimensions are approximate. ss: single stranded, ds: double stranded.

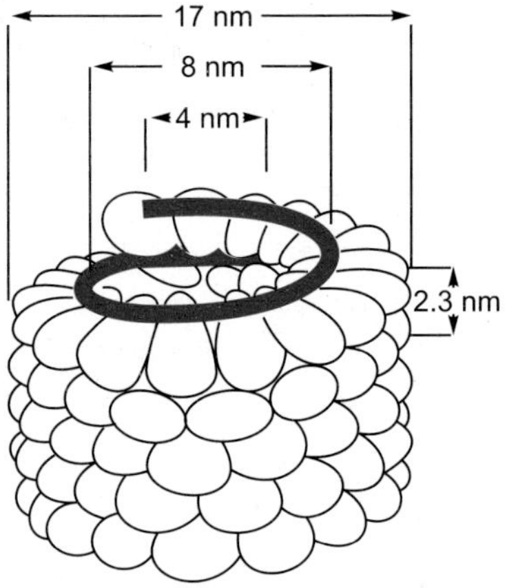

Figure 1.3 The TMV particle. On this drawing of a segment of the helix, 5 turns (out of 132) are shown. The RNA is indicated as a ribbon jammed between the subunits, shown here as simple volumes (Cornuet, 1987).

(Fig. 1.4). These predictions of the secondary structure are not entirely reliable, and they must be confirmed by referring to proteins already known by means of crystallographic methods. The tertiary structure of a polypeptide chain results essentially from the spatial organization of helices and sheets; it is stabilized by non-covalent interactions between amino acids: hydrogen bridges and saline bonds between the polar groups, hydrophobic interactions between non-polar groups. These forces act within subunits as well as between subunits.

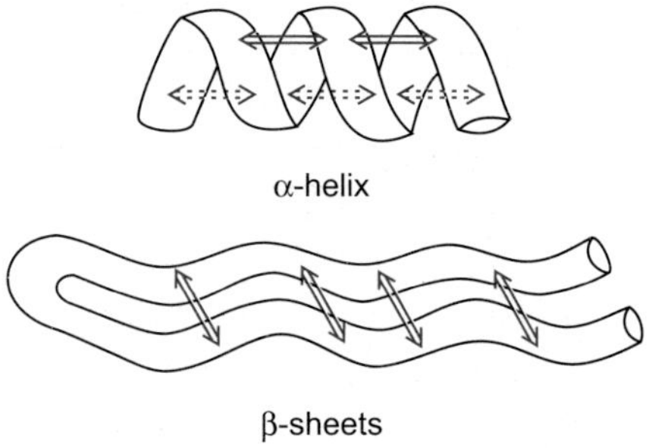

Figure 1.4 Structural conformations of the polypeptide chain. Depending on the sequence, the chain may roll into itself to form an α-helix (intra-chain linkages) or fold to form β-sheets (inter-chain linkages). These structures are stabilized by non-covalent bonds represented by arrows (Cornuet, 1987).

The subunit of the TMV capsid has a high proportion of secondary structures: 50% in α-helices, 10% in β-sheets as well as loops. Four antiparallel helices (denominated LS, RS and LR, RR) form the basis of the structure. Nearer the outside, two short helices are close to the NH_2 and COOH ends, which are found on the outer surface. On the lateral faces, there is an alternation of hydrophilic and hydrophobic zones (Fig. 1.5).

Interactions between the subunits

Interactions between the subunits stabilize the viral particle either through lateral bonds between subunits contiguous on the helix or between superimposed units on two successive turns.

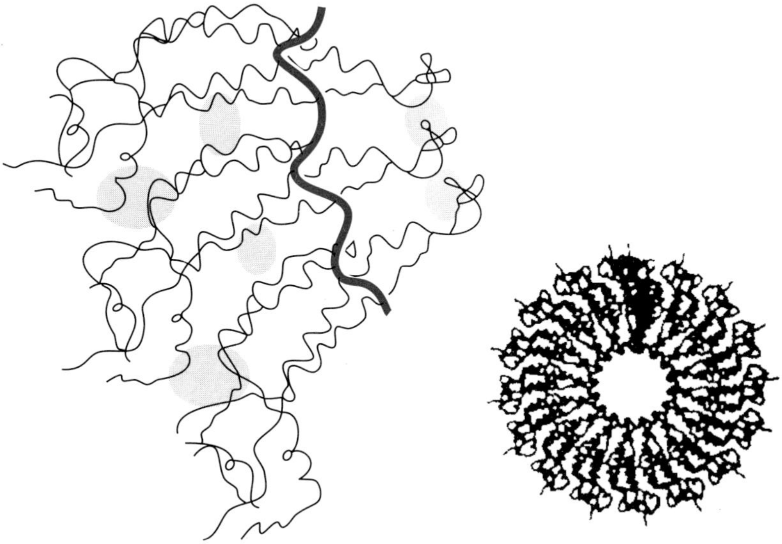

Figure 1.5 Spatial arrangement of TMV subunits, interaction between them and with RNA. At left, three adjacent subunits in a turn of the helix. The axis of the helix is perpendicular to the page. The polypeptide chain is a fine line, the viral RNA is represented by a heavy line. The zones of hydrophobic interactions are indicated in grey (Namba et al., 1989). At right, a complete gyre of the helix is represented; it contains 16.3 subunits (photo by T.M.A. Wilson).

— The lateral bonds are essentially non-polar. They are realized by aromatic and aliphatic amino acids that form a hydrophobic band close to the outer surface; two other highly hydrophobic regions are located towards the inside of the particle.
— The bonds between two superimposed subunits are mostly electrostatic (Fig. 1.6). A large zone of non-specific interactions is found towards the inside of the helix, a second less extensive zone is located at a radius of around 5.5 nm. A third zone located towards the outside of the helix stabilizes the subunit.

Protein-RNA interactions

Viral RNA is associated at each subunit by the intermediary of three bases, in two different ways:
— The bases are linked to amino acids of the LR helix of the subunit located below by very close interaction between the phosphates of the negatively charged RNA and the basic arginines 90 and 92, but there are other bonds between each of the bases and several amino acids (Fig. 1.6).
— Other electrostatic interactions occur between the RNA and the proteins, especially with amino acids of the RR segment of the subunit located below.

The interaction between RNA and the protein is ultimately realized by three types of bonds (see Fig. 9.13):

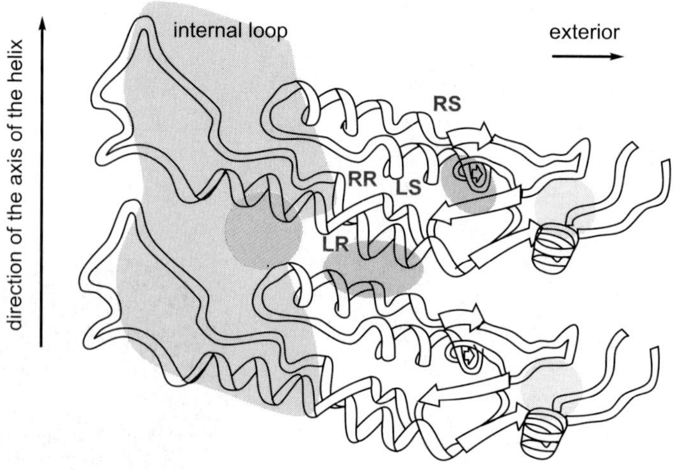

Figure 1.6 Two subunits located one below the other. The four principal α-helices are designated as RS, LS, RR, and LR. The zones of electrostatic interactions are indicated in dark grey. The external hydrophobic band perpendicular to the page is in light grey. The volume in which the viral RNA is found is indicated in red (Namba et al., 1989).

- — Electrostatic forces between arginines and phosphate groups of the RNA.
- — Specific hydrogen bonds, including with the group 2'-hydroxyl of ribose.
- — Non-specific hydrophobic interactions with the LR helix, which allows encapsidation of the complete sequence. Several amino acids implicated directly in the RNA bonding are conserved between strains.

Despite the stability of the TMV virion, the RNA must be able to decapsidate and encapsidate

The structure of the particle must ensure protection of the RNA as well as permit its disassembly during infection; these two states are possible because of a metastable equilibrium whose energy is close to the minimum energy state. Apart from the forces that stabilize the particle, there are repulsive electrostatic forces between the TMV subunits, due to carboxyl groups that are neutralized in the particle by protons or calcium ions. During entry into the cell, the change in pH and calcium concentration allows repulsive forces to destabilize subunits of the last turns of the helix. Starting at the 5' end of the viral RNA, the process of decapsidation is linked with the attachment of ribosomes to RNA (Lu et al., 1998). Decapsidation proceeds from the 5' end to the 3' end with the ribosomal complex (Chapter 2, Fig. 2.2) and very quickly covers two-thirds of the RNA, i.e., the end of the replication genes. In a second step, decapsidation proceeds inversely from the 3' end to the 5' end and terminates in the region that contains the origin of assembly (described below). The newly synthesized replication proteins participate in this second step (Wu and Shaw, 1997). Decapsidation is thus an active phenomenon, linked to translation and replication.

If the TMV particle can dissociate into subunits and RNA *in vitro*, it can also reassemble by simply mixing the RNA and the subunits and by adjusting the pH, ionic force, and concentrations (Fraenkel-Conrat and Williams, 1955; Lebeurrier et al., 1977). These assembly experiments have shown that the information necessary for encapsidation, a highly specific phenomenon, is contained in the components of the structure. The starting point of encapsidation is the formation of a polarized double disc (2 × 17 subunits) that recognizes the encapsidation origin of the viral RNA, a structure in three successive hairpins located in the gene of the 30 kDa protein about 900 residues from the 3' end (Jonard et al., 1977). This secondary structure and a sequence with repeated AAG motifs (including guanine 122, which is at the beginning of the process) are recognized by one or two subunits of the double disc, which leads it to its development into a helical structure. At this stage, the 3' and 5' ends of the RNA are found on the same side of the growing rod (Fig. 1.7). The elongation occurs by the positioning of subunits next to each other, chiefly on the 5' end, i.e., on the side of an RNA loop that slides to the extent that the RNA chain gets established between the subunits, without specificity of sequence. Encapsidation is completed in the same fashion on the side of the 3' end. The hydrogen bonds that are established between the subunits make it a highly cooperative process (Namba et al., 1989). In another rod-shaped virus, PCV (*Peanut clump virus, Pecluvirus*), two different sequences drive the encapsidation of RNA-1 and two other unrelated sequences the encapsidation of RNA-2 (Hemmer et al., 2003).

Through experiments in assembling viral particles by mixing RNA and capsidial subunits of two strains of TMV, Fraenkel-Conrat and Singer established in 1957 that the viral RNA codes for its capsid protein, thus demonstrating for the first time that RNA can carry genetic information.

Flexuous viruses also have helical symmetry

Helical symmetry allows the construction of rigid particles (*Tobravirus, Tobamovirus, Furovirus, Hordeivirus*, etc.) or flexuous particles (*Potexvirus, Potyviridae, Carlavirus, Closteroviridae*, etc.). The encapsidated genomes are of widely varying size; some genomes are segmented into several molecules (*Tobravirus, Hordeivirus, Furovirus, Benyvirus, Bymovirus*, etc.). The length of particles may reach 2000 nm in the genus *Closterovirus*. The flexibility is linked to a greater pitch of the helix than in rigid particles as well as lesser interactions between the subunits towards the periphery of the helix. Unlike the rigid particles, the flexuous particles generally do not have a clearly visible central cavity and they are less stable than those of the TMV.

The potyviruses have flexuous particles of 600 to 900 nm length and 11 nm diameter composed of around 2000 subunits. The pitch of the helix is 3.3 nm; 7 to 8 subunits form one turn. The N and C ends of the capsidial subunit are on the outside of the particle (Fig. 1.8) (Shukla and Ward, 1989). The secondary and tertiary structure of the subunit is comparable to that described for TMV and other viruses of helical symmetry. The origin-of-assembly sequence of TVMV (*Tobacco vein mottling virus, Potyviridae*) is located near the 5' end of viral RNA (Wu and Shaw, 1998).

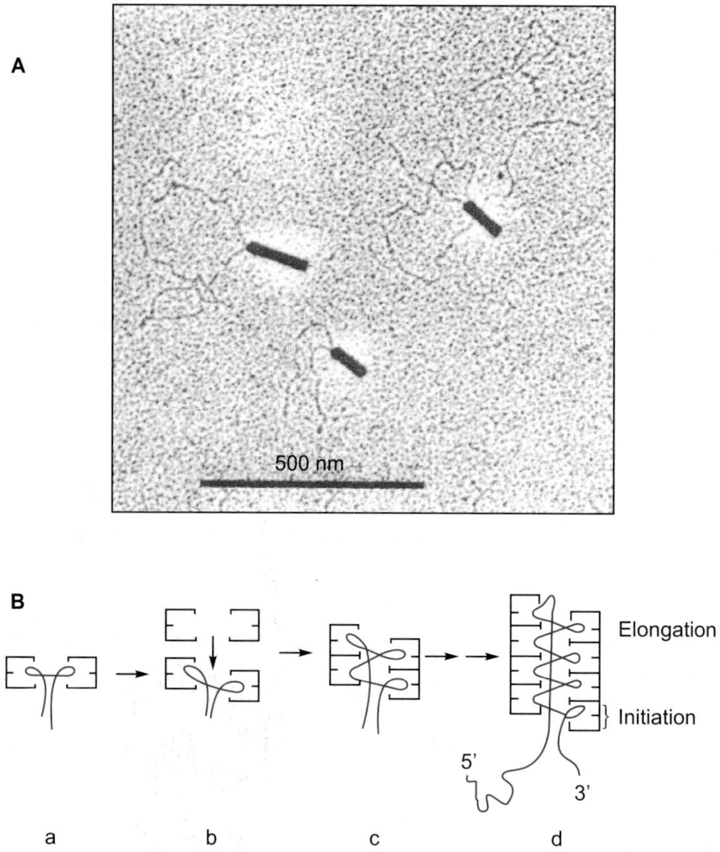

Figure 1.7 Encapsidation of the RNA of TMV. A: Rods in the course of encapsidation showing the two ends of the free RNA on the same side of the particle. B: Encapsidation. The viral RNA is inserted at the level of the origin of assembly in a double disc of proteic subunits (a, b). The double disc is thus destabilized into a helical structure (c); elongation of the encapsidated form occurs towards the two ends (d). Nicolaieff and Hirth in Francki, Milne and Hatta, 1987.

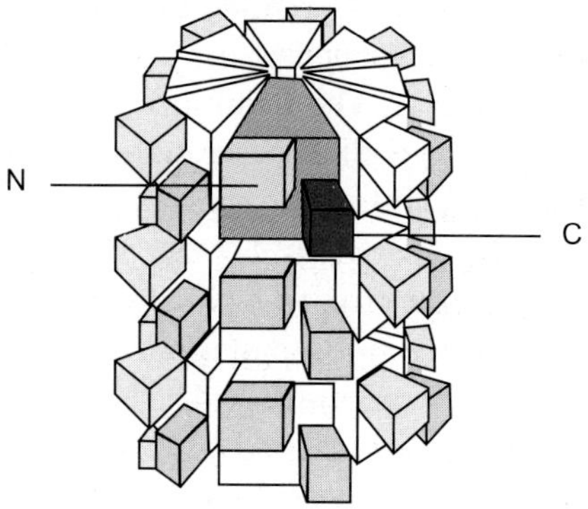

Figure 1.8 Fragment of the particle of a potyvirus. The subunits are arranged in a helix and the NH_2 (N) and COOH (C) ends are clearly exposed on the outside (Shukla and Ward, 1989).

The enveloped viruses have a nucleocapsid of helical symmetry

Unlike animal viruses, only few viruses infecting plants have an outer coat. These viruses (*Rhabdoviridae, Bunyaviridae*) have the common characteristic of multiplying in the cells of their insect vector. The virions of *Rhabdoviridae* are bacilliform. The inner part constitutes the nucleocapsid, formed by an RNA linked to a basic protein. This ribonucleoprotein is arranged in a helix. The outer coat is derived from cellular membranes; it contains two viral proteins one of which is directed towards the exterior. The virions of tospovirus have an analogous structure. Their general form is irregular, approximately spherical; the nucleocapsids have helical symmetry (Fig. 1.9).

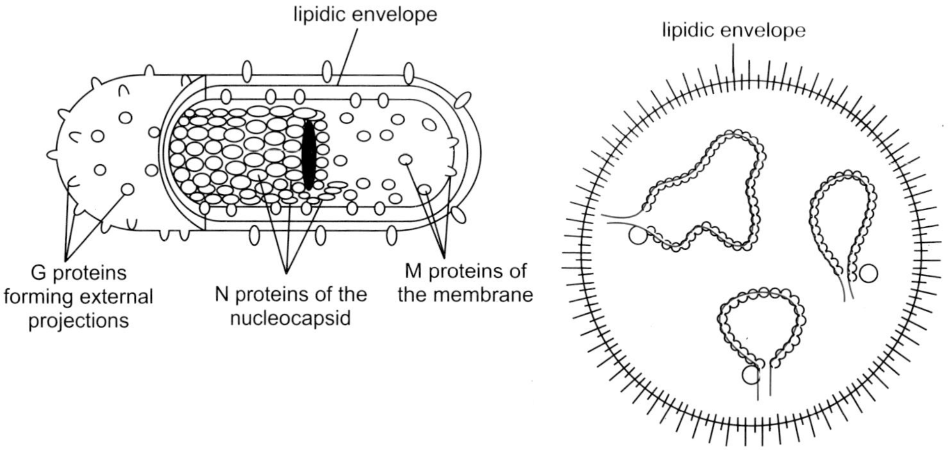

Figure 1.9 Virus with envelope: viral particles of a rhabdovirus (at left) and a tospovirus showing three nucleocapsids inside the lipidic membrane (at right).

Isometric virions

Isometric virions have icosahedral symmetry

Many viruses form isometric particles with a central cavity in which the nucleic acid is positioned. These particles do not always have the form of an icosahedron, but their form is derived from the icosahedron and they have elements of such symmetry. The icosahedron is a solid with 20 faces, which are equilateral triangles, forming among them 12 vertices. This solid presents three series of axes of rotation generating symmetries: 5-fold axes pass through the vertices; 3-fold axes are located at the centre of the triangles and are perpendicular to the plane of the triangle; 2-fold axes pass through the midpoint of edges and are perpendicular to the edges (Fig. 1.10). Each face is made up of three identical asymmetrical units, and this type of structure allows 60 subunits, each occupying the place of an asymmetrical unit on the icosahedron, to constitute a closed surface, all having the same environment.

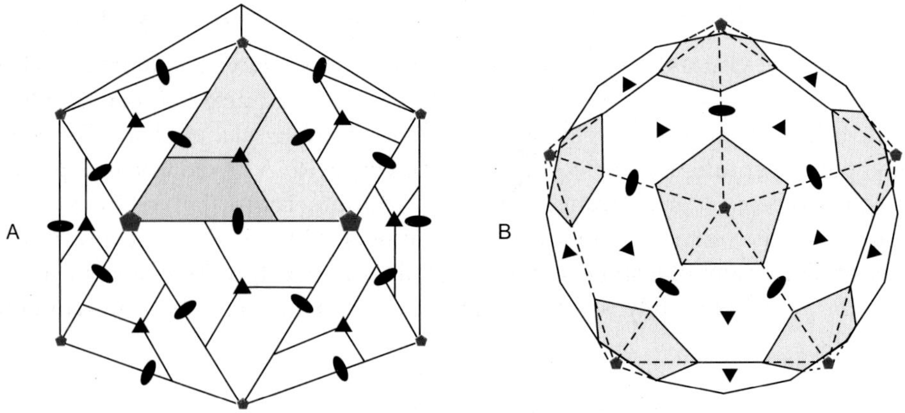

Figure 1.10 Drawing of an icosahedron; its surface is made up of 20 faces that are equilateral triangles and 12 vertices. Three series of symmetry axes (5-fold vertices are shown as red pentagons, the centres of 3-fold faces are shown as triangles; midpoints of edges are shown as ovals). Each triangle contains 3 asymmetrical units (one is shaded in dark grey). Here one capsidial subunit occupies the place of an asymmetrical unit of the icosahedron (T = 1), and the capsid is formed of 60 subunits. B: Truncated icosahedron at the level of vertices; its surface is made up of 20 hexagons and 12 pentagons. The pattern formed is that of a soccer ball.

Icosahedral symmetry allows the assembly of the maximum number of subunits in entirely identical environments to constitute an isometric particle. These 60 subunits correspond through the 2-fold, 3-fold, and 5-fold axes of rotation; they surround at the centre of the viral particle a fixed internal volume that can shelter a polynucleotide of around 1.5 kb at most.

This volume is sufficient for certain very simple viral structures. The capsid of the satellite of TNV (*Tobacco necrosis virus, Necrovirus*), which has a diameter of 17 nm, is composed of 60 subunits; they surround a cavity of around 6 nm diameter containing RNA of 1.2 kb that code only for this subunit of 195 amino acids. The presence of an assistant virus is necessary for its replication. Another example of a particle formed of 60 subunits is that of *Nanoviruses*, which have fractionated their genetic information in about 10 circular DNA of around 1 kb, each encapsidated in an icosahedral particle.

Apart from these few rare cases, the icosahedron made up of 60 elements delimits an internal space whose volume is insufficient for most viral genomes. One solution is that adopted in the family *Geminiviridae*: their particles are formed of two adjacent icosahedrons to which a morphological unit is added to realize a junction, which gives the virion a twinned appearance. The cavity thus formed contains a circular DNA molecule of 2.5 to 3 kb. Another solution to increase the volume, adopted by numerous viruses, consists of subdividing each asymmetrical unit of the icosahedron to place several subunits in it. In total a number of subunits that is a multiple of 60 is obtained, between which the contacts are no longer identical. Caspar and Klug (1962) established that in these more complex edifices, the subunits could form hexamers, pentamers, or dimers, which are the morphological subunits of the capsid. The contacts between subunits are "quasi-equivalent"; in addition, the subunits must be able to effect the molecular changes needed to build different structures and to organize them spatially, and thus to present a polymorphism of conformation (Johnson, 1996). These changes differ according to the virus. The final number of subunits is 60 × T, where T (number of triangulation) is the number of subunits in the asymmetrical unit of the icosahedron. Many plant viruses have structures of T = 3 with 180 identical subunits; this is true of TYMV (*Turnip yellow mosaic virus, Tymovirus*) (Fig. 1.11A) and TBSV (*Tomato bushy stunt virus, Tombusvirus*) (Fig. 1.12A). When the subunits are not identical, the assembly is more complex but always responds to the symmetries of the icosahedron. This is the case with CPMV (*Cowpea mosaic virus, Comovirus*), which has two types of capsidial subunits of 43 and 24 kDa; the small subunits form pentamers, while the large ones, grouped in threes, occupy the position of hexamers (Fig. 1.11B).

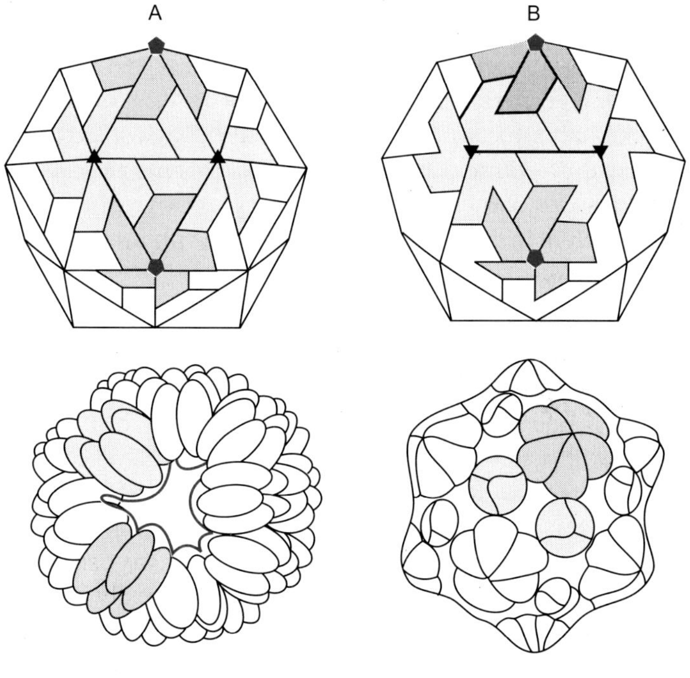

Figure 1.11 Morphological units on capsids of icosahedral symmetry. A: Virus with a single capsid protein (e.g., TYMV). Above, only some axes of symmetry are shown. Two hexamers and two pentamers are identified (Rossmann and Johnson, 1989). Below, the subunits separated into pentamers and hexamers surround the RNA (in red), made visible by the removal of some subunits (Cornuet, 1987). B: Virus with two capsid proteins (CPMV). Above, two trimers and two pentamers are identified as well as certain axes of symmetry (Rossmann and Johnson, 1989, with modifications). These morphological units are also drawn below (Crowther et al., 1974).

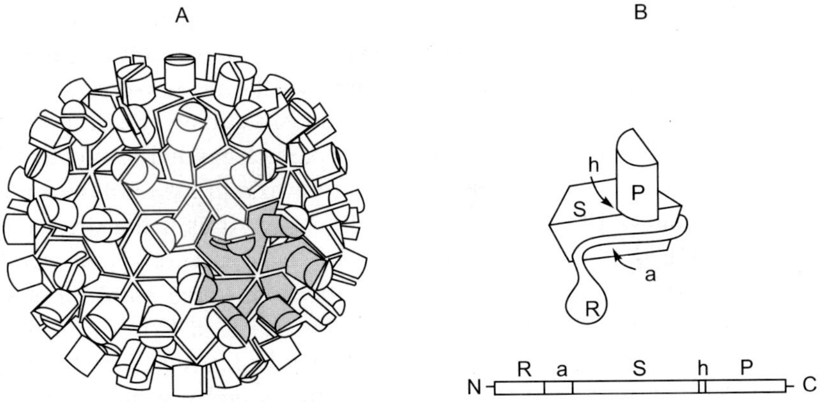

Figure 1.12 A particle of TBSV (*Tomato bushy stunt virus*), 32-35 nm in diameter, the capsid of which is made up of 180 subunits. They are grouped into 90 dimers forming protuberances. One hexamer is shaded in grey, one pentamer is shown in red. B: Polypeptide chain with the different domains, represented linearly and folded. Two adjacent P domains form the protuberances. The S domain presents the "β sandwich" pattern (Olson et al., 1983).

The triangulation number T takes defined values (1, 3, 4, 7, 13 in plant viruses, sometimes much higher in animal viruses). The external capsid of the reoviruses has a structure of T = 3. The capsid of CaMV (*Cauliflower mosaic virus*, *Caulimovirus*), the genome of which is a double-stranded DNA of 8 kbp, has a structure of T = 7, with 420 subunits distributed into 12 pentamers and 60 hexamers. Analysis under electron cryomicroscopy shows that the hexamers are not regular but present two trimers slightly different in their interaction with the neighbouring subunits (Cheng et al., 1992).

Variations on the theme of "180 subunits"

The TYMV is highly concentrated in the cells that it infects, where it can form paracrystals. This is one of the first isometric viruses in which the structure was studied by X-ray diffraction showing that the capsid (diameter about 30 nm) is formed of 180 subunits that form protuberances on the surface of the particle. Its triangulation number is T = 3. The observations under electron microscope show 32 morphological groups. These units of structure are 12 convex pentamers and 20 flat hexamers in which the axes of symmetry coincide respectively with the 5-fold and 3-fold axes of the icosahedron (Fig. 1.11A). The pseudo-spherical structure obtained is reminiscent of a soccer ball, formed of 12 black pentagons and 20 white hexagons. It is derived from the icosahedron by truncation of the vertices (Fig. 1.10B).

The TBSV, which was crystallized by Bawden and Pirie in 1936 and the architecture of which was studied by Harrison in 1978, presents a capsid formed of 180 subunits. The electron microscope reveals 90 regularly distributed protuberances that result in the juxtaposition of the outer domain "P" of two subunits forming dimers (Fig. 1.12). Each subunit has two globular domains, P and S, linked by a flexuous connection of 5 amino acids, and the angle between these two domains varies with the position of the dimer in the structure. The NH_2 end, strongly basic, is located within the capsid, where it interacts with the viral RNA.

Most simple viruses with icosahedral symmetry have capsidial subunits of the same structure

The common element is a structural domain of approximately trapezoidal form resulting from the folding of the polypeptide chain (Fig. 1.13). The two β-sheets, each made up of 4 antiparallel segments, make up the sides of the subunit, and two α-helices are inserted between them, the whole being connected by numerous loops. It may be noted that the β-sheets predominate, while in the rod structure there are mostly α-helices. With some exceptions, this compact structure, the "eight-stranded β sandwich", is present in all the isometric plant viruses. It is sensitive to its environment: divalent ions, pH, and the presence of RNA influence the changes of conformation required by the quasi-equivalence of subunits. The capsidial subunit of CCMV (*Cowpea chlorotic mottle virus*, *Bromovirus*) is presented as a ribbon of 190 amino acids folding according to the "β sandwich" pattern (Fig. 1.14). The residues 27 to 42 are ordered when subunits form hexamers and not when they form

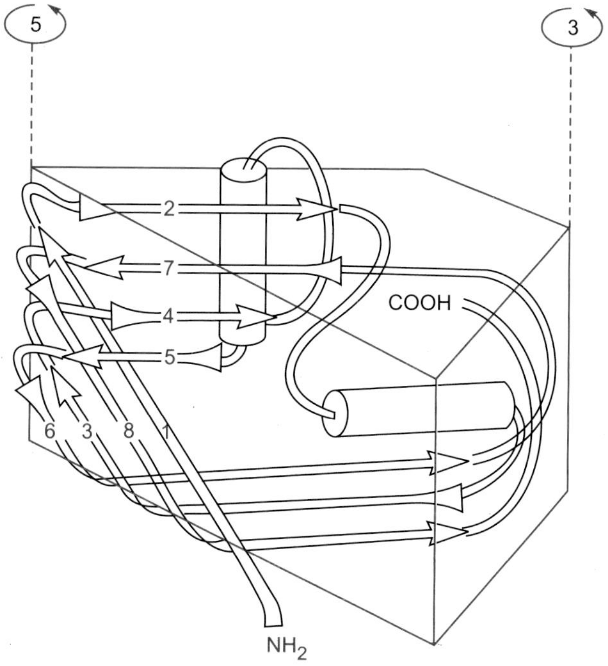

Figure 1.13 Folding of the polypeptide chain to form the "β sandwich" of viruses with icosahedral symmetry. The two β-sheets are formed by segments [1, 8, 3, 6] and [2, 7, 4, 5]. The 3-fold and 5-fold axes are indicated by arrows. The upper surface is on the outside of the particle. The ends of the chain emerge on the outside in a variable manner.

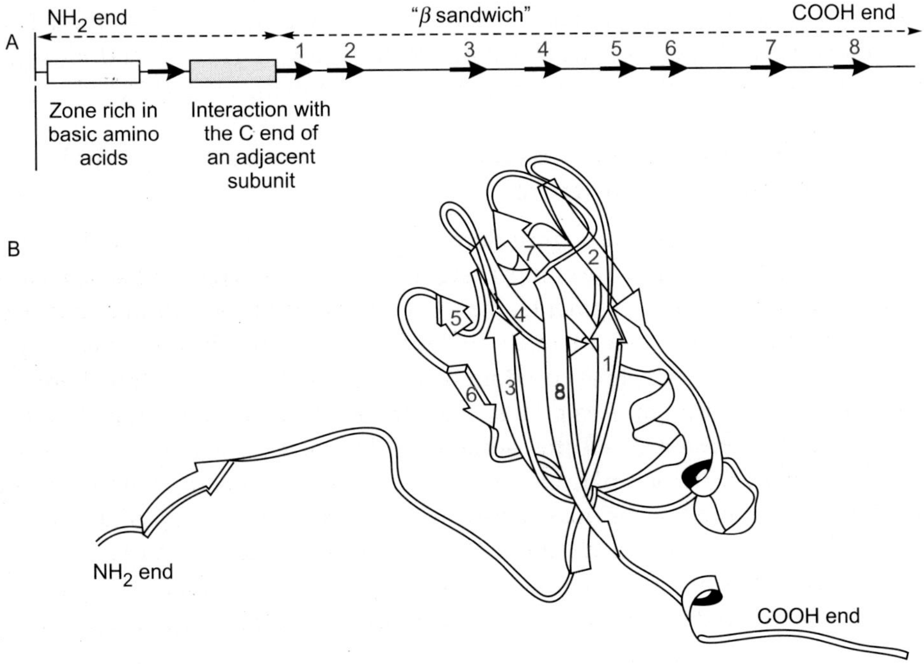

Figure 1.14 Subunit of CCMV (*Cowpea chlorotic mottle virus, Bromovirus*). A: The polypeptide chain (190 amino acid residues) is represented in a linear fashion. Three principal domains: NH_2 end, β sandwich, COOH end. The eight sheets of the β sandwich are numbered 1 to 8. B: The chain is shown folded. The ends spread out to link the subunit to its neighbours (Zhao et al., 1995).

pentamers; these residues modulate the stability of the quaternary structure. This conformational polymorphism is necessary for particle assembly (Johnson and Speir, 1997).

Packaging process is specific

After synthesis of multiple copies of capsid subunits and replication of viral nucleic acid, the components interact in the process of virus assembly, or packaging, and form new particles. The encapsidation process needs the neutralization of charges of the nucleic acid; this neutralization is often realized by the N-terminal end of the capsidial protein that contains a high proportion of basic amino acids. When this condition is not met (*Tymovirus, Comovirus*), the neutralization results from the presence of polyamines in the virion.

The four RNAs of bromoviruses (three genomic and one subgenomic) are separately encapsidated into three particles containing either RNA1, RNA2, or RNA3 plus the subgenomic RNA coding for the capsid protein; the three icosahedral particles have the same size and morphology. The packaging of the bromovirus CCMV has been studied *in vitro* and *in vivo*. The components of the virus can be packaged *in vitro*; at a pH higher than 7, the viral particles form in a few seconds, while at pH less than 6 and low ionic force, empty particles are formed without RNA. The capsidial protein is fixed preferentially on its own RNA in a mixture with non-viral RNAs. The two ends of the polypeptide form two "arms" that move far from the centre of the subunit. To ascertain their exact role, experiments of *in vitro* packaging were carried out with capsidial proteins expressed in *Escherichia coli* (Zhao et al., 1995). The edification of the virion begins with the spontaneous formation of dimers by the intermediary of C-terminal "arms" that extend towards the exterior and are associated with an N end of a nearby subunit (Fig. 1.14). The flexibility is sufficient to realize all the quasi-equivalent reactions, because the dimer can be flat or curved. The formation of the capsid begins with a flat double hexamer made up of six dimers, then the structure grows and curves by the addition of dimers in the presence of RNA and Ca ions. The final appearance of the surface is that of a truncated icosahedron with 20 flat hexamers and 12 convex pentamers. The cohesion of the particle is ensured partly by protein-protein bonds (which allow the edification of empty capsids) and partly by RNA-protein interactions. Viral RNA might promote the formation of the final structure from dimers through the binding of N-terminal region of the capsid protein, containing a basic arginine-rich motif conserved among plant viruses (Choi and Rao, 2003). In TYMV and SBMV (*Southern bean mosaic virus, Sobemovirus*), numerous particles without RNA are observed in the preparations: the bonds between proteins dominate.

The relative proportion of hexamers and pentamers may be different from the 20/12 ratio of the truncated icosahedron when the triangulation number is 3. The hexamers formed by the capsid of AMV (*Alfalfa mosaic virus, Alfamovirus*) are particularly stable; they constitute the dominant form and their interaction with RNA determines the size of the particle, which could take on a bacilliform appearance. The hexamers form a cylindrical surface that curves and closes at the ends by the insertion of pentamers.

RNA of icosahedral viruses has an origin of assembly

The specific recognition between viral RNA and its capsid protein is observed *in vivo*; non-viral RNAs are rarely encapsidated, and the same is true of foreign viral RNAs in the case of co-infection. The determinants of encapsidation of TCV (*Turnip crinkle virus, Carmovirus*) have been precisely recorded *in vivo* by observing the aptitude for encapsidation of mutant viral RNA. A region of 186 nucleotides, at the 3' end of the capsid gene, is essential; it contains an essential motif of 28 nucleotides forming a hairpin (Qu and Morris, 1997). In the multicomponent BMV (*Brome mosaic virus, Bromovirus*), the 3'-terminal structure of viral RNAs (201 nt assuming a tRNA-like conformation) plays a crucial role in *in vitro* RNA packaging, functioning as a nucleating element to initiate virus assembly. But in addition to the tRNA structure, the capsid protein interacts with selective signals located elsewhere on each BMV RNA. The selective packaging signal was identified for RNA3: a 187 nt sequence in the 5' movement protein. So encapsidation of BMV RNA3 is mediated through a bipartite signal (Choi and Rao, 2003).

To be infectious, the virion must undergo a conformational change

It has been mentioned that the end of the TMV virion, upon its entry into the cell, is destabilized by the action of pH and ions. The decapsidation thus primed is pursued by the ribosomes. A process of destabilization of viral particles has been observed in various conditions for isometric viruses the particles of which change conformation and swell when the forces of cohesion diminish. This swelling makes the RNA end accessible to the ribosomes (Brisco et al., 1986). The process of decapsidation linked to translation has been observed, for isometric viruses, in the genera *Alfamovirus, Sobemovirus,* and *Bromovirus* (Brisco et al., 1986).

Viral nucleic acids

■ DNA or RNA

In viruses, genetic information is carried by either DNA or RNA. Double-stranded DNA carries genetic information in the cell. Highly stable in its double-stranded form, it is complemented, for the transfer of information to the protein, by messenger, transfer, and ribosomal RNAs. Viral genomes are present in more varied forms; they are made up of DNA or RNA and either of these supports can be single-stranded or double-stranded.

In plants, only viruses in families *Caulimoviridae, Geminiviridae,* and *Nanoviridae* have DNA genomes. The other viruses have RNA genomes, most of them single-stranded. Single-stranded RNA can be "positive-sense" (RNA(+)), i.e., directly messenger, or "negative-sense" (RNA(–)), the complementary copy of which is messenger. Viral RNA can also be double-stranded (*Reoviridae*). In 1956, it was found that viral RNA could on its own carry information, unlike RNA in the cell, where the primary support is DNA (Fraenkel-Conrat, 1956; Gierer and Schramm, 1956). In evolutionary terms, the success of RNA genomes indicates that it is advantageous for a virus to use this support. Its chemical fragility in the single-stranded form, and especially its sensitivity to ribonucleases, is probably counterbalanced by the formation of secondary double-stranded structures and the close association with proteins. All viruses need a messenger RNA to express their genes. Those in which the genome is not directly messenger use different mechanisms to transcribe their genome into mRNA (Chapter 2).

■ Cellular mRNA ends

Because of the process of synthesis, the chain of nucleotides that constitute RNA is oriented (Fig. 1.15A). The first triphosphate nucleoside positioned carries three

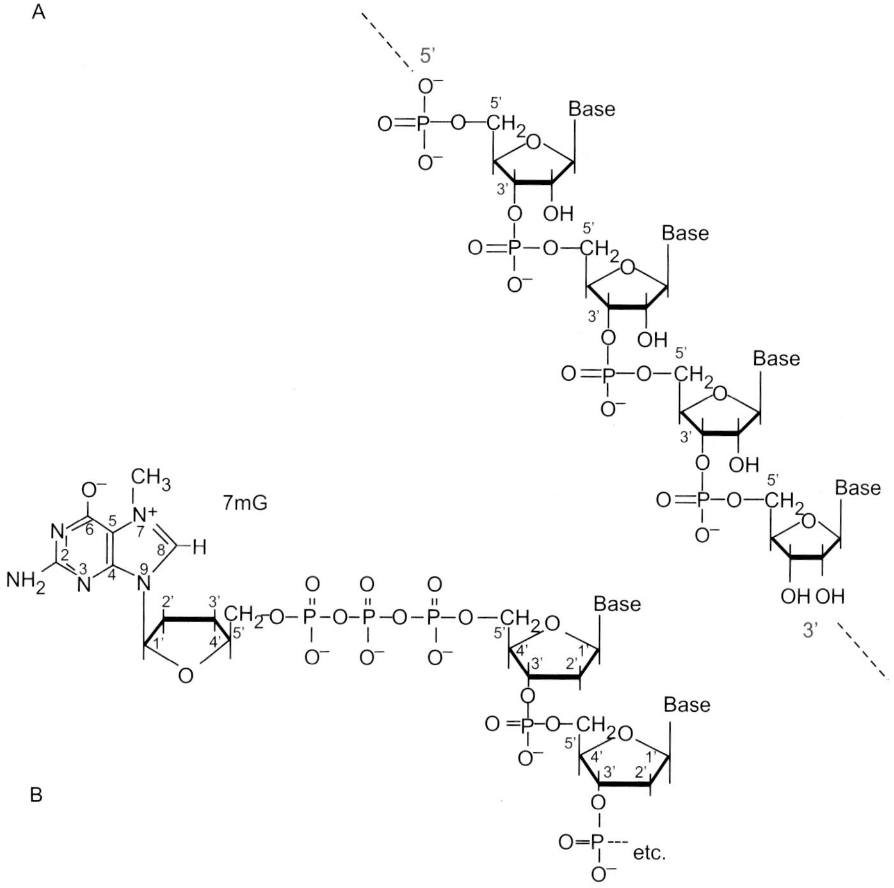

Figure 1.15 A: A segment of oriented RNA chain. B: 5' end carrying a cap.

phosphate residues on the 5' carbon of ribose, and this end is called 5'. The chain grows by the formation of phospho-diester linkages between the 3' carbon and the 5' carbon of the subsequent triphosphate nucleoside. The last nucleoside has a ribose carrying an OH on the 3' carbon; this is the 3' end. These ends carry structures that protect the RNA from the action of exonucleases and directly intervene in the initiation of protein synthesis (Chapter 2). The 5' end of cellular mRNA carries a cap made up of a methylated guanosine in position 7 (m7G) and inverted, forming a triphosphate bond with the first nucleotide of the chain m7G$^{5'}$ ppp^{5}XpYp (Fig. 1.15B). The first two bases X and Y are also methylated. The 3' end carries a polyA of variable length (20 to 400 A). Certain viral transcripts have the same terminal structures as the cellular messengers.

The cellular mRNA begins and ends with non-coding regions of variable length that flank the coding region in which an open reading frame (ORF) opens, beginning generally with the initiation codon AUG in a favourable context, and ending in a termination codon (UAA, UAG, UGA).

Extremities of genomic viral RNA

The ends of single-stranded genomic viral RNAs of positive polarity carry diversified structures.

At the 5' end, there is a cap (Fig. 1.15) (but here X and Y are not methylated), or a viral protein linked to RNA by a covalent bond (called VPg), or even any particular structure (pppX-). The VPg (genome-linked viral protein) is coded by the viral genome. It is not necessary for the translation of RNA. It is cleaved from a polyprotein in which it is part of the replication module (Fellers et al., 1998). Because of its role as a primer in replication, it is linked to the 5' end of the RNA by a covalent bond with a tyrosine (*Potyviruses* and poliovirus) or a serine (*Comoviruses*). Its size varies from 3.5 to 24 kDa depending on the virus.

At the 3' end is a poly A of variable length, or a tRNA-like structure, or any particular structure (-Y). The tRNA-like structure was discovered on the RNA of TYMV (Pinck et al., 1970). The sequence of 159 nucleotides of the 3' end leads to the folding of the chain (Fig. 1.16) by several series of base pairings that give it a part of the structure and functionality of the valine tRNA (Yot et al., 1970; Giégé, 1996). The 3' region of the viral RNA is the place where the replication enzymes recognize it specifically to make a copy (Chapter 3).

Six combinations are found between the various structures of the 3' and 5' ends:

— 5' cap----3' polyA: *Potexvirus, Trichovirus, Benyvirus*

— 5' cap----3' tRNA: *Bromovirus, Cucumovirus, Furovirus, Hordeivirus, Tobamovirus, Tobravirus, Tymovirus*

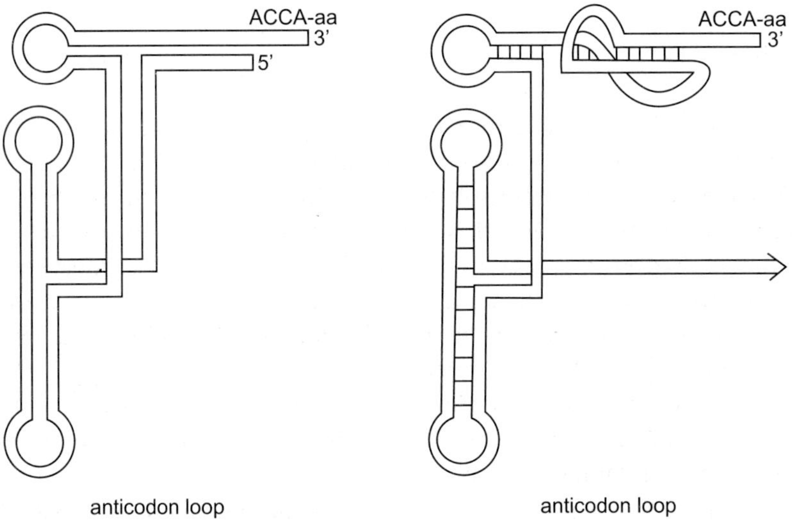

Figure 1.16 3' end of a cellular tRNA, which does not carry a pseudo-knot (at right); 3' end in tRNA-like structure of a viral RNA carrying a pseudo-knot (at left) (Giegé, 1996).

— 5' cap----3' Y: *Alfamovirus, Carmovirus, Closterovirus, Machlomovirus*
— 5' VPg----3' polyA: *Comoviridae (Comovirus, Nepovirus), Potyviridae (Bymovirus, Potyvirus)*
— 5' VPg----3' Y: *Sobemovirus, Polerovirus, Enamovirus*
— 5' X----3' Y: *Luteovirus, Necrovirus*

(where X and Y represent structures that are still not precisely described).

The genomes of plant viruses code for 4 to 12 proteins, and many mechanisms are used by the viruses to express a large amount of information in a minimum of sequences. Their genes are always very close to one another; it is not rare to find overlapping or sometimes superimposition in different reading frames, and some translation strategies allow a still greater increase in the number of proteins synthesized (Chapter 2). The genetic information is carried by a single RNA molecule (undivided or monopartite genome) or several molecules (two or three, divided or multipartite genome). In this last case, the terminal structures of different RNAs are most often identical or very similar.

Secondary and tertiary structures

Some elements of RNA structures can be predicted by the observation of sequences, by mutation experiments, and by phylogenetic analyses. Statistical programs can be used to visualize folds and suggest optimal or sub-optimal structures.

The secondary structures in hairpin or stem-loop form are constituted when a single chain presents complementary inverse sequences, which in pairing will form a stem in a double helix, separated by a few nucleotides forming a loop. These are sites of interaction with cellular or viral proteins, and with other nucleotide sequences. A pseudo-knot is formed when the loop of a hairpin pairs with a nearby or more distant nucleotide sequence, forming a tertiary structure (Fig. 1.17). The tRNA-like structures carried by certain viral RNAs at the 3' end are formed of hairpins and pseudo-knots, the latter being absent in cellular tRNA (Fig. 1.16).

Hairpins and pseudo-knots are present in different locations on the RNA, in relation with translation, replication, encapsidation, and other processes. The

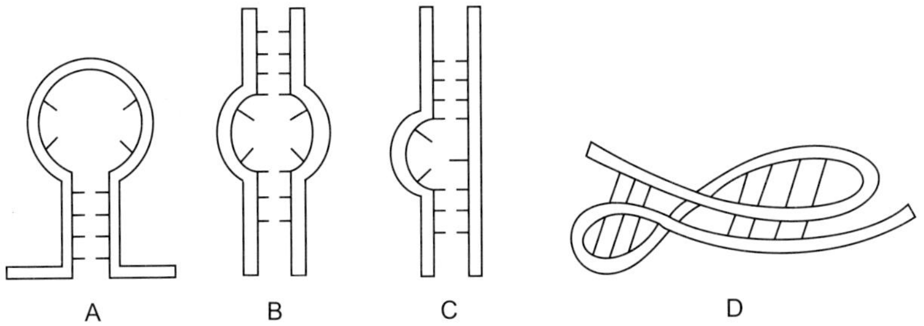

Figure 1.17 Secondary and tertiary structures of RNA. A: Hairpin structure (or stem-loop). B: Internal symmetrical loop. C: Internal asymmetrical loop. D: Pseudo-knot.

ultimate three-dimensional structure of the viral RNA is thus determined by its sequence, then by the secondary structures, and finally by the relationships between secondary structures themselves and between them and the immediate or general environment (nucleic acid, viral or cellular proteins, metallic ions).

■ Structural peculiarities of viral DNA

In the family *Caulimoviridae*, the genome is a circular double-stranded DNA. Each of the strands presents discontinuities with a redundant segment. At the time of infection, the discontinuities are paired and the DNA appears in the form of a mini-chromosome (Chapter 3). The *Geminiviridae* has a genome formed of a circular single-stranded DNA that has a specific hairpin, the origin of bidirectional transcription (Chapter 3).

Viral information: a protected message

The security of the viral message is ensured in two kinds of environments:

- Outside the cell and up to entry into the cell, the capsid packages and transports the viral genome.
- Inside the cell, cellular and viral proteins, as well as the secondary structures, protect the viral genome during the phases of its cycle: translation, replication, and translocation. Then, at the end of the cycle, the capsidial subunits surround the newly synthesized nucleic acids to form new viral particles.

2

CHAPTER

Infection of the Cell: Synthesis of Viral Proteins

To illustrate the process of viral multiplication, we propose to use a military metaphor. An invader (the virus) barges into a peaceful territory (the cell). With an extremely small army and resources (a nucleic acid protected by its capsid), the invader will convert part of the cell's means of production (the synthesis of proteins and nucleic acids) to support its own infinite multiplication. This takeover can succeed only with some mastery over information, by introducing into the general current (the system of protein synthesis and replication) a new piece of information (the viral genetic message) that will consequently redirect the economy (the cellular metabolism). Some collaborators (certain cellular proteins) must be exploited for the purpose. If such collaborators are found, the cell becomes an unwitting host for the invading virus. It will produce mechanisms (e.g., replicase) that escape the control of the cell and multiply the subversive information. Depending on the nature of the information, this process will take various routes. The continuation of the story depends on the capacity of the virus to integrate itself irreversibly into the cellular life cycle, while the cell develops defensive actions and the virus develops counter-defensive actions, selected for in the process of coevolution of the virus and the host.

Viral multiplication is a phenomenon that occurs first on the cellular scale, and necessarily involves a cycle in the cell (Fig. 2.1). After the viral particle enters the cell, the nucleic acid is uncoated and viral information becomes accessible. The viral nucleic acid is often a messenger, as in the case of positive-sense RNA viruses. If it is not a messenger, it is transcribed into messengers. Subsequently translation takes place, during which the viral messenger expresses its information in proteins; it is followed by replication, in which the viral genome serves as a template from which

Viruses cited

BMV (*Brome mosaic virus, Bromovirus*); BYDV (*Barley yellow dwarf virus, Luteovirus*); CPMV (*Cowpea mosaic virus, Comoviridae*); CMV (*Cucumber mosaic virus, Cucumovirus*); HIV (*Human immunodeficiency virus, Retroviridae*); PPV (*Plum pox virus, Potyvirus*); PVX (*Potato virus X, Potexvirus*); SbDV (*Soybean dwarf virus, Luteovirus*); TCV (*Turnip crinkle virus, Carmovirus*); TEV (*Tobacco etch virus, Potyvirus*); TMV (*Tobacco mosaic virus, Tobamovirus*); TNV (*Tobacco necrosis virus, Necrovirus*).

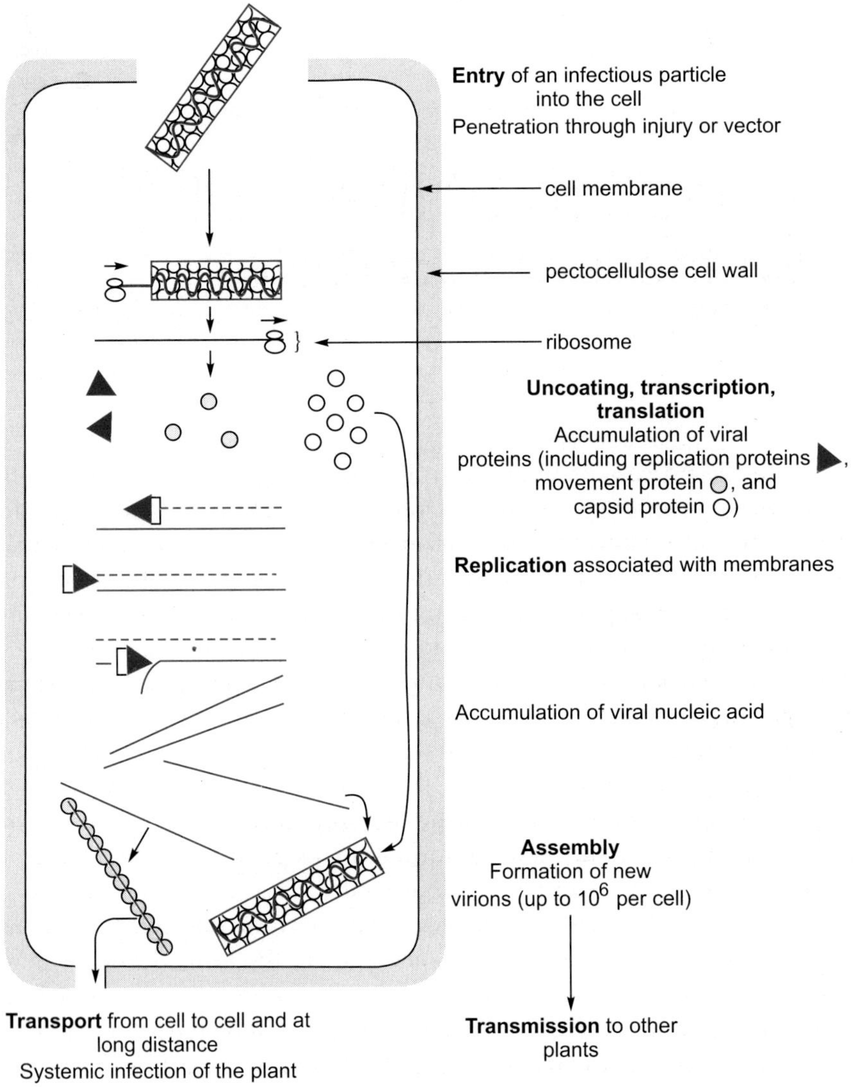

Figure 2.1 Cycle of an RNA virus inside the cell.

numerous copies are made. In fact, these steps are not always distinct: uncoating is related to translation and the first proteins synthesized act immediately and participate in the other phases of multiplication. The viral components thus multiplied will be assembled, disseminated gradually to the neighbouring cells, then transported and replicated throughout the plant.

Entry of viral genetic information into the cell

■ The penetration of a virus into the plant is similar to a break-in

Plants resist the penetration of viruses by means of their cuticle and pectocellulose wall; an intact tobacco leaf soaked in a suspension of TMV (*Tobacco mosaic virus, Tobamovirus*) or in a suspension of viral RNA does not become infected. For the inoculation to succeed, fresh lesions must be created on the leaf surface. In experiments, such lesions are made with carborundum powder (silicon carbide), which is sprinkled on the leaf before the virus suspension is deposited and carefully spread out (the leaf must not be deeply cut). The abrasive is supposed to remove the cuticle locally, as well as scraps of the cell wall and undoubtedly also of the cytoplasmic membrane. It has been shown that an immediate rinsing of the leaf after this scraping increases the number of effective penetrations into the epidermis, probably by eliminating the ribonucleases released by the abrasion.

Most plant viruses are selected constantly in nature to be transmitted from plant to plant by a vector (e.g., aphids, leafhoppers, nematodes, fungi) that, by the breakage it causes in the plant tissues, ensures the penetration of the virus (Chapter 9). Is there an event of specific recognition to trigger the infection process?

An initial recognition that determines the penetration as well as the disassembly of the nucleic acid has been indicated in the case of animal viruses and bacteriophages. For example, HIV (*Human immunodeficiency virus, Retroviridae*) uses receptor sites, CD4s, to bind on macrophages and certain lymphocytes, target cells of this virus in the organism. This fixation on the CD4 receptor does not always suffice for the virus to penetrate, and several cofactors including chemokine receptors make it possible for the viral coat to fuse with the cellular membrane, which will introduce and release the RNA. In the case of the bacteriophage Qβ, the recognition site is located on the sexual hairs of *Escherichia coli*: the fixation of the bacteriophage on this site determines the injection of its RNA into the bacterium.

Such sites have never been located for plant viruses. Nevertheless, older studies report that a "pre-inoculation" of the tobacco leaf by a preparation of TMV capsid protein followed by an overinoculation by the complete virus prevents infection. If the overinocuation is effected with TMV RNA or with another virus, such as CMV (*Cucumber mosaic virus, Cucumovirus*), the infection is normal. This experiment has been interpreted as indirect evidence of the existence of specific sites in the cell that are necessary for the uncoating of TMV particles.

Nevertheless, it has been demonstrated that when the TMV particle is treated at pH 8.2, the RNA can be translated *in vitro* in rabbit reticulocyte lysate. At pH 8.2, the architecture of TMV is sufficiently destabilized (for at least a segment of 49 nucleotides at the 5' end of the RNA) to become accessible to ribosomes, which as they progress displace each subunit of the capsid, one by one. This process is called "disassembly". We can also observe under electron microscope (Fig. 2.2) shortened particles extended by a polysome (Wilson, 1984). The observation of these particles *in vitro* as well as *in vivo* (Shaw et al., 1986) has thus led to the concept of co-translational disassembly. The translation of RNA does not require a previous specific interaction between the virus and a cellular site. It is clear that the information available leaves the debate open on the nature of the very first virus-plant interaction.

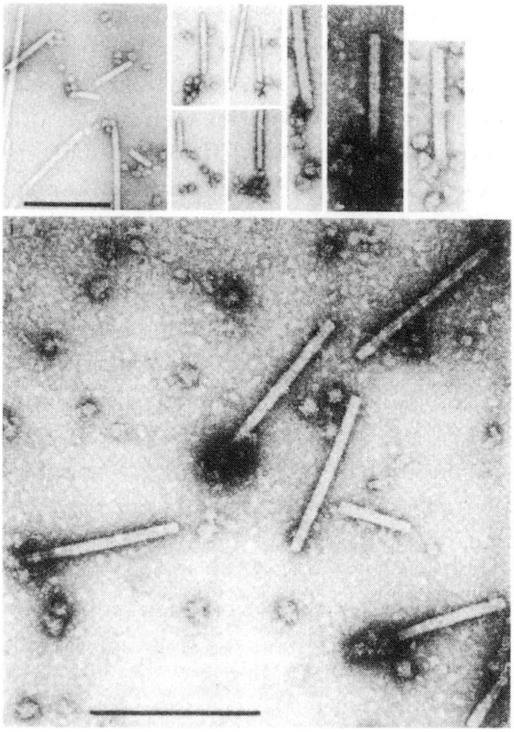

Figure 2.2 TMV particles partly uncoated *in vitro*. In a reticulocyte lysate, the RNA of complete particles treated at pH 8.2 is partly engaged in a polysome. The bar represents 300 nm, the length of a complete particle (Wilson, 1984).

Synchronous infection of protoplasts for the analysis of a viral cycle: advantages and limitations

The protoplast, a cell that survives without a cell wall, can be inoculated directly in the presence of a polycation, polyornithine, using viral RNA (Aoki and Takebe, 1969) or virus particles (Takebe and Otsuki, 1969). Observations under electron microscope indicate a local disorganization of the cytoplasmic membrane by the accumulation of charges of the polycation. This disorganization mimics the breaking of the cell membrane that naturally precedes the infection of the first cell. The penetration can also be realized by means of very brief electric shocks (electroporation).

The protoplast suspension is—unlike in the leaf, where one out of 10,000 cells is initially infected—a synchronous system of infection: the replication of the virus is measured on the protoplasts infected at time 0, and there is no infection of new protoplasts during the incubation. The phases of the infection proceed at the same time in each cell, which favours the analysis of kinetics and mechanisms.

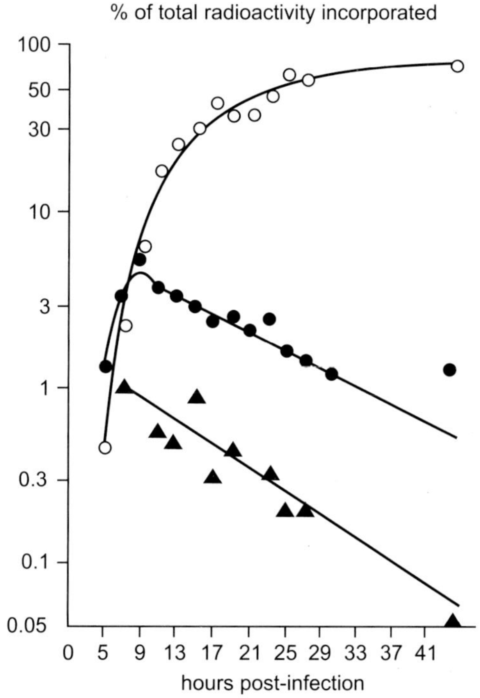

Figure 2.3 Time-course of production of three viral proteins in TMV-infected protoplasts. The Y-axis indicates the percentage of the total incorporation of radioactivity in each of three proteins during the incubation of protoplasts in the presence of ^{3}H-leucine. The X-axis indicates the time between inoculation and beginning of labelling. ▲, 183 kDa; ●, 126 kDa; ○, 17 kDa (capsid).

For example, the infection of tobacco protoplasts by TMV has made it possible to analyse the kinetics of viral protein synthesis (Fig. 2.3) by incorporation of radioactive amino acids over a short period: in this way, early viral proteins 126 kDa and 183 kDa, which have a maximum of synthesis 9 h after inoculation, can be differentiated from a late protein, the capsid protein, which is synthesized over 30 h in much greater quantity than proteins involved in replication. Moreover, the duration of the TMV cycle in the cell can be evaluated at 30 h (Siegel et al., 1978). The kinetics of synthesis of the 30 kDa protein has also been determined (Watanabe et al., 1984).

This system made it possible to understand *in vivo* certain phenomena such as the uncoating of TMV, to determine the time required and to show that this uncoating is bidirectional. The uncoating occurs first from the 5' end to the 3' end, releasing 70% of the viral RNA in 2 to 3 min., then more slowly from the 3' end to the 5' end. The liberation of the origin of assembly is the end of this process, 20 to 25 min. after the penetration of the virus (Wu and Shaw, 1996). The concept associates translation and uncoating from the 5' end to the 3' end. Moreover, the synthesis of 126 and 183 kDa replication proteins is necessary for the realization of the second phase of the uncoating; the newly synthesized viral replicase is implicated *in vivo* in the uncoating from the 3' end to the 5' end (Wu and Shaw, 1997).

The infection of protoplasts is very useful in a comparative analysis of the cellular phenomena of infection in various situations, according to the genetics of the plant and/or the genetics of the virus, for example, when a recessive gene of *Arabidopsis* coding for a cofactor of the replication of TMV-Cg (TMV strain that multiplies in *Arabidopsis*) is mutated (Oshima et al., 1998) or when potato protoplasts having a dominant Rx gene induce a blockage in viral replication of some strains of PVX (*Potato virus X, Potexvirus*) (Kohm et al., 1993).

The system of infection of protoplasts cannot be applied to the study of events occurring in plant tissues (movement, hypersensitivity). Note also that the stress of the preparation under high osmotic pressures makes this cellular context greatly different in its physiology from that constituted by the cell in the plant.

Viral messengers

Viral messengers and cytoplasmic ribosomes

The TMV particles are made up of RNA and capsid proteins. These constituents can be separated for various strains, and they can be inoculated by mixing RNA and proteins of different strains. It is thus observed that the descendants are those of the strain that provides the RNA and that the RNA alone is infectious (Fraenkel-Conrat and Singer, 1957). This experiment was the first demonstration of the infectious potential of viral RNA as the carrier of a specific genetic message.

If we compare the quantity of information carried by the genome of bacteria and that carried by viruses, we observe a change in the order of magnitude resulting chiefly from the absence in viruses of genes involved in energy production, the synthesis of precursors, and the protein synthesis system. The virus carries only the information that allows it to redirect the cellular mechanism; its genome, DNA or RNA, gives the following order to the cell: "multiply me". The coded message is very precisely adapted to the execution of this process. It carries the information for those proteins essential to replication, movement, and transmission that the virus cannot find in the cell. For the reading of its information, the virus depends totally on the cellular mechanism of protein synthesis; its genetic message must be recognized and read by eukaryotic ribosomes. Experiments aimed at blocking infection *in vivo* by specific inhibitors—cycloheximide inhibiting translation by 80S ribosomes (cytoplasmic ribosomes) and chloramphenicol inhibiting translation by the 70S system (in the plasts and mitochondria)—have shown that the 80S system is used in all studied cases. Since cellular messages are generally monocistronic, this system expresses only the first gene present at the 5' end. Consequently, the viruses are adapted to this situation by using various strategies to express all the genes of their message.

When the viral genome is not directly messenger, it is transcribed into messenger RNA

No matter what the nature of the nucleic acid, the genomic organization, and size of particles may be, viruses express their proteins through the intermediary of an RNA recognized by the cell as a messenger.

The simplest case seems to be that of viruses with positive-sense RNA, which constitute the large majority of plant viruses: the viral RNA is translated directly, and consequently the viral RNA without its capsid is infectious. The genetic material of certain viruses with positive-sense RNA, easy to purify in large quantities, has been used *in vitro* to understand the mechanisms of the translation of eukaryotic messengers; note, however, that while cellular messengers generally have a cap at the 5' end and a polyA at the 3' end, viral messengers very frequently have a different structure at one or both ends (Chapter 1).

DNA viruses use the cellular RNA polymerase to transcribe their genomes into messenger RNAs. Double-stranded DNA viruses (*Caulimoviridae*) generate two messenger RNAs by transcription effected by the cellular RNA polymerase II, in the nucleus in which the virus is found in the form of a minichromosome. The viral DNA has two regions with typical sequences (TATA-box) that are recognized by this RNA polymerase, the 19S promoter and 35S promoter (named by the sedimentation constant of their respective transcripts). The 35S promoter is particularly effective, and the transcription is the genome amplification phase. The viral DNA also has terminators recognized by the RNA polymerase II. The transcription gives rise to two major capped and polyadenylated transcripts, 19S and 35S. The ability of the 35S promoter to allow a very high transcription level of a gene placed downstream, in a constitutive manner and in the absence of viral infection, is widely used in genetic engineering. Single-stranded DNA viruses (*Geminiviridae, Nanovirus*) are bidirectionally transcribed from an intergenic region, where the origin of replication is also found along with two back-to-back promoters comprising sites of recognition by the cellular RNA polymerase II (see Chapter 3, Fig. 3.18).

Except in positive-sense RNA viruses, in which the RNA is directly messenger, and DNA viruses that use the cellular mechanism of transcription, the virus must generate its own messenger RNAs. Since the cell does not have enzymes copying long RNA, this function is realized by a viral transcriptase encapsidated in the virion with the genomic RNA in the cases of double-stranded RNA viruses (*Partitiviridae, Reoviridae*) and negative-sense RNA viruses (*Rhabdoviridae, Bunyaviridae*). In the family *Reoviridae*, the genome is segmented into 10 double-stranded RNAs that are transcribed as monocistronic messenger RNAs. The *Tospovirus* and *Tenuivirus* have ambisense RNA, i.e., the negative-sense strand and its positive-sense complementary strand each have an open reading frame in their 5' region.

The initiation of translation requires many partners

Cap-dependent initiation (Fig. 2.4A)

The translation of messengers is a process with several steps: initiation, elongation, and termination. It results from protein-protein, RNA-protein, and RNA-RNA interactions. Initiation is a particularly complex step (Browning et al., 1996). Eukaryotic messengers are read by ribosomes from their 5' end, where the presence of a hypermethylated base in the cap (m^7Gppp-) helps recognition by the eukaryotic initiation factor eIF4F (a complex of several proteins, notably eIF4E, which is attached to the cap, eIF4A, which has helicase activity, and eIF4G, which is attached to mRNA and to the ribosome). The small ribosomal subunit 40S associated with the eIF3 factor and carrying the methionine initiator tRNA binds to this set through an eIF4F-eIF3 interaction; it scans the messenger RNA until the recognition of an initiation codon AUG in a favourable context. The large ribosomal subunit 60S thus joins the small subunit, forms the 80S ribosome and protein synthesis can begin. It proceeds up to the first stop codon (UAA, UAG, or UGA), where the ribosomes are disassembled. Generally, in a eukaryotic cell, each gene is transcribed in the form of a monocistronic messenger RNA (i.e., containing a single open reading frame (ORF)).

Even though the cap and polyA are found at two ends of the molecule of cellular messenger RNAs, many observations have shown that the translation initiation depends on their synergy and their functional interaction (Gallie, 1996, 1998). In plants and yeast, the agents of this interaction are notably eIF4G (which is part of eIF4F) and a protein that binds specifically to polyA (called polyA binding protein, pAbp); these two proteins are physically associated (Tarun and Sachs, 1996). The result is a circularization of the messenger RNA by protein-protein interaction, which is visible under microscope (Wells et al., 1998). This arrangement allows the selection of intact messenger RNA.

Cap-independent initiation (Fig. 2.4B and C)

Since the cap and the polyA act in concert to initiate the translation, how are translation factors recruited when one of the elements is missing, or both? In fact, the viral RNAs that are directly messengers do not all have caps. Some have at the 5' end a protein linked covalently (VPg), others have neither cap nor VPg. These RNAs are highly active messengers, which indicates that the stimulating activity of polyA is exerted also in the absence of the cap, as has been demonstrated in yeast (Preiss and Hentze, 1998). The mRNAs that are naturally not capped are viral RNAs.

The situation of RNA carrying VPg at the 5' end and a polyA tail at the 3' end was specifically studied in the family *Picornaviridae* of animal viruses (poliovirus). The presence of VPg is not essential for their translation. The non-coding 5' region is very long, folded into numerous secondary structures, and it contains an internal

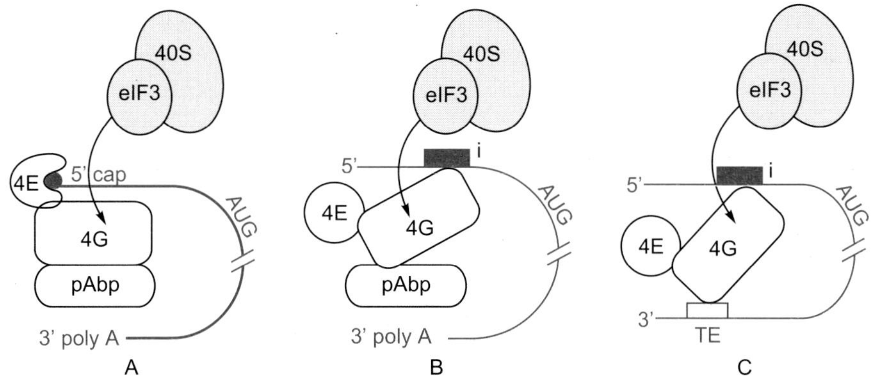

Figure 2.4 Initiation of cellular and viral RNA translation. Interaction of translation factors with RNA ends. A: RNA messenger with cap and polyA. B: RNA without cap, initiation independent of cap, i = initiation zone. C: RNA without cap and polyA (example of BYDV). TE = translational enhancer.

ribosome entry site (IRES), to which the ribosomal subunit 40S binds directly. The IRES region replaces the cap to locate the subunit 40S on the messenger. Like that of *Picornaviruses*, *Potyvirus* RNA lacks cap structures at the 5' end and has a polyA tail at the 3' end (Fig. 2.4B).

In PPV (*Plum pox virus, Potyvirus*), the ribosomes run through the 5' region from its end to an AUG codon placed in a favourable context (especially a puric base in position -3 with respect to the AUG initiation codon), having ignored an initiation codon placed in a less favourable context. This is called "leaky scanning" (Simon-Buela et al., 1997). In TEV (*Tobacco etch virus, Potyvirus*), the non-coding 5' region of 143 nucleotides contains two centrally located regulatory elements required to direct cap-independent translation (Niepel and Gallie, 1999); both elements are required to interact functionally with the polyA through a mechanism involving specifically eIF4G (Gallie, 2001). In several potyviruses, the capacity of the non-coding 5' region placed between two reporter genes to promote the expression of the second gene has been demonstrated (Levis and Astier-Manifacier, 1993; Basso et al., 1994; Niepel and Gallie, 1999).

Some RNA have neither cap nor polyA (Fig. 2.4C). The RNA of BYDV (*Barley yellow dwarf virus, Luteovirus*) comprises a short sequence (translational enhancer, TE) located in the 3' region, at more than 4.5 kb from the 5' end, which strongly stimulates translation (100 times). However, if a cap is added to the viral RNA, this sequence is not necessary. Thus, there is a mechanism that permits initiation in the absence of the cap. The initiation factor eIF4F recognizes this region, which suggests the role of TE in the fixation of the translation complex at the 5' end (Wang et al., 1997; Allen et al., 1999). Moreover, the genomic and subgenomic RNAs of several BYDV isolates, SbDV (*Soybean dwarf virus, Luteovirus*), and TNV (*Tobacco necrosis virus, Necrovirus*) have conserved stem-loop structures in their 3' and 5' terminal sequences allowing direct base-pairing and formation of a closed loop mRNA necessary for translation (Guo et al., 2001). A strong stimulation of the translation by a functional cooperation between the non-coding 3' and 5' regions has been described for the satellite of TNV and for TCV (*Turnip crinkle virus, Carmovirus*) (Meulewaeter et al., 1998; Qu and Morris, 2000).

The cap-independent translation represents several advantages for the viral RNA. The initiation phase is not closely dependent on eIF4F and eIF4E, the quantity of which is often limiting (Leonard et al., 2000). Moreover, methyl-transferase and guanyl-transferase functions necessary for the addition of the cap become useless (Wang et al., 1999). Several IRES were identified in cellular mRNAs involved in stress response or in the control of cell proliferation and differentiation.

Translation of the viral messenger

Competition between viral and cellular messengers

Translation of the viral messenger is often favoured over that of cellular messengers

Role of 5' structures

The presence of a cap results from the expression of a viral function (methyl-guanyl-transferase) carried by the C-terminal part of the 126 kDa protein of TMV or the protein 1a of BMV (*Brome mosaic virus, Bromovirus*), for example.

Before the initiator AUG is the non-coding 5' region of variable length, also called the "leader". Some leaders have been particularly studied. The TMV leader, 68 nucleotides long and carrying a cap, is particularly effective in promoting the translation of the 126 kDa protein *in vitro* and *in vivo* in protoplasts. It can also be used to stimulate the translation of other proteins in constructions of chimeric genes (Gallie et al., 1987). It is highly rich in adenine and uracil and contains no G. It has no evident secondary structure and contains repeat CAA motifs that play an important role in the efficiency of translation of downstream genes (Gallie and Walbot, 1992). Other leaders are proved to be good stimulators of translation, even outside their natural context. Leaders capable of realizing internal initiation contain nucleotide motifs present in the IRES that, in synergy with other factors, give them this property (Niepel and Gallie, 1999).

Role of 3' structures

The non-coding 3' region acts in synergy with the non-coding 5' region following interactions between these two regions and translation factors. For example, the 5' region of the RNA of TMV, which has a cap, interacts with the 3' region, which has a secondary structure of a pseudo-knot type just upstream of the terminal structure in tRNA (Leather et al., 1993). These two elements act in conjunction to stimulate the translation. Generally, the messenger RNAs, cellular or viral, must now be represented as ball-shaped structures in which the ends are very close, allowing multiple protein-protein interactions (Fig. 2.4). These interactions are seen *in vivo* by the double hybrid method in yeast, and *in vitro* by co-precipitation and by immunological detection. The cellular factors fixed by the 3'-polyA ends begin to be determined (polyA binding protein, pAbp). In the case of the luteovirus BYDV mentioned above, it seems that the TE region directly recruits the translation factors; these factors are carried near the 5' end (Fig. 2.4). There is total functional substitution since the TE region serves its function even when inserted in the 5' region (Wang et al., 1997). The TE segment is also present in the subgenomic RNAs 1 and 2; it may play a central role in the regulation of translation of various viral RNAs (Wang et al., 1999).

Role of IRES, internal ribosome entry site

In the family *Picornaviridae,* in which the RNA carries a VPg, there is simultaneously stimulation of the translation of viral RNA and repression of that of cellular RNAs by proteolysis of the eIF4G factor necessary to the translation of capped messengers; a viral protease cleaves the cap-binding factor eIF4E from eIF4G (Sachs et al., 1997). This aggressive strategy allows competition with the cellular messengers that is favourable to the virus.

Activation of translation by a viral protein

In *Caulimovirus*, the DNA is first transcribed into two RNAs; the 35S RNA, which is a complete copy of the genome, is polycistronic and carries 7 ORFs; the 19S RNA is the monocistronic messenger of the P6 protein. In an attempt to explain the low *in vitro* translation efficiency of 35S RNA, it has been observed that the P6 protein acts in *trans* as a translation activator of the 35S transcript (Bonneville et al., 1989). This protein, called transactivator, plays a major role in the translation of different ORFs of the 35S RNA that are very close or overlapping, helping the ribosomes at the end of the ORF to begin the reading of the next ORF (termination-reinitiation) (Jacquot et al., 1997). The mode of action of P6 will involve a close interaction with ribosomal proteins; one of them has been identified as a component of the ribosomal subunit 60S (Leh et al., 2000). Phenomena of splicing have also been described in the caulimoviruses and could explain the expression of certain internal ORFs (Kiss-Laslo et al., 1995).

Expression of all the viral genes

The viral genomes present an extreme compactness (the very large majority having a genome less than or equal to 10 kb) that often leads them to express several proteins from a single segment. Their messenger RNAs code most often for several genes while the eukaryotic cell can usually read only the first of them. Various strategies to overcome this constraint aim to ensure the synthesis of all viral proteins while retaining the possibility of regulating their expression. Some of these mechanisms are rarely used by cellular mRNA but are exploited effectively by viral messengers. The viral messenger RNA is not passively translated by cellular ribosomes; the mechanisms necessary to secure its expression are inscribed in its primary, secondary, and tertiary structures.

Fragmentation of message

A simple mechanism to accommodate the limitations of monocistronic translation is to fragment the message by either of the following means:

— *By segmentation of the genome.* Many genera (40%) have their genetic information divided among two or more nucleic acid molecules. The various segments can be assembled in a single particle (*Tospovirus, Reoviridae*). More frequently, each segment is coated alone in a particle; these viruses are termed "multi-component" viruses (*Begomovirus, Partitiviridae, Comoviridae, Bromoviridae, Dianthovirus, Idaeovirus, Tobravirus, Furovirus, Pomovirus, Pecluvirus, Benyvirus, Hordeivirus, Bymovirus, Crinivirus, Ourmiavirus* ...). Each messenger RNA has a 5' end accessible to ribosomes.

— *By segmentation in several transcripts.* In the minus-sense RNA *Rhabdoviridae*, the monopartite genome is transcribed in the form of several monocistronic messengers.

Transcription in subgenomic RNA

The TMV RNA carries three ORFs; only the 5' ORF is translated directly on the genomic RNA, the other two are silent in the translation process of genomic RNA. They are expressed by the intermediary of two collinear RNAs that are partial copies of the genomic RNA. In these copies, called subgenomic RNAs, a gene is found in position 5' and thus susceptible of being read (Fig. 2.5). Generally, the copy is realized by replicase (see Chapter 3) from an internal transcription promoter located on the negative-sense strand, and it proceeds up to the 3' end in the absence of a termination signal. The result is that the subgenomic RNAs are always co-terminal with the 3' end.

Many viral genera have subgenomic RNAs to express the genes close to the 3' end (Miller et al., 1985; Smith and Harris, 1990). Generally, a small number of subgenomic RNAs are produced, but when the genome is large, many subgenomic RNAs are produced (up to 10 in *Closteroviridae*). They are coated if they carry the capsid protein recognition sequence; they may be capped, which gives them a high translation efficiency. For example, the subgenomic RNA coding for the TMV capsid is capped while that coding for the 30 kDa protein is not. The intensity of the subgenomic transcription is an element that regulates the expression of proteins. The 30 kDa protein (movement protein) is expressed in small quantity and in a transient fashion at the beginning of infection; the 17.5 kDa protein (capsid) is expressed strongly during the infection (Fig. 2.3).

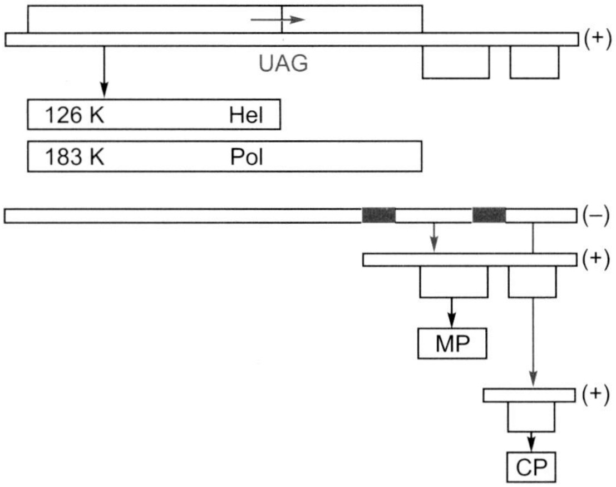

Figure 2.5 Translation and transcription of TMV RNA. The ORF 1 contains a UAG codon that allows the readthrough and synthesis of two proteins 126 and 183 kDa. The internal promoters of transcription (in red) located on the negative strand trigger the synthesis of subgenomic RNAs, messengers of the movement protein (MP) and the capsid (CP).

It must be emphasized that expression through subgenomic RNAs leads to two cycles of translation: early translation of replication proteins, then translation of other proteins (e.g., movement protein, capsid). The principal groups involved are: *Bromoviridae, Closterovirus, Dianthovirus, Enamovirus, Furovirus, Hordeivirus, Idaeovirus, Luteovirus, Polerovirus, Machlomovirus, Nucleorhabdovirus, Potexvirus, Sobemovirus, Tenuivirus, Tobamovirus, Tobravirus, Tombusviridae, Tospovirus, Tymovirus.*

Facultative reading of AUG initiation codon (leaky scanning)

When several AUG are present in the 5' region, the efficiency of recognition of one of them depends on its flanking sequences and possible secondary structures. Some viral RNAs have two AUG, the first in a suboptimal context and the second in an optimal context for the translation initiation. The first may be ignored by some of the ribosomes in favour of the second. This mechanism presents different forms that all allow the simultaneous synthesis of two proteins from the same RNA segment. Two proteins with the same C-terminus and different N-terminus (the 95 kDa and the 105 kDa proteins of CPMV (*Cowpea mosaic virus, Comoviridae*)) are initiated from two AUG in the same reading frame. In the genus *Tymovirus*, two AUG close to the 5' end open two overlapping reading frames in different phases. The same is true in genera *Luteovirus* and *Polerovirus* for ORF 3 and 4 expressed through the intermediary of a single subgenomic RNA (Dinesh-Kumar et al., 1992), as well as in genera *Enamovirus* and *Tombusvirus*.

The mRNA of caulimoviruses begins with an exceptionally long leader (600 nucleotides), which carries several very small ORFs. The ribosomes normally scan

A famous example: *Tobacco mosaic virus*

TMV (*Tobacco mosaic virus*) has a special place in the history of virology. It was the first virus to be described, purified, crystallized, and obtained in the form of an infectious nucleic acid. Its RNA was the first messenger introduced in the *in vitro* protein synthesis system developed by Nierenberg and Mathaei (1961); with the TMV RNA as messenger, they hoped to observe the synthesis of the capsid protein. In fact, the results showed the synthesis of two proteins of 126 and 183 kDa, and the absence of the capsid of 17 kDa, which is the most abundant *in vivo*. The nucleotide sequence (Goelet et al., 1982) has allowed for an explanation of this phenomenon and a description of the translation of this messenger.

Genomic RNA (Fig. 2.5) carries a cap at the 5' end. A non-coding region of 68 nucleotides is followed by the initiator AUG codon of the reading frame, which ends with a UAG stop codon. This ORF corresponds to a protein of 126 kDa. A few of the ribosomes pass over the stop codon (this is called read through), continue up to the UAA stop codon, and synthesize the 183 kDa protein (Pelham, 1978). These two proteins are involved in replication. The rest of the genomic RNA is not read and remains silent. The corresponding proteins are read on two messengers that are found in the infected cell. These are partial copies of genomic RNA, called subgenomic RNAs, which each carry at the 5' end an ORF accessible to ribosomes, one coding for the 30 kDa movement protein and the other for the 17 kDa capsid. This example shows how the virus uses several tricks to express, in a eukaryote cell, four proteins with a genome that has only around 6400 nucleotides. The high density of coding information correlated with a very efficient expression of this information is a general feature of small viral genomes.

this leader and then short-circuit the central part and are transferred close to the beginning of the first large ORF (Scharer-Hernandez and Hohn, 1998).

Stop-codon suppression or readthrough

In some circumstances, the presence of a stop codon may not lead to the end of the synthesis of the protein. Some of the ribosomes may continue the reading and thus allow the highly differentiated expression of two open reading frames co-terminal at the 5' end. This mechanism was discovered and studied specifically in TMV (Pelham, 1978). If the TMV RNA is translated in an *in vitro* protein-synthesizing system, two proteins are obtained of 126 and 183 kDa, initiated at the same AUG codon. These proteins have distinct roles in replication. The termination of the 126 kDa protein occurs at the level of an "amber" UAG stop codon (Fig. 2.5). There are rare tRNAs called "suppressors", capable of effecting the codon-anticodon linkage needed to pursue the synthesis even in the presence of a stop codon. The presence of a tRNA suppressor of the "amber" codon has been verified in tobacco, where it is found with a frequency of around 10^{-3}. It has, moreover, been demonstrated by mutagenesis (Valle et al., 1992) that the two codons downstream of an "amber" codon are essential for the readthrough of the RNA of TMV, which is an infrequent event (the ratio of 126 kDa to 183 kDa has been assessed at 20/1 *in vivo*). This process of polymerase synthesis by readthrough is found in genera *Tobamovirus, Tobravirus, Tombusvirus, Carmovirus,* and *Furovirus* (Skuzecski et al., 1991).

The readthrough strategy is used by many bacterial, animal, and plant viruses. In some cases, signals that favour readthrough have been identified on the RNA. In BYDV-PAV, the stop codon terminating the capsid gene can be ignored (in around 1% of cases) and a fusion protein is thereby obtained that will be part of the virion and intervene in transmission. This phenomenon seems to depend simultaneously on the RNA sequence immediately downstream of the stop codon and a second more unexpected signal located 700 bases downstream (Brown et al., 1996). The readthrough of the capsid gene is found in genera *Luteovirus, Polerovirus, Enamovirus,* and *Benyvirus.*

Frameshift

An alternative method for reading the genetic message is frameshift. In several genes, the protein synthesis coded by the ORF nearest the 5' end (P1) is accompanied by the synthesis of a second protein (P1-2, which is a fusion protein of ORF 1 and 2) in a smaller quantity (Prufer et al., 1992; Brault and Miller, 1992). This second protein is very often polymerase, and its synthesis results from a slipping, before the stop codon of P1, of some of the ribosomes that shift their reading frame by retreating (–1 frameshifting) one nucleotide (*Dianthovirus, Enamovirus, Luteovirus*) or advancing (+1 frameshifting) one nucleotide (*Closterovirus*) (Fig. 2.6).

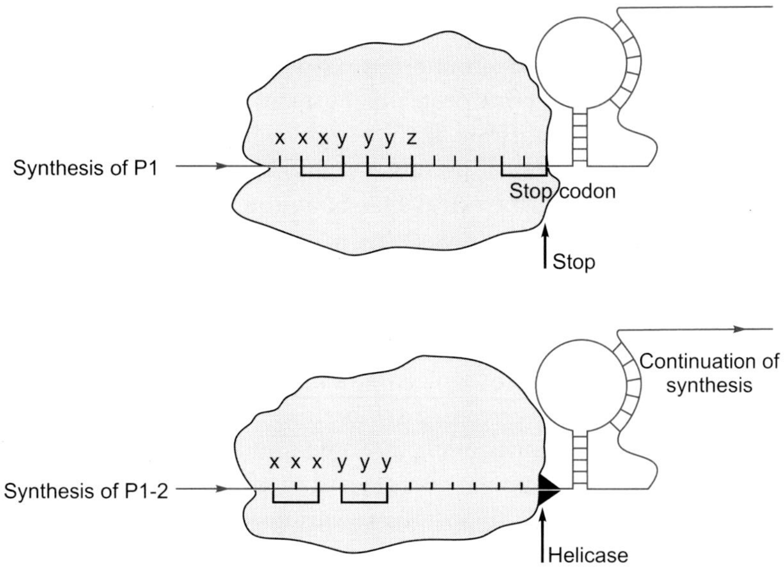

Figure 2.6 Frameshift. Above, synthesis of P1. Below, the two tRNAs associated with triplets XXY and YYZ shift by retreating by one base and reassociate with the triplets XXX and YYY. The ribosomes (in grey) continue the reading and synthesize the protein P1-2 (Farabaugh, 1996).

The (–1) frameshift may operate by means of the presence of a slippery sequence, a heptamer made up of repeated nucleotides (X XXY YYZ, where Y is A or U), followed by a stop codon and a hairpin secondary structure or pseudo-knot, which favours a pause of ribosomes. The majority of ribosomes read X XXY YYZ and then find the stop codon of protein P1. Some pause at the approach of a pseudo-knot; the two tRNAs of peptide site and amino acid site of ribosome slip simultaneously by one nucleotide and reassociate with the RNA in the XXX YYY phase. The ribosome escapes the stop codon and continues synthesis after having destroyed the secondary structure by means of helicase activity of translation factors, thus synthesizing P1-2 (Fig. 2.6). The polymerase gene in the genera *Luteovirus, Dianthovirus*, and *Enamovirus* is expressed by a (–1) slipping of the reading frame by some of the ribosomes. This slipping programmed by a favourable sequence and secondary structure is known presently in about 10 genera of plant viruses as well as in animal viruses (especially *Retroviridae*) (Dinman et al., 1998). It also occurs in some cellular messengers (bacteria, yeast, man) (Alam et al., 1999).

Many functional analogies can be noted between readthrough and frameshift, both of which involve a reading "error" and a "recoding". Some of their signals carried by RNA are sometimes interchangeable. The result is a coordinated and regulated synthesis of two partly overlapping proteins, one of them in a small quantity.

Cleavage of a polyprotein

Many RNA viruses have messengers that carry a single ORF corresponding to several proteins. Their translation results in a large protein called "polyprotein". The functional proteins appear in the course of a process of proteolysis realized by viral proteases. These proteases (serine proteases, cysteine proteases, aspartic proteases) are highly specific to their substrate. In the genus *Potyvirus*, three different protease activities carried by viral proteins have been identified (nuclear inclusion NIa, helper component HC-Pro, and P1); they cleave the polyprotein of around 350 kDa into 10 functional proteins. NIa can effect cleavages in *cis* and in *trans*, while HC-Pro and P1 self-cleave at their COOH end (Fig. 2.7).

This process allows at least two levels of regulation: regulation of the rate of proteolysis by each of the proteases and regulation of the efficiency of cleavage site recognition. The different sites recognized by a single protease are not hydrolysed at the same rate. Slow processing produces stable intermediates playing a role during the virus infection cycle (Merits et al., 2002). Despite these regulations, the final products are produced in equimolar quantities; the excessive non-structural proteins accumulate after use and are sequestered in various cellular compartments in the form of inclusions (cytoplasmic, nuclear inclusions) (Riechmann et al., 1992). The viral proteins that effect proteolysis have other functions in the viral cycle (Schaad et al., 1996). The post-translational proteolysis is realized by viral enzymes, and it is

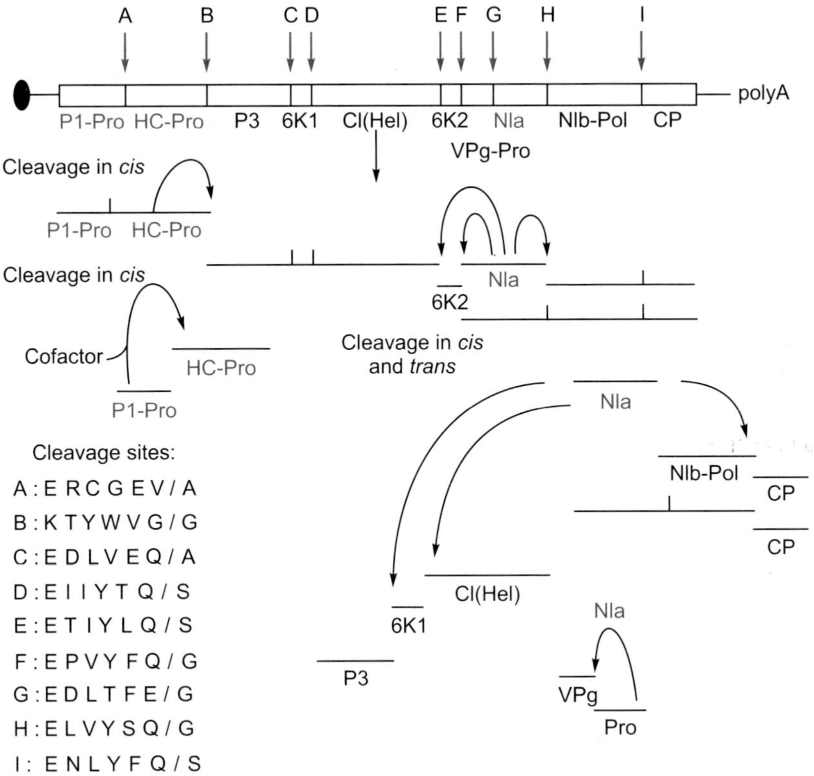

Figure 2.7 Cleavage of the potyvirus polyprotein. The arrows A through I indicate sites of proteolysis on the genetic map of TEV; the sequence of cleavage sites is given. The respective proteolytic activities of NIa, HC-Pro, and P1-Pro are shown.

sufficiently specific to constitute an antiviral target, as in the case of antiproteases against HIV.

This strategy involves the following viruses to varying degrees: *Potyvirus* (S, C), *Bymovirus* (S, C), *Comoviridae* (S), *Sequiviridae* (S), *Tymovirus* (C), *Closterovirus* (C), *Benyvirus* (C), *Marafivirus* (C), *Polerovirus* (S) (S = serine protease; C = cysteine protease).

Multiple strategies

Most viruses use several strategies of expression and regulation. From genome maps of type members of different viral genera, it is easy to observe that a virus or group of viruses use several mechanisms. For example, we find the following:

— Subgenomic RNA and multipartite genome: *Bromovirus*, *Hordeivirus*
— Polyprotein processing and bipartite genome: *Bymovirus*
— Subgenomic RNA and readthrough: *Tobamovirus*

- — Polyprotein processing, 2 AUG, bipartite genome: *Comovirus*
- — Polyprotein processing, 2 AUG, subgenomic RNA: *Tymovirus*
- — Subgenomic RNA, frameshift, readthrough, internal initiation: *Luteovirus*

It is often observed that a strategy or set of strategies is common to different viruses of a genus or family (Zaccomer et al., 1995). For example, translation into polyprotein is common to the family *Potyviridae*.

The use of host ribosomes prevents virus control by antibiotics

Control against pathogenic bacteria has made great progress with the use of antibiotics, which specifically inhibit the protein synthesis system of the bacterium, allowing the cellular system to function. A similar mechanism cannot exist in the case of viruses, except if the processes used by the viruses are very infrequent in the translation of cellular messengers.

3

CHAPTER

Infection of the Cell: Replication of the Viral Nucleic Acid

Replication is generally understood to be the cellular process by which the viral genomic nucleic acid is amplified; the term "transcription" denotes the synthesis of messengers, for example the subgenomic RNAs. From the 1960s onward, there have been many studies devoted to replication. From these studies, the principal aspects of the process have been precisely described. However, many points remain unknown, and the results obtained with a virus can only cautiously be extrapolated to viruses having the same genomic organization. The objective is to define the viral and cellular proteins involved in the process as well as the polynucleotide structures on which they act, for the different types of viral genomes.

Replication of positive-sense RNA viruses

Positive-sense RNA viruses represent circa 70% of known plant viruses; their study has provided many insights into the successive stages of replication. Translation of

Viruses cited

AMV (*Alfalfa mosaic virus, Alfamovirus*), BMV (*Brome mosaic virus, Bromovirus*); BNYVV (*Beet necrotic yellow vein virus, Benyvirus*); BSV (*Banana streak virus, Badnavirus, Caulimoviridae*); BYDV (*Barley yellow dwarf virus, Luteovirus*); CaMV (*Cauliflower mosaic virus, Caulimovirus*); CMV (*Cucumber mosaic virus, Cucumovirus*); CPMV (*Cowpea mosaic virus, Comovirus*); FBNYV (*Faba bean necrotic yellow virus, Nanovirus*); LNYV (*Lettuce necrotic yellow virus, Rhabdoviridae*); PPV (*Plum pox virus, Potyvirus*); PSbMV (*Pea seed-borne mosaic virus, Potyvirus*); PVCV (*Petunia vein clearing virus, Caulimoviridae*); PVX (*Potato virus X, Potexvirus*); RHBV (*Rice hoja blanca virus, Tenuivirus*); RSV (*Rice stripe virus, Tenuivirus*); SYNV (*Sonchus yellow net virus, Rhabdoviridae*); TBRV (*Tomato black ring virus, Comoviridae, Nepovirus*); TCV (*Turnip crinkle virus, Tombusviridae, Carmovirus*); TEV (*Tobacco etch virus, Potyvirus*); TGMV (*Tomato golden mosaic virus, Geminiviridae*); TMV (*Tobacco mosaic virus, Tobamovirus*); TRSV (*Tobacco ringspot virus, Comoviridae*); TuMV (*Turnip mosaic virus, Potyvirus*); TVCV (*Tobacco vein clearing virus, Caulimoviridae*); TYLCV (*Tomato yellow leaf curl virus, Begomovirus*); TYMV (*Turnip yellow mosaic virus, Tymovirus*); WDV (*Wheat dwarf virus, Mastrevirus, Geminiviridae*).

the parental infectious RNA gives rise to viral replication proteins. The parental RNA then forms a replication complex with newly synthesized viral proteins and cellular proteins of partly unknown nature. This complex recognizes the 3' end of the parental RNA which is first copied and then incorporated in the replicative form constituted by the positive strand and its complete complementary copy. Under the action of the replicase, the replicative form evolves into an intermediary replication form made up of the negative strands and several growing positive strands. The positive strands are released and constitute the descendants of the parental RNA. There are thus two main distinct stages: in the first a positive strand is copied into a negative one, and in the second the negative strand is copied into numerous positive ones (box and Fig. 3.1).

Which genes govern replication?

Is replication governed by viral genes or plant genes? This is the question that has been posed in the earliest studies on replication of RNA viruses. Here the study of the replication of RNA bacteriophages has played a pioneering role. A semi-purified extract from bacteria infected by the bacteriophage Qβ presented a replicase activity *in vitro*, that is, it amplified the viral RNA. This extract was already sufficiently purified to determine its composition in polypeptides and to observe that it contained one subunit of viral origin and three cellular proteins (Blumenthal and Carmichael, 1979). Following this example, researchers attempted to purify, from plants infected by viruses, replication complexes functioning *in vitro* (Fraenkel-Conrat, 1983). Several host-virus models, refined over years, gave a coherent picture of the replication of RNA viruses. This process is effected by a viral enzyme, polymerase, with the help of other viral proteinic factors (notably helicase, methyl-transferase, and others) as well as cellular factors, all of which constitute the replicase. Replicase is closely associated to the viral RNAs it copies, and together with them forms the replicative complex.

However, in non-infected cells there is an endogenous RNA polymerase that copies DNA or RNA templates. It was first described in chinese cabbage extracts and then in other plants, in which an RNA polymerase of 130 kDa copying various template RNAs *in vitro* was very strongly stimulated after infection by positive-sense RNA viruses (Astier-Manifacier and Cornuet, 1978). The product of the reaction is quite small, which suggests that this endogenous polymerase plays no direct role in viral replication (Schiebel et al., 1993, 1998). This activity copurifies with the poorly purified replicative systems and has complicated the analysis of the first isolation assays of replication systems. It is now known that it is not included in the replication complex and that its role must be looked for at other levels of the host-pathogen relationship (Chapter 6).

Viral polynucleotides in the infected cell

Following studies of RNA bacteriophages, viral RNAs of sequence complementary to infectious RNA were looked for and found to be present in cells infected with plant viruses. They are conventionally called negative-sense RNA (or (–) sense RNA). After extraction, the viral RNA is found in a double-stranded structure composed of a complete positive strand and a complete negative strand, called *replicative form*. There are also branched forms consisting of a negative strand linked to positive strands of variable length; this construction is called the *intermediary replication form* (Fig. 3.1).

These polynucleotides can be distinguished by means of their different sensitivity to degradation by pancreatic ribonuclease because of their structure. The positive strands and negative strands are joined by hydrogen bonds. These bonds are reinforced when the ionic force is high. In the presence of 0.2 M NaCl, the double-stranded RNAs are resistant to hydrolysis by pancreatic ribonuclease, while the single-stranded RNAs are hydrolysed. Because of these differences, it is possible to follow the progression of a radioactive RNA precursor incorporated during a very short time to infected cells. The radioactive labelling is found first in structures totally or partly resistant to hydrolysis by ribonuclease, which are replicative forms and intermediary replication forms. In the next stage, nearly the entire labelling is found in the single-stranded viral RNA, which is sensitive to RNase hydrolysis (Jackson et al., 1972). *In vivo*, the replicative and intermediary forms of replication are probably double-stranded on a small part of their length (where replication is realized) and during extraction for analysis the positive and negative strands are joined in a zippered manner.

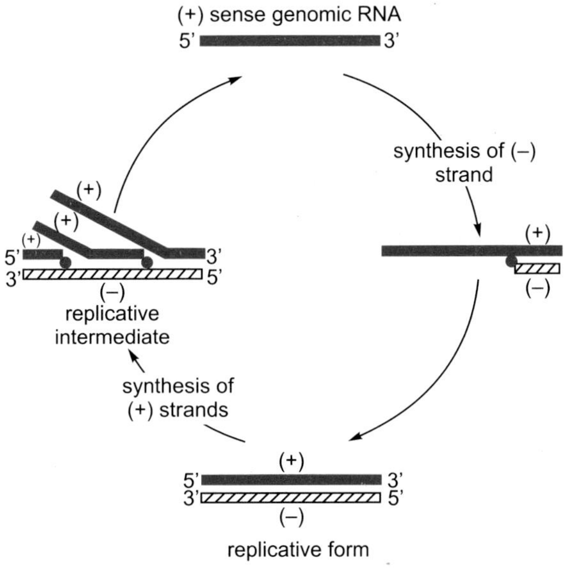

Figure 3.1 Replication cycle of genomic RNA and its complementary strand. The viral and cellular proteins involved directly in the synthesis are represented by a red dot.

A model virus for the study of replication: BMV

A model virus, BMV (*Brome mosaic virus, Bromovirus*), offers several favourable circumstances for the study of replication. Its genome is divided and the proteins coded by each RNA are clearly identified. Its replication can be studied *in vitro* using extracts of infected plants and *in vivo* through the infection of barley protoplasts. Infectious transcripts derived from complementary DNA in which mutations can be introduced are available. A more recent advance is the possibility of achieving replication of BMV in yeast, which opens access to the genetics of the host and to the determination of cellular functions that are involved in replication.

The BMV genome is divided into three capped RNAs (called RNA 1, 2, 3) that are messengers (Fig. 3.2). The RNAs 1 and 2 code for proteins 1a (109 kDa, helicase) and 2a (94 kDa, polymerase). The three RNAs are necessary for systemic infection, but RNAs 1 and 2 are sufficient to infect the protoplasts. Proteins 1a and 2a are part of the replicase which also contains cellular factors. It recognizes viral RNA at the 3' end and synthesizes the negative strand on the model of the infectious positive strand. Then it recognizes the end of the negative strand and synthesizes the positive strands. The RNA 3 negative strand has an inner site that is recognized by replicase, which allows the synthesis of a subgenomic RNA coding for the capsid (reviewed by Noueiry and Ahlquist, 2003).

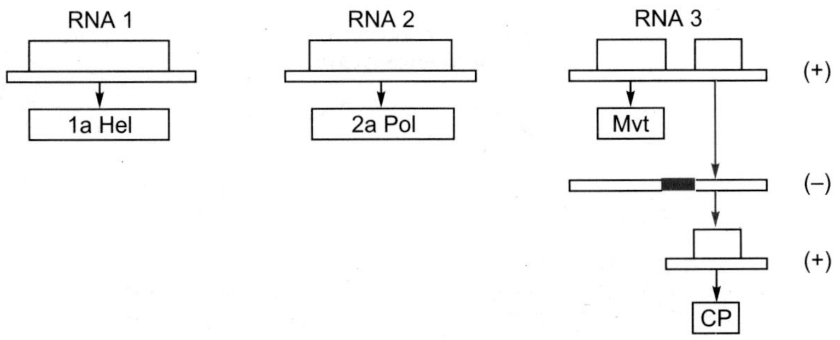

Figure 3.2 BMV RNAs, ORFs, and synthesized proteins. The transcription promoter (in red) located on the negative RNA 3, allows the synthesis of the RNA4, messenger of the capsid (CP). Hel, helicase; Pol, polymerase; Mvt, movement protein.

All the positive-strand RNA viruses have a polymerase function and a majority have a helicase function. These two proteins can be associated with other viral proteins for replication. Significant sequence homologies in the polymerase gene and in the helicase gene allow a classification into several distinct groups called supergroups or superfamilies, within which the processes of replication have common characteristics (see Fig. 3.16). The two principal supergroups are described successively in this chapter:

- Alpha-like superfamily (for example, BMV and TMV) in which the RNA has a cap structure.
- Picorna-like superfamily (for example, CPMV and TEV). Their RNA has a VPg protein.

Membrane sites of replication

Is the replicative complex associated with cellular structures? From the earliest studies, it has been observed that it is linked to host cell membranes, and that each viral genus realizes its replication on a specific membrane site.

Cytological studies have shown that infection by many positive-sense RNA viruses induces proliferation with formation of numerous membranous vesicles and modification of different cellular membranes. Cells infected by TYMV (*Turnip yellow mosaic virus, Tymovirus*) exhibit typical modifications of the external chloroplast membrane, which develops spherical invaginations (Fig. 3.3); viral RNA synthesis takes place in these invaginations (Garnier et al., 1986). This external chloroplast membrane is also very likely the replication site of AMV (*Alfalfa mosaic virus, Alfamovirus*) (de Graaf et al., 1993).

The replication sites of TMV (*Tobacco mosaic virus, Tobamovirus*) are found associated with clusters of tubules in the endoplasmic reticulum (Hills et al., 1987). The same is true in the genus *Potyvirus* (Martin et al., 1995). The sites of synthesis are in spherical invaginations of the endoplasmic reticulum for BMV (Restrepo-Hartwig

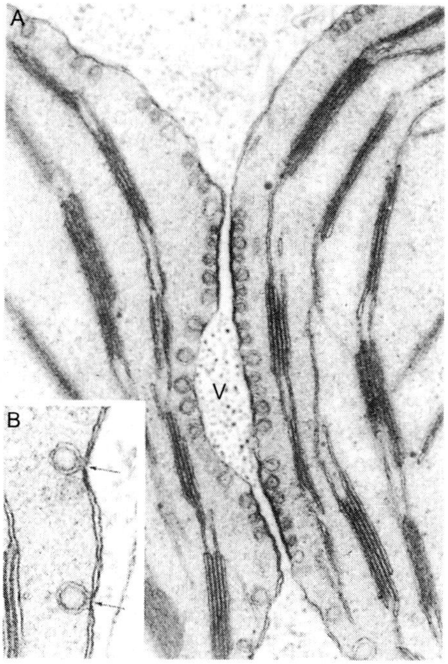

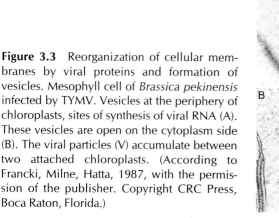

Figure 3.3 Reorganization of cellular membranes by viral proteins and formation of vesicles. Mesophyll cell of *Brassica pekinensis* infected by TYMV. Vesicles at the periphery of chloroplasts, sites of synthesis of viral RNA (A). These vesicles are open on the cytoplasm side (B). The viral particles (V) accumulate between two attached chloroplasts. (According to Francki, Milne, Hatta, 1987, with the permission of the publisher. Copyright CRC Press, Boca Raton, Florida.)

and Alhquist, 1996), in an invagination of the membrane of vacuoles or tonoplast (*Cucumovirus*), in an invagination of the mitochondria membrane (*Tobravirus*), in clusters of vesicles in a rosette derived from the endoplasmic reticulum or Golgi apparatus (*Nepovirus*).

The determinants that target a replicative complex at a type of membrane are still little known. Several viral helicases have been shown to act as membrane anchor. A fragment of the BMV helicase protein is buried within a membrane structure (Dohi et al., 2002). The helicase of TRSV (*Tobacco ring spot virus, Comoviridae*) is a transmembrane protein with a transmembrane domain near the VPg location (Han and Sanfaçon, 2003). The assembly of replication complexes takes place in membrane vesicles, where the polymerase protein is driven through interaction with the helicase protein. In the case of BMV, the 1a helicase protein mediates the membrane localization of 2a polymerase (Schwartz et al., 2002). In a similar manner, the TYMV 140 kDa protein is able to interact with cellular membranes in the absence of other viral factors. The expression of the sole 140 kDa is sufficient to promote the clumping of chloroplasts, a characteristic cytological abnormality induced by TYMV infection. The cytoplasmic 66 kDa polymerase is located to membranes through specific interaction with the 140 kDa (Prod'homme et al., 2003). The TYMV replication complex is then assembled in the cytoplasmic vesicles resulting from the invagination of the chloroplast envelope.

It is observed that the membrane anchorage is a permanent feature. The expression of the BMV RNA in a non-natural host, yeast, results in the synthesis of a replicative complex linked to membranes as in a plant cell (Quadt et al., 1995; Restrepo-Hartwig and Ahlquist, 1999). In potyviruses, a small protein of 6 kDa realizes the anchorage of replication proteins in the endoplasmic reticulum by means of an intramembrane hydrophobic domain (Restrepo-Hartwig and Carrington, 1994). The endoplasmic reticulum forms vesicles (Schaad et al., 1997a).

Reorganization of cellular membranes provides to replication complexes a protected compartment, in which viral RNA is isolated from competing cellular RNAs and where viral double-stranded RNA is protected from host defence response. This new, viral-induced compartment functions as a virus-specific organelle (Ahlquist et al., 2003).

Isolation and solubilization of replicases

The advantage of an *in vitro* replication system is that it allows to characterize the components of the enzymatic system responsible for RNA replication. This advantage justifies the numerous and prolonged tasks of purification of viral replicases.

If an extract of infected cells is fractionated, the enzymatic activity is associated with the membrane fraction. It can be liberated from membranes by the action of detergents, the choice of which is critical for the conservation, always partial, of the activity. The polynucleotides still linked to the extract are separated by physical methods (separation of phases) or enzymatic methods (action of a nuclease). At various stages of purification, the synthesis activity of the enzymatic extract is measured by viral RNA synthesis in the presence of the four triphosphate nucleosides that are precursors of RNA, in a buffered medium at pH 8 containing Mg^{2+} ions. The purification protocols are applied to healthy plants as a control experiment. The synthesis functions by consuming the precursors, the triphosphate nucleosides, which are rich in energy.

When the replicase is still attached to its template, its *in vitro* activity consists of terminating the strands begun *in vivo*. After elimination of the endogenous template, the enzymatic activity of RNA synthesis becomes dependent on the template added in the experiment. The polynucleotides synthesized are isolated, quantified, and compared with the viral RNA in terms of size, sequence, and polarity. The RNA replicase activity that is expressed in the presence of template RNA shows a specificity for the RNA of the infecting virus. The product of the reaction is the full-length copy of the viral RNA, as well as the subgenomic RNA in the case of BMV and CMV (*Cucumber mosaic virus, Cucumovirus*). The continuation of the reaction, that is the synthesis of full-length positive strands from this product, has been obtained in plant systems for the first time with the CMV replicase. This replicase is obtained after a highly effective purification, which probably allows elimination of inhibitors of the synthesis of positive strands (Hayes and Buck, 1990). An extraction protocol for replicase that can synthesize full-length positive and negative strands has also been developed for TMV (Osman and Buck, 1997).

Replicase contains viral proteins: polymerase and helicase

The viral proteins present in the replication complex, polymerase and helicase, have been precisely identified for several viruses. For example, TYMV replicase contains a polypeptide of 115 kDa that is part of the *in vitro* translation products of viral RNA, as has been shown with an anti-replicase serum. Moreover, a serum against the 115 kDa protein partly inhibits the *in vitro* replicase activity (Candresse et al., 1986).

With viruses having a divided genome, protoplast infection can be obtained with just part of the viral genome: RNA 1 and 2 of BMV, CMV, and AMV, RNA 2 of CPMV (*Cowpea mosaic virus, Comovirus*). It can be concluded that these RNAs carry the information for replication.

The polypeptides present in the replicative complex have been studied only in some favourable cases (BMV, CMV, TYMV, CPMV, TMV). The replicase activity results most often from the presence of two viral proteins. One is always present and contains motifs characteristic of cellular or viral polymerases. The other is not always essential and contains one or several motifs characteristic of helicases.

Polymerase

The study of purified replicase activity shows that new RNA strands are synthesized through complementary copy of homologous viral RNA. To function, the replicase must simultaneously effect the fixation of the template RNA and that of precursor triphosphate nucleosides, and then realize the covalent phosphodiester bond between the 3′OH end of the growing strand and the α phosphate of the following nucleoside 5' triphosphate. The hydrolysis of the two other β and γ phosphate bonds provides the energy (Fig. 3.4).

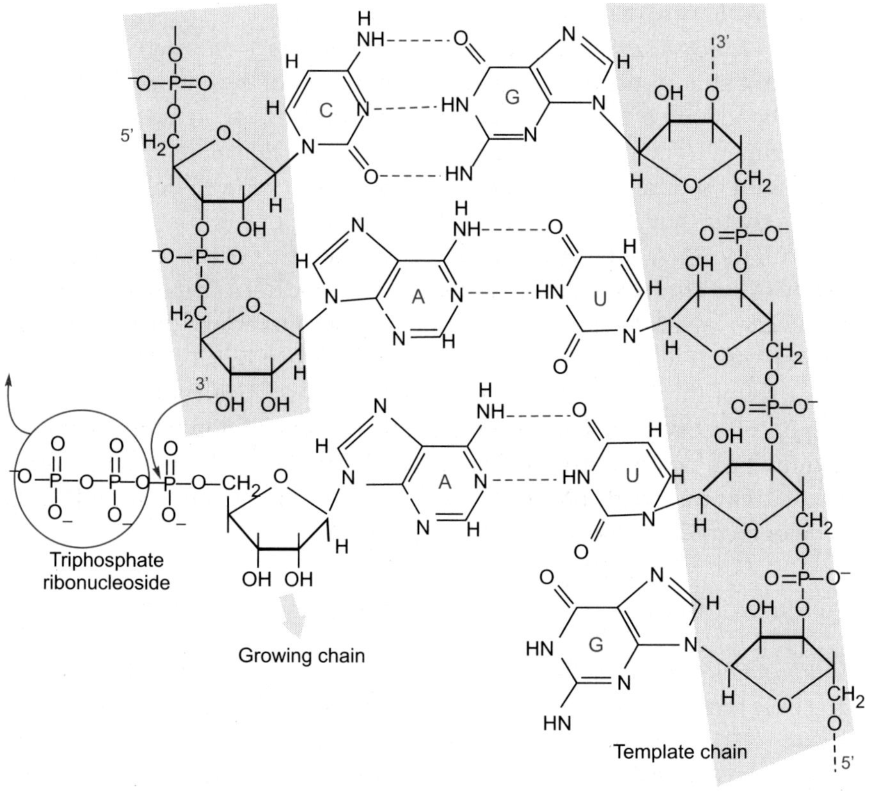

Figure 3.4 Complementary copy of an RNA and formation of a phosphodiester bond in the strand being synthesized. The ribose-phosphate skeleton of the polynucleotide chain is shown in grey, the hydrogen bonds between complementary bases in broken red lines.

In searching for the motifs common to plant and animal viral RNA polymerases and the polymerase of poliovirus, which was already well known, Kamer and Argos (1984) drew attention to the glycine–aspartic acid–aspartic acid motif (GDD) surrounded by hydrophobic residues present in polymerases of all these viruses. If mutations are made at this motif by transforming it into ADD, GED, or GAD in the RNA of PVX (*Potato virus X, Potexvirus*), the RNA loses its infectivity (Longstaff et al., 1993). The same applies for the mutation of an aspartic acid of the GDD motif in TYMV RNA (Weiland and Dreher, 1993). It is now considered that the viral proteins containing this motif are involved in replication by expressing an activity of polymerization of nucleotides, even though the polymerase activity has been biochemically demonstrated only for some viruses. This motif is present in the 2a proteins of BMV, CMV, AMV, in the 183 kDa protein of TMV, and in the 110 kDa protein of CPMV and is identified in all the positive-sense RNA viruses to the extent their genome is sequenced. The GDD triplet and its hydrophobic environment form a structure of two antiparallel chains linked by hydrogen bonds exposing a short loop containing GDD on the outer side. Other motifs are strongly conserved in the polymerases of positive-sense RNA viruses and certain double-stranded RNA viruses (Koonin, 1991a) (see box).

The polymerase function

The polymerase function is the only one that is present in all RNA viruses. In positive-sense RNA viruses, the RNA-dependent RNA polymerases (RdRp) are the proteins that present the clearest similarities of sequence (Kamer and Argos, 1984; Argos, 1988). These similarities are limited to very short motifs. Eight have been identified, always aligned in the same order (Koonin and Dolja, 1993), which constitute the "polymerase signature" (Candresse et al., 1990).

This concept has opened up new avenues of investigation in various domains. On the basis of similarities found in these eight motifs, the polymerases of positive-sense RNA viruses have been compared. The resulting phylogenetic tree (Koonin, 1991a) shows that they can be classified into three major groups, 1, 2, and 3, each containing plant and animal viruses as well as bacterial viruses for group 2 (Fig. 3.5). This element is important in view of a unified classification. In connection with other characteristics (helicase, translation strategies), it has led to the definition of several supergroups of positive-sense RNA viruses (cf. Figs. 3.13, 3.14, 3.15).

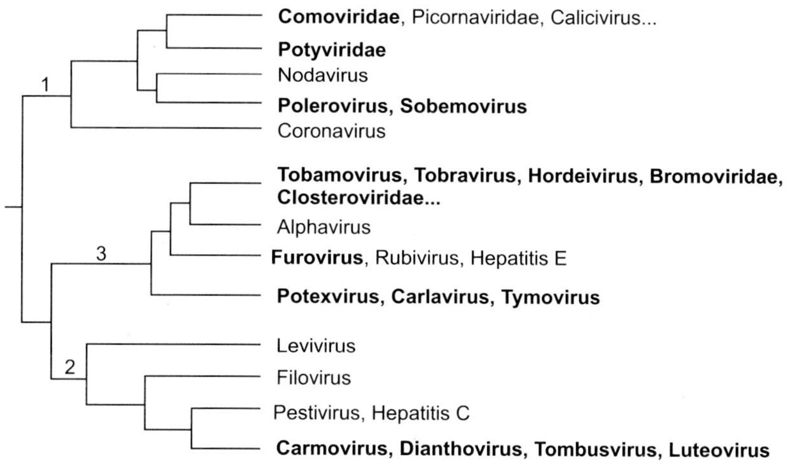

Figure 3.5 Phylogenetic tree of RNA polymerases of positive-strand RNA viruses: three groups (polymerases of type 1, 2, and 3) within which the homologies are significantly greater than between the groups (according to Koonin, 1991a). The plant viruses are in bold type.

When comparisons are extended to all polymerases for which the sequence is known, similarities are found in the sequences of motifs conserved and more particularly three successive motifs referred to here as A, B, and C (Fig. 3.6, top), the direct implication of which in the fixation of nucleotides and the reaction of polymerization (Fig. 3.6, bottom right) has been demonstrated by the observation of mutants.

Other very interesting comparisons involve the secondary and tertiary structures. Despite limited sequence homologies, the secondary structures (the succession and length of β-sheets and α-helices) are highly conserved for the polymerases of six viral genera including *Bromovirus, Tombusvirus, Tobamovirus,* and poliovirus (O'Reilly and Kao, 1998).

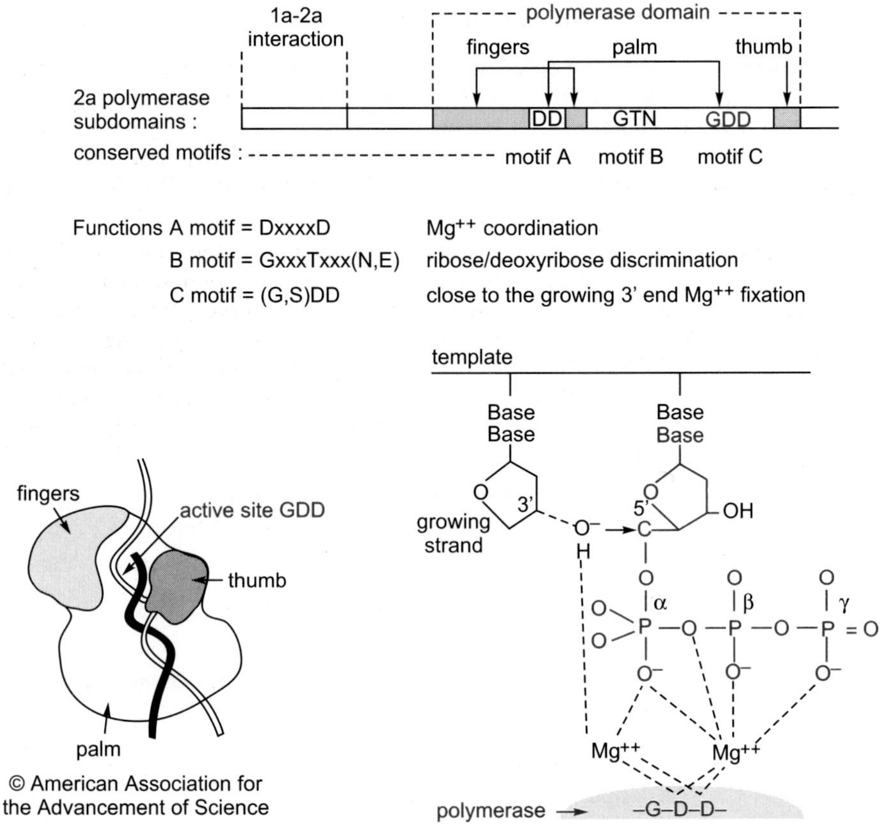

Figure 3.6 Domains of the polymerase gene (2a) of BMV, and remarkable motifs (top). G, glycine; D, aspartic acid; S, serine; T, threonine; N, asparagine; E, glutamic acid. This arrangement is also observed in TMV, TBRV, and poliovirus (according to O'Reilly and Kao, 1998). Bottom left, spatial representation of a polymerase, where the active site is found in the central groove between the fingers and the thumb (according to Kholstaedt et al., 1992). Bottom right, polymerization reaction at the GDD site indicating the role of Mg^{++} ions and phosphates (Steitz, 1998).

The tertiary structures are known in rare cases in which an X-ray crystallographic diffraction pattern of the protein can be obtained: DNA polymerase I, reverse transcriptase of the human immunodeficiency virus (HIV), RNA polymerase of phage T7, 3D polymerase of poliovirus, sometimes in co-crystal with the template.

These studies have shown that the activity of these proteins results from the interaction of several modular domains. These enzymes present astonishing similarities in folding and are all described as a half-open right hand with several domains, in particular the thumb and fingers forming a deep groove adjacent to the palm (Fig. 3.6, bottom left). On the bottom of the groove is a catalytic site of the phosphodiester bond, with very similar motifs A and C fixing the two essential Mg^{2+} ions. The template strand is introduced from the outside, kept in place by the different domains of the enclosed hand; direct contact is established between the polymerase and the template at the level of the active site, where the aspartate residues (DD) that fix two Mg^{++} ions are found. As shown by the high resolution crystal structure of hepatitis C virus polymerase, movement of the thumb might take place on template binding, and translocation of the nascent molecule might result in concerted movements of thumb and fingertips (Bressanelli et al., 1999). The double-stranded product emerges on the wrist side (Kohlstaedt et al., 1992). The similarities of the secondary structures suggest that all the polymerases function according to this diagram, with variants.

Helicase

Several amino acid motifs are conserved between the 1a proteins of BMV, CMV, and AMV, and the 126 kDa protein of TMV. One of these motifs (motif I: GxGKS, T) is characteristic of proteins having a nucleotide fixation activity, detected experimentally for the 126 kDa protein of TMV (Evans et al., 1985). Another motif (motif II: DE or DD preceded by several hydrophobic amino acids) makes possible the chelation of Mg^{2+} necessary for the hydrolysis of nucleotides. The cytoplasmic inclusion (CI) protein of potyvirus PPV (*Plum pox virus*) has these motifs and manifests *in vitro* two aspects of helicase activity (Lain et al., 1991):

- ATPase activity. It can effect *in vitro* the hydrolysis of ATP in the presence of single-stranded RNA. This activity is totally dependent on Mg^{2+} ions; it is linked to the presence of motifs I and II.
- Unwinding of double-stranded RNA structures. The experiment consists in placing the potential helicase in the presence of a triphosphate nucleoside, Mg^{++} ions, and a duplex structure in which one of the strands is radioactive. Electrophoresis can be used to see whether the strands have been separated. It is observed that if the RNA substrate has a free 3' end not engaged in a duplex structure, the two strands are separated and the reaction progresses from the 3' end to the 5' end. The region that confers the property of RNA binding is a conserved motif (motif VI) rich in basic amino acids (Fernandez et al., 1995).

Many cellular and viral proteins have these motifs and have a helicase activity of destabilization of duplex structures, demonstrated or supposed (Kadaré and Haenni, 1997).

The helicase function

Helicases form a class of enzymes capable of unwinding double-stranded RNA or DNA by destroying the hydrogen bonds that stabilize their structure (Kadaré and Haenni, 1997). This action is coupled with hydrolysis of triphosphate nucleosides that provides energy. Sequence comparisons (Gorbalenya et al., 1990) show that, in all the RNA viruses, the viral proteins that have motifs of nucleotide fixation present up to seven homologous regions (called I, Ia, II, III, IV, V, and VI) that are more or less conserved. Often, the helicase function is associated with the presence of these motifs, while the activity has been demonstrated in a small number of examples. On the basis of sequence homologies, the plant viral helicases can be phylogenetically clustered in three groups (that also include cellular helicases and helicases carried by DNA viruses) (Gorbalenya and Koonin, 1989):

- group 1, Alpha-like viruses (*Bromoviridae, Tobamovirus*, etc.);
- group 2, *Potyviridae*;
- group 3, *Comoviridae*.

What is the role of helicases in viral replication? Helicase is present in purified replicases and it is a necessary component of the replicative complex. It may help the displacement of the strand already synthesized in the replication intermediary by destroying the secondary structures that hinder the progression of the replicase. Finally, helicases have a tendency to form multimers and notably hexamers surrounding the nucleic acid like a ring. Koonin (1991b) formulated the hypothesis that helicase, in replication complex, could increase the fidelity of the RNA-RNA copy which does not benefit from any system of control or repair. This makes it necessary for replication of genomes of more than 5 to 6 kb. It seems established that not all RNA viruses have a helicase function, only those in which the genome exceeds 6 kb (Koonin and Dolja, 1993).

Polymerase and helicase interact closely in the replication complex

The yeast two-hybrid system is a powerful tool to investigate protein-protein relationships *in vivo*: the expression of a reporter gene is determined by a close interaction between the two proteins studied. This system used for proteins 1a and 2a of BMV shows that these two proteins interact in the replication complex and that two proteins 1a interact by their N-terminal region (O'Reilly et al., 1998b). The replicative complex thus involves multiple protein-protein relationships; these results suggest that, in the complex, the two proteins could form multimers in a stoichiometric ratio that remains to be identified.

The replicative complex of TMV capable of synthesizing the negative strands is composed of one molecule of the 183 kDa protein (polymerase) and one molecule of the 126 kDa protein (helicase) as well as cellular proteins (Watanabe et al., 1999). One function of the 126 kDa protein is to bind in *cis* on the viral RNA; the 183 kDa protein then joins it (Lewandowski and Dawson, 2000). The 126 kDa protein assembles in oligomeric complexes, probably hexameric ring-like structures, in which the 183 kDa protein is included (Goregaoker and Culver, 2003).

A great diversity of additional functions is found in proteins with helicase domain. The BMV protein 1a exhibits a C-ter helicase activity and an N-ter methyltransferase domain involved in capping function. Viruses with a "triple gene block" (TGB, see Chapter 4) possess two helicase domains, one involved in replication and the other in movement of the virus in the plant.

Specific cooperation with cellular factors

The purification of replication complexes has clearly shown that cellular proteins are part of them. The first replicase studied, that of the bacteriophage Qβ, is made up of one viral subunit, polymerase, and three subunits from the bacterial host, which are implicated in protein synthesis: a protein S1 of the ribosomal particle 30S and two elongation factors of polypeptide chains, EF-Tu and EF-Ts. This enzymatic complex can direct the synthesis of positive-sense RNA from the negative-sense RNA of Qβ but does not direct the synthesis of negative-sense RNA from a positive-sense RNA template unless another factor HF1 (another protein associated with the ribosome) is linked to the 3' end of the genomic RNA. The four components of the replicase are indispensable and play different roles, some of which have been elucidated. Synthesis in the presence of competitor RNAs has shown that two independent sites of replicase, one situated on S1, the other on EF-Tu, are able to bind respectively to the 3' end of positive-sense and negative-sense RNA (Brown and Gold, 1996). These results allow the proposal of a rather simple model for the bacteriophage Qβ replication: the polymerase is brought to the 3' end of the RNA by the intermediary of S1 and it effects the synthesis of the negative strand. The EF-Tu factor thus recognizes the 3' end of the negative strand and the polymerase is fixed there by its intermediary. Replicase effects forward and back movements on the positive and negative templates to which it is fixed via an interaction with the cellular proteins (Lai, 1998). In brief, the cellular proteins realize the specific binding with the viral RNA, convey the enzymatic complex to the end of the strands, and introduce an asymmetry in the synthesis of positive and negative strands. The presence of subunits of the elongation factor EF_1 has been indicated in the replicase of poliovirus and vesicular stomatitis virus. These examples show how cellular proteins are subverted from the host machinery.

Plant proteins are associated with the viral proteins

The cellular proteins engaged in the replication complexes are identified by biochemical and genetic methods. The replication complex is linked to membranes, and the use of detergents to solubilize it for the *in vitro* synthesis of viral RNA has very frequently led to its denaturation and loss of activity, except in some favourable cases. The molecular weight of cellular proteins that copurify with replicase has been determined in the case of TYMV (45 kDa) (Mouches et al., 1984) and CMV (50 kDa)

(Hayes and Buck, 1990). The replicase activity is lost if this protein is eliminated. What is the identity of these factors essential to RNA synthesis? The replicase of BMV, isolated from barley leaves, contains a cellular protein (Quadt et al., 1993) that is serologically linked to the 41 kDa subunit of the initiation factor eIF3 (constituted of 10 subunits) that can be found for example in wheat seed; this protein interacts specifically with polymerase 2a. In another example, that of the replicative complex of TMV extracted from tomato, a cellular protein is linked to the P126 and P183 proteins; by immunology, it is observed that it is essential to their activity and that it is very close to the 56 kDa subunit of the initiation factor eIF3 carrying an RNA binding motif (Osman and Buck, 1997). These eIF3 subunits associated with replication complexes stimulate negative strand RNA synthesis possibly in relation with the aminoacylatable tRNA-like structures at the 3' end where negative strand RNA synthesis begins (Ahlquist et al., 2003).

Replication thus necessitates the intervention of cellular proteins that give to the replicase its efficiency and specificity. To clarify their role further, it is necessary to study a host-virus system in which the molecular genetics of the host is very well known, for example, *Arabidopsis thaliana* or yeast. The BMV can be replicated in yeast after having been introduced in the form of plasmids (Ishikawa et al., 1997a). Several mutants affected in the replication of RNA 3 have been isolated, they are distributed in three (or more) groups of complementation, each corresponding to a cellular function necessary for replication (Ishikawa et al., 1997b). Because the yeast genome has been entirely sequenced, identification of implicated proteins progresses rapidly. OLE1 is a yeast gene encoding an integral membrane protein essential in lipid metabolism; mutations in OLE1 block BMV replication after assembly of the replication complex but before initiation of negative strand RNA synthesis (Lee et al., 2001). YDJ1 yeast gene encodes a chaperone protein localized on endoplasmic reticulum membrane; mutations in YDJ1 block initiation of negative strand RNA synthesis (Tomita et al., 2003).

TMV replication in *Arabidopsis thaliana* is reduced to a low level in mutants named tom1 and tom2. TOM1 and TOM2 genes encode transmembrane proteins and the TOM1 protein can interact with the helicase domain of TMV replication proteins. Inactivation of TOM1 completely inhibits TMV multiplication. Gene products of TOM1 and TOM2 are essential constituents of the TMV replication complex (Tsujimoto et al., 2003).

A complementary approach is the yeast two-hybrid system for the study of protein-protein interactions: it has been found in the case of TuMV (*Turnip mosaic virus, Potyvirus*) that the protein VPg, implicated in replication, interacts specifically with four proteins of its host *Arabidopsis thaliana*, one of them being the translation initiation factor eIFiso4E (Wittmann et al., 1997). Yeast two-hybrid analysis demonstrated reproducible and specific interactions between ZYMV (*Zucchini yellow mosaic virus, Potyviridae*) polymerase and host polyA-binding protein; this protein is directly implicated in translation of polyadenylated RNAs.

Thus, different host factors (already described or to be discovered) participate in several functions needed for viral replication, in particular localization of replication complexes and initiation of negative and positive strand RNA synthesis.

Being asymmetrical, replication mostly produces positive chains

Highly purified replicase of CMV produced *in vitro* one negative strand for every seven positive strands (Hayes and Buck, 1990). This asymmetry is much more marked *in vivo* (French and Ahlquist, 1987). Some aspects of the regulation process between positive and negative strands are known in the case of BMV. If protoplasts are infected with RNA 1 and 2 in the absence of RNA 3, the positive/negative ratio is 1; in the presence of RNA 3, it is 100 (Marsh et al., 1991). By deletions and mutations, it can be shown that the intercistronic region of RNA 3 (Fig. 3.2) effects this regulation in *trans*, by a mechanism that is still to be determined. This region also determines the global level of replication.

The accumulation of negative strands of TMV in infected protoplasts reaches its maximum at the end of 6 h after infection and then ceases; the ratio of positive to negative strands is then around 100. The accumulation of positive strands continues vigorously beyond 18 h (Ishikawa et al., 1991). The increase of the positive/negative ratio during the cellular infection thus results from the asymmetrical functioning of the replication intermediary form, as well as from the cessation of synthesis of negative strands. It can be hypothesized that the replicative complex is not exactly the same for the synthesis of positive strands and for that of negative strands, and that a cellular factor is found in limiting quantity.

Structural and sequence requirements for replicase promoters

Promoters for negative strand RNA synthesis

Replicase must specifically recognize "its" viral RNA among all the polynucleotides of the cell. This recognition occurs at the 3' end, where nucleotide polymerization begins and progresses from the 3' end to the 5' end. The 3' untranslated region presents a tRNA-like structure, a polyA, or a sequence without known particularity. Viral sequences indispensable to the initiation of negative strand synthesis have been identified *in vivo* and *in vitro* in the 200 3'-terminal nucleotides of positive strand RNA of several viruses. For example, BMV replicase can function *in vitro* with a 3'-terminal fragment of 153 bases as a template, in which the folded structure tRNA ("cloverleaf") is present (Lahser et al., 1993). The template activity of this fragment or the viral RNA is abolished when a segment of 39 bases is hybridized at its 3' end. It is interesting to note that the hybridization of an internal segment of the 3' region, not at the end, does not stop the replicase. It thus seems essential that the CCA-OH 3' end be a single-stranded form.

The resemblances between the tRNAs and the 3' end of certain viral RNAs are long-standing (Pink et al., 1970); they have several properties in common, the main one being the capacity of carrying an amino acid (valine for TYMV RNA). Several genera have this structure: *Tymovirus, Bromovirus, Cucumovirus, Tobamovirus, Hordeivirus*; in these cases, its presence is essential to replication (Miller et al., 1986). In TYMV, this region comprises three elements: a 5' segment that resembles a tRNA with a branch acceptor of valine, a segment comprising pseudo-knots, and finally the 3' end in single strand of ACCA sequence. *In vitro* studies make it possible to determine the role of each element. The principal determinant for the initiation of the synthesis of negative strands is the terminal sequence ACCA. Mutations or deletions cannot be introduced here (particularly in the last A) without considerably diminishing the activity. The segment with pseudo-knots intervenes by its structure rather than by a specific sequence (Deiman et al., 1998). As for the acceptor branch of an amino acid, an aminoacylation must be undergone by the host aminoacyl tRNA synthetases for the initiation of synthesis to be effective. This aminoacylation is not specific. By mutation the valine may be replaced by methionine. This suggests that a cellular protein having the properties of an elongation factor could be implicated, what is known for the bacteriophage Qβ (Giegé, 1996). The new strand begins with a G complementary to the C preceding the terminal A. The terminal A will be added after synthesis by host tRNA nucleotidyl-transferase. An important role of the tRNA-like structure is to present 3'CCA in a highly accessible but moderately specific conformation; specificity may be enhanced by the strong *cis* preferential replication exhibited by TYMV (Weiland and Dreher, 1993) (Fig. 3.7). Regulation of negative strand synthesis may be mediated through 3' aminoacylation host proteins that bind at or near the 3' end (Dreher, 2002).

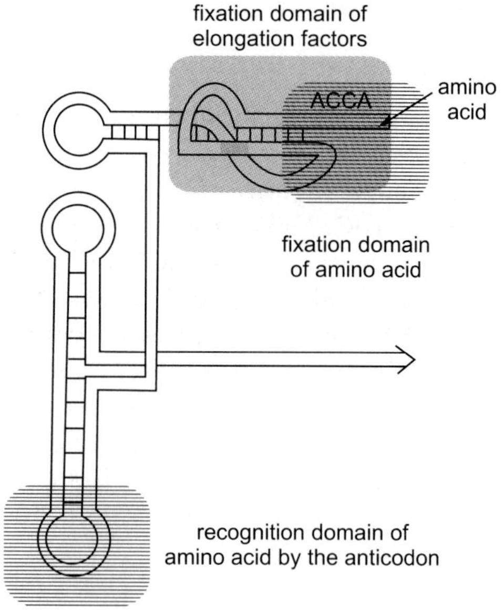

Figure 3.7 3' end of BMV RNA. This RNA is aminoacylated by valine. The drawing highlights the structural analogies with the cellular tRNA, with the recognition sites by synthetases (shown in grey) and by factors of elongation (in red). In this last site is the pseudo-knot region, which does not exist in the tRNA (according to Giegé, 1996).

The 3' end of *Alfamovirus* and *Ilarvirus* does not stably present a tRNA-like structure. By a series of secondary hairpin structures, it specifically recognizes a motif of the coat protein and the presence of a small quantity of the coat protein is necessary for the viral cycle to begin: this is called activation by the coat protein (Bol, 1999). But it seems established that the 3' end can adopt an alternative conformation including a pseudo-knot needed for replication, which gives it a tRNA-like structure and function analogous to those of BMV and CMV. Each of these conformations corresponds to a phase of the cycle (translation for the first and replication for the second), and the passage from one to the other will be a means of modulating the successive roles of viral RNA (Olsthoorn et al., 1999).

The roles of terminal 3' regions of RNA having a polyA are also being studied. Several results suggest that the secondary structures of RNA are necessary. The 3' end of the RNA of TEV (*Tobacco etch virus, Potyvirus*) (the 300 nucleotides of the end of the capsid gene and the non-coding 3' region) has a set of four stem-loops (5'-A-B-C-D-3'), the structure of which is necessary for the amplification of the genome because the mutations that destroy the structure greatly diminish the replication (Haldeman-Cahill et al., 1998); moreover, the sequence of the loop A is indispensable in *cis* for the replication of RNA (Fig. 3.8). The same authors have observed that it is also necessary that the ribosomes progress up to the capsid gene, i.e., close to loop A. To explain these different results, it may be proposed that the ribosomes, having progressed up to loop A, carry in the downstream 3' region (corresponding to the end of the capsid gene and the non-coding 3' region) viral and cellular proteins active in *cis* in the replication. This coupling of translation and replication prevents the replication of defective genomes.

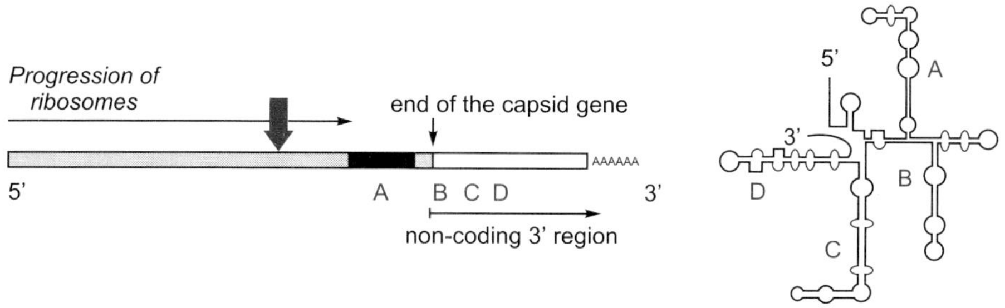

Figure 3.8 3' end of TEV RNA, promoter of plus strand synthesis. The four stem-loops are labelled A, B, C, D. The arrow indicates the zone the ribosomes must reach in order to let the replication to start (according to Haldeman-Cahill et al., 1998).

Promoters for positive strand RNA synthesis

Promoters for positive strand RNA synthesis are found in the 3' end of the negative strand, complementary to the 5' end of the positive strand. Interesting determinations on the synthesis of positive strands of BMV have resulted from the observation of sequence homologies between the 5' end of the viral RNA and cellular elements, that is, the internal control regions of promoters of genes coding for the tRNA (called ICR 1 and 2)

The homologies observed have suggested that these motifs could play a role in the control of replication. To test their functionality, mutations were introduced in the ICR motifs or close to them, then the rate of replication of mutants was measured (Pogue et al., 1992). In the absence of mutation, the 5' end of the positive strand RNA adopts a stem-loop structure, and this folding creates a free single-stranded 3' end on the negative strand RNA. In contrast, in the mutants, the synthesis of the positive strands is observed to be diminished (from 70 to 97%) to the extent that mutation affects the folding. A similar observation was made for poliovirus. Pogue and Hall (1992) concluded that the stem-loop structure of the 5' end is necessary and they constructed a model that takes into account the role of this end in the synthesis of positive strands and integrates a large number of data on the replication of positive strand RNA viruses (Fig. 3.9). The presence and importance of a stem-loop close to the 5' end of the viral RNA has been highlighted in BNYVV (*Beet necrotic yellow vein virus, Benyvirus*), TYMV, PVX (Miller et al., 1998), and AMV (Vlot et al., 2003).

Transcription promoters for subgenomic RNA synthesis

In many virus groups, the genes located in the 3' part of the genome are expressed through the intermediary of subgenomic RNAs, which are partial copies of the viral genome resulting from an RNA-RNA transcription. How are these copies made?

The study of the *in vitro* synthesis of BMV RNA (Miller et al., 1985) clearly shows that the synthesis of subgenomic RNA is the result of the action of the replicase. The same is true *in vivo* for the tymovirus TYMV (Gargouri et al., 1989). Replicase recognizes an internal promoter, the extent of which has been determined on the RNA 3 of BMV. It is located in the intergenic region of the negative strand and essentially comprises 17 bases upstream of the starting point of the synthesis. The adjacent sequences upstream and downstream play a more indirect role. Some nucleotides of the promoter are particularly important in recognition by the replicase. They are conserved in the homologous promoters of *Bromoviridae* and Alphavirus of animals (Siegel et al., 1997).

The nucleotide sequence of the promoter is decisive, but the secondary structure also seems very important for its activity. The promoter of two subgenomic RNAs of TCV (*Turnip crinkle virus, Carmovirus*), defined *in vivo*, comprises 30 nucleotides; a segment of 9 nucleotides immediately upstream of the starting point of the synthesis is preceded by a hairpin structure of 21 nucleotides. The structure as well as the adjacent sequence are essential for the regulation of the synthesis (Wang et al., 1999). Similarly, the promoter of the subgenomic RNA 1 of BYDV (*Barley yellow dwarf virus, Luteovirus*) has two hairpin structures, one of which contains the initiation site (Koev et al., 1999). This promoter, located on the negative strand, is formed of two loops (Fig. 3.10):

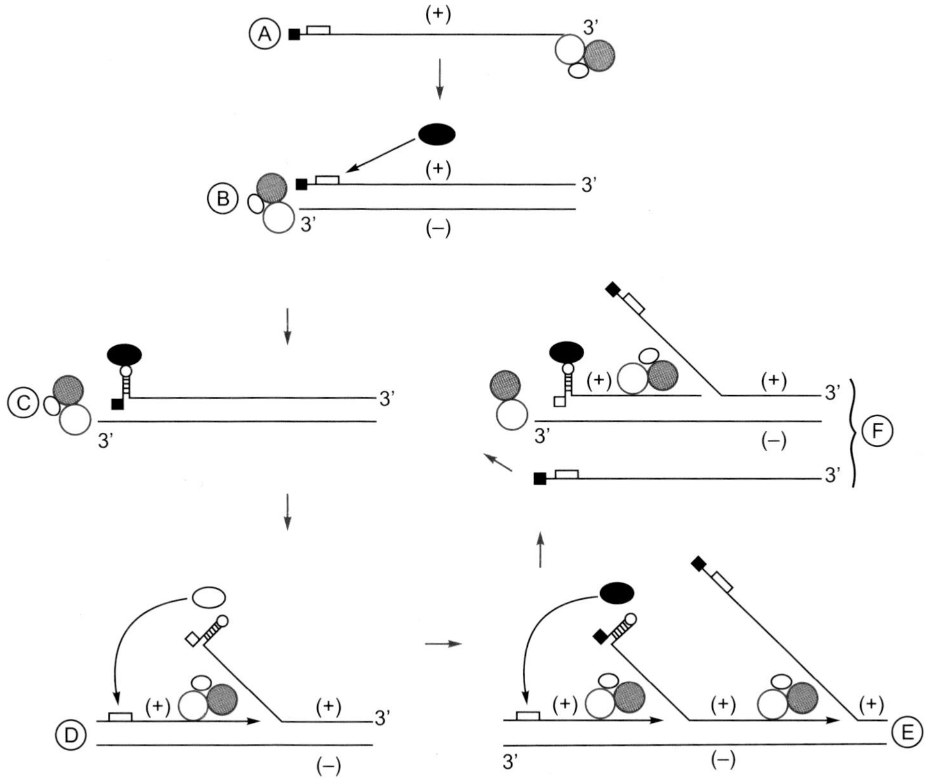

Figure 3.9 Initiation of positive-strand RNA synthesis of BMV and the replication cycle. A. Replication proteins (polymerase and helicase ⚬, cellular factor ○) recognize the 3' end and synthesize the negative strand. Cap: ■; ICR: ▭. B. Another cellular factor (⬬) recognizes the 5' end of the positive strand by means of the ICR motif; a secondary structure is established on the positive strand. The 3' end of the negative strand becomes free. C, D, E. The replicative intermediate can function with several growing positive strands. F. A complete positive strand is released (according to Pogue and Hall, 1992).

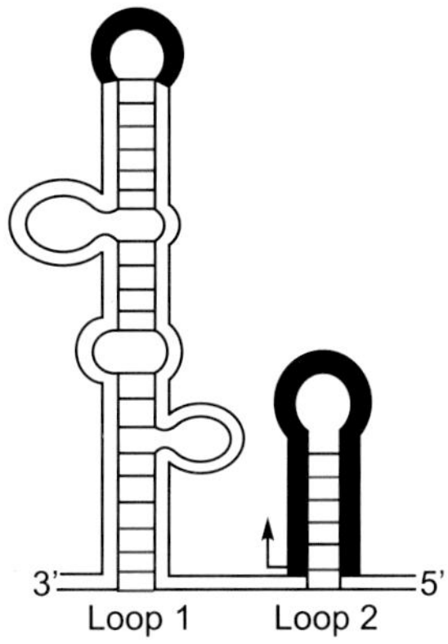

Figure 3.10 Promoter for subgenomic RNA 1 synthesis of BYDV (according to Koev et al., 1999).

- Loop 1 is the replicase recognition site. The sequence of the loop (in black) and the secondary structures underlined in red are essential.
- Loop 2 contains the nucleotide in which the transcription begins, marked by an arrow. Its sequence (in black) is indispensable. The secondary structures underlined in red play a regulatory role (transcription increases when they are suppressed).

The promoters for synthesis of the three subgenomic RNAs of BYDV have been mapped and they show surprisingly little sequence or structural similarities (Koev and Miller, 2000).

■ Module of replication of viruses in which RNA has a VPg

The viruses with a protein linked to the viral RNA by a covalent bond (VPg) at the 5' end and a polyA at the 3' end have modalities of replication that differentiate them from viruses with a cap and tRNA-like structure. They belong to families *Potyviridae* and *Comoviridae*. These viruses express their message in the form of one or two polyproteins that are then cleaved by viral proteins carrying a protease function (Riechmann et al., 1992; Vos et al., 1988). The availability of proteins intervening in replication depends thus on a supplementary parameter, the place and moment at which the protease activities involving these proteins are expressed. This remark here concerns the "replication module", which contains several contiguous proteins in the polyprotein.

From 1984 onward, analogies have been discovered between the CPMV genome and that of poliovirus: sequence homologies, strategy of expression, continuous arrangement of genes directly related to replication (see Fig. 3.13). Polymerase of poliovirus is well known; its homologue in CPMV (Fig. 3.11) is the 87 kDa protein, which bears the GDD motif surrounded by hydrophobic residues. Two motifs characteristic of the helicase function are carried by the 58 kDa protein and the infectious capacity is lost if a mutation is introduced in one or the other of these motifs (Peters et al., 1994). These elements designate the proteins 87 and 58 kDa for the polymerase and helicase functions respectively (Fig. 3.11). By analogy with poliovirus, it is believed that the initiation of the synthesis of viral strands requires a primer. The role of VPg in the initiation of replication is suggested by the presence of this protein at the ends of the positive and negative strands of the replicative forms. Purified poliovirus polymerase can use *in vitro* VPg as a primer and a polyA template to synthesize VPg-polyU, which is then elongated into negative strand RNA (Paul et al., 1998). Positive strand RNA synthesis is initiated with a large pool of VPg-pUpU synthesized on a specific hairpin of the viral RNA; VPg-pUpU functions as a primer to initiate positive strand RNA synthesis by using the AA sequence at the end of the negative strand template, resulting in an asymmetrical mode of replication (Morasco et al., 2003) (see also p. 143).

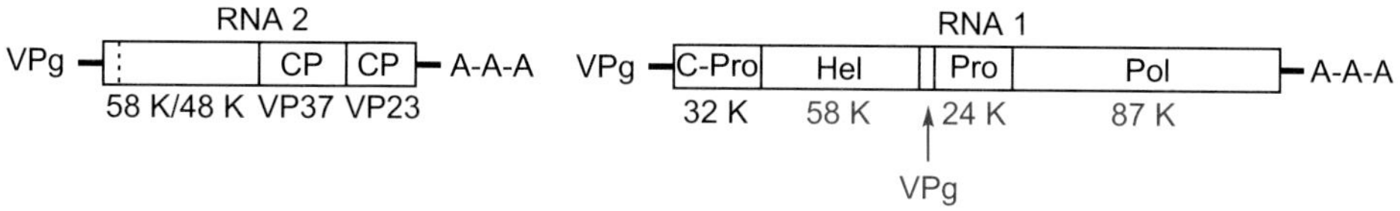

Figure 3.11 Genetic map of CPMV RNA. In red are the proteins implicated in replication.

A model, still partly speculative, has been proposed for CPMV taking into account all these elements: the 200 kDa polyprotein derived from RNA 1 is cleaved by internal autoproteolysis into two proteins of 32 and 170 kDa. The 170 kDa protein, linked to membranes, will be progressively cleaved liberating polymerase, helicase, and VPg protein (Peters et al., 1995). The synthesis of negative strand starts off with uridylated VPg as primer. VPg-polyU is associated by complementarity with polyA of the template positive strand that the polymerase copies by elongation of the primer. A proteolysis between VPg and Pro then releases the VPg protein linked to the neosynthesized chain (Fig. 3.12). Therefore, proteolysis is closely linked to the initiation of the replication. Experimentally, it is observed that the 32 kDa protein remains closely associated with the 170 kDa protein, and that it slows down the proteolysis as a function of the need for formation of the replication complex.

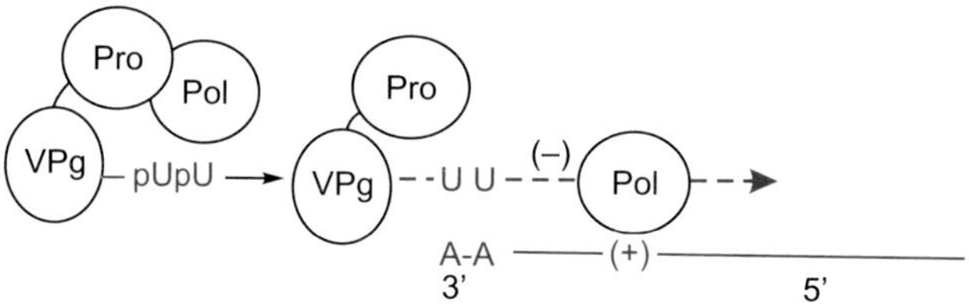

Figure 3.12 Diagram proposed for the initiation of negative strand RNA synthesis of a virus in which the RNA carries a VPg protein, indicating the possible roles for VPg-Pro and Pol proteins. The VPg protein serves as a primer and is linked to the end of the RNA (according to Li et al., 1995).

In the genus *Potyvirus*, the structural analogies of the genome and the mode of expression with the poliovirus also suggested functional analogies of proteins. The polyprotein is cleaved by three viral proteases (see Fig. 2.7). Among the resulting proteins, several may intervene directly in replication:

- NIb (nuclear inclusion b) carries polymerase motifs. Its polymerase activity was demonstrated *in vitro* in the case of TVMV NIb expressed in *E. coli* (Hong and Hunt, 1996).
- CI carries helicase motifs. Its activity of unwinding of the double strand was demonstrated for PPV (Lain et al., 1991).
- NIa (nuclear inclusion a) is constituted of the VPg protein in its N-terminal region and by a protease in its C-terminal region.

- 6K2, a small protein necessary for replication, is located in membrane proliferations. It could be an element of transmembrane fixation of the complex (Schaad et al., 1997).

These four proteins are arranged side by side in the polyprotein: -CI-6K2-NIa (VPg-Pro)-NIb- and their cleavage is effected in *cis* by NIa (Fig. 2.7). The slowest cleavage is the internal cleavage of NIa, which liberates VPg (Schaad et al., 1996). The proteins NIa, NIb, and CI are part of the replicative complex isolated from tissues infected by PPV (Martin et al., 1995). Regulation of RNA synthesis results from time-course of proteolytic processing at each proteolytic junction. Fully processed proteins as well as intermediates play a role during the infection cycle (Merits et al., 2002).

Protein-protein interactions have been discovered *in vivo* in yeast between NIb and NIa (Li et al., 1997). A strong physical interaction can also be detected *in vitro* between NIb and the VPg domain of NIa (Fellers et al., 1998). It is believed that these interactions link NIb to membranes by the intermediary of NIa-6K2 and locate the polymerase close to the VPg. The initiation of viral strands occurs through the VPg region of NIa used as a protein-primer in the form VPg-pU (Fig. 3.12).

It is clear that in viruses in which the RNA has a VPg, translation, proteolysis, and replication are interwoven in a closely regulated process. *In vivo*, the replication of RNA 1 of CPMV can only be effected in *cis*, that is, the replicative complex is assembled on the same strand from which the proteins were translated (Bokhoven et al., 1993). The RNA 2 is replicated in *trans* by replication proteins coded by RNA 1 (Fig. 3.11). The replication complex binds to the viral RNA associated with ribosomes by the intermediary of the 58 kDa protein coded by RNA 2, which carries a motif for RNA binding. Translation and replication are here closely coupled, which could explain the difficulties of isolating a replicase effecting the entire replication cycle *in vitro*.

■ RNA elongation

We have just seen that the initiation of the replication of RNA linked to VPg needs a VPg-pU primer. Is the need for a primer universal? It is known that the different classes of cellular polymerases copy or transcribe DNA from internal promoters by using primers. To copy an RNA from an end presents an additional difficulty: the stable fixation of the replication complex at the end of the template strand.

The replication complex comprises replicase, the template RNA, and the precursor triphosphate nucleosides. After studying the functioning of the BMV replication complex of BMV in the presence of limiting quantities of precursors (Sun and Kao, 1997a) and measuring the capacity of the replicases to use specific primers (Kao and Sun, 1996), the authors described complexes of increasing stability.

- The first stage is the specific recognition of replicase and template.

- The second stage is the formation of primary phosphodiester bonds that give rise to numerous oligonucleotides complementary to the 3' end. If their length is less than 8 nucleotides, they are released and the complex dissociates (abortive synthesis).
- To go from initiation to elongation, the complex must be stable. This occurs from around 8 nucleotides (Sun and Kao, 1997b). This mode of functioning is analogous to that of DNA-dependent RNA polymerases.

Inhibition of the expression of cellular genes

It is very difficult to analyse the mechanisms by which the early inhibition action of the virus is exerted on the metabolism of its host. Histochemical observations made on cotyledons of peas in the course of infection by PSbMV (*Pea seed-borne mosaic virus, Potyvirus*) showed that the infection (measured by *in situ* hybridization of negative and positive strand RNAs) progresses according to a front along which the expression of certain genes of the host is greatly inhibited. In the cells that become infected and in which the virus actively multiplies, the accumulation of mRNA of 10 cellular genes is reduced to a very low level. This phenomenon is transitory, it coincides with replication and the protein synthesis is restored in the cells located behind the front along which the infection advances, which is formed of 4 to 8 rows of cells (Wang and Maule, 1995). This point distinguishes it from the shutoff observed in the animal cells, which persists up to the end of the cycle. Not all the genes are affected, which could be explained by a specific degradation of certain messengers (Aranda et al., 1996).

Positive-sense RNA virus supergroups

The nucleotide sequence of numerous positive-sense RNA viruses of bacteria, plants, and animals is now known, which makes it possible to compare even distant viruses. It has become obvious that different viruses (in terms of morphology, pathogenicity, animal or plant host) have similarities with respect to the genes, arrangement of genes, and modes of expression of genes (Haseloff et al., 1984). Because of the high mutation rate of RNA genomes and the recombination frequency, it is probable that the conservation of some even very short motifs, involved in major functions, and their arrangement on the genetic map have a high significance.

The homology between motifs of replication proteins, first demonstrated between poliovirus (*Picornavirus*) and CPMV (*Comovirus*), was extended to an entire supergroup (or superfamily) called *Picorna*-like supergroup (Fig. 3.13). The common characteristics of the genera and families of this supergroup are as follows:

- homologies in the polymerase gene (polymerase of type 1, see Fig. 3.5);
- presence of a VPg (protein linked covalently at the 5' end); frequent polyA at the 3' end of the RNA;

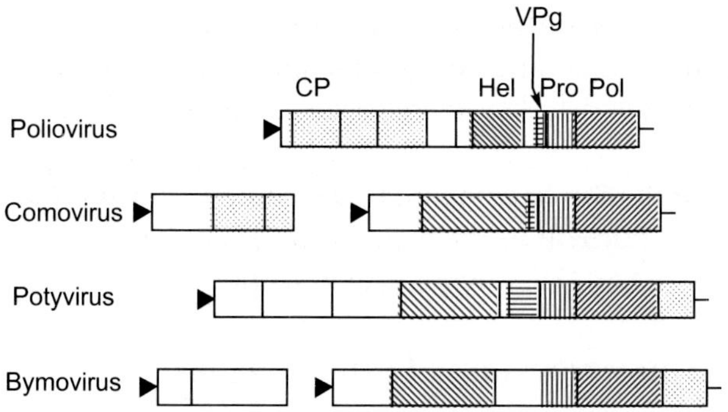

Figure 3.13 Picorna-like supergroup (according to Goldbach and Wellinck, 1988). Domains: CP (capsid), Hel (helicase), VPg, Pro (protease), Pol (polymerase).

- translation strategy: polyprotein and proteolysis;
- the replication module on the viral RNA conserves the order 5'-helicase-VPg-protease-polymerase-3'.

This supergroup comprises animal viruses (*Picornaviridae, Caliciviridae, Waikavirus*, etc.) and plant viruses (*Potyviridae, Comoviridae*).

In the same way, significant homologies have been found between the 126 and 183 kDa proteins of TMV (*Tobamovirus*), the 206 kDa protein of TYMV (*Tymovirus*), the proteins coded by RNA 1 and 2 of *Bromoviridae*, and the Alphaviruses of animals (Goldbach and Wellink, 1988; Goldbach et al., 1991). These viruses are now grouped in the Alpha-like supergroup (Fig. 3.14). The common characters of the genera and families of this supergroup are the following:

- sequence homologies in the polymerase gene (type 3 polymerase);
- sequence homologies in the helicase gene (type 1 helicase);

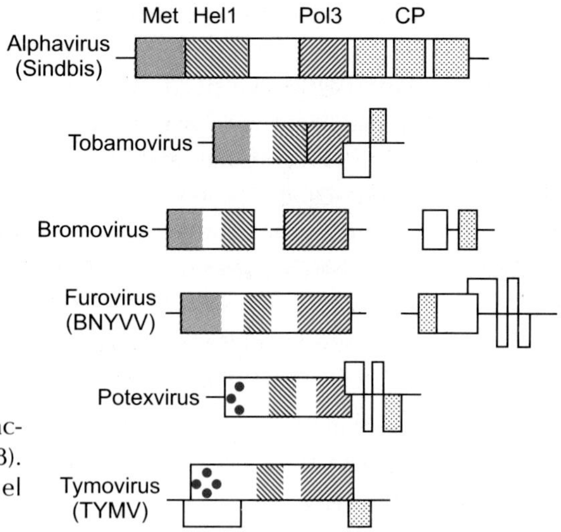

Figure 3.14 Alpha-like supergroup (according to Goldbach and Wellinck, 1988). Domains: methyltransferase (Met), Hel (helicase), Pol (polymerase), CP (capsid).

- conserved arrangement of replication genes;
- translation strategies: cap at the 5' end, subgenomic RNAs, frequent readthrough.

This supergroup comprises animal viruses (*Alphavirus (Sindbis), Rubivirus, Arterivirus*, etc.) and plant viruses (*Bromoviridae, Tobamovirus, Tymovirus, Tobravirus, Furovirus, Hordeivirus, Potexvirus, Closteroviridae, Carlavirus, Trichovirus*, etc.).

Two other supergroups, Carmo-like and Sobemo-like, have also been proposed (Fig. 3.15) (Gorbalenya, 1995). Not all the positive-sense RNA viruses are categorized into supergroups.

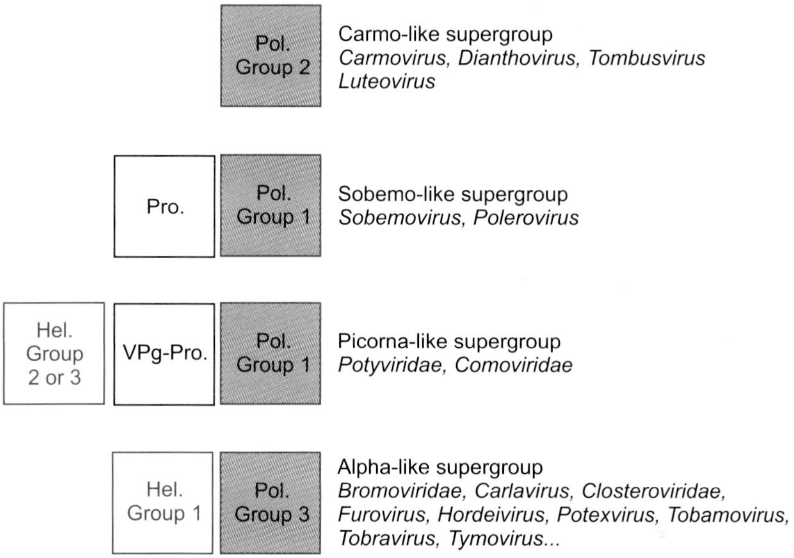

Figure 3.15 Positive-sense RNA viruses: modular organization of the replication proteins in the supergroups (Gorbalenya, 1995). The phylogenetic origin of replication proteins of each of the supergroups proposed is indicated. The other proteins are not shown.

It must be emphasized that supergroups are not actually a classification category but rather a grouping based on criteria of genome sequence and structure. It is observed that this grouping brings to light other common points (e.g., structure of RNA ends, translation strategies) and divergences (e.g., movement protein, capsid). To explain the similarities, we can resort to various non-exclusive phenomena: common ancestry, interviral recombination, acquisition of host genes (examples of each of these phenomena have been described). It is observed that the similarities in the replication genes do not imply similarities in other genes, for example the capsid gene. The concept of modular evolution (Fig. 3.15) makes it possible to understand that the genes or gene modules could have been acquired from different origins, and that the grouping of viruses must rely preferably on the fundamental character of the replication strategy.

Replication of negative-sense RNA viruses

Two families of this group comprise viruses that infect plants: *Rhabdoviridae* (*Cyto-* and *Nucleorhabdovirus*) and *Bunyaviridae* (*Tospovirus*) as well as the genera *Tenuivirus* and *Ophiovirus*. Many of these viruses have the common characteristic of replicating in plant cells as well as in the cells of invertebrates. They also have in common the organization of their genome and the particular structure of their RNAs, which have inverted complementary sequences at their ends. These RNAs can thus form circles closed by non-covalent bonds. The terminal regions, where the transcription and replication promoters are found, are conserved for the different RNAs of the same virus.

■ *Rhabdoviridae*

In the *Rhabdoviridae*, all the viral proteins are present in the virion. The virion is protected by a lipoproteic coat and contains at least 5 proteins (Chapter 1, Fig. 1.19). The genome is constituted of a single molecule of negative-sense single-stranded RNA of around 13 kb. From the 3' end to the 5' end, the 5 ORFs code for the nucleocapsid N, a phosphoprotein M2, a protein of unknown function, the matrix protein M1, a glycoprotein G, and the polymerase-transcriptase L (Chapter 15 No. 53, 54, profile of family *Rhabdoviridae*). The replication of *Rhabdoviridae* was first studied in viruses infecting animals (*Vesicular stomatitis virus, Rabies virus*). In plants the principal events of viral multiplication have been described for the nucleorhabdovirus SYNV (*Sonchus yellow net virus*): to become infectious and messenger, the genomic RNA must be transcribed by the L transcriptase incorporated in the virion. It is transcribed into capped and polyadenylated monocistronic mRNAs (5 for the SYNV), each coding for a viral protein. The replication proteins then copy the negative strand genomic RNA into a complete positive strand, which serves in turn as a template to synthesize the descendants. The viral RNA then successively serves as template for transcription and replication (Fig. 3.16).

These events can be followed *in vitro*: in fact, the SYNV multiplies essentially in the nucleus, which swells and develops inclusions called "viroplasms" in the infected cells. This has allowed the preparation from infected nuclei of an enzymatic extract that can transcribe viral RNA *in vitro* into messenger RNA and synthesize strands of genomic polarity. The nuclear extract contains proteins N and M2, and protein L of around 200 kDa, the sequence of which includes the polymerase signature (GDNQ) present in all the negative-sense RNA viruses. An antiserum against the protein L inhibits enzymatic activity. This activity is double: one, the viral RNA is observed to be transcribed into messenger RNAs sequentially, the messengers being decreasingly abundant from the 3' end to the 5' end, and two, the extract can be used to synthesize segments of strands of genomic polarity (Wagner and Jackson, 1997).

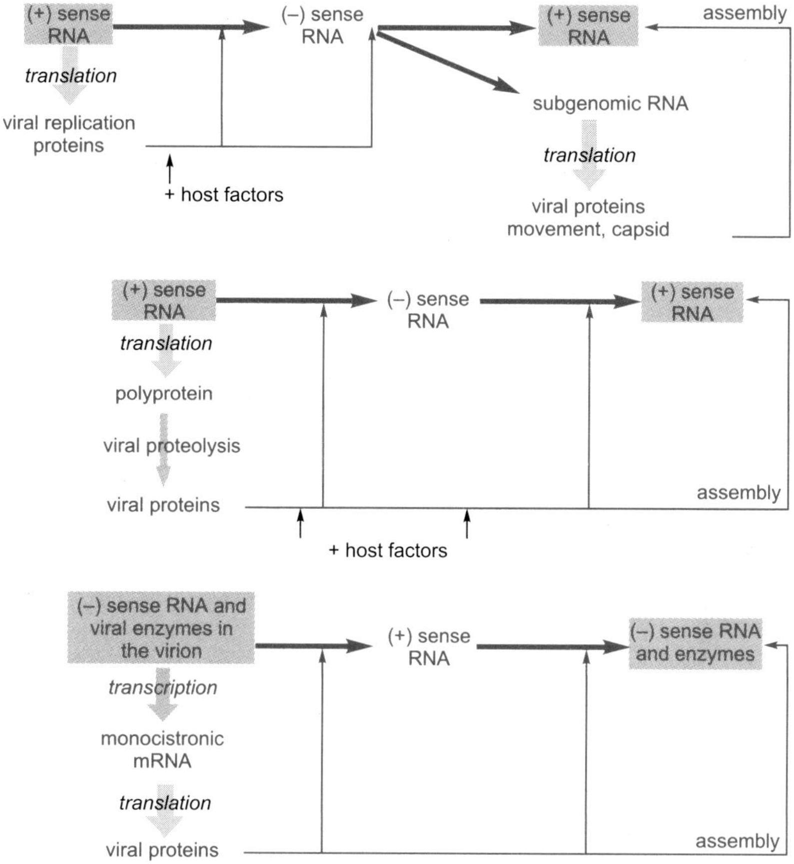

Figure 3.16 Comparison of replication cycles of RNA viruses. Top, positive-sense RNA viruses and synthesis of subgenomic RNA. Centre, positive-sense RNA viruses and polyproteins. Below, negative-sense RNA viruses. The functions coded by the viral genome are shown in red, the host functions in black.

In cells infected by SYNV, a series of events is observed; the viral particle fuses its envelope with the endoplasmic reticulum and the nucleocapsid reaches the nucleus, where it establishes a viroplasm that transcribes and replicates the viral RNA. Since the translation occurs in the cytoplasm, the process involves numerous transfers between the cytoplasm and the nucleus, and sequences of nuclear localization have been indicated in protein N. The newly formed nucleocapsids bud through the internal nuclear membrane and the surrounded particles accumulate in the perinuclear space in vesicles containing virions.

Two genera are presently recognized: the *Cytorhabdovirus* (LNYV, *lettuce necrotic yellow virus*) replicate in the cytoplasmic viroplasms, a synthesis preceded in some cases by a nuclear phase. The *Nucleorhabdovirus* (SYNV) multiply in the nucleus, where the viral proteins will be transported by means of the nuclear localization signal they carry.

Tenuivirus and *Tospovirus*

Tenuivirus and *Tospovirus* have negative-sense RNA and ambisense RNA.

Tenuivirus have segmented genomes (up to 6 segments); their polarity is either negative or ambisense. The longest of the RNAs of RSV (*Rice stripe virus*, *Tenuivirus*) has negative polarity. Its complementary strand codes for a protein L that is incorporated in the virion. This RNA polymerase has a transcription and polymerization activity demonstrated for RHBV (*Rice hoja blanca virus*, *Tenuivirus*). During the transcription, the RNAs acquire a cap by fixing a short sequence 5' of a cellular messenger (Nguyen et al., 1997; Ramirez and Haenni, 1994).

The virions of *Tospovirus* are formed of a lipidic membrane in which two glycosylated viral proteins G1 and G2 are inserted. It encloses three ribonucleoprotein complexes (viral RNA + nucleocapsids + enzymatic protein) (Fig. 3.17). Two of the RNAs are ambisense (S and M) and the third, L, has a negative polarity and its complementary strand codes for an RNA polymerase incorporated in the virion (Richmond et al., 1998) (Chapter 15, no. 52). This enzyme transcribes the genomic RNAs into messenger RNAs and plays the role of viral replicase.

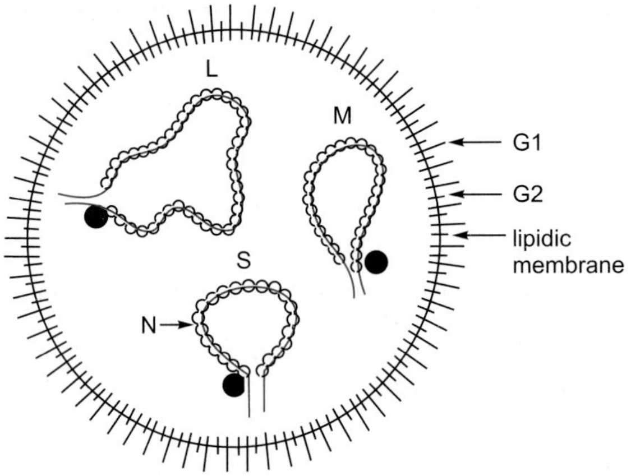

Figure 3.17 Viral particle of a tospovirus. Three viral RNAs (in red) are surrounded by N proteins. The outer lipidic membrane contains two viral proteins, G1 and G2. ●: RNA polymerase incorporated in the virion.

Replication of single-stranded DNA viruses: *Geminiviridae* and *Nanoviridae*

The genome of *Geminiviridae* is constituted of either one or two circular single-stranded DNAs less than 3 kb in size. This relatively small genome encodes no polymerase gene, and it relies extensively on host factors for its multiplication. In the first stage of multiplication, the single-stranded DNA is converted into double-stranded DNA by the action of host enzymes. This dsDNA is transcribed in two opposite directions from divergent promoters located in the intergenic region (Fig. 3.18).

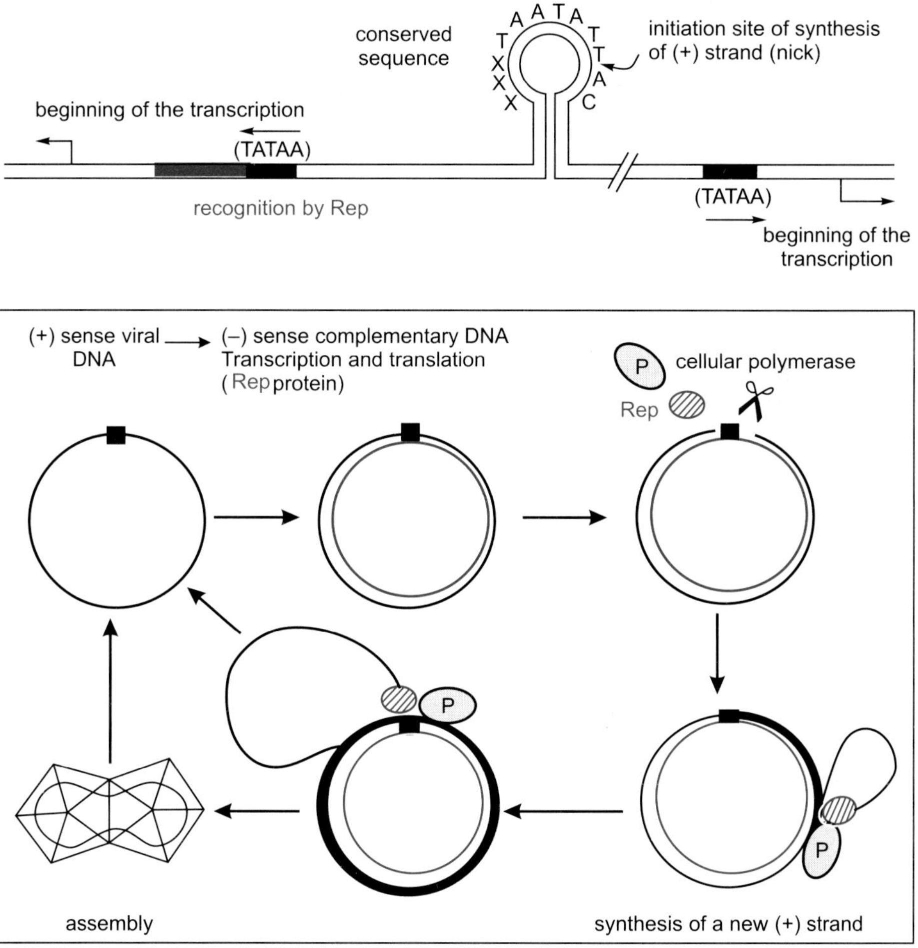

Figure 3.18 Multiplication cycle of *Geminiviridae*. The parental chain is a black circle, the complementary copy a red circle, and the newly synthesized chain is a thick black line. Top: the intergenic region contains several signals: the potential stem-loop and its restriction site, the TATA-box of recognition for the transcription and the starting sites of transcription in the opposite directions, the recognition motif of Rep (according to Bisaro, 1996).

In the second stage, cellular DNA replication proteins and the viral protein Rep (replication associated protein) initiate the "rolling circle" replication process on the intermediate double-stranded circular DNA: plus strands are continuously copied on the minus-strand template and shifted from the dsDNA.

Two viral elements are indispensable for replication

By mutagenesis and complementation, it has been demonstrated that, apart from cellular enzymes, two viral structures directly intervene in this synthesis, the protein Rep and the viral sequence recognized by Rep:

- The protein Rep associated with replication (also called AL1 or AC1 or C1-C2, molecular weight around 40 kDa) is translated from the complementary-sense transcript. It is the only viral protein indispensable for the infection of protoplasts. It has no polymerase motif, and its sequence resembles the sequence of Rep proteins that initiate the rolling circle replication of bacterial plasmids. One of its functions is an endonuclease activity that specifically cleaves a DNA site at the beginning of synthesis and effects a ligation at the end of synthesis (Laufs et al., 1995).
- Rep specifically recognizes a sequence of the intergenic region with high affinity (Orozco and Hanley-Bowdoin, 1996) and creates downstream a cleavage of positive-sense DNA in an adjacent motif in the potential stem-loop (TAATATTAC) exactly conserved in all the *Geminiviridae* (Fig. 3.18). This cleavage allows the priming of the synthesis of positive strands. The bipartite *Geminiviridae* have intergenic regions with identical specific recognition sites.

Rolling circle replication occurs in several phases:

- The viral DNA is transported into the nucleus by means of a nuclear localization signal carried by the capsid (Liu et al., 1999a). In the nucleus, the single-stranded DNA is converted into double-stranded DNA by cellular enzymes.
- The double-stranded DNA serves as a template for transcription and replication. Rep nicks parental circular DNA at the specific conserved site and remains covalently linked to the 5' end. The free 3' OH end serves as a primer for the synthesis, by cellular enzymes, of a new positive strand on the template of the negative strand, which displaces the parental strand (Fig. 3.18). Rep then exerts its ligation function to render the displaced strand circular.

The intergenic region contains the origin of replication as well as the promoters of the bidirectional transcription. By recognizing this region, Rep represses the transcription. This role is added to those already indicated of endonuclease and ligase. The specificity for origin recognition of TYLCV (*Tomato yellow leaf curl virus, Begomovirus*) by Rep has been mapped to the N-terminal 116 amino acids (Jupin et al., 1995).

The Rep protein of *Geminiviridae* acts on the cellular cycle

For their replication, the *Geminiviridae* depend on cellular proteins that replicate the host DNA. The replication of the cellular DNA is the result of a set of coordinated events, regulated by positive and negative cellular factors. In mammals, the proteins of the retinoblastome family (Rb) act as negative regulators of the cell cycle by neutralizing transcription factors of the family EF2 necessary for the transcription of numerous genes implicated in the replication of DNA. An early protein of mammal oncoviruses forms a stable complex with Rb proteins by the intermediary of an LxCxE motif; the negative regulator is thus rendered inoperative and the transcription factors become available, which favours the entry of the cell into the S-phase of division and the synthesis of DNA replication enzymes necessary for the virus.

It seems established that in the family *Geminiviridae,* the virus is capable of interfering with proteins that control the plant cellular cycle, in a manner analogous to the mammal oncoviruses. The replication of viral DNA is realized in the nucleus by cellular enzymes normally absent from the quiescent nucleus of a differentiated cell. During infection by TGMV (*Tomato golden mosaic virus, Begomovirus*), the nucleus becomes hypertrophied; by immuno-localization, the viral Rep as well as a cellular factor (called PCNA) associated with replication of the cellular DNA can be detected. This factor is normally detected only in dividing cells, but it is found also in cells of transgenic *Nicotiana benthamiana* expressing the Rep protein. This strongly suggests that one of the functions of Rep is to induce the de-repression of DNA replication. The protein RepA of WDV (*Wheat dwarf virus, Mastrevirus*) carries a motif LxCxE that, in a yeast two-hybrid system, recognizes a mammal Rb protein as well as a homologous protein of maize. It seems that viruses of the family *Geminiviridae* infecting differentiated cells all have a protein Rep that recognizes an Rb protein. This strategy is thus analogous to that of animal oncoviruses in which a viral protein blocks the action of a regulator of the cell cycle (Xie et al., 1995; Liu et al., 1999b). However, unlike in infections by oncoviruses, the process does not continue past the S-phase, and the infected cells do not divide.

In sum, replication does not require viral polymerase but a protein associated with replication that induces the synthesis of cellular enzymes having a DNA polymerase activity, which the virus uses for its multiplication (Fig. 3.19). It must be noted that the neutralization of proteins of Rb type has multiple effects on the cell and plant metabolism (Gutierrez, 2000).

Nanoviridae also deregulate the cell cycle

The *Nanoviridae* genome is composed of several single-stranded circular DNAs of around 1 kb. Eleven distinct DNAs have been found in an isolate of FBNYV (*Faba bean necrotic yellow virus, Nanovirus*) and five different Rep proteins. Each Rep is carried by one of the DNAs of which it specifically controls the replication, except Rep2 which controls the replication of six other DNAs. The component coding for Rep2 is the only one that is common to all the isolates of FBNYV. The activity of these Rep proteins is the cleavage of the DNA at the specific site and its ligation as in the *Geminiviridae* (Timchenko et al., 1999).

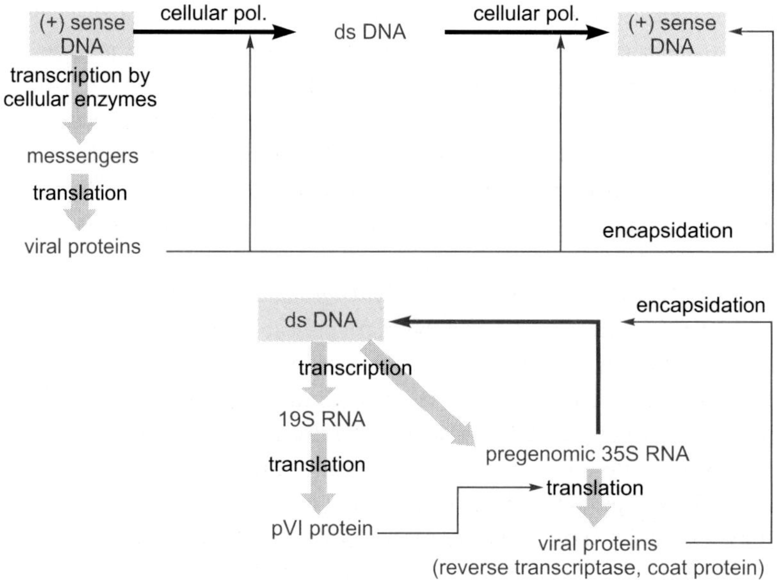

Figure 3.19 Comparison of replication cycles of DNA viruses. Top, single-stranded DNA viruses (*Geminiviridae*); bottom, double-stranded DNA viruses (*Caulimoviridae*). The viral functions are indicated in red, the host functions in black.

In contrast, the cell cycle deregulation activity is not ensured by Rep. In fact, the motif LxCxE already revealed in *Geminiviridae* is carried by a viral protein called Clink (cell cycle link). Clink is the stimulator of viral replication by its capacity to link and interact with proteins regulating the cell cycle (Aronson et al., 2000).

Replication of double-stranded DNA viruses: *Caulimoviridae*

Plant viruses with double-stranded DNA genome belong to six genera assembled in the family *Caulimoviridae*. In this family, replication of viral DNA is achieved by a process of reverse transcription: viral DNA is first transcribed into mRNA by cellular enzyme. After translation of viral proteins, viral RNA is reverse transcribed by a viral protein, reverse transcriptase.

The 8 kb genome of CaMV (*Cauliflower mosaic virus, Caulimovirus*) is constituted of a circular double-stranded DNA molecule (Fig. 3.20) comprising three discontinuities (or gaps) with redundant termini: two (Δ2 and Δ3) on the positive strand (transcribed strand), one (Δ1) on the negative strand (see Fig. 3.22). The discontinuity Δ1 is conserved in all the *Caulimovirus*; it is adjacent to the site on which the $tRNA^{met}$ that will serve as a primer in the replication is attached by complementarity.

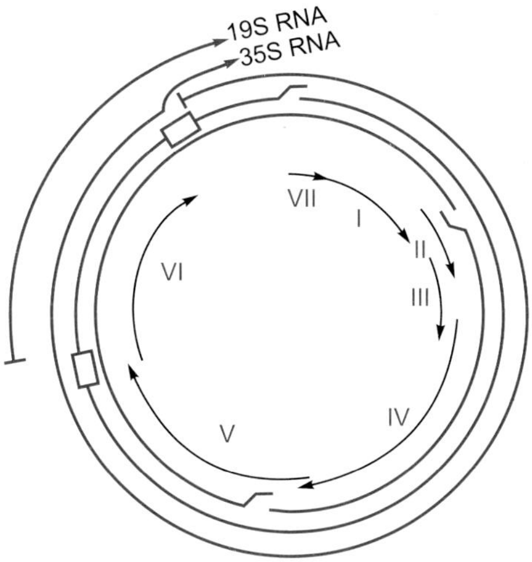

Figure 3.20 CaMV circular double-stranded DNA with position of gaps. Promoters of transcription are shown as boxes and transcripts as thin outer lines. Proteins are numbered I to VII (Jacquot et al., 1998).

After penetration of the virus into the cell, the DNA migrates into the nucleus, where the gaps are filled, and is surrounded with histones to form a supercoiled minichromosome (Fig. 3.21). The transcription by the host DNA-dependent RNA polymerase II, which recognizes the viral promoters 35S and 19S, is the only genome amplification phase. It gives rise to two major transcripts called 35S (pregenomic RNA) and 19S (subgenomic RNA), which have the same 3' end. Their translation in the cytoplasm gives rise to viral proteins.

The 19S RNA is monocistronic, it codes for a protein PVI that forms cytoplasmic inclusion bodies called viroplasms, in which the viral particles accumulate (Mazzolini et al., 1989). Viroplasms are the site of viral DNA synthesis and assembly of virions.

The 35S RNA is a complete copy of the genome, increased by a redundant segment (this is possible on circularized DNA) of 180 nucleotides corresponding to the region Δ1. The protein PVI stimulates the expression of proteins coded by the 35S RNA: PI (movement), PII (transmission by aphids), PIII and PIV (protein associated with the capsid and capsid protein), and PV (polymerase reverse transcriptase) (Jacquot et al., 1997) (Fig. 3.22).

The 35S transcript plays two roles: it is messenger for the expression of viral proteins (except PVI) and it is a template (pregenomic RNA) in the reverse transcription by which the minus sense DNA is synthesized. Near its 5' end, it has a sequence of 13 bases complementary to the end of $tRNA^{met}$ that will serve as primer for the viral reverse transcriptase synthesizing the viral DNA. The redundant sequences of 180 nucleotides of its ends allow the reverse transcriptase to "jump" from the 5' end to the 3' end and continue up to the complete synthesis of the DNA

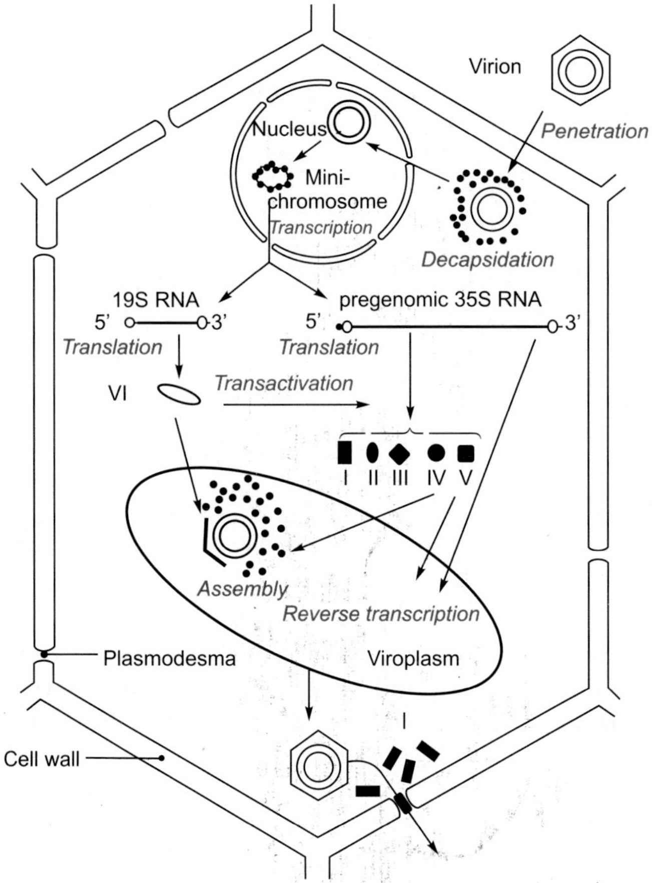

Figure 3.21 Multiplication cycle of CaMV (Jacquot et al., 1998).

strand. This template function can be carried on only once, because the RNase H activity associated with reverse transcriptase immediately digests the RNA, except for two small segments that will serve as primers for the synthesis of the complementary positive strand. This double-stranded DNA will be either newly transcribed into RNA or assembled (Figs. 3.19 and 3.21).

The mode of replication of CaMV is characteristic of the Pararetrovirus supergroup, which includes *Caulimoviridae* (*Badnavirus* and *Caulimovirus* of plants) and *Hepadnaviridae* (hepatitis B virus of vertebrates); a pregenomic RNA serves as template for synthesis by reverse transcription of the encapsidated viral DNA, which is not integrated in the host genome. In the *Retroviridae* infecting vertebrates, the pregenomic RNA is encapsidated, while the viral DNA is integrated into the host genome. We notice a functional analogy between these two groups: the genome exists alternatively in two forms, DNA and RNA, and the passage between the two forms is effected by a viral enzyme, the reverse transcriptase. There are also structural analogies: in the *Caulimovirus*, equivalents of the genes *gag* (capsid), *pol* (reverse transcriptase), and *env* (viroplasm protein) of *Retroviridae* are present. The sequence of amino acids of the CaMV reverse transcriptase has motifs conserved with those of animal retroviruses of murine leukaemia and Rous sarcoma.

Reverse transcriptase is the common element of pararetroviruses and *Retroviridae*, as well as retrotransposons (Fig. 3.23). This multifunction protein has several known activities: aspartate-proteinase in the N-ter domain, reverse transcriptase and RNase H in the C-ter domain (Takatsuji et al., 2003). In the *Retroviridae* and retrotransposons, it also has an integrase-endonuclease function allowing insertion into the host DNA, a function absent in pararetroviruses. However, it must be noted

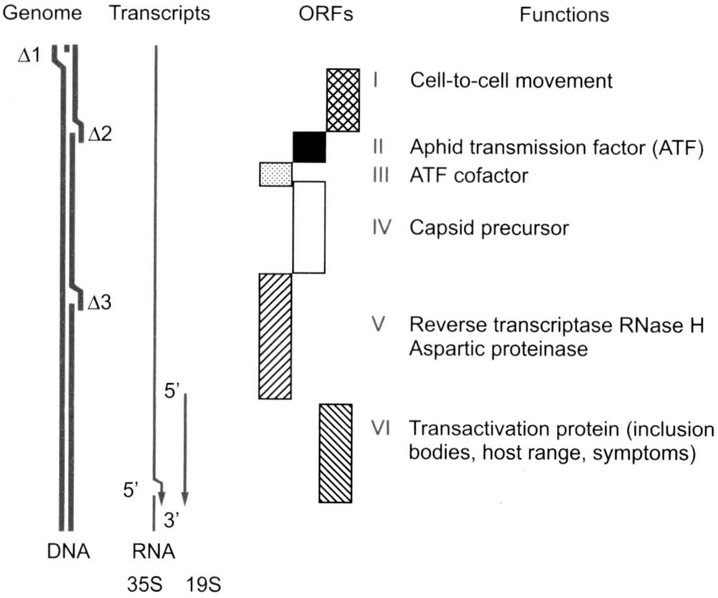

Figure 3.22 Principal functions of proteins coded by CaMV DNA. The DNA is represented linearly (Jacquot et al., 1998).

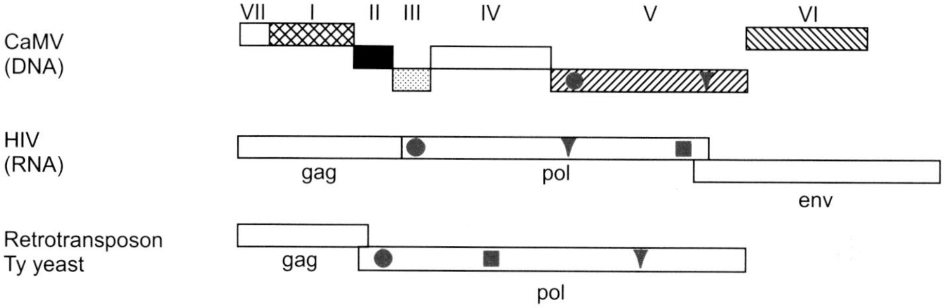

Figure 3.23 Virus genomes having a reverse transcriptase. Protease motifs ●, reverse transcriptase ▼, integrase ■.

that cytogenetic and molecular observations have shown that BSV (*Banana streak virus, Badnavirus, Caulimoviridae*) can be integrated in the genome (Harper et al., 1999). Every *Musa* species contains BSV sequences, partial sequences or multiple copies of the complete genome (Geering et al., 2001). Multiple copies arranged in tandem are a potential source of BSV infections observed in tissue culture of interspecific hybrids between *Musa acuminata* and *Musa balbisiana* (Chapter 13).

Integrated sequence of the complete genome can therefore be activated to give episomal infections under certain conditions. This situation is observed with BSV as well as with PVCV (*Petunia vein clearing virus, Caulimovirus*) and TVCV (*Tobacco vein clearing virus, Caulimovirus*). PVCV entire genome is present in many cultivars of *Petunia hybrida* (Richert-Pöggeler et al., 2003). In the polyprotein encoded by PVCV DNA, putative integrase motifs have been identified (Richert-Pöggeler and Shepherd, 1997). TVCV is transmitted only vertically. Its genome exists as multiple partial sequences in the host genome and becomes episomally active (Lockhart et al., 2000, 2002).

In the group of reverse transcribing viruses, two new families (with genera containing, among others, members of the plant kingdom) have been proposed by the International Committee on the Taxonomy of Viruses, at the frontier between viruses and retrotransposons: *Pseudoviridae* (retrotransposons of barley, *Arabidopsis*, etc...) and *Metaviridae* (retrotransposons of *Arabidopsis*, peas, soybean, etc...) (Lockhart et al., 2002).

Conclusion

As a comparative study of viral cycles shows (Figs. 3.16 and 3.19), viruses use various strategies to insert the viral multiplication into the host metabolism. The part ensured directly by the host enzymes varies according to the strategy. In all cases, however, the virus exploits the resources of the host cell at several levels:

- It draws from cellular reserves to find the energy and precursors required for the synthesis of its proteins and nucleic acids.
- It borrows the host metabolic mechanisms and especially the protein synthesis system of the host.
- Because of the limited quantity of information of the genome, resulting from the constraints of encapsidation, the virus completes its replication device by using host proteins, which it deflects from their usual function.

In all these domains, the virus is a competitor for the cellular metabolic pathways. To take the lead, the viral genome has various structures that can catch the cellular proteins the virus needs.

The capacity of a given virus to infect for example a pea cell and not a tobacco cell results partly from the possibility of obtaining this cooperation of molecules and cell structures; these specific interactions are a significant factor in the host range of a virus.

Many viruses do not use the host transcription systems but promote the synthesis of their own enzymes (RNA polymerase and helicase for RNA viruses, reverse transcriptase), which form replicative complexes with cellular proteins. These specific devices allow them to escape cellular regulatory systems that tightly control the multiplication of nucleic acids. When they have no specific replication system, they use a cellular system by removing the obstacle of its regulation (e.g., *Geminiviridae, Nanoviridae*).

4

CHAPTER

Plant Virus Movement

While the expression of the viral message (Chapter 2) and the replication of viral nucleic acid (Chapter 3) are important phases of the infection cycle, if the infection stops within the first infected cells, which are often small in number, the plant will remain healthy. The infection is called subliminal. Some cases of subliminal infection were discovered when it was realized that protoplasts from plants considered to be resistant to a virus were found to be able to support the replication of that virus (Furusawa and Ono, 1978). Subliminal infection of cotton or cowpea by TMV (*Tobacco Mosaic virus, Tobamovirus*) was also observed (Sulzinski and Zaitlin, 1982). In susceptible plants, such as tomato or tobacco, TMV infection propagates from the first infected cells towards other organs. This kind of infection is called systemic. The capacity of a virus to move within a plant is therefore an important component of a successful infection. Virus movement was first studied in the TMV-tomato and TMV-tobacco systems before being studied for other viruses.

TMV movement: a model system

Kinetics of virus spread in the plant and in the leaf

In 1934, a classic plant virology experiment was carried out by Samuel (1934). Several tomato plants were inoculated on a single leaf with TMV (Fig. 4.1), then at

Viruses cited

AMV (*Alfalfa mosaic virus, Alfamovirus*); BCMNV (*Bean common mosaic necrosis virus, Potyvirus*); BCTV (*Beet curly top virus, Curtovirus*); BDMV (*Bean dwarf mosaic virus, Begomovirus*); BGMV (*Bean golden mosaic virus, Begomovirus*); BMV (*Brome mosaic virus, Bromovirus*); BNYVV (*Beet necrotic yellow vein virus, Benyvirus*); BSMV (*Barley stripe mosaic virus, Hordeivirus*); BWYV (*Beet western yellows virus, Polerovirus*); BYDV-PAV (*Barley yellow dwarf virus-PAV Luteovirus*); CaMV (*Cauliflower mosaic virus, Caulimovirus*); CCMV (*Cowpea chlorotic mottle virus, Bromovirus*); CMoV (*Carrot mottle virus, Umbravirus*); CMV (*Cucumber mosaic virus, Cucumovirus*); FHV (*Flock housel virus, Nodavirus*); GFLV (*Grapevine fanleaf virus, Nepovirus*); LMV (*Lettuce mosaic virus, Potyvirus*); MSV (*Maize streak virus, Mastrevirus*); ORSV (*Odontoglossum ringspot virus, Tobamovirus*); PCV (*Peanut clump virus, Pecluvirus*); PEMV (*Pea enation mosaic virus, Enamovirus*); PLRV (*Potato leafroll virus, Polerovirus*); PPV (*Plum pox virus, Potyvirus*); PVX (*Potato virus X, Potexvirus*); PVY (*Potato virus Y, Potyvirus*); RCNMV (*Red clover necrotic mosaic virus, Dianthovirus*); SHMV (*Sunn-hemp mosaic virus, Tobamovirus*); TAV (*Tomato aspermy virus, Cucumovirus*); TBSV (*Tomato bushy stunt virus, Tombusvirus*); TEV (*Tabacco etch virus, Potyvirus*); TMV (*Tobacco mosaic virus, Tobamovirus*); TuMV (*Turnip mosaic virus, Potyvirus*); TVMV (*Tobacco vein mottling virus, Potyvirus*); TYLCV (*Tomato yellow leaf curl virus, Begomovirus*); WClMV (*White clover mosaic virus, Potexvirus*).

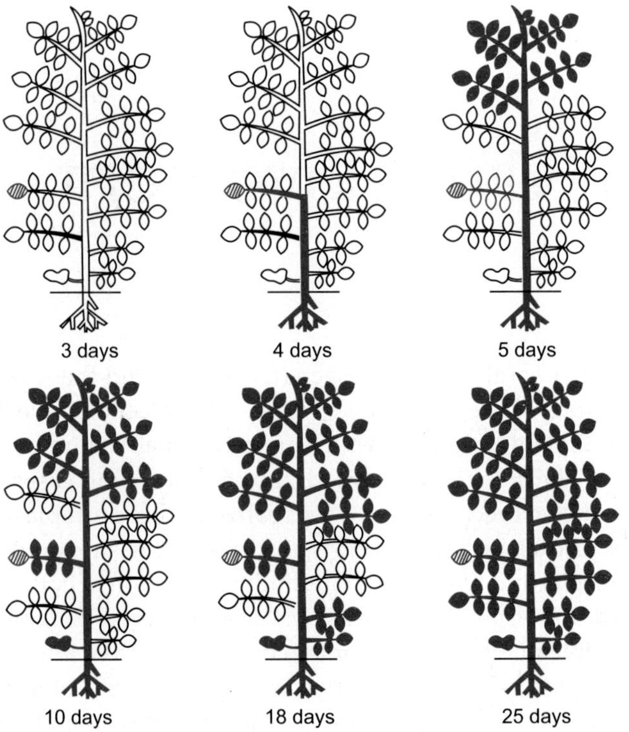

Figure 4.1 Movement of TMV in tomato (Samuel 1934). The inoculated leaflet is shown in hatched red and the infected organs in red.

different times, plants were divided and cuttings were made to allow the infection to develop. The presence of the virus in each plant part was monitored using a local lesion host (hypersensitive host, see Chapter 6). The number of local lesions is related to the titre of the virus in the inoculum.

This experiment showed that TMV stays for three days in the infected leaf, then is detectable the fourth and fifth days post-infection (p.i.) in the roots and in apical leaves before invading the rest of the plant. According to Samuel (1934), these observations were considered to be consistent with the theory of slow cell-to-cell movement of the virus through the plasmodesmata, combined with a rapid distribution of the virus throughout the plant through the phloem.

The infection of a tobacco leaf by TMV at 25°C was followed over 14 days after a mechanical infection, the same amount of leaf tissue was harvested at defined times and then ground in a standard volume of buffer, and TMV was assessed using a local lesion host as above (Dawson and Painter, 1978). In this experiment (Fig. 4.2) three phases can be observed. On the first two days no virus is detectable, then, between the second and the tenth days, the number of infectious particles steadily increases to reach a plateau on day ten. It is estimated that about 1/10,000 cells was initially infected and that 5 to 6 hours are needed for the virus to be detectable in adjacent cells, then the number of infected cells increases until the seventh day.

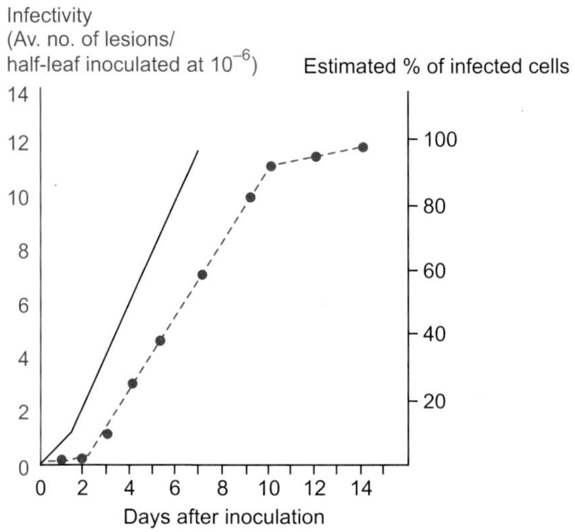

Figure 4.2 Kinetics of infection of a tobacco leaf inoculated on day 0 with TMV. -●- represents the evolution of infectivity (in number of local lesions) after mechanical inoculation; —— represents an estimation of the evolution of the percentage of infected cells (Dawson and Painter, 1978).

■ Cell-to-cell movement of TMV

The movement protein mobilizes the viral RNA

Neither the TMV particle (300 × 18 nm) nor the viral RNA (estimated at 10 nm when stastistically folded as a ball) can diffuse freely across the 2.5 nm lumen of the plasmodesmata (Gibbs, 1976) (Fig. 4.3). However, TMV RNA can move from cell to cell, as shown by the inoculation of a viral genome from which the coat protein is deleted. Although no virus particle can form, the RNA infects the mesophyll cells. It was shown that a thermosensitive mutant of TMV, named Ls1, can replicate but not spread outside of infected cells at the non-permissive temperature (Nishiguchi et al., 1978). The mutation was mapped at the gene coding for the 30K protein (Ohno et al., 1983) and thermosensitivity was introduced into the wild-type 30K protein by the substitution of Pro153 by Ser (Meshi et al., 1987). Moreover, the expression of the wild-type 30K protein in transgenic tobacco can complement the Ls1 deficiency for cell-to-cell movement at the non-permissive temperature (Deom et al., 1987). This protein was therefore called the movement protein (MP). Movement proteins were subsequently discovered in many plant viruses.

To promote virus movement, it can be anticipated that the TMV 30K protein must bind to the viral RNA and mobilize it toward and through the plasmodesmata.

The TMV 30K protein has two RNA binding domains

Gel shift experiments demonstrate that the TMV MP has a strong affinity for single-stranded RNA or DNA in a cooperative manner and without sequence specificity.

Binding of the first protein to the first site (4-7 nucleotides) stimulates second-site binding thanks to the establishment of protein-protein interactions; this proceeds until the nucleic acid molecule is completely covered by the MP. Binding of the MP to RNA disrupts intramolecular secondary structure to produce a long and thin (1.5 to 2 nm) ribonucleoprotein complex (Fig. 4.4). Two RNA binding domains were characterized. Domain A (amino acids 112-185) and domain B (amino acids 185-268) are independently able to bind RNA (Citovsky et al., 1990, 1992) (Fig. 4.5).

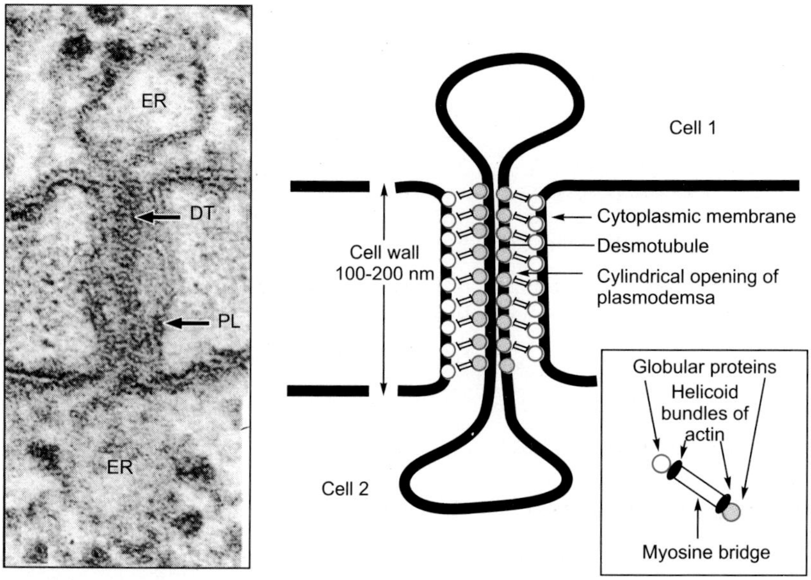

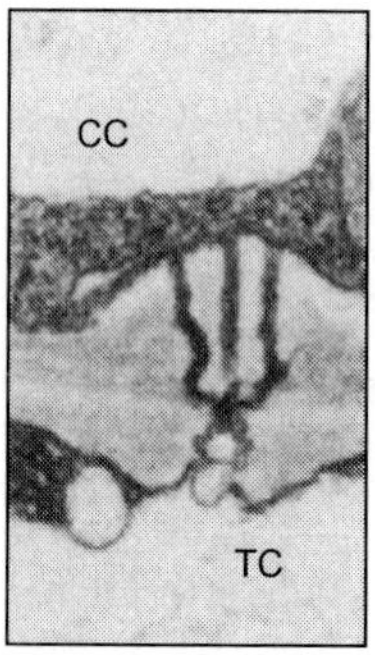

Figure 4.3 Linear plasmodesma (× 240,000) and delta-branched plasmodesma (× 18,000) from Esau and Thorsch (1985). Drawing of plasmodesma showing globular proteins on the surface of desmotubule and plasmalemma; on these globular proteins are anchored helicoid bundles of actin connected by myosine bridges (Ding et al., 1992b; Stussi Garaud et al., 1998). ER, ergastoplasm; PL, plasmalemma; TC, Sieve tube; DT, desmotubule; CC, companion cell.

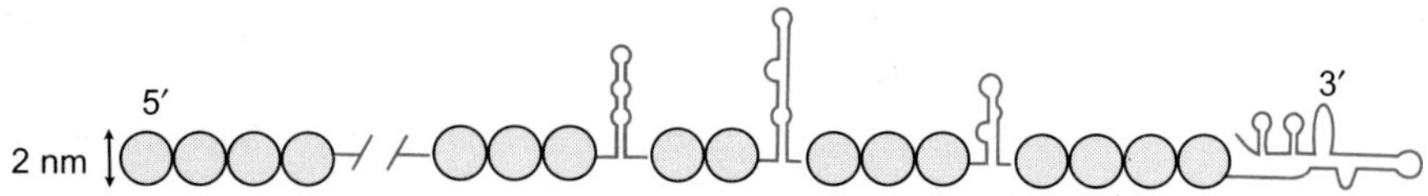

Figure 4.4 Structure of TMV MP-RNA complex (Citovsky et al., 1990; copyright Cell Press, Cambridge, Mass.). The MP molecules are shown as grey circles. They continuously protect the first 5400 nucleotides of RNA. In the remaining sequence of 1000 nucleotides, MP cannot connect to secondary structures where are located the origin of assembly and the tRNA-like 3′ end (see Chapter 1).

The TMV 30K protein interacts with plasmodesmata and the cytoskeleton

TMV MP fused to the green fluorescent protein (GFP) was found to localize next to microtubules at the proximity of plasmodesmata, associated with actin microfilaments.

TMV MP has also been found within the plasmodesmata of infected cells (Tomenius et al., 1987) and of MP-transgenic plants (Atkins et al., 1991) and was found to carry a fused GUS protein in the plasmodesmata (Waigmann and Zambryski, 1995). Amino acids 9 to 11 are necessary for this localization.

Increase of plasmodesmata exclusion size by the TMV 30K protein

Transgenic tobacco plants expressing the TMV 30K protein allow the free diffusion of 10 kDa fluorescent dextran, whereas in wild-type plants the exclusion limit is at approximately 1 kDa. Co-injection of dextrans together with the 30K protein confirms that 10 kDa molecules can diffuse between cells in young leaves and that molecules up to 20 kDa can move between cells of developed leaves. These experiments also demonstrate movement of the 30K protein itself and of a co-injected 2000 kDa viral RNA (Waigmann et al., 1994). The 2 nm width RNA-protein complex can thus cross the plasmodesmata.

Since the injection of profilin, an actin-depolymerizing protein, allows the passage through the plasmodesmata of 20 kDa dextran, it has been suggested that actin microfilaments may be a target of TMV MP (Ding et al., 1996a). Amino acids 126-224 (domain E) are involved in the increase in plasmodesmatal permeability (Waigmann et al., 1994).

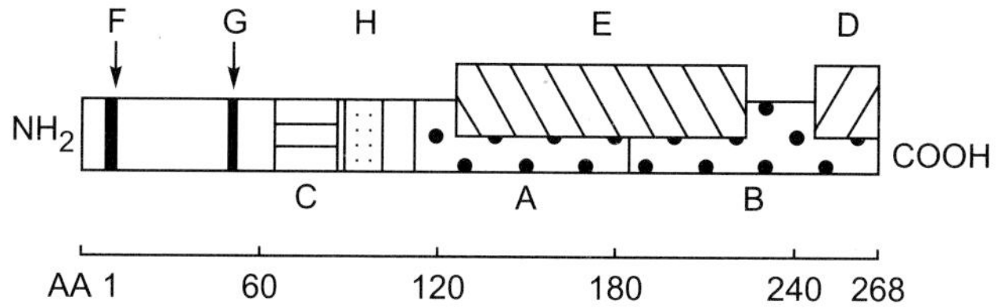

Figure 4.5 Functional domains in the TMV MP protein (redrawn from Waigmann et al., ©1994, National Academy of Sciences USA)

Domain A	aa 112-185	RNA binding
Domain B	aa 185-268	RNA binding
Domain C	aa 65-86	folding
Domain D	aa 247-268	phosphorylation
Domain E	aa 126-224	increase of plasmodesmatal exclusion size
Domain F	aa 9-11	plasmodesmatal localization
Domain G	aa 49-51	alignment with microtubules
Domain H	aa 88-101	microtubule binding

During infection by a TMV bearing an MP-GFP fusion, it was found that a fluorescent halo is associated with the expansion of each infection site. When a 20 kDa dextran is injected in this zone, its diffusion is limited to the border of the zone. Therefore, the increase in plasmodesmatal exclusion size is a transient phenomenon linked with the progression of the viral infection (Heinlein et al., 1995; Oparka et al., 1997). However, the MP protein is associated with plasmodesmata in the whole infected zone, suggesting that its activity is regulated by inactivation.

Indeed, the C terminus of the protein (amino acids 247-268) (Fig. 4.5) is phosphorylated by a serine/threonine kinase associated with the cell wall of mature leaves. Phosphorylated MP is thought to be sequestered in the central cavity of the plasmodesmata, possibly limiting its detrimental physiological activities such as non-specific binding to host cell RNA (Citovsky et al., 1993).

At the N terminus, a stretch of hydrophobic amino acids (65 to 86, domain C) (Fig. 4.5) are important for correct folding of the 30K protein.

Long distance movement of TMV

From the initially infected mesophyll cells, the virus enters the vascular tissue to propagate systemically and move out of this tissue to infect the whole plant. It enters the vascular tissue through small veins. The virus successively infects a cell of the perivascular parenchyma, cells of the phloem parenchyma, companion cells, and then the sieve tube.

The critical point in virus movement appears to be the entry into the sieve tubes and exit out of the sieve elements; this is linked to the particular structure of plasmodesmata linking the companion cells to the sieve tubes (Fig. 4.6).

The vein network and delta plasmodesmata

Plant vascular tissue comprises different forms of veins. Class 1 veins correspond to the primary vein, which divides into secondary veins (class II). From class II veins a network of class III veins delimits small islands of limb tissue, which is irrigated by the smallest class IV and V veins. These are completely embedded in limb tissue and account for about 90% of the vascular network in mature leaves.

Plasmodesmata linking the vascular tissue to companion cells have different properties than those of mesophyll cells. They form a typical delta-like branching on the side of the companion cell and a pore on the side of the sieve tube (Fig. 4.3). They also have a larger mass exclusion limit (10 to 20 kDa) than plasmodesmata of mesophyll cells (about 800 Da, Kempers et al., 1993).

The coat protein is essential for TMV movement in phloem tissue

A TMV mutant unable to produce the coat protein cannot systemically invade the plant. The virus accumulates in phloem parenchyma cells but is not detected in

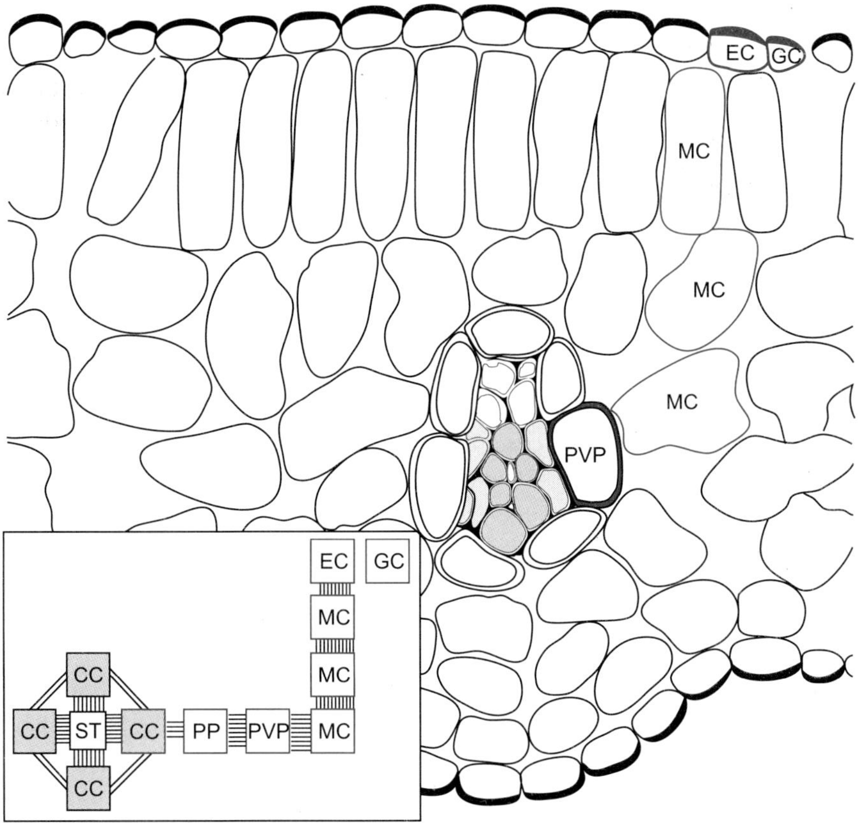

Figure 4.6 Different leaf cell types that can be infected by viruses: GC = stomata guard cell; EC = epidermal cell; MC = mesophyll cell; PVP = perivascular parenchyma; PP = phloem parenchyma; CC = companion cell; ST = sieve tube. Inset: relative number of plasmodesmata linking cell types.

companion cells (Ding et al., 1996b). *Odontoglossum ringspot virus* (ORSV) is a tobamovirus that is limited to the infected cell in *Nicotiana tabacum*. If the TMV coat protein gene is replaced by that of ORSV, the resulting chimeric virus can move from cell to cell but cannot invade the phloem (Hilf and Dawson, 1993). This suggests that the coat protein is necessary not just to protect the viral RNA in phloem tissue, but also to interact with host components of the vascular plasmodesmata.

The role of the MP in long distance movement is difficult to study independently of its role in cell-to-cell movement. The MP does not increase the size exclusion limit of the plasmodesmata of phloem cells but a chimeric virus where the TMV MP is replaced by ORSV MP is defective for phloem transport (Ding et al., 1992a).

Replication proteins are also involved in systemic movement

Some TMV strains accumulate at low levels in leaves distant from the inoculated leaf. The corresponding mutations were found not to affect the coat protein or MP protein genes but the gene for the 126 kDa protein, which is involved in replication. These mutants replicate at the same rate as a wild type in protoplasts or inoculated leaf, but in the latter very few companion cells were found to be infected, suggesting that replication proteins play a role in the transport through phloem plasmodesmata (Ding et al., 1995c; Derrick et al., 1997).

The virus moves with the flux of photosynthesis products

The sieve tube is composed of living elongated cells (20×250 μM) with two compartments: one is a fixed cytoplasmic compartment forming a thin layer along the cell wall, composed of membranes, plastids, and mitochondria but devoid of nucleus, ribosome, and vacuole, and the other is a large mobile compartment in which fluid transit can occur. Sieve elements communicate by large pores (200-400 nm or more). The sieve tube complex allows the transport of carbon metabolites from "source" carbon autotrophic leaves to "sink" immature leaves. In source leaves, given the high concentration in sugars, the hydrostatic pressure of the sieve element can reach 3000 kPa. The distribution of metabolites between source and sink leaves can take as little as 30 minutes. A sieve element is traversed by the flux in about two seconds (Sjölund, 1997). Therefore, the viral transport complex when inoculated in source tissue can rapidly reach developing tissues and cells.

In young developing leaves, class IV and V veins are not functional, therefore the virus is unloaded at the level of class III veins, as demonstrated by the propagation of the infection of a chimeric *Potato virus X* (PVX, *Potexvirus*) containing the GFP gene (Roberts et al., 1997). Class IV and V veins are infected at a later stage, through cell-to-cell transport from infected mesophyll cells.

In young developing leaves, the viral infection is propagated uniformly, as seen following GFP fluorescence, although the apical part can be more severely infected. During development, leaves undergo an irreversible physiological transition from importer to exporter of photosynthesis products. The source-sink transition is materialized by a narrow band, where the small veins involved in the export of carbohydrates are differentiated, moving from the apical part of the leaf (source) to the basal part (sink). This transition zone acts as a barrier to virus movement and the infection is then confined to the basal zone. This can be conveniently monitored by the observation of mosaic symptoms (Chapter 5; Plate I.6). The behaviour of the transition zone, from the apical to the basal part of the leaf, is unaffected by the viral infection (Roberts et al., 1997). The cells in the sink (basal) zone of young leaves progressively mature into source cells from which the virus can slowly invade other, older leaves as observed in the experiment of Samuel (1934). When the plant is young, all the leaves are eventually infected, but in older plants some leaves can escape the infection, although their sensitivity to experimental virus infection persists. This part of the viral movement needs further study.

TMV, like many viruses, can infect young leaves but does not enter meristems of infected plants. This was first demonstrated by Limasset and Cornuet (1949) by the inability to infect healthy plants with extracts from excised meristems of infected tobacco. From these observations, methods were developed to cure crop plants of viruses (see Chapter 10). The ectopic expression of a potexvirus MP in the meristem allows the propagation of the virus and of post-transcriptional gene silencing (PTGS) signals to the meristem (see Chapter 5). This suggests that an RNA surveillance system is protecting the shoot apex and the cell that gives rise to the reproductive structures and to the gametes from the entry of viruses (Foster et al., 2002).

Plasmodesmata allow the trafficking of macromolecules in healthy plants

In healthy plants, several proteins and RNA were found to cross the plasmodesmatal barrier to drive cellular processes at a distance from the cell that synthesized them. The first example was the maize transcription factor KN1 (*knotted-1*), which is involved in the maintenance of undifferentiated state of meristem cells. This protein can increase plasmodesmatal permeability, move from cell to cell through the plasmodesmata, and promote the movement of its own mRNA (Lucas et al., 1995). Other proteins were found to move from cell to cell to promote cell behaviour during development (Mezitt and Lucas, 1996; Nakajima et al., 2001).

Many proteins and RNA were also found to move to the sieve tubes from the companion cells through the vascular plasmodesmata (Fisher et al., 1992, Ruiz-Medrano et al., 1999). To study the molecules present in the phloem sap, cucurbits were used as experimental system, because it is easy to purify phloem sap from these species. Several phloem-located mRNA were identified, coding for various regulatory proteins (Ruiz-Medrano et al., 1999). Two RNA binding proteins were purified and characterized. *Cucurbita maxima* Phloem Protein 16 kDa (CmPP16) can interact with plasmodesmata and move from cell to cell. CmPP16 mRNA is found only in the companion cell, whereas the protein is found in the sieve tube (Xonocostle-Cazares et al., 1999). *Cucurbita maxima* Phloem Small RNA Binding Protein 1 (CmPSRP1, 20 kDa) was found to bind to 25 nucleotide single-stranded small RNA species, such as those involved in PTGS in response to viruses (siRNA) and in regulation of endogenous gene expression (microRNA). Cell-to-cell movement of small RNA was also promoted by co-microinjection with CmPSRP1 (Yoo et al., 2004).

Therefore, the virus hijacks different plant processes to promote its movement and systemic infection. The microtubule and microfilaments network moves the MP-RNA complex from the replication site to the plasmodesmata, then a system allowing supracellular regulation by macromolecular exchange moves it from cell to cell through the plasmodesmata, and a systemic regulatory system acts at the interface between the sieve tube and companion cells.

The movement of other viruses

Movement proteins have been identified in most plant virus families studied. As viruses are thought to have acquired cellular genes during their evolution, MPs might derive from host proteins involved in crossing the plamodesmata (Koonin and Dolja, 1993). Indeed, a partial sequence similarity has been found between the MP of *Red clover necrotic mosaic virus* (RCNMV, *Dianthovirus*) and the CmPP16 protein (see above). Although viral MPs generally lack primary sequence similarities, they display similar properties, indicating that they may have common three-dimensional motifs. As will be shown in the following examples, the various functions necessary for virus movement can also be provided by several cooperating polypeptides.

■ The triple gene block (TGB) is used by eight viral genera

In *Allexivirus, Benyvirus, Carlavirus, Foveavirus, Hordeivirus, Pecluvirus, Pomovirus,* and *Potexvirus,* a block of three overlapping reading frames (TGBp1, TGBp2, TGBp3) is required for cell-to-cell movement. These triple gene blocks are clustered in two classes depending on the size of TGBp1 and TGBp3.

Class 1 TGB in *Benyvirus* (Fig. 4.7), *Hordeivirus* and *Pecluvirus*

TGBp1 is of a size between 42 kDa and 63 kDa and harbours an RNA helicase consensus domain; the N-terminal domain is rich in positively charged amino acids and is involved in a non-specific interaction with RNA. TGBp2 and TGBp3 are highly hydrophobic and are associated with membranes and the cell wall.

The use of chimeras between the TGB of different hordeiviruses shows that TGB proteins interact functionally and that these interactions require the presence of host factors (Solovyev et al., 1999).

Another indication of interactions is that TGBp1 of *Peanut clump virus* (PCV, *Pecluvirus*) is localized in the vicinity of plasmodesmata in infected plants, but not in transgenic plants expressing the protein alone. The localization is restored when the transgenic plant is infected with a PCV mutant unable to produce TGBp1 (Erhardt et al., 1999).

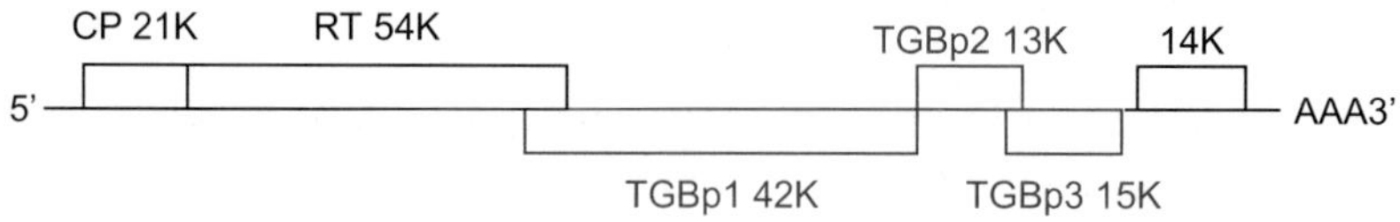

Figure 4.7 Genomic organization of the RNA 2 of BNYVV.

Class 2 TGB in *Potexvirus*

In Potexviruses the coat protein is involved in cell-to-cell movement together with TGB class 2 proteins.

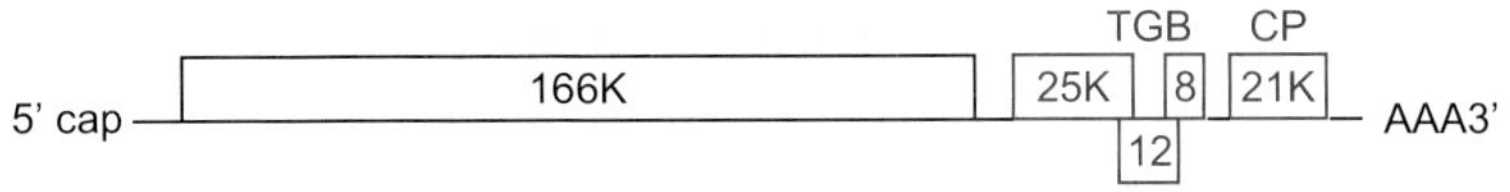

Figure 4.8 Genomic organization of PVX.

TGB class 2 proteins, such as those of PVX (*Potatovirus X, Potexvirus*) are shorter than TGB class 1 proteins. TGBp1 (25 kDa) lacks the N-terminal extension of class 1 proteins. Experiments with PVX and *White clover mosaic virus* (WClMV, *Potexvirus*) in *Nicotiana benthamiana* show that TGBp1 can increase plasmodesmatal permeability, can move from cell to cell, and can bind RNA in association with TGBp2, TGBp3, and the coat protein (CP). In the absence of TGBp2, TGBp3, and CP, TGBp1 can interact with plasmodesmata to induce an increase in size exclusion limit, but it is dysfunctional in terms of potentiating its own cell-to-cell transport. In the case of PVX, cell-to-cell movement occurs as a TGBp1-RNA-CP complex.

■ CMV movement involves movement proteins together with structural and replication proteins

Cell-to-cell movement

Proteins encoded by RNA3 are involved in cell-to-cell movement in all CMV (*Cucumber mosaic virus, Cucumovirus*) hosts (Fig. 4.9).

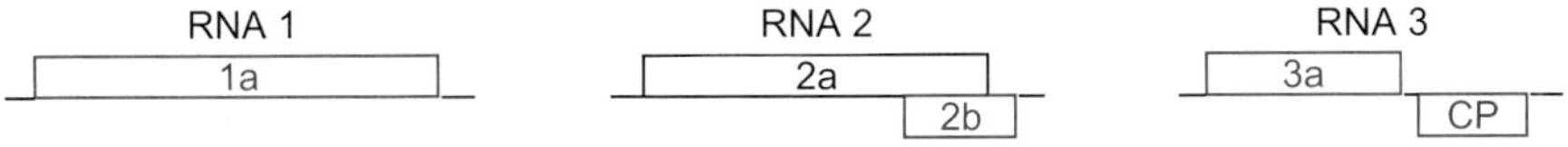

Figure 4.9 Genomic organization of CMV.

CMV 3a protein is a MP that binds RNA, increases the permeability of mesophyll and epidermal plasmodesmata, and leads to RNA movement into neighbouring cells (Ding et al., 1995a; Kaplan et al., 1995; Li and Palukaitis, 1996). The coat protein is also involved in movement (Canto et al., 1997), although the capacity to form virions is not required for cell-to-cell movement (Kaplan et al., 1998). However it has been suggested that virions are able to cross some plasmo desmata.

Long distance movement

Proteins from the three genomic RNA are involved in long distance movement:

— A domain of the coat protein is required that is different from that required in cell-to-cell movement (Suzuki et al., 1995; Taliansky and Garcia-Arenal, 1995; Boccard and Baulcombe, 1993).

— The helicase coded by RNA 1 is involved in the systemic accumulation of CMV in zucchini squash (Gal-On et al., 1994).
— The 2b protein, encoded in a reading frame overlapping that of the polymerase, in RNA2 is involved in systemic movement in some hosts such as cucumber (Ding et al., 1995b).

In potyviruses, the coat protein and HC-Pro cooperate with the cytoplasmic inclusion and the VPg for movement

Cell-to-cell movement

In *Bean common mosaic necrosis virus* (BCMNV, *Potyvirus*) and *Lettuce mosaic virus* (LMV, *Potyvirus*), both CP and HC-pro increase plasmodesmatal permeability when microinjected to *Nicotiana benthamiana* cells. HC-Pro induces the movement of 20-39 kDa dextrans, whereas CP allows only for 10 kDa dextran permeability. HC-Pro also moves more rapidly than CP and can mobilize the viral RNA (Rojas et al., 1997). *Potato virus Y* (PVY, *Potyvirus*) HC-Pro also displays an RNA-binding activity (Maia and Bernardi, 1996). Genetic analysis of *Tobacco etch virus* (TEV, *Potyvirus*) shows, however, that the CP is essential for cell-to-cell movement through its conserved central core domain, which is necessary for RNA encapsidation. These two activities, movement and encapsidation, appear independent (Dolja et al., 1994). Mutations within the N- and C-terminal regions of *Tobacco vein mottling virus* (TVMV, *Potyvirus*) HC-Pro were clearly found to affect cell-to-cell movement (Klein et al., 1994).

P1-Pro	HC-Pro	P3	6K	CI Helicase	6K	NIa-VPg	NIa-Pro	NIbPolymerase	Capsid
32	52	50		71	6	21	27	58	30

Figure 4.10 Polyprotein of a potyvirus. Order and molecular mass (kDa) of viral proteins (TEV).

The cytoplasmic inclusion (CI) protein

The potyviral CI protein has helicase and ATPase activities. N-terminal mutants in the TEV CI are unable to move from cell to cell, although viral replication is unaffected in isolated cells (Carrington et al., 1998). The CI does not modify plasmodesmatal permeability but is the monomeric component of cytoplasmic inclusions (pinwheels) (Fig. 4.11), cone-shaped structures formed with a central cylinder and triangular radial surfaces. Pinwheels have long been suspected to be involved in potyviral intercellular movement since early electron microscopy studies showed that they are located on both sides of the plasmodesmata at the beginning of the viral infection and within the plasmodesmata (Lawson and Hearon, 1971; Rodriguez-Cerezo et al., 1997; Roberts et al., 1998). Viral RNA and CP are also detected in the same plasmodesmatal crossing structures. It has been proposed that CI structures could replace the cytoskeleton and guide the CP-RNA complex towards the plasmodesmata in an energy-dependent process.

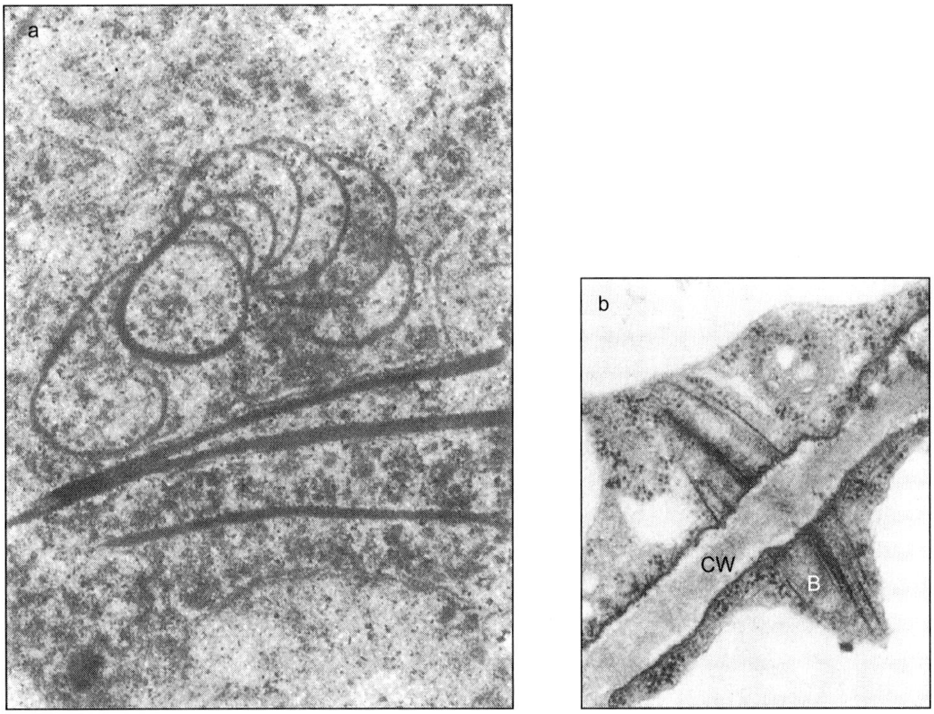

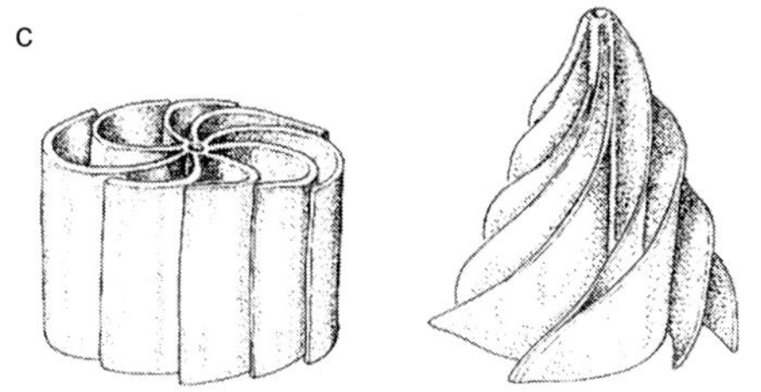

Figure 4.11 Cytoplasmic inclusion body of potyviruses. a, cross-section; b, longitudinal section on both sides of the cell wall (CW) (Lawson and Hearon, copyright Academic Press, 1971); c, three-dimensional drawing (Kurstak, 1981; Andrews and Shalla, 1974).

The VPg protein

The VPg (viral protein genome-linked) is thought to be involved in virus replication (see Chapter 3). Tobacco plants bearing the *va* gene restrict the cell-to-cell movement of PVY. A virus strain with a single amino acid change in the VPg coding sequence was found to overcome *va* resistance. VPg interacts with at least two host proteins, and one of unknown function is called PVIP (potyviral VPg interacting protein) (Dunoyer et al., 2004). Mutation of the *Turnip mosaic virus* (TuMV, *Potyvirus*) VPg at position 12, which is thought to interact with PVIP, reduced cell-to-cell (and systemic) movement but not viral replication rate, and reduction of PVIP levels reduces susceptibility to TuMV. The other VPg interactor is the eukaryotic initiation factor 4E (eIF4E), which has a probable role in viral translation, and hence replication, but was also found to promote cell-to-cell movement by an unknown mechanism (Gao et al., 2004).

Long distance movement

Both the central domain of HC-Pro and the C-terminal domain of the coat protein have been involved in potyviral long distance movement (Cronin et al., 1995; Kasschau et al., 1997; Dolja et al., 1995). *Plum pox virus* (PPV, *Potyvirus*) is restricted to the inoculated leaves of wild-type *Nicotiana tabacum* but can successfully invade transgenic plants expressing the 5′ terminal half of the TEV genome, which systemically infects tobacco. A mutation in the TEV HC-Pro coding sequence abolishes the complementation. In addition, it has been shown that co-infection with the unrelated CMV can complement PPV movement in *N. tabacum* (Saenz et al., 2002). HC-Pro appears to have related activities in amplification and long distance movement of potyviruses. It is possible that some viral proteins such as HC-Pro are specifically required for viral movement toward the companion cells of the sieve tube or for viral genome amplification within these cells, perhaps through the inhibition of RNA silencing (Kasschau and Carrington, 2001).

Mutations in the TEV cytoplasmic inclusion protein were also found to reduce long distance movement (Carrington et al., 1998).

The VPg was also found to influence systemic movement. In the *Nicotiana tabacum* cultivar V20, two recessive genes prevent the TEV systemic infection from reaching the sieve tubes and restrict it to phloem companion cells. TEV-Oxnard is a strain that can infect V20. The construction of chimeric viruses makes it possible to map this ability in the VPg coding region (Schaad et al., 1997b).

Phloem-restricted viruses

Luteovirus, Polerovirus, Tenuivirus, Marafivirus, Reovirus and some members of the *Closteroviridae* and *Geminiviridae* are restricted to the phloem of infected plants. They cannot be mechanically inoculated and a vector is necessary for infection. This requirement can be circumvented by the use of cloned viral DNA or cDNA copy transferred to the plant genome by *Agrobacterium tumefaciens* (Leiser et al., 1992).

Phloem restriction is not due to the inability of the virus to multiply in other cell types, as demonstrated by the infection of mesophyll protoplasts by several poleroviruses. Co-inoculation of *Potato leafroll virus* (PLRV, *Polerovirus*) with a potyvirus (PVY) or an umbravirus (CMoV, *Carrot mottle virus*) allows partial infection of mesophyll cells by the polerovirus (Barker, 1987, 1989). This suggests an assistance to the polerovirus for transport from the phloem to the mesophyll. Such assistance is permanent in the case of the *Pea enation mosaic virus* (PEMV, *Enamovirus*). This virus has two genomic RNA: RNA1 is similar to a polerovirus genome lacking the MP and RNA2 is similar to an umbravirus that lacks a coat protein gene. Therefore, PEMV appears to be a stable association between two defective genomes (Demler et al., 1996). The umbravirus moiety allows the replication in mesophyll cells and the mechanical transmission of the complex; this happens also in co-infection between PLRV and some umbraviruses (Mayo et al., 2000).

In the phloem, *Barley yellow dwarf virus* (BYDV, *Luteovirus*), *Beet western yellow vein virus* (BWYV, *Polerovirus*), and PLRV are localized in the sieve tubes, companion cell, and occasionally the phloem parenchyma (Shepardson et al., 1980). Systemic infection therefore requires only movement between the companion cells and the sieve tube and rapid movement through the vascular system. Therefore, phloem-restricted viruses have an MP to cross the phloem plasmodesmata. In PLRV, this protein is coded by ORF4 and has a molecular mass of 17 kDa (Fig. 4.12). The protein displays properties similar to the TMV MP, including the binding of RNA (Tacke et al., 1991). In infected plants or in transgenic plants transformed with the whole viral genome, this protein is preferentially located with the plasmodesmata linking the sieve tube to companion cells rather than with those linking the mesophyll to the perivascular bundle. However, in transgenic plants expressing the PLRV p17, the protein is found associated with all types of plasmodesmata. This suggests that other viral proteins are required for the proper localization of p17 (Schmitz et al., 1997).

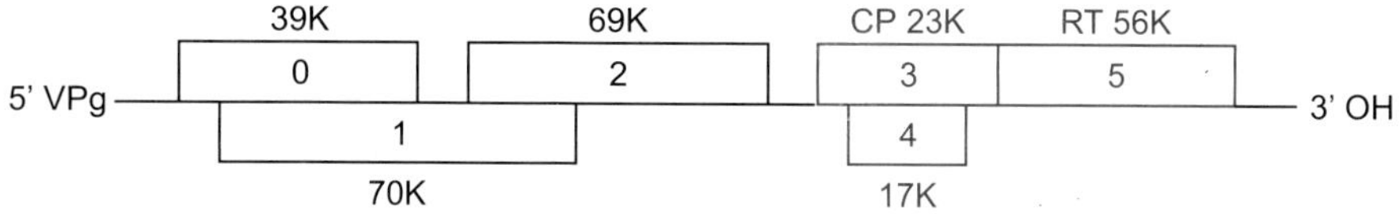

Figure 4.12 Molecular organization of a *Polerovirus* genome: PLRV.

The role in movement of the protein coded by the ORF4 of Luteoviruses and Poleroviruses has been confirmed for BYDV (Chay et al., 1996) but not for BWYV; in the latter case it has been suggested that the virion is the form moving from cell to cell (Ziegler-Graff et al., 1996).

■ Viruses replicating in the nucleus

Two proteins are involved in the movement of bipartite begomovirus DNA through the nuclear pores and plasmodesmata

Since the coat protein of *Bean dwarf mosaic virus* (BDMV, bipartite *Begomovirus*) is not involved in systemic movement, it implies that it is the viral DNA that is moving intercellularly and systemically. As replication occurs in the nucleus, the DNA must first move into this organelle before cell-to-cell movement. Genetic studies with BDMV show that DNA B codes for movement functions, whereas DNA A codes for replication and encapsidation functions. Proteins B-C1 and B-V1 are required for the systemic movement of both DNA. When produced in *Escherichia coli* and microinjected in mesophyll cells of bean and *N. benthamiana*, BC-1 moves rapidly from cell to cell, increases plasmodesmatal permeability, and allows double-stranded DNA to cross the plasmodesmata. Double-stranded DNA is the replicative intermediate in begomovirus replication and is thus also probably the movement form. BV-1 protein is required for the export of single-stranded and double-stranded viral DNA through the nuclear pore, a function not required in the case of RNA viruses. BV-1 and BC-1 are therefore both required to promote the movement of the viral nucleic within and out of the infected cells (Noueiry et al., 1994).

The coat protein is essential for movement of monopartite *Geminiviridae*

Monopartite *Geminiviridae* lack the DNA B component, suggesting that they use other mechanisms for movement. Indeed, some of them are restricted to the phloem, such as the *Beet curly top virus* (BCTV, *Curtovirus*) (Hormuzdi and Bisaro, 1993) and *Tomato yellow leaf curl virus* (TYLCV, *Begomovirus*) (Chérif and Russo, 1983). In these viruses, the coat protein is an essential element of movement and nuclear transport. The CP of *Maize streak virus* (MSV, *Mastrevirus*) has a 22 amino acid N-terminal domain rich in basic amino acids, which binds circular single-stranded and double-stranded DNA in a non-specific manner; moreover, this domain behaves like a nuclear localization signal. Co-injection of CP with single-stranded viral DNA shows that CP can reach the nucleus, where it is required for the formation of the viral particle, but can also trigger the entry of DNA into the nucleus (Liu et al., 1999a). Similar observations were made with the CP of TYLCV.

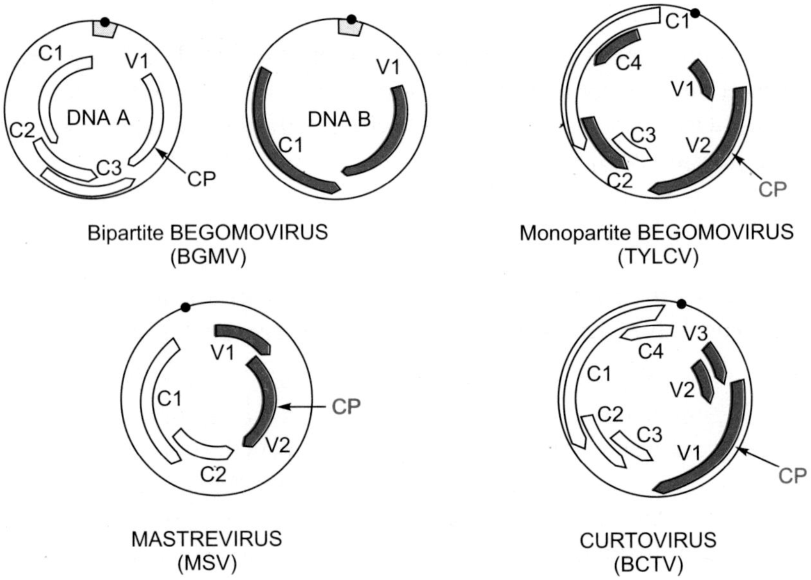

Figure 4.13 Genomic organization of *Geminiviridae*. The name of genes depends on the viral DNA strand (v) or complementary DNA (c) that codes them. CP indicates the gene by which the capsid protein is coded. The genes involved in the movement are in red.

The movement of monopartite *Geminiviridae* also involves several small molecular weight viral proteins (10-15 kDa). Movement of MSV requires the V1 protein (Boulton et al., 1993) and movement of BCTV requires the synthesis of proteins V2 and V3 (Stanley et al., 1992; Hormuzdi and Bisaro, 1993). TYLCV systemic movement requires, in addition to CP, the protein V1 in *Nicotiana benthamiana* or the proteins V1, C2, and C4 in tomato (Jupin et al., 1994; Wartig et al., 1997).

Another form of cell-to-cell movement: the viral particle

The movement protein forms intercellular tubules

In some spherical viruses, the MPs polymerize in intracellular tubules that are used by the virus to move from cell to cell. Comoviruses, nepoviruses, and caulimoviruses were found aligned as beads within tubules with an internal diameter equal to that of the viral particle. In tospoviruses, the tubules contain nucleocapsids. These tubules cross the cell wall and replace the desmotubule of the plasmodesmata (van Lent et al., 1990; Perbal et al., 1993; Wieczorek and Sanfaçon, 1993; Kasteel et al., 1996) (Fig. 4.14). Tubules were found to be formed almost exclusively by the MP (Perbal et al., 1993; Kasteel et al., 1997). Indeed, the expression of *Grapevine fanleaf virus* (GFLV, *Nepovirus*) MP in transgenic plant and cell lines leads to the formation of tubules located at the plasmodesmata. In that case, it was found that the formation of tubules involves the secretory pathway and their location at the plasmodesmata requires an interaction with the cytoskeleton (Laporte et al., 2003).

Tubules were also found to form across the plasma membrane of protoplasts and in cultured insect cells, suggesting that a functional plasmodesmata is not required for their assembly (Fig. 4.14).

The movement of the viral particle is still debated for some viruses

As the coat protein is involved in the cell-to-cell movement and systemic movement of many viruses, it is still unclear in some cases whether cell-to-cell movement involves the viral particle itself or another type of complex containing the viral nucleic acid, CP, and MP.

Cell-to-cell movement with tubule formation in CMV?

In the family *Bromoviridae*, cell-to-cell movement is complex; the CP is not required for some viruses such as *Cowpea chlorotic mottle virus* (CCMV) but it is required in association with MP for CMV, *Alfalfa mosaic virus* (AMV), and *Brome mosaic virus* (BMV). However, for CMV and AMV the formation of virions is not necessary for movement and tubules composed of MP are formed in infected protoplasts

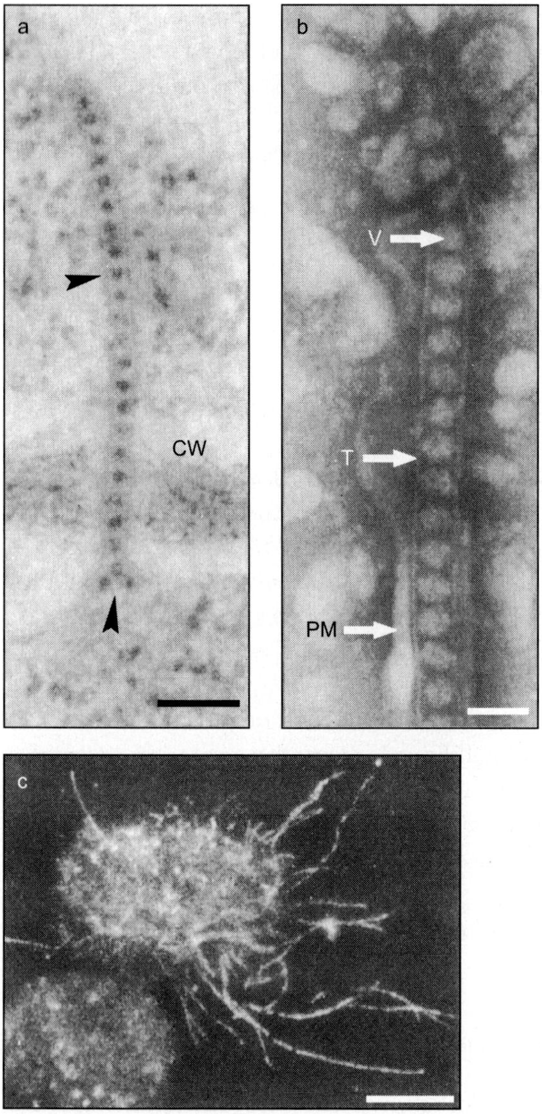

Figure 4.14 *Cowpea mosaic virus* (CPMV, *Comovirus*) movement. a: Virus containing tubules across the plasmodesmata of cowpea cells (scale bar: 0.1 μm). b: Tubular structures emerging from a protoplast (V = virus; T= tubule; PM = plasma membrane; scale bar: 0.05 μm). c: Confocal laser scanning microscopy of an infected protoplast labelled with fluorescent anti-58K/48K antibodies (scale bar: 10 μm).

(Kaplan et al., 1998). A point mutation in CMV MP leads to the failure to form tubules but the virus can still invade *N. tabacum* and *N. benthamiana*. The capacity to form tubules appears to be required for cell-to-cell movement in epidermal tissues only (Canto and Palukaitis, 1999). CMV tubules may be only transient structures in cells where they were observed (Stussi-Garaud et al., 1998); they may also be required to invade certain hosts although movement in *Nicotiana* does not require them.

The role of the virion in cell-to-cell movement in *Potyvirus* and *Potexvirus*

In potyvirus CP, the domains involved in movement and encapsidation are genetically separated, suggesting that formation of the particle is independent of movement. However, in plasmodesmata crossed by cytoplasmic inclusion, it is possible to identify fibrils similar to virus particles and reacting with anti-CP antibodies, suggesting that the cytoplasmic inclusion targets the viral particle to the plasmodesmata (Roberts et al., 1998).

In PVX, a similar observation can be made where viral particles are found within the plasmodesmata of infected cells. Moreover, this observation was made in plasmodesmata at the front of infection, making it more significant than in already infected cells. To allow the passage of the PVX particle (radius 13 nm), plasmodesmata must be more modified than for the passage of the TMV MP-RNA complex (2 nm radius), the desmotubule cannot be recognized, and the internal membranes and proteins are apparently lost (Santa Cruz et al., 1998).

Long distance movement

The coat protein is involved in the systemic movement of many virus genera: *Bromovirus, Carmovirus, Comovirus, Furovirus, Luteovirus, Polerovirus, Potexvirus, Potyvirus, Tobamovirus*, and monopartite *Geminiviridae*. However, as the CP is not necessary for long distance movement of other genera—*Enamovirus, Hordeivirus, Tombusvirus, Umbravirus* and bipartite *Geminiviridae*—the result is that the protection of viral nucleic acids during systemic invasion of the plant does not require the formation of a virus particle. Indeed, the viroids that are naked RNA can invade plants systemically (see Chapter 7) and several plant mRNA were found to circulate in the sieve tube (Citovsky and Zambryski, 2000; Ruiz Medrano et al., 1999).This latter observation supports the idea that the integration and identity of plant cells depends largely on macromolecular information trafficking within the whole plant (Lucas and Lee, 2004).

Although the first studies on TMV long distance movement suggest that it involves a nucleoprotein containing viral RNA, CP, and MP (Dorokhov et al., 1984), later studies show that in the phloem sap there are many infectious particles and the viral RNA can only be detected as virus particles (Simon-Buela and Garcia-Arenal, 1999). It remains to be clarified whether a ribonucleoprotein complex can form in companion cells while the viral particle forms within the residual cytoplasm of the sieve tube or, whether the CP interacts with specialized plasmodesmata of the phloem.

The viral particle of some poleroviruses can cross phloem plasmodesmata

The cell-to-cell movement of phloem-restricted *Luteoviruses* and *Poleroviruses* is limited to the passage between the companion cells and the sieve tube and occasionally between companion cells and the phloem parenchyma. BWYV viral particles can be observed in the plasmodesmata linking the companion cells to the sieve tubes, and mutants lacking the CP are unable to move from cell to cell, suggesting that the viral particle is the form moving from cell to cell. Additionally, in BWYV, the P74 readthrough protein, which is a C-terminal extension of the CP (Fig. 4.12), facilitates the movement between nucleated phloem cells and the sieve tube (Mutterer et al., 1999). This would explain how the viral particles that are injected without non-structural proteins into the sieve tube by the aphid vector can reach the companion cell, where they replicate.

Within *Polerovirus* coexist two different mechanisms to cross the branched plasmodesmata linking the sieve tube to companion cells, a nucleoprotein distinct from the viral particle for PLRV, and the viral particle itself for BWYV. Whether these two mechanisms can be used by the same viral species or in different hosts is currently unknown.

Virus movement: a paradigm for macromolecular trafficking within plants

■ Superfamilies

Movement proteins were grouped into four superfamilies on the basis of structural features: TGB proteins, *Tymovirus* MP, small (< 10 kDa) polypeptides coded by carmo-associated viruses and by some *Geminiviridae*, and the 30K superfamily. In the 30K superfamily, secondary structure predictions reveal a common core structure composed of β sheets, with α helices on both sides. A subgroup in this family includes the tubule forming MP and phloem proteins such as CmPP16 (Melchers, 2000).

■ Movement proteins can be genetically exchanged

The movement of a virus or a viral strain unable to move within a given host can be complemented by co-infection with another virus (Atabekov and Talianski, 1990). For example, TMV can infect barley if it is inoculated with BSMV (*Barley stripe mosaic virus, Hordeivirus* or BMV (Hamilton and Nichols, 1977). Co-infection does not show, however, what the complementing components are. The exchange of the MP between a rod-like virus *Sunn-hemp mosaic virus* (SHMV, *Tobamovirus*) and an icosaedral virus CCMV was demonstrated by building a hybrid virus (de Jong and Alhquist, 1992). Movement-defective BNYVV and BSMV can be similarly complemented by the TMV 30K MP (Lauber et al., 1998; Solovyev et al., 1996). In that case, the movement is restored in hosts common to the TGB-bearing viruses and TMV.

As many viruses are able to replicate within single plant cells, the restriction for a full infection is often a limitation to movement. Indeed, *Flock house virus* (FHV, *Nodavirus*), an insect virus that was found to replicate in several types of plant cells, can be successfully mobilized from cell to cell and systemically in transgenic *Nicotiana benthamiana* expressing the TMV 30K MP or the RCNMV MP (Dasgupta et al., 2001).

Plasmodesmata are multiple barriers to virus movement

The movement of viruses can be restricted by their inability to cross the plasmodesmata between different plant tissues.

Restriction between the perivascular parenchyma and phloem parenchyma

Tomato aspermy virus (TAV, *Cucumovirus*) and CMV have some hosts in common. Both can infect cucumber, but TAV infection is restricted to the inoculated leaf and CMV coat protein allows it to systemically infect that host. The block in TAV movement appears to be at the interface between the perivascular parenchyma and the phloem parenchyma preventing it from reaching the vascular tissue (Thompson and Garcia-Arenal, 1998).

This barrier also operates in the opposite direction as in *Polerovirus* movement, which is limited to the phloem tissues. Immunocytochemical analysis shows that BWYV was indeed restricted between the phloem parenchyma and perivascular parenchyma (Sanger et al., 1994).

Restriction between the phloem parenchyma and companion cells

An analysis of *Tobamovirus* and *Potyvirus* movement in the infection of small veins of *Solanaceae* and *Fabaceae* shows that companion cells adjacent to infected phloem parenchyma cells are very rarely infected (Ding et al., 1998). In the other direction, this barrier appears to exist to limit the movement of PLRV in infected potato leaves (Schmitz et al., 1997).

Restriction between the companion cells and the sieve tubes

In the case of TEV, there is restriction between the companion cells and the sieve tubes. The HC-Pro protein is involved in the ability to cross this barrier.

Concluding remarks

These examples suggest that the plasmodesmata linking different types of tissues have different properties. Viruses must acquire the capacity to interact with these different plasmodesmata to invade the plant.

The extent of virus movement therefore depends on many interactions with the cytoskeleton and with each type of plasmodesmata and also on the ability of the virus to replicate in the different cell types (see Chapter 3). The two processes, replication and movement, are sometimes difficult to distinguish experimentally and it is often difficult to determine whether a lack of movement does not result in fact from an absence of replication in certain tissue types. Nevertheless, if we refer to TMV as an example, it is clear that the entire viral genome is involved in both processes. Given the number of interactions required with host components and the small size of most viral genomes, this may explain why the host range of most viruses is relatively narrow and an increase in host range may require an increase in the viral genetic material.

This appears to happen in some viruses with an exceptionally large host range where new reading frames have evolved to this end. For example, CMV acquired the 2b gene (overlapping the 2a gene), which together with the CP is required for long distance movement in some hosts. *Tomato bushy stunt virus* (TBSV, *Tombusvirus*) has the gene coding for the P19 protein overlapping with the MP gene; this protein is required for systemic movement in spinach and pepper but not in *N. benthamiana* or *N. clevelandii* (Scholthof et al., 1995).

A deeper understanding of the molecular basis of viral movement requires the identification of host components interacting with viral proteins and genomes to allow movement. Genetic analysis in host plants reveals host functions involved in movement. For example, *Arabidopsis thaliana* ecotypes were used to screen for the ability to support TEV long distance movement leading to the isolation of the RTM1 gene encoding a lectin-like protein (Whitham et al., 1999; Chisholm et al., 2000). The structure of the different plasmodesmata, the nature of their components, and their regulation are also explored (for reviews see: Aaziz et al., 2001; Ehlers and Kollmann, 2001, Heinlein and Epel 2004; Oparka and Santa Cruz, 2000).

Since some plant proteins belong to the same protein families as viral movement, it is likely that viruses acquired their movement functions from their hosts.

It is also becoming increasingly evident that the exchange of macromolecules such as RNA and proteins through the plasmodesmata plays an important role in the integration of plant physiology and development and that viruses hijack these pathways (for reviews see: Heinlein and Epel, 2004; Lucas and Lee, 2004; Oparka, 2004).

5

CHAPTER

The Defence Reaction of the Infected Plant

Viral infection triggers an array of defence mechanisms in the plant, and viruses have evolved to counteract these defences to various extents. For successful purification of viruses, virologists know that, during the infection, there is a peak of virus concentration (about 1 μg/g of leaf tissue for a polerovirus, 100 μg/g for a potyvirus, 1-5 mg/g for a tobamovirus). Following this peak, the viral concentration usually decreases, as shown in a study of AMV (*Alfalfa mosaic virus, Alfamovirus*) concentration in tobacco leaf (Fig. 5.1). The highest concentration is reached 12 days post infection (p.i.), then the viral load decreases to very low levels (Ross, 1941; Gibbs and Harrison, 1976). In the case of *Potyviruses*, the viral concentration in a leaf increases over about three weeks, then, even at higher temperatures (20-25°C), it decreases rapidly and, in leaves developing strong symptoms, the virus can be undetectable using ELISA. This sequence of events is reproducible in each infected leaf, as demonstrated in tobacco infected with PVA (*Potato virus A, Potyvirus*) (Bartels, 1954). This drop in concentration also affects TMV (*Tobacco mosaic virus, Tobamovirus*), a very stable virus, when plants are maintained at high temperatures. Between 28°C and 32°C, the virus accumulation lasts for seven to eight days and is followed by a sharp decline (Lebeurier and Hirth, 1966).

These examples suggest that during many virus infections plants develop a defence reaction that limits virus accumulation.

Viruses cited

AMV *(Alfalfa mosaic virus, Alfamovirus)*; CaLCuV *(Cabbage leaf curl virus, Begomovirus)*; CaMV *(Cauliflower mosaic virus, Caulimovirus)*; CMV *(Cucumber mosaic virus, Cucumovirus)*; CTV *(Citrus tristeza virus, Closterovirus)*; PVA *(Potato virus A, Potyvirus)*; PVX *(Potato virus X, Potexvirus)*; TBRV *(Tomato black ring virus, Nepovirus)*; TBRV *(Tomato black ring virus, Nepovirus)*; TEV *(Tobacco etch virus, Potyvirus)*; TGMV *(Tomato golden mosaic virus, Begomovirus)*; TMV *(Tobacco mosaic virus, Tobamovirus)*; TRSV *(Tobacco ringspot virus, Nepovirus)*; TRV *(Tobacco rattle virus, Tobravirus)*; TSV *(Tobacco streak virus, Ilarvirus)*;

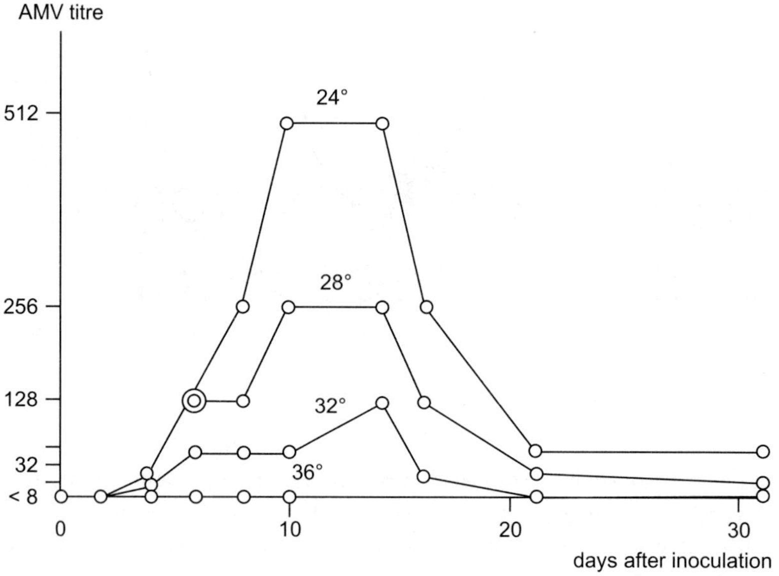

Figure 5.1 Evolution of AMV titre in tobacco leaves as a function of days after inoculation (Gibbs and Harrison, 1976). The plants were maintained at the temperatures indicated after inoculation. The virus concentration was estimated by serology.

Recovery from viral infection is due to an RNA silencing mechanism similar to post-transcriptional gene silencing (PTGS)

Recovery of developing organs in virus-infected plants is often observed with nepoviruses and caulimoviruses. When *Nicotiana clevelandii* is infected by TBRV (*Tomato black ring virus, Nepovirus*), the plant develops severe symptoms, but soon newly growing leaves appear healthy and become resistant to superinfection. This resistance is specific for viral strains with close sequence similarity to the infecting strain.

If a fragment of the TBRV genome is inserted into a PVX (*Potato virus X, Potexvirus*) vector, the hybrid virus cannot infect the TBRV-infected plant, whereas wild-type PVX causes successful infection, showing that the TBRV RNA is the target of the recovery mechanism (Ratcliff et al., 1997). Similarly, the recovery mechanism in plants infected by CaMV (*Cauliflower mosaic virus, Caulimovirus*) appears to target the accumulation of 19S and 35S RNA, although the transcription rate is not modified (Covey et al., 1997; Al Kaff et al., 1998) .

The establishment of the resistant state is not linked to a peculiar property of the viral RNA. If TRV (*Tobacco rattle virus, Tobravirus*) is used as a vector to express the Green Fluorescent Protein (GFP), after a strong initial GFP expression in infected leaves until about 8-10 dpi (days post-infection), the fluorescence disappears. At this stage, the plants are resistant to the inoculation by a PVX vector containing GFP sequences, but not to inoculation by a PVX carrying β-glucuronidase (GUS) sequences (Ratcliff et al., 1999).

This superimmune state is similar to that observed in experiments in which a viral genome fragment is expressed in transgenic plants. In some cases, the plant is initially susceptible to the virus, but as the infection progresses new leaves recover from the infection and the recovery is associated with a sharp drop in the virus-derived transgene RNA level (see Chapter 12). Moreover, it was found that the transcription rate of the transgene was unaffected, leading to the hypothesis that a post-transcriptional degradation mechanism specific to the transgene and also targeting the homologous viral RNA was involved. Recovering leaves are also resistant to superinfection by the same virus (Lindbo et al., 1993).

The expression of transgene sequences derived from the plant's own genes has also often been shown to lead to the silencing of both the transgene and the target gene at the post-transcriptional level by an essentially similar mechanism. This phenomenon was given various names: post-transcriptional gene silencing (PTGS), homology-dependent gene silencing (HDGS), cosuppression, RNAi (Vaucheret et al., 2001; Zamore, 2001; Ahlquist, 2002; Baulcombe, 2004). For clarity It will be called RNA silencing in this chapter. Following its discovery in plants, RNA silencing was found to exist in a broad range of eukaryotic organisms (Mello and Conte, 2004)

The mechanism of RNA silencing

The process of RNA silencing has been determined from multiple genetic and biochemical studies in plants (*Arabidopsis thaliana, Zea mays*), animals (*Caenorhabditis elegans, Drosophila melanogaster, Homo sapiens*), and fungi (*Neurospora crassa, Schizosaccharomyces pombe*).

RNA silencing is initiated by double-stranded RNA, which is an obligate intermediate in the replication of RNA viruses. The mechanisms of RNA silencing have been reviewed by Meister and Tuschl (2004).

Long double-stranded RNA is initially processed by a class of dsRNA-specific RNAse-III-type endonucleases, collectively known as Dicer, into small double-stranded RNA of 21 to 24 nucleotides with two 3′ nucleotide overhangs. These small siRNA (short interfering RNA) are then unwound in a protein complex known as RISC (RNA-induced silencing complex) containing RNA helicases hydrolysing ATP and Argonaute (Ago) type of proteins with RNA-binding activities. The single-stranded siRNA can then guide the RISC complex to its target RNA, which is then cleaved in the middle of the complementary region (Fig. 5.2). In that way, once activated by double-stranded RNA (or sometimes by aberrant RNA of unknown structure), RNA silencing will target and destroy all RNA with sufficient sequence similarity with the initiator RNA. The threshold of sequence similarity is defined by the size of the siRNA.

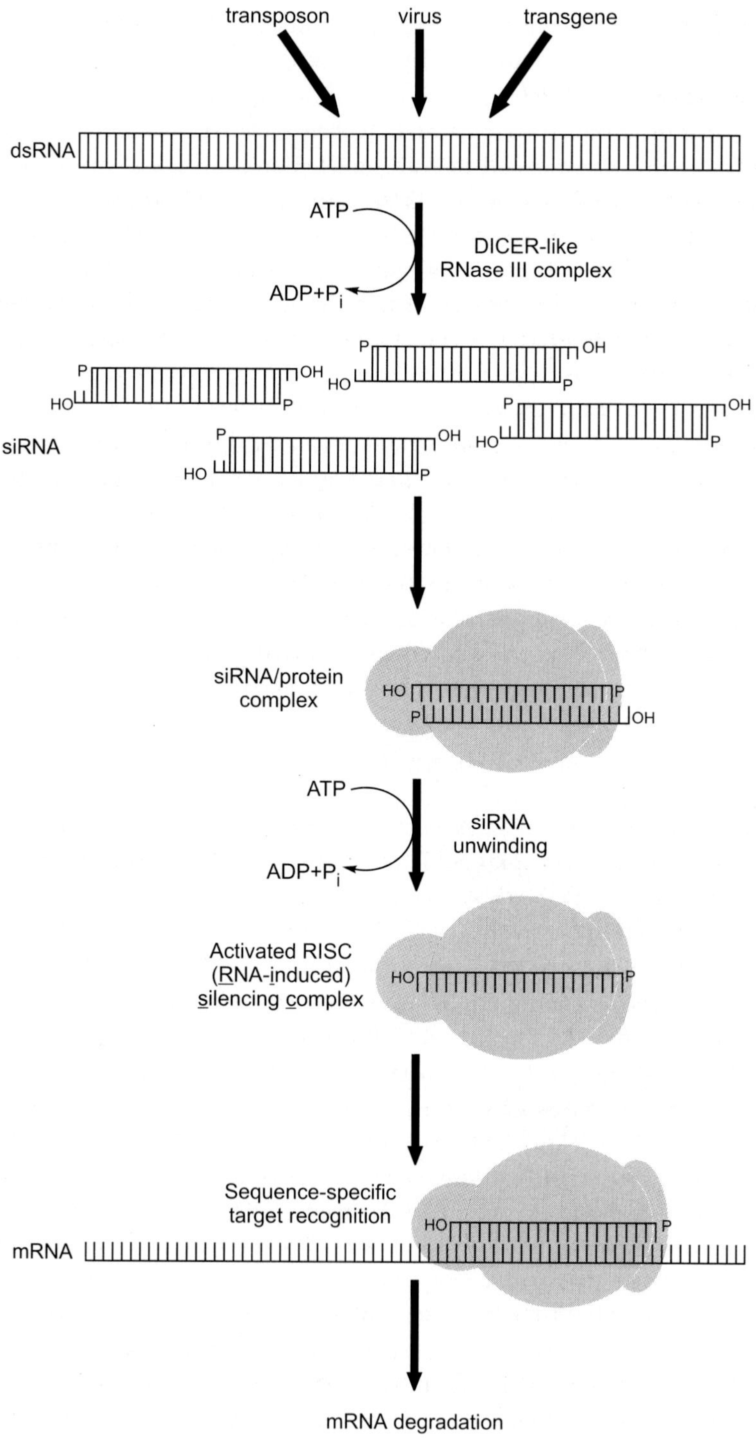

Figure 5.2 The mechanism of RNA silencing (Redrawn from Roth et al., 2004, with permission).

The role of RNA-dependent RNA polymerases

Other components of RNA silencing in plants are the RNA-dependent RNA polymerases (RdRp) discovered as early as 1971 by Astier-Manifacier and Cornuet (Astier-Manifacier and Cornuet, 1971, 1978). This enzyme appears to be required for the formation of double-stranded RNA from some RNA substrates. The DNA genome of CaLCuV (*Cabbage leaf curl virus, Begomovirus*) cannot initiate RNA silencing in an *Arabidopsis thaliana* mutant lacking one of the RdRp encoding genes, suggesting that this RdRp is involved in the formation of double-stranded RNA from the virally encoded mRNA (Muangsuan et al., 2004). RdRp is also important for the control of some RNA virus infections, since the same RdRp mutant plant also shows stronger CMV (*Cucumber mosaic virus, Cucumovirus*) symptoms than the wild type, although these plants are not altered in their susceptibility to other viruses (Mourrain et al., 2000; Dalmay et al., 2000). Further insights on the role of RdRp in the control of virus infection in the wild-type plant came from the observation that *Nicotiana benthamiana*, a plant that is highly susceptible to many viruses, lacks one form of RdRp and transgenic plant expressing a heterologous RdRp gene from *Medicago truncatula* have stronger virus resistance (Yang et al., 2004). Some RdRp encoding genes are also transcriptionally activated by salicylic acid, therefore linking RNA silencing to the general plant response to pathogens known as systemic acquired resistance (SAR, see Chapter 6).

Systemic propagation of RNA silencing

One of the hallmarks of RNA silencing in plants is that it can propagate systemically (Palauqui et al., 1997). Although the precise nature of the systemic signal is unknown, its sequence specificity strongly suggests that it is derived or composed of siRNA or derivatives that are small enough to cross the plasmodesmata (Jorgensen et al., 1998; Hamilton and Baulcombe, 1999). A role of the RdRp appears to be the relay amplification of the signal required for efficient long distance signalling (Himbert et al., 2003). A model would be that RdRp can initiate the synthesis of new double-stranded RNA from the target RNA using either internal initiation or siRNA as primers; this new double-stranded RNA will in turn be substrate for Dicer to generate new siRNA. Indeed, the generation of new siRNA from both the 5′ and 3′ of the initial target RNA sequence has been observed.

Altogether, the recovery mechanism observed in some plant virus combinations may reflect the outcome of a race between the virus and the systemic RNA silencing toward the young developing tissues. Efficient recovery would be obtained if the RNA silencing signal won the race, the virus will then invade a tissue where the RNA silencing mechanism is already built up to degrade its genomic RNA or mRNA.

Different states of activation of the RNA silencing were observed in transgenic plants expressing transgenes derived from the genome of viruses that do not efficiently trigger plant recovery, such as potyviruses. Within a set of independent transformants, three phenotypes can be observed: immunity, recovery, or full susceptibility to the virus (Lindbo et al., 1993). Immune plants result from transgene insertion events causing the formation of high levels of double-stranded RNA. This is associated with multiple head-to-tail insertion in the plant genome or ambisense transcription from endogenous plant promoters. In this case, RNA silencing complexes are ready to function in all cells and the viral RNA is immediately destroyed following entry. In recovery plants, low levels of double-stranded RNA or aberrant RNA (such as untranslatable RNA) are produced and the RdRp activity possibly activated by the viral infection is required to produce the double-stranded RNA levels required for a complete RNA silencing response. The virus can develop in the infected leaves, but the propagation of the mobile silencing signals prevents it from infecting newly developed leaves. In fully susceptible plants, the transgene is correctly inserted and expressed and does not produce any substrate leading to RNA silencing activation.

However, the recovery observed in a transgenic plant expressing virus-derived transgenes differs from recovery occurring during natural virus infection. While the former can lead to the complete disappearance of the virus, the latter is generally not associated with elimination of the virus. Also, whereas both TRV and PVX can initiate an RNA silencing response in *N. benthamiana*, only the TRV-infected plant can recover, suggesting that other factors play a role in the initiation of recovery. This was further strengthened by the isolation of a non-recovery mutant of TSV (*Tobacco streak virus, Ilarvirus*), which is known to induce a strong recovery response in *Nicotiana tabacum*. The mutation maps in an intergenic region of the viral genome (Xin and Ding, 2003).

RNA silencing as a defence against invasive nucleic acids

The RNA silencing mechanism has the capacity to respond to virus infection as well as to invasive DNA potentially leading to the formation of double-stranded RNA, such as transposons. Indeed, RNA silencing was found to limit the expression of retrotransposons and to be linked in an unknown manner to transcriptional DNA silencing by the chromatin modification of repeated DNA structures (Rudenko et al., 2003; Martienssen, 2003).

The apical meristems are symplasmically isolated from the rest of the plant body and many viruses are unable to invade the meristems. Viruses able to enter the meristem are often also able to induce a strong recovery. For example, tobacco develops a very strong recovery from infection with TRSV (*Tobacco ringspot virus, Nepovirus*), which is able to invade the stem cells of the apical meristem (Valleau, 1941; Roberts et al., 1970). Therefore, the plant-virus coevolution may have selected some interactions in which viruses causing very mild symptoms and therefore not greatly detrimental to the plant fitness because of recovery can gain a propagative advantage in accessing the meristem and increase their chance to be vertically propagated by seed (see Chapter 8).

Virus-induced gene silencing (VIGS)

In 1996, an elegant experiment demonstrated that the RNA degradation process affecting some transgene sequences and the viral resistance observed in transgenic plants expressing virus-derived transgenes were two sides of the same coin. Tobacco plant lines in which the reporter transgene coding for β-glucuronidase (GUS) was silenced were inoculated with an RNA virus vector (PVX) carrying GUS sequences. In that situation the plants were found resistant to the recombinant PVX virus but not to wild-type PVX (English et al., 1996). Further experiments show that the infection of a plant carrying a non-silenced gene or transgene with a PVX vector carrying sequences similar to that gene leads to the RNA silencing of that gene. Similar observations were made using TGMV (*Tomato golden mosaic virus, Begomovirus*) and TRV, and in the model plant *Arabidopsis thaliana* with CaLCuV. This process was called virus-induced gene silencing (VIGS) and has been used in several experiments to rapidly down-regulate endogenous plant genes (reviewed in Robertson, 2004).

Viruses can suppress RNA silencing

The discovery that many viruses produce inhibitors of the various steps of RNA silencing was a strong proof that it has evolved as antiviral mechanism (reviewed in Roth et al., 2004).

The synergistic effect of potyviruses in double infections

The coinfection of TEV (*Tobacco etch virus, Potyvirus*) with unrelated viruses such as PVX, CMV, or TMV causes stronger symptoms than either virus alone and leads to an increase in the multiplication of the non-potyvirus when compared with singly infected plants. The element involved in this synergistic effect has been mapped on the potyviral genome and was found to be the central domain of the multifunctional HC-pro protein, which is also involved in proteolytic processing, aphid transmission and long distance movement (Pruss et al., 1997; Maia et al., 1996).

Further experiments in which plants expressing HC-pro were crossed with plants with an RNA silenced reporter transgene demonstrate that HC-pro reactivates transgene expression and is thus a suppressor of RNA silencing (Anandalakshmi et al., 1998; Kasschau and Carrington, 1998). Many other viral suppressors of RNA silencing were subsequently discovered, in different viral groups, using the same type of experiments as above or transient assays in which viral vectors, such as PVX, were used to express silencing suppressors in transgenic plants silenced for reporter genes (Voinnet et al., 1999). Silencing suppressors were identified in both RNA viruses and DNA viruses such as geminiviruses (Table 5.1). Up to three distinct suppressors of silencing were detected in the large RNA genome (20 kb) of CTV (*Citrus tristeza virus*, *Closterovirus*) (Lu et al., 2004). CTV is propagated in trees and the evolution of multiple suppressors may be required for long-term survival in the host.

Viral inhibitors of RNA silencing can act at various steps of the RNA silencing pathway

Since RNA silencing is a multi-step process, it is not surprising that silencing suppressors were found to act at different steps of the pathway. They can be classified as (1) suppressors interfering with the metabolism of small RNA and (2) suppressors interfering with the systemic movement of the silencing signal (Table 5.1). However, depending on the type of assay (reverse silencing of transgenes or of transiently expressed genes), conflicting results have sometimes been observed. Suppressors are also tools to study the silencing and related pathways. The search

Table 5.1 Silencing suppressors identified in different genera of plant viruses.

Group	Genus	Virus	Suppressor	Mode of action
(+) RNA	*Carmovirus*	*Turnip crinkle virus* (TCV)	CP	local, systemic
	Closterovirus	*Beet yellows virus* (BYV)	P21	local
		Beet yellow stunt virus (BYSV)	P22	
		Citrus tristeza virus (CTV)	CP	systemic
			P20	systemic
			P23	local, systemic
	Cucumovirus	*Cucumber mosaic virus* (CMV)	2b	systemic
	Furovirus	*Beet necrotic yellow vein virus* (BNYVV)	P14	?
	Hordeivirus	*Barley stripe mosaic virus* (BSMV)	γb	?
	Pecluvirus	*Peanut clump virus* (PCV)	P15	local, systemic
	Polerovirus	*Beet western yellows virus* (BWYV)	P0	local
	Potexvirus	*Potato virus* X (PVX)	P25	systemic
	Potyvirus	*Potato virus* Y (PVY)	HC-Pro	local, systemic ?
		Tobacco etch virus (TEV)		
	Sobemovirus	*Rice yellow mottle virus* (RYMV)	P1	?
	Tombusvirus	*Tomato bushy stunt virus* (TBSV)	P19	?
(–) RNA	*Tenuivirus*	*Rice hoja blanca virus* (RHBV)	NS3	local
	Tospovirus	*Tomato spotted wilt virus* (TSWV)	NS_s	local, systemic
ssDNA	*Begomovirus*	*African cassava mosaic virus* (ACMV)	AC2	?
		Tomato yellow leaf curl virus-China (TYLCV-Ch)	C2	

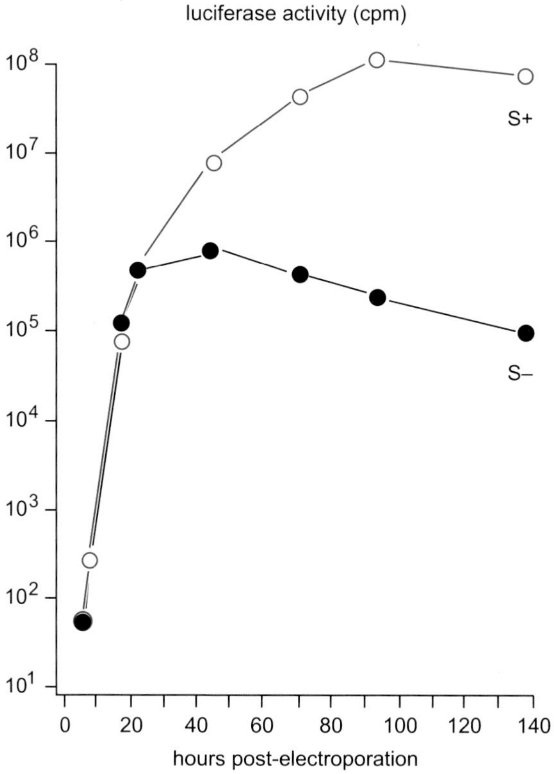

Figure 5.3 The effect of TEV HC-Pro on PVX replication rate. PVX replication was measured by replacing the coat protein gene, which is dispensable for replication in single cells, by a luciferase reporter gene (PVX-luc). The red graph represents PVX-luc expressing a wild-type HC-Pro from TEV, the black graph represents PVX-luc expressing a mutant HC-Pro unable to promote synergism (modified from Pruss et al., 1997).

for proteins interacting with silencing suppressors reveals potential novel components or unsuspected relations between silencing and other cell processes. Indeed, a calmodulin-related protein was found to interact with potyviral HC-pro and to suppress silencing when it is overexpressed (Anandalakshmi et al., 2000). The microRNA (miRNA) pathway is an endogenous gene regulation mechanism by which small RNA produced from the genome can regulate genes located at distinct loci at the translational or RNA stability level. Like siRNA, miRNA are processed through the Dicer complex from double-stranded precursors. Silencing suppressors were found to interfere with the miRNA pathway, and their expression, independently of the viral infection, produced a range of morphological phenotypes related to deregulation of genes involved in various developmental aspects (Mallory et al., 2002; Chapman et al., 2004). It is therefore likely that at least part of the symptoms associated with viral infections come from the action of silencing suppressors on host gene regulation.

Mosaic symptoms are due to local silencing

Systemic infection of tobacco by TMV leads to mosaic symptoms on young leaves with dark green, pale green, and yellow islands of tissue. This mosaic develops on leaves that are less than 1.5 cm long when the infection reaches them (Atkinson and Matthews, 1970) (Plate I.2). The virus can be found in yellow and pale green areas, whereas the dark green islands are free of virus and resistant to superinfection with the same virus but not to other viruses. Plants regenerated through tissue culture from the dark green islands are also free of virus. This phenomenon is reminiscent of some of the recovery phenotypes observed when transgenic plants expressing virus-derived transgenes are challenged by the virus. In young leaves, dark green islands appear on chlorotic infected tissue, the virus accumulation and transgene RNA are strongly reduced (Guo and Garcia, 1997; Moore et al., 2001). In addition, accumulation of a viral vector containing sequences of another virus is severely reduced in dark green islands if co-infected with the virus. These experiments strongly suggest that the cellular basis of mosaic symptoms is the establishment of local RNA silencing. Interestingly, dark green islands are not composed of clonal cells deriving from a single cell acquiring a superimmune state in the primordia and transmitting it to daughter cells. Thus, a specific signal must move within a group of cells during leaf development to trigger the silenced state (Jorgensen et al., 1998).

Concluding remarks: RNA silencing, the tip of the iceberg?

Following its discovery as a bizarre phenomenon occurring in transgenic plants, the field of RNA silencing has revealed major mechanisms governing virus-host interactions and gene expression not only in plants but also in most other eukaryotes. However, some organisms such as the yeast *Saccharomyces cerevisiae* seem to lack most of the molecular components of the RNA silencing pathway. The modifications of chromatin structure occurring upon the establishment of cytoplasmic RNA silencing raise new questions on the integration of functions between the cytoplasm and nucleus in eukaryotic cells. Highly structured RNA such as those of viroids and satellite RNA appear to escape the degradation step of RNA silencing although they are able to initiate it, as shown by the accumulation of siRNA (Wang et al., 2004). This suggests that RNA silencing might have a role in shaping the structure of viral and subviral RNA and possibly also of the cell's own RNA.

The long story of the discovery of the elusive RNA-dependent RNA polymerases

RNA-dependent RNA polymerases, enzymes that copy RNA into RNA, were first discovered as viral enzymes involved in RNA virus replication. They were thought to be absent from non-virus-infected cells until 1971, when Astier-Manifacier and Cornuet discovered an RdRp activity in healthy cabbage plants that was increased by RNA virus infection. Further experiments by other groups suggest that RdRp activity may exist in other plant cells (Fraenkel-Conrat, 1986). Research on these mysterious enzymes and about their molecular nature was reduced for nearly 30 years and in 1998 the cloning of an RdRp encoding gene was reported by Wasseneger team (Schiebel et al., 1998). Its function came to light when the research teams of Baulcombe (Dalmay et al., 2000) and Vaucheret (Mourrain et al., 2000) independently identified genes involved in suppression of RNA silencing and were surprised to discover that they were related to the sequence previously published by Schiebel et al. (1998). RdRp activity is now known to be central to the generation of short interfering RNA during the antiviral RNA silencing response in plants. However, RdRp genes are not confined to plants and were found in animals such as *Caenorhabditis elegans* and in fungi such as *Neurospora crassa* and *Schizosaccharomyces pombe*. In *S. pombe*, RdRp was found to be required for the maintenance of DNA silencing at the transcriptional level, therefore linking RNA to epigenetic inheritance, potentially opening up a new aspect of evolutionary processes in eukaryotes (Martienssen, 2003).

6

CHAPTER

Resistance with Hypersensitivity Reaction and Extreme Resistance

In the plant world, numerous species have developed a surveillance system to recognize the aggressor virus and immediately block infection. These plants are sensitive to the virus but become resistant in response to the infection itself. In most cases, necrosis is associated with the phenomenon, and this is *resistance with hypersensitivity reaction* (HR); when no necrosis appears, the resistance is an *extreme resistance*.

The manifestation of these two types of resistance requires in fact the existence of a complementary pair of genes, a resistance gene (R) of the plant and an avirulence gene (Avr) of the virus. The specific recognition of the product of the R gene and of a ligand dependent on the Avr gene (which in the simplest case is the product of Avr itself) activates a cascade of signals that induce different defence responses appropriate for blocking the progression of the virus. Loss or alteration of one of these two genes leads to the disease.

The *high specificity* required to *induce* the resistance (we will see that a single amino acid substitution in the sequence of the protein coded by a viral gene for avirulence can control the aptitude of this protein to trigger resistance) contrasts with the *non-specificity of the defence mechanism mobilized* that is active against a wide range of pathogenic agents.

Viruses cited

ClYVV (*Clover yellow vein virus, Potyvirus*), CMV (*Cucumber mosaic virus, Cucumovirus*), LMV (*Lettuce mosaic virus, Potyvirus*), PVMV (*Pepper veinal mottle virus, Potyvirus*), PVX (*Potato virus X, Potexvirus*), PVY (*Potato virus Y, Potyvirus*), TCV (*Turnip crinckle virus, Carmovirus*), TEV (*Tobacco etch virus, Potyvirus*), TMV (*Tobacco mosaic virus, Tobamovirus*), TNV (*Tobacco necrosis virus, Necrovirus*), TuMV (*Turnip mosaic virus, Potyvirus*).

This simple description applies to numerous cases of resistance to viruses. It applies also to plant resistance to other types of pathogenic agents, such as fungi, bacteria, or nematodes. This classic "gene for gene" concept was formulated by Flor (1942) from a study on resistance of flax (*Linum usitatissimum*) to a rust fungus (*Melampsora lini*).

In this chapter, resistance with HR and extreme resistance are briefly described, followed by induction of resistance, signal transduction, and finally the expression of resistance.

Description of resistance

■ Resistance with hypersensitive response

Two dominant genes, N from *Nicotiana glutinosa* or N′ from *Nicotiana sylvestris*, give to these species resistance with hypersensitivity to TMV (*Tobacco mosaic virus, Tobamovirus*).

The genes N and N′ were independently introduced in certain cultivars of *Nicotiana tabacum* (Holmes, 1938; Valleau, 1943). The mechanical inoculation of these "hypersensitive" cultivars with TMV caused local necrotic lesions to form on the treated leaf; these lesions consist of a necrotic disc of about 4 mm diameter, surrounded by a light green ring of 1 to 1.5 mm. It is possible to detect TMV by infectivity tests, in the necrotic disc and in the light green ring, but not in the remainder of the leaf blade, which remains green (Martin and Gallet, 1966; Antoniw and White, 1986) (Fig. 6.1).

A comparative kinetics at 22°C of the multiplication of TMV on the sensitive tobacco leaf (nn) and the hypersensitive tobacco (NN) records a similar evolution of the level of TMV during the first 30 h. Around 33 to 36 h, local necrotic lesions appear. From this time, the TMV multiplication rates decrease in the hypersensitive host (Konate and Fritig, 1984). The development of resistance leads to a limitation of the TMV to cells of the light green ring, while in the sensitive tobacco the viral multiplication leads to a systemic infection. This experiment clearly demonstrates that *the resistance did not exist beforehand but was induced by the infection*.

This resistance is heat-sensitive: at 30°C, a hypersensitive tobacco behaves like a sensitive tobacco. It is possible to vary the time during which the infected hypersensitive tobacco is kept at high temperature to cause, on return to 22°C, an extension of necrosis to the tissue in which the TMV is multiplying. For example, after the formation of local lesions at 22°C, a transfer to 30°C leads to resumption of viral multiplication; if this system is maintained for 3 days at 30°C, a return to 22°C leads to the formation of lesions of 10 to 15 mm diameter (Martin and Gallet, 1966).

The resistance induced in response to the infection by TMV is not confined to the immediate proximity of the necrotic lesion. It extends to other parts of the plant. A

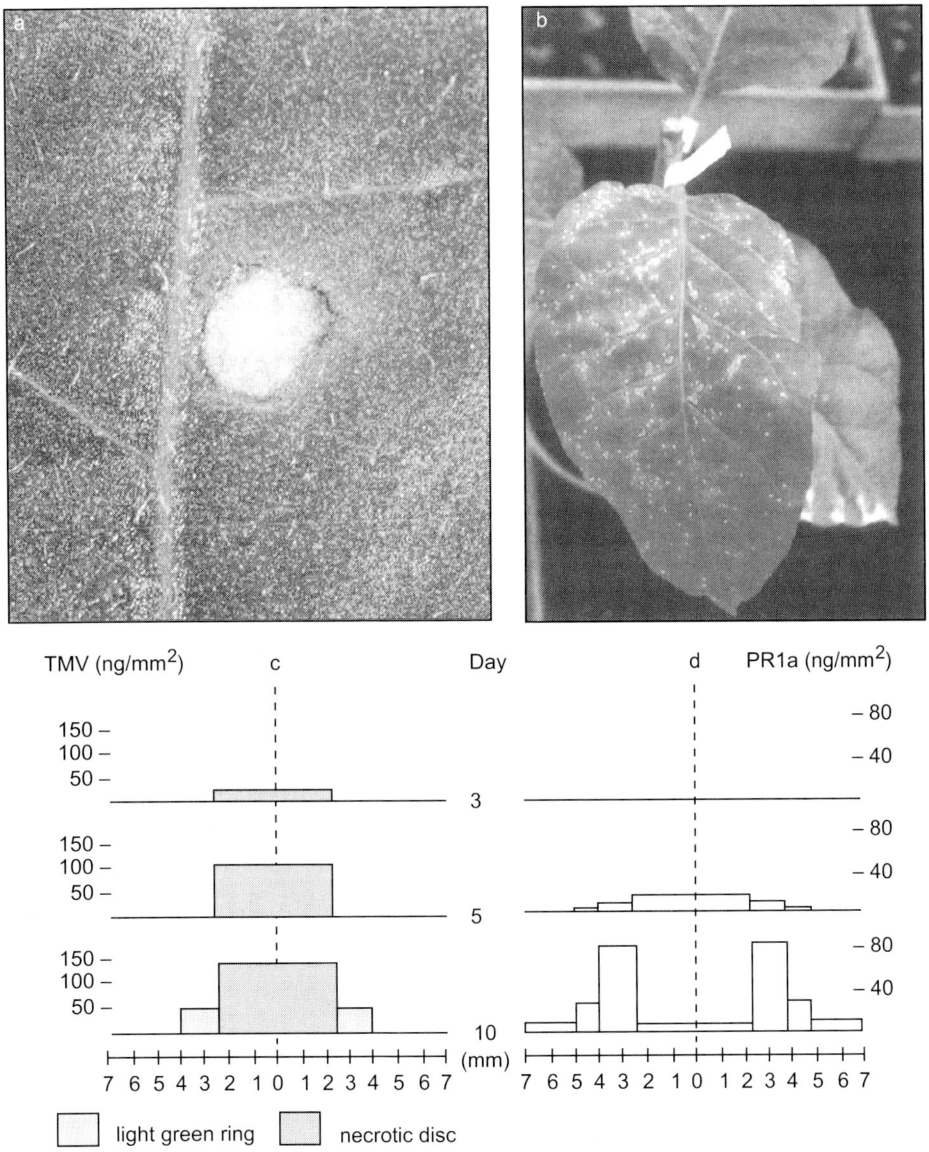

Figure 6.1 (a and b) Local necrotic lesions induced by TMV on *Nicotiana tabacum* cv. Xanthi (NN). (c) Concentration of TMV in a local lesion at different times after inoculation. (d) Concentration of the PR1a protein at these same times (the broken line indicates the centre of the lesion) (Antoniw and White, 1986; with the kind permission of Kluwer Academic Publishers).

tissue ring around the necrosis becomes completely resistant to reinoculation; Ross (1961a) called this property *localized acquired resistance* (LAR). The reinoculation of distant tissues leads to the formation of necrotic lesions that are much smaller and sometimes fewer in number; Ross (1961b) called this *systemic acquired resistance*

(SAR). The leaves that are at the early stage at the time of the first inoculation are themselves the site of this induction, which suggests that a signal has been produced during the first infection, a signal that could propagate itself throughout the plant so that the plant responds more effectively to a new infection. Note that the resistance is non-specific: localized resistance and systemic acquired resistance are effective not only against the inducing TMV but also against very different viruses such as TNV (*Tobacco necrosis virus, Necrovirus*) (Ross, 1961a, b) (Fig. 6.2).

Extreme resistance

Resistance may be induced and not give rise to necrosis, as in the case of the Rx/PVX system.

The Rx gene gives potato extreme resistance against common strains of PVX (*Potato virus X, Potexvirus*). The leaves do not manifest local lesions after inoculation, which, in the first place, suggests that this is a case of direct resistance.

The system has been studied by comparing the synthesis of an avirulent strain of PVX and a virulent strain (PVX-HB) in protoplasts of sensitive potato (rx) and Rx potato. Over the course of the first 8 h, the synthesis of viral RNA is of the same order in all cases. Between 8 and 16 h after infection, this synthesis declines and stops for the PVX/Rx combination, while it proceeds normally in the other cases.

This study recalls what happens on the leaf with TMV. Obviously, the resistance is not expressed at time 0, it is induced some hours after infection (Kohm et al., 1993).

Moreover, although the recognition phase is specific to certain strains of PVX, the resulting response phase inhibits the accumulation of PVX as well as that of other viruses (CMV, *Cucumber mosaic virus, Cucumovirus*, or TMV). This confirms that extreme resistance reveals mechanisms analogous to hypersensitive resistance, with respect to resistance, even if local lesions do not appear.

Induction of resistance

Resistance with HR of tobacco to TMV

The presence of a resistance gene does not guarantee resistance to all strains of a particular virus, as shown in the following table:

	Host plant		
	N. tabacum cv. Samsun nn	*N. sylvestris* N'N'	*N. tabacum* cv. Xanthi NN
Common strain of TMV	mosaic	mosaic	local lesions
Aucuba strain of TMV	mosaic	local lesions	local lesions

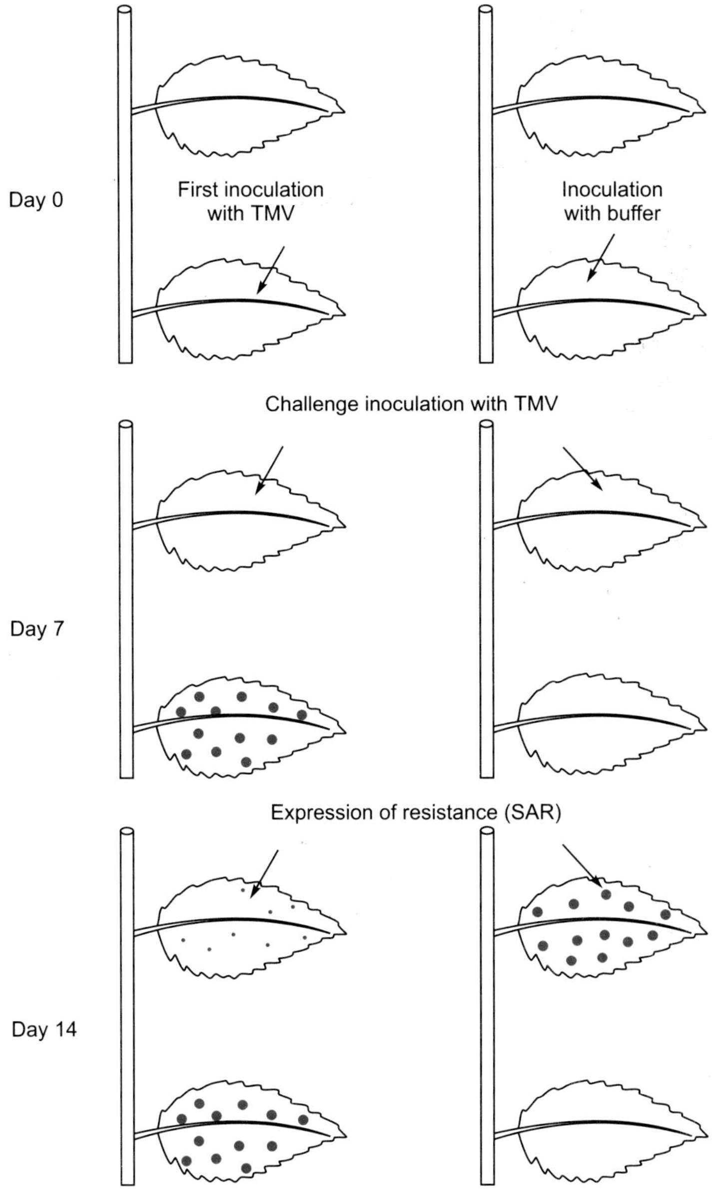

Figure 6.2 Systemic acquired resistance (SAR). On day 7, local lesions are visible on the leaf that was inoculated on day 0. The induction of SAR in other leaves is revealed by a challenge inoculation the effects of which are recorded on day 14: reduction of local lesions in size and number with respect to the control.

If the behaviour of TMV strains is considered *with respect to the gene* N′, the aucuba strain is *avirulent* and the common strain is *virulent*. (In plant pathology, *virulence* is a component of pathogenic capacity defined with respect to a *resistance*. Another component is *aggressiveness* of a strain, which relates to the intensity of the symptoms it causes on a sensitive line; see Chapter 11.)

The avirulence gene with respect to N′ is the capsid gene

An experiment to construct chimeric genomes by exchange of fragments between the cDNA of the common strain and the aucuba strain made it possible to show that the property of avirulence or virulence with respect to the gene N′ is determined by the capsid gene. The capsid protein is the direct elicitor of resistance; constitutive expression of this gene by calluses or transgenic plants of *N. sylvestris* is sufficient to activate the hypersensitivity reaction (Culver and Dawson, 1991). A more precise analysis showed that the substitution of *a single amino acid* at certain precise points of the capsid protein suffices to transform a virulent strain into an avirulent strain (Knorr and Dawson, 1988). When the capsid is compared within a range of avirulent strains for the N′ gene, it is observed that, even though the similarity between the different protein sequences varies from 82% to 34%, all these proteins conserved the same tertiary structure and probably a potential site of interaction with the product of N′ gene (Taraporewala and Culver, 1997).

The avirulence gene related to N is the helicase gene (126 kDa)

If the capsid gene is deleted from a strain that is avirulent for N′, local lesions are no longer observed on *N. sylvestris* N′N′ but are observed on *N. tabacum* cv. Xanthi NN. Thus, the determinant of avirulence against gene N is not the capsid gene.

The selection of a virulent strain of TMV (Csillery et al., 1983) made it possible to construct chimeric genomes between the cDNA of this virulent strain and the cDNA of an avirulent strain and to identify the gene coding for the protein of 126 kDa as an avirulence gene; more specifically, it is the zone between the amino acids 692 and 1116 (Padgett et al., 1997). The transient expression of the 126 kDa protein or of the fragment involved in the avirulence induces necrosis in tobacco NN and not in tobacco nn, which demonstrates that this sequence directly elicits the hypersensitivity reaction in the absence of other viral proteins or RNA replication (Abbink et al., 1998).

Gene N is the first gene of resistance to a virus that has been cloned

Gene N has been detected in the genome of *N. tabacum* Samsun NN using the activator element Ac of the maize transposon. Increased production of mutants that have Ac inserted in the N gene and have thus lost the property of resistance has been facilitated by the use of the absence of hypersensitive response at high temperature. In varying the duration for which plantlets are kept at 30°C, systemic infection by TMV can be favoured; a transfer to 22°C induces hypersensitivity reaction in all the tissues in which TMV multiplies, i.e., total necrosis of a plantlet having a functional N gene, which allows an easy selection of insertion mutants (Whitham et al., 1994).

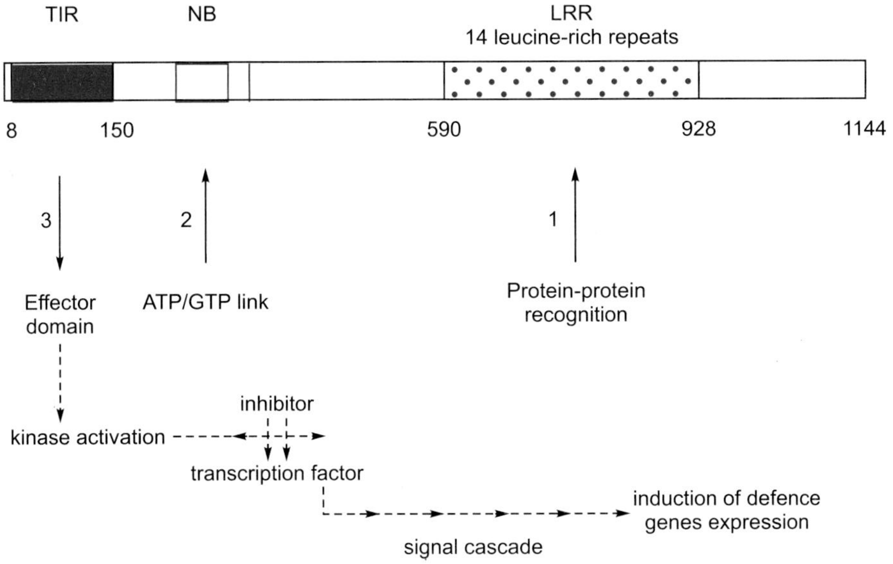

Figure 6.3 Domains of N protein and hypothetical diagram of induction of the defence mechanism.

Domains of N protein

The N gene codes for a protein of 1144 amino acids (131.4 kDa) in which several particular domains have been detected.

C-terminal recognition domain containing 14 imperfect repeats of a motif of 26 amino acids rich in leucine (leucine-rich repeats, LRR)

The importance of LRR domains in resistance is attested by their conservation in most resistance genes known and by experiments in which a single amino acid change makes the gene non-functional (Bent, 1996).

The LRR are generally involved in protein-protein interactions. This domain, found in about 30 resistance gene products, constitutes the principal determinant of specificity for Avr products (van der Biezen and Jones, 1998a). In the precise case of the N protein, LRR could be involved in the specific interaction with the viral helicase zone responsible for avirulence; this hypothesis, attractive in the context of "gene for gene relationship", can now be tested experimentally.

Central nucleotide binding (NB) domain

A consensus domain, defined from various proteins capable of binding ATP or GTP phosphates, is present in the N protein and is highly conserved in several known resistance gene products. This suggests that triphosphate nucleoside binding is essential for the function of these proteins. Loss of function was observed after experimental point mutations in this domain.

Toll and interleukin-1 receptors (TIR)

The amino-terminal part of N presents a homology (20% of amino acids identical, 42% of amino acids similar) with the effector domain of a receptor of *Drosophila*. This receptor, Toll, triggers a *signal cascade* involved in the dorsoventral polarity of the *Drosophila* embryo. The interaction with a ligand allows Toll to activate a serine-threonine kinase that itself phosphorylate a transcription factor, releasing it from a complex with an inhibitor.

The amino-terminal part of N also presents a homology (16% of amino acids identical, 41% of amino acids similar) with the effector domain of 1L-1R, interleukin receptors in mammals. In the immune system, the perception of signals produced by a pathogenic agent leads to binding of the interleukin-1 to the cytoplasmic domain of the 1L-1R receptor. This interaction leads to the activation of a kinase protein that releases the transcription factor NF-kB from a complex with an inhibitor; the transcription factor migrates into the nucleus, where it activates the transcription of genes for inflammatory response and for immune response.

This TIR domain is found in the product of gene L6 of flax for resistance to the fungus *Melampsora lini* and in the protein coded by the gene RPP5, which gives *Arabidopsis* resistance to *Peronospora parasitica*.

The NB-LRR proteins appear as adapting molecules that respond to recognition by a signal emission. Schematically, the recognition of a product of the gene for avirulence by LRR could control a hydrolysis of triphosphate nucleosides by the NB domain. This would be the source of energy necessary for a modification of conformation that allows exposure of the amino-terminal part TIR and the triggering of a cascade of phosphorylations by various kinases and phosphatases and lead to the activation of a transcription factor that then transcribes the range of genes for the defence mechanism.

■ Extreme resistance: the Rx-PVX system in potato

The avirulence gene of PVX related to Rx is the capsid gene

Constructions of chimeras between virulent and avirulent PVX strains have made it possible to reveal the implication of the capsid gene and the importance of amino acids threonine (T121) and lysine (K127) in avirulence. Infection of protoplasts carrying the Rx gene with a chimera of TMV expressing the capsid gene of PVX in place of the capsid gene of TMV shows that production of the capsid protein of PVX is necessary and sufficient for elicitation of the Rx resistance, but that the production of viral particles is not required (Bendahmane et al., 1995).

The central part of the capsid is essential for its elicitor activity, particularly position 121, at which substitutions lead to the following range of elicitor activity: PVX-TK > PVX-SK > PVX-PK > PVX-VK > PVX-KK = 0.

The Rx gene is the second gene for resistance to a virus that has been cloned and sequenced

The locus Rx, in potato cultivar Cara, was genetically defined in a clone of bacterial artificial chromosome (BAC). The method adopted to demonstrate the presence of Rx in a BAC clone can be used to isolate other genes of extreme resistance. It consists of integrating, in each BAC to be tested, the genome of an avirulent strain of the virus; this genome has already had a gene removed that determines cell-to-cell movement and is labelled with the gene uidA coding for GUS. If a BAC contains Rx and PVX simultaneously, its inoculation by biolistics into a host sensitive to PVX generates no expression of GUS in the affected cells (in the absence of Rx, PVX multiplies and blue spots that reveal GUS are easily observed 48 h after inoculation).

The fact that this BAC contained Rx was confirmed by mutating the genome of avirulent PVX into a virulent PVX (capsid $K_{121}R_{127} \rightarrow T_{121}K_{127}$). This mutation re-established the appearance of blue spots. This approach, used also for subcloning, made it possible to specify the location of Rx in a fragment of DNA of 11 kb. Sequence analysis showed that Rx has 3 exons and 2 introns of 234 and 111 bp (Bendahmane et al., 1999).

The stable transformation of potato of genotype rx by such a BAC gave it extreme resistance to PVX. The transient experiments show, moreover, that Rx is functional in heterologous species *N. tabacum* and *N. benthamiana*.

The Rx protein is also a NB-LRR protein

The Rx protein is composed of 937 amino acids with a molecular mass of 107.5 kDa.

Two out of three domains of protein N are found:

- One C-terminal recognition domain (amino acids 473 to 868) comprising 14 to 16 imperfect leucine-rich repeats (LRR).
- A central nucleotide binding (NB) domain.

The Rx protein also presents the following:

- Unknown motif at the C-terminal end.
- An N-terminal effector domain presenting a consensus similar to leucine-zipper (LZ). The N-terminal domain of LZ type exists in all the proteins coded by genes for resistance to different pathogenic agents that do not have a TIR domain. This LZ domain presents contiguous repeats of a sequence (I/R)xDLxxx that facilitates the formation of coiled-coil structure and the formation of dimers or trimers.

Extreme resistance controlled by Rx is epistatic on resistance with HR related to N

The two R-Avr systems of resistance, N-126 kDa of TMV (which produces local lesions) and Rx-CP of PVX (which does not produce them), were solicited in the same plant in a concomitant way. This experiment was carried out with *N. tabacum* NN that is transgenic for the Rx gene. It was infected with a construction of TMV in which the capsid gene of PVX replaced that of TMV; *this construction thus codes for each of two genes of avirulence relative to N and Rx.*

This double solicitation leads to a state of resistance without formation of local lesions, while the local lesions appear on the tobacco plants that have only the gene N. The interpretation is that extreme resistance (Rx-PVX CP) is activated much more rapidly than resistance with HR (N-TMV 126 kDa). The Rx-dependent resistance is effective at the cellular level against the chimera of TMV with CP of PVX and hinders the progress of an induction by N-126 kDa of TMV.

This result is coherent with the kinetics of the appearance of resistance in the N-TMV system. Konate and Fritig (1984) propose that a period of interaction of 30-36 h between the product of the N gene and that of the 126 kDa avirulence gene is necessary to trigger necrotic stress and induce the expression of resistance. In the protoplasts, during the viral cycle, the N system therefore has no time to induce resistance, while the Rx system is functional.

Cell death is a secondary phenomenon and not the cause of resistance with HR

Extreme resistance does not cause local lesions to appear. Nevertheless, when agrobacteria transformed by the CP gene of PVX are infiltrated into Rx transgenic tobacco, the tissues become necrotic. This shows that Rx resistance can also lead to cell death. In these tissues, CP is produced constantly, which is not the case in the infection experiment, in which the efficacy of the induction quickly blocks in return the input of the avirulence factor.

These experiments show thus that resistance does not depend on the appearance of local necrotic lesions. Necrosis is a secondary phenomenon (see below).

Signal transduction

After infection, recognition of a virus—or any other pathogen—is followed by a series of intracellular, intercellular, and systemic signals that coordinate resistance at the plant level. These signals involve kinases, phosphatases, G-proteins, and changes in the levels of Ca^{2+}, H_2O_2, and salicylic acid (Hammond-Kosack and Jones, 1996).

Intra- and intercellular signal transduction

The data relate on the one hand to the tobacco N–TMV system, which allows, by transfer from a temperature of 30°C to 22°C, induction of the hypersensitivity reaction in a synchronous fashion in a large number of cells, and on the other hand to cultures of parsley and tomato cells treated by an elicitor.

A reactive oxygen species (ROS, including the superoxide ion O^{2-} and hydrogen peroxide H_2O_2) appears 2 min. after induction (Doke and Ohashi, 1988). In mammals, H_2O_2 plays the role of secondary intracellular messenger; it activates the transcription factor NF-*k*B implicated in the inflammatory immune response. In the plant, the genes for defence mechanism are activated in response to an H_2O_2 treatment (Levine et al., 1994).

The appearance of the ROS supposes the induction of the assembly at the level of membranes of a NADPH oxidase complex constituted of four protein subunits, two membranary and two cytoplasmic. The appearance of the ROS is dependent on the ion flow (K^+, Cl^-) across the cytoplasmic membrane and an influx of Ca^{2+}. The scenario could bring in the effect of the phosphorylation of proteins required for the activation of Ca^{2+} channels in the membranes and of the G-proteins (proteins linked to guanine, known to transfer a signal produced by a ligand-receptor interaction with an effector molecule) that stimulate the activation of Ca^{2+} channels. These G-proteins also activate a Ca^{2+}-dependent protein kinase that favours the translocation of cytoplasmic subunits of the NADPH oxidase complex towards the membrane binding site (Higgins et al., 1998).

The activation of transcription of defence genes may be manifested 5 to 30 min. after the induction of hypersensitivity reaction in parsley cells.

Transduction leading to systemic acquired resistance (SAR)

The significant range of defence responses induced rapidly and in a coordinated manner in different cell compartments involves a small number of regulators of low molecular weight, salicylic acid and derivatives, jasmonic acid and derivatives, and ethylene (C_2H_4).

Salicylic acid plays an important role in signal transfer leading to SAR

In the TMV–tobacco NN system, salicylic acid (SA) is a key molecule acting as secondary signal in the induction of local as well as systemic resistance. The induction of SAR coincides with a significant increase in the synthesis of this acid not only at the infection site but also in distant non-infected tissues. The key role of SA has been indicated in analysing transgenic plants expressing the nahG gene of *Pseudomonas putida* coding for the enzyme salicylate hydroxylase, which converts SA into inactive catechol. These plants do not accumulate SA and are incapable of developing a SAR, which indicates that the accumulation of SA is required for the expression of SAR (Gaffney et al., 1993).

The establishment of SAR in non-infected tissues is, moreover, correlated with the induction of defence genes "PR" (see p. 140). These PR genes have been cloned and characterized and have been chosen as molecular markers of SAR: 9 families of PR genes in tobacco and the genes PR-1, PR-2, and PR-5 in *Arabidopsis*; PR-1 is the widely predominant marker protein of SAR in tobacco and *Arabidopsis*.

When TMV is the inducing pathogen, the range of PR proteins synthesized overlaps the range of PR proteins induced by the experimental application of exogenous SA; in transgenic plants expressing the gene nahG, PR proteins do not appear and resistance to TMV is greatly diminished (Ryals et al., 1996).

Salicylic acid is not likely to be the primary signal that migrates long distances from infected tissues to activate SAR elsewhere in the plant. This conclusion results from a grafting experiment in which a tobacco NN scion was grafted on to a nahG transgenic NN stock. The inoculation of the stock by TMV induces a negligible accumulation of SA in comparison to the 185 times higher level of non-transgenic controls; in these conditions, the migration of the primary signal outside of the stock does not seem to be affected, the level of expression of SAR marker genes and the resistance induced in the graft being equivalent to those observed in non-grafted NN tobacco (Vernooij et al., 1994). The reciprocal experiment to graft a nahG scion on to a non-transgenic stock did not lead to induction of SAR in the graft, which indicates that SA is an essential signal for SAR and that it is required downstream of the signal at long distance.

The analysis of promoters of certain PR genes has indicated a common "W" box of 25 base pairs the presence of which is correlated with potential activation by SA. Several proteins capable of binding to this sequence have been cloned. One of these factors, SRRB2, is phosphorylated by a MAP-kinase activated by SA, which greatly increases its affinity for DNA (Klessig et al., 1998).

Genetic analysis of SAR

Genetic analysis makes it possible to trace the signal transduction pathway in *Arabidopsis* (Ryals et al., 1996).

Mutants constitutively expressing SAR

- *lsd* (*lesions stimulating disease*): recessive mutants *lsd1, lsd3, lsd5* and dominant mutants *lsd2, lsd4, lsd6, lsd7*.
- *acd* (*accelerated cell death*): recessive mutants *acd1, acd2*.

These mutants are spontaneously displaying cell death in the absence of a pathogen. They express the marker genes of SAR at a high level, have high SA concentrations, and strong resistance to pathogens. These properties are suppressed

Secondary signals

Salicylic acid (SA), which accumulates either in free form or in the form of glucoside in the immediate proximity of the infection site, derives from phenylpropanoid pathways. However, its synthesis is not regulated at the PAL level. It is produced from a pre-formed conjugate of benzoic acid. The BA2-H enzyme that converts benzoic acid into salicylic acid is actively induced before the appearance of HR.

Jasmonic acid (JA) and *methyl jasmonate* (MeJA) are cyclopentanone derivatives of linolenic acid. They are widespread in the plant world, where they serve as growth regulators intervening in vegetative development, fruit development, and pollen viability.

Ethylene influences different growth and development processes, particularly germination, cell elongation, senescence, abscission, and fruit maturation as well as responses to stress. The biosynthesis pathway is known and the genes of key enzymes, ACC synthase and ACC oxidase, have been cloned.

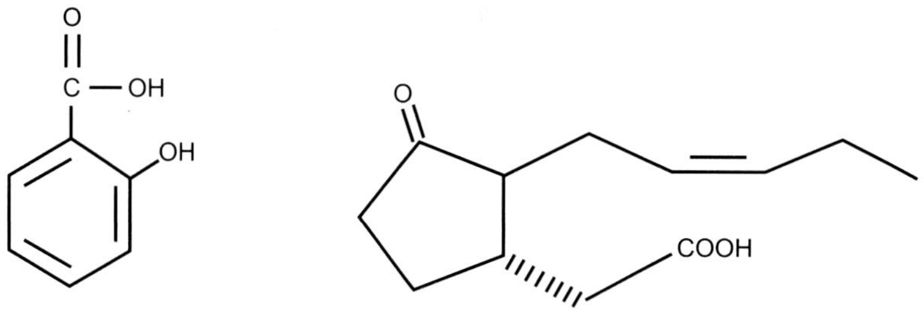

Figure 6.4 Salicylic acid (left) and jasmonic acid (right).

by nahG in *lsd6* and *lsd7*; the formation of necrosis is not suppressed by nahG in *lsd2* and *lsd4*.

- *cim* (*constitutive immunity*): dominant mutant *cim3*.

This mutant presents high SA levels and constitutively expresses SAR marker genes (PR1, PR2, PR5) in the absence of cell death; it is greatly resistant to pathogenic agents. The inhibition of SA accumulation by expression of nahG suppresses resistance as well as the expression of SAR marker genes.

Mutants defective in the expression of SAR

- *nim* (*noninducible immunity*).
- *npr* (*nonexpresser of PR genes*): recessive *npr* mutant.

These mutants do not respond to an exogenous application of SA or of a synthetic activator of SA such as INA (2,6-dichloro-isonicotinic acid) or even to a powerful activator used to protect various crops against a wide range of pathogens, BTH (S-methyl ester of benzo-1,2,3 thiadiazole-7-carbothioic acid). INA and BTH induce in

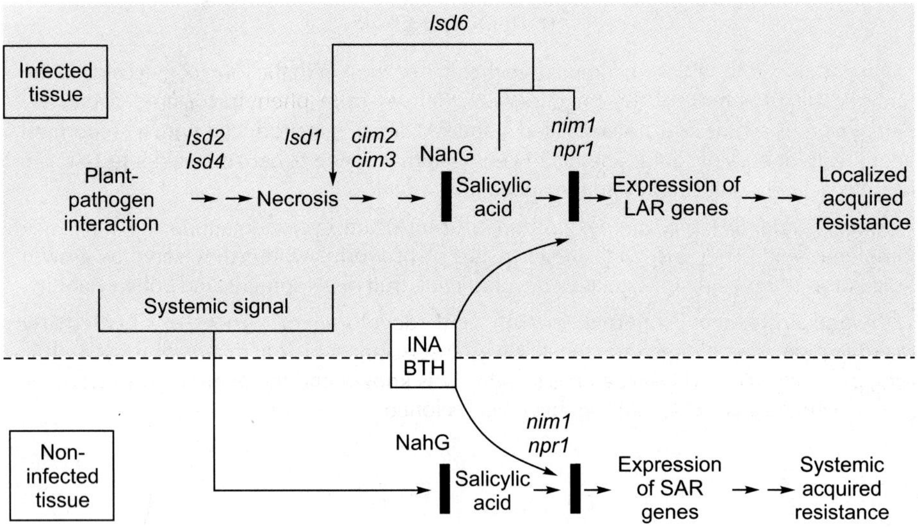

Figure 6.5 Signal transduction leading to LAR and SAR (Hunt and Ryals, 1996). The necrosis induced by the pathogen triggers the activation of localized resistance (upper part of figure) or SAR (below). The formation of the lesion is anterior to the accumulation of salicylic acid. It is not inhibited in *Arabidopsis* nahG, which expresses the salicylate hydroxylase. The two signal cascades depend on SA; they are blocked in nahG plants (Hunt and Ryals, 1996; with the permission of the publisher, © CRC Press, Boca Raton, Florida).

tobacco the same SAR marker genes as SA, even on nahG plants. They thus act downstream of SA on the same signal transduction pathway.

The hypersensitivity reaction: a programmed cell death that leads to signal amplification

The N protein shares a signalling motif with regulators of cell death in animals. The Apaf-1 gene in humans codes for an adaptor between a defender of the cell (Bc12) and a killer (cysteine proteases called caspases). The product of Apaf-1 also has three domains, a carboxy-terminal domain of protein recognition, an amino-terminal effector domain, and a central domain called NB-ARC: this is the central domain that is common to plant and animal proteins (ARC = Apaf in humans, R in plants, Ced4 in a nematode), and its activation leads to cell death (Van der Biezen and Jones, 1998b).

Hypersensitivity reaction is a genetically programmed cell death, which requires active participation of the host. The triggering of HR is probably due to the burst of active oxygen species; several observations indicate a very close correlation. For example, the infiltration of superoxide dismutase in leaves of tobacco NN infected by TMV compromises the development of HR (Doke and Ohashi, 1988).

The flows of K^+ and Ca^{2+} observed also play a role in the activation of DNA endonucleases. A typical event of apoptosis is the breaking up of DNA into fragments of around 50 kDa having 3′OH ends characteristic of endonucleolytic cleavages. These 3′OH terminations in DNA have been observed in the case of the induction of HR by TMV (Mittler et al., 1995).

The induction of PR and SAR has been found invariably to be linked to infections generating local necrotic lesions. This notion is reinforced by the characterization of mutants of *Arabidopsis* lcd and acd, which clearly shows the link between cell death and SAR.

Other signal transfer pathways do not involve salicylic acid

Transgenic nahG plants also made it possible to reveal SA-independent pathways of defence. Several defence reactions could be activated in plants without an increase in SA level. Jasmonic acid and ethylene now seem to be important signals in these alternative defence pathways. The exogenous application of these signal molecules induces a subset of defence genes, for example, in *Arabidopsis*, the gene for thionine Thi 2.1 and the gene for a defensine PDF 1.2, basic proteins with antimicrobial activity; these genes are not induced by SA.

Hypersensitivity reaction with infection by a pathogen could trigger an expression of SA-dependent defence genes as well as SA-independent expression. These pathways have undoubtedly a common part at the beginning of the induction, then diverge. Numerous mutants have been used to trace the common part and the diverging parts in *Arabidopsis* (Pieterse and van Loon, 1999). In certain cases, the divergence could be highly precocious: two different signal transfer pathways have been observed in *Arabidopsis* depending on whether the R protein is of the TIR-NB-LRR type or of the LZ-NB-LRR type (Aarts et al., 1998).

Since the structure of the R protein does not vary fundamentally with the type of pathogen, it may be supposed that the defence strategy relies on the induction of a set of responses directed against any type of pathogen and not on a response targeted against the recognized pathogen. Experiments show that this idea must be modulated. The first signals are relayed and amplified by the production of molecules that act as secondary signals controlling the different induction pathways of defence genes. In preferentially activating the appropriate routes or combination of routes, the plant is thus capable of adapting its defence to some extent against the aggressor pathogen.

The expression of resistance

■ The activation of numerous genes

The initial interaction is immediately followed by signal cascades that activate certain transcription factors. Finally, a wide range of genes involved in the defence mechanism are induced. Their number is difficult to estimate because several of these genes also have physiological functions (Reymond and Farmer, 1998).

The defence proteins are briefly presented here according to their distribution in the cytoplasmic, parietal, and extracellular compartments.

Cytoplasmic proteins

Enzymes of the phenylpropanoid pathways lead from phenylalanine to coumaroyl-CoA; the very first, phenylalanine ammonia lyase (PAL), also controls the synthesis of benzoic acid, precursor of salicylic acid.

From coumaroyl-CoA, the enzymes *chalcone synthase* and *chalcone isomerase* lead to the synthesis of phytoalexins; the enzymes *coumaric alcohol dehydrogenase* and *peroxidase* control the synthesis of polyphenols such as lignin.

When tobacco plants expressing an antisense PAL are inoculated with TMV, the kinetics of local lesions formation and their number and sizes are unchanged. Only the centre of the lesion is white and not brown because of the absence of oxidized phenolic products (Pallas et al., 1996). These observations suggest that the products of the phenylpropanoid pathway other than salicylic acid are not implicated in resistance to TMV.

Cell wall proteins

Hydroxyproline-rich glycoproteins (HRGP), glycine-rich proteins (GRP) and proline-rich proteins (PRP) are excreted at the cell wall. They form there by formation of di-tyrosine bonds under the effect of H_2O_2, a three-dimensional network reinforced around the cell, to which bind polyphenols such as lignin, the synthesis of which is regulated by PAL and OMT (orthodiphenol-O-methyltransferase). This transformation interferes with the degradation of cell walls by enzymes of pathogenic bacteria or fungi.

Extracellular proteins

The PR proteins, acidic proteins soluble at pH 3 and resistant to proteases, are secreted in the intercellular space of the leaf. They are part of the large family of stress proteins, having nevertheless the peculiarity of accumulating (Fig. 6.1) to up to 1% of the total soluble proteins of the leaf (Gianinazzi and Ahl, 1983; Kauffmann et al., 1987; Niderman et al., 1995; Van Loon, 1997; Legrand et al., 1987; Bol et al., 1990). These PR proteins are grouped into 11 families including, in particular:

- three families of chitinases (PR-3, PR-4, PR-11) and one family of *β-1, 3-glucanases* (PR-2), which disorganize the cell wall of fungi;
- one family of chitinases with lysozyme activity (PR-8) with possible effect against bacteria;
- two families of *antifungal molecules* (PR-1, PR-5 similar to thaumatine), active against oomycetes (which have no chitin in their cell wall);

— one family of *protease* or *α-amylase inhibitors* (PR-6), which prevents leaf tissue digestion by insects;
— one specific *peroxidase* (PR-9) implicated in lignin formation and reinforcement of cell walls.

Basic PR proteins, very similar in sequence and function to acidic PR proteins, have been identified. They appear in low quantity in vacuoles, not in response to the infection but in a regulated manner during the course of leaf, flower, and root development. They have a physiological as well as defence role.

The genes of numerous defence proteins present a common motif (H box) at their promoter. The transcription factor *myb1* binds to H boxes and it has been shown in parsley and tobacco that the gene coding for *myb1* is specifically expressed during the course of the induction of resistance (Yang and Klessig, 1996). In tobacco, a significant homology exists between the regulator sequences of the LTR-U3 region of the retrotransposon Tnt1A and the H box. The induction of this retrotransposon Tnt1A has been clearly demonstrated in tobacco NN inoculated by TMV, correlated with the activation of defence genes, more particularly genes activated early, such as PAL. The significance of this phenomenon with respect to plant defence remains a subject of debate (Moreau-Mhiri et al., 1996; Grandbastien, 1998).

The mechanism of resistance to viruses is not yet elucidated

The PR-10 family, the only family constituted of cytoplasmic proteins, has sequence relationships with ribonucleases. A ribonuclease having for target a structure common to numerous viruses such as the replication intermediate double-stranded RNA could rapidly block the multiplication of viruses, as shown in experiments of constitutive expression by tobacco of the *pac1* gene of *Schizosaccharomyces pombe* coding for a double-stranded RNA specific ribonuclease. Plants expressing this transgene are not affected physiologically and are partly protected against infection by CMV or PVY (*Potato virus Y, Potyvirus*) (Watanabe et al., 1995). But at present there is no evidence that members of the PR-10 family are active against viruses.

Certain PR proteins have been discovered by immunocytochemistry in the plasmodesmata (Murillo et al., 1997). This observation suggests possible inhibition of virus movement. Other authors suppose a blockage of plasmodesmata by callose as an essential factor of resistance to viruses in the case of resistance with HR (Beffa et al., 1996).

The localization of TMV in tobacco NN plants has also been associated with a replication-inhibiting protein (IVR) active against different viruses. This protein does not have the properties of PR proteins. A protein of 21.6 kDa capable of reducing TMV replication in preliminary assays on tobacco nn leaf discs has been identified by differential screening of cDNA libraries (Akad et al., 1999). Components of the RNA silencing pathway, such as RNA dependent RNA polymerases were also found activated during SAR (see p. 117).

Recessive resistance genes

Recessive resistance genes against viruses have been known for a long time and are found to be prevalent in some plant species. The fact that a resistance gene behaves in a dominant or recessive manner suggests different molecular bases for its action. Dominant genes are typically those that trigger the hypersensitive response and their mode of action, although not perfectly understood, involves the recognition of a pathogen-specific component followed by an array of signalling events culminating in the establishment of the resistance. In that model of receptor-ligand interaction a single functional copy of the gene is sufficient to establish the resistance, which behaves in a dominant manner during crosses. In contrast, two hypotheses can be drawn for the molecular basis of a recessive behaviour; these genes can code either for negative regulators of defence mechanisms and cell death or for components required for the virus life cycle. Since viruses recruit many different host components to perform their replication cycle, the latter hypothesis predicts that one of these components is missing or mutated to ensure resistance. As an example of the former hypothesis, the *rrt* gene for resistance to *Turnip crinkle virus* (TCV, *Carmovirus*) in *Arabidopsis thaliana* is dependent on salicylic acid signalling and therefore the *rrt* mutation is likely to enhance the hypersensitive response and the RRT allele behaves dominantly as a repressor of defence mechanisms (Chandra-Sekara et al., 2004).

eIF4E proteins as determinants of resistance/susceptibility

Recessive resistance genes are overrepresented in the interaction between plants and potyviruses, where about half of the resistance genes behave in a recessive manner whereas that proportion falls to 20% for other pathogens. The *pvr2* gene controlling resistance to *Potato virus Y* (PVY, *Potyvirus*) in pepper was the first to be identified and was found to code for the eukaryotic initiation factor 4E, a protein binding the cap structure present at the 5′ end of most eukaryotic messenger RNA during the initiation of translation (Ruffel et al , 2002; Chapter 2, Fig. 2.4). Point mutations in the N terminus of this protein prevent virus replication. Interestingly, the eIF4E protein was previously found to interact with the viral protein genome-linked protein (VPg), which is covalently bound to the 5′end of the genome of viruses of the picorna-like superfamily, including potyviruses (Wittmann et al., 1997; Schaad et al., 2000). The eIF4E protein is encoded by a small gene family in most plants and the fact that mutation of a single member of the family is sufficient to ensure resistance suggests that other members of the family are not able to sustain viral replication. This was further supported by a knock-out mutation in the *Arabidopsis thaliana* eIF(iso)4E gene preventing the replication of *Turnip mosaic virus* (TuMV, *Potyvirus*), *Lettuce mosaic virus* (LMV, *Potyvirus*), and *Tobacco etch virus* (TEV, *Potyvirus*) without preventing infection by *Clover yellow vein virus* (ClYVV, *Potyvirus*), whereas the inactivation of eIF4E allows TuMV but not ClYVV replication (Duprat et al., 2002; Lellis et al., 2002; Sato et al., 2005). In addition, in pepper, while the replication of PVY is prevented by the single *pvr2* mutation, the resistance to *Pepper veinal mottle virus* (PVMV, *Potyvirus*) requires both a mutation in the eIF4E isoform corresponding to *pvr2* and another mutation in the eIF(iso)4E isoform corresponding to another locus known as *pvr6* (Ruffel et al., 2006). Therefore, it appears that while some viruses appears to require specifically one eIF4E isoform to perform their replication cycle, others can use several of them.

The involvement of eIF4E in the infection cycle of plant RNA viruses has been further documented by the discovery that the recessive resistance genes *mo1* to LMV in lettuce and *sbm* to *Pea seed-borne mosaic virus* (PSbMV, Potyvirus) also correspond to mutated alleles of eIF4E (Nicaise et al., 2003; Gao et al., 2004). Although prevalent for potyviruses since it was found to be evolutionarily selected to give rise to natural resistance alleles, the role of eIF4E in virus cycle is not specific to this group, since chemically induced mutations for resistance to CMV (*Cucumber mosaic virus, Cucumovirus*) in *Arabidopsis* maps to eIF4E and eIF4G another protein of the eukaryotic translation initiation complex (Chapter 2, Fig. 2.4). In addition, a domain in the 3′ region of the satellite RNA of *Tobacco necrosis virus* (TNV, *Necrovirus*) binds to eIF4E proteins to promote efficient translation (Gazo et al., 2004). The precise biochemical role of eIF4E in potyvirus infection has still to be defined. As eIF4E binds VPg and mutations in the potyviral VPg allow some virus strains to overcome the eIF4E-mediated resistance, it is likely that the interaction with the VPg is required at some step of the viral replication cycle (Moury et al., 2004). The VPg might function as a cap mimic and recruit the translation initiation complex to the viral RNA. However, as TEV RNA possesses an internal ribosome entry site (IRES, see Chapter 2), the benefit in recruiting the cap-binding initiation complex is not clear. Also, VPg-eIF4E interaction can play a role in genome replication. In the replication model of picorna-like viruses, the urydylated VPg (VPgpUpU) serves as a primer for complementary strand synthesis; as eIF4E is connected to polyA-binding protein through the eIF4G bridge, the VPg primer might then be positioned near the viral polyA tract. In potyviruses, VPg urydylation by the NIb polymerase has been demonstrated (Pusstinen and Makinen, 2004) as well as the existence of a complex containing VPg, the viral polymerase, eIF4E and PABP (Leonard et al., 2004; Wang et al., 2000). In the case of the *sbm* gene for resistance to PSbMV, it has also been proposed that eIF4E-mutated alleles hinder viral cell-to-cell movement (Gao et al., 2004).

Conclusion

Evolution constantly selects in each virus the forms best adapted to each host plant, capable of cooperative interactions at the level of replication and movement. In the plant population, evolution has selected three mechanisms to limit infection by viruses:

— Silencing is induced in the sensitive plant by perception of an aberration at the level of nucleic acids, double-stranded RNA in the case of RNA viruses, after an intensive virus multiplication phase; this perception leads to specific degradation of the viral RNA, genomic or transcribed.

- The second mechanism is a surveillance system to perceive a protein coded by the virus and to give the alert for the triggering of a general programme of control against pathogenic agents. The plant thus manifests *extreme resistance* or, more frequently, *resistance with hypersensitivity reaction* in which necrosis serves to amplify signals and allows the development of systemic acquired resistance.
- The third mechanism is a change in host proteins (such as elongation initiation factors) so that they will not interact anymore with viral proteins in order to achieve essential steps in viral multiplication.

The surveillance system involves R genes, the ancestors of which come very likely from endogenous systems capable of linking a signal emission with recognition (Hammond Kosack and Jones, 1997). A significant number of NB-LRR proteins related to R proteins control development in mammals, yeasts, and insects: some hundreds of NB-LRR proteins are undoubtedly also present in *Arabidopsis*, and two of them control size and form of floral organs. The structural homology of R proteins with the transcription activator of class II of the major human histocompatibility complex suggests that the R genes of plants and genes involved in immunity in mammals could have a common evolutionary origin.

The R genes must have a capacity for rapid evolution of specificity. It must be considered that the constraints to which R genes are subjected in their evolution are not only selection for recognition of pathogenic agents but also selection against the recognition of endogenous proteins so as not to trigger a disadvantageous phenomenon inappropriately. The prevalence of NB-LRR in various plant species is coherent with the function attributed to them by surveillance molecules adaptable to pathogens with rapid evolution. The differential specificities of R proteins are often associated with duplication, deletion, and exchange of sequences in the regions coding for LRR. Comparative analysis of NB-LRR proteins shows that β-sheets of LRR comprise hypervariable zones that are subject to diversification pressure. The LRR domains appear thus as rapidly adaptable surfaces for recognition of Avr signals (van der Biezen and Jones, 1998a). In the genus *Solanum*, three genes similar to Rx have been sequenced, two that confer resistance to PVX and one that determines a resistance to a nematode: the LZ and NB domains are highly conserved, and the LRR domain alone is less conserved in the protein corresponding to the resistance to the nematode. This study opens up prospects for experiments designed to use a gene for specific resistance to obtain a broad-spectrum resistance by creating *in vitro* variability at the LRR domain (Bendahmane et al., 2000).

After the initial R/Avr recognition, the signal cascades that are triggered seem to be conserved in different plant species. In controlling the transformation of the tomato *Lycopersicum esculentum* by the N gene derived from *Nicotiana glutinosa* and that of *N. benthamiana* by the Rx gene derived from *Solanum tuberosum*, the two resistance genes remain perfectly functional. A better understanding of signal transduction pathways and the nature of the long distance signal is important for theoretical reasons and silencing by VIGS appears to be a rapid method for determining the implication of a given gene in a transduction pathway. The manipulation of this pathway could have interesting practical consequences; preliminary results in attempts to use one of the genes of the transduction pathway, i.e., the *npr*1 gene, to confer to plants a broad-spectrum resistance are encouraging. This gene exists in all species in which it has been looked for. A moderate overexpression (2 to 3 times) of this gene suffices to lead to activation of genes downstream, the synergistic effect of which leads to resistance against various pathogens without plant growth and development being affected (Cao et al., 1998).

In the constant battle between viruses and plants, the expression of this rapidly adaptable surveillance system is an important element of restriction of the host range. It is accompanied by constitutive resistance, which is linked either to a lack of the host co-factor capable of specific interaction with the viral factors in the course of replication and movement or to the production of an inhibitor of these viral functions. The agronomic exploitation of these different types of resistance is developed in Chapter 11.

7

CHAPTER

Subviral Pathogenic RNAs: Satellites and Viroids

Viruses are intracellular obligate parasites. They share this characteristic with other pathogenic agents the nucleic acid of which is smaller than that of the simplest viruses. These are satellites, which need a helper virus, and viroids, which multiply without recourse to a helper virus.

Satellite viruses and satellite RNAs

All satellites closely depend on a virus and are defined by several characters:

- They cannot replicate without a helper virus.
- They are not necessary for the cycle of their helper virus.
- They have no sequence homology with their helper virus, except in very limited regions (often at termini), which differentiates them from defective interfering RNA.

Satellites have a single-stranded RNA of 200 to 1400 nucleotides and behave like molecular parasites of their helper viruses, which number about 30. In some cases, they greatly modify the symptoms. Hence, the possibility of attenuating the symptoms and the incidence of viral diseases using the expression of certain satellites in crops has been explored. It presently appears that satellites are restricted to the plant kingdom, even though it is believed that the Δ virus found in humans is a satellite of the hepatitis B virus.

The modalities of the virus-satellite association are various. There are satellite viruses coding for their own coat protein and RNA satellites that do not encode a capsid protein and are encapsidated in the coat proteins of their helper virus.

Satellite viruses

Satellite viruses have no polymerase but code for their capsid. The RNA of satellite viruses (around 1 kb) codes for a protein, in which it is encapsidated to form icosahedral particles of around 17 nm. There is no serological relationship between the helper virus and its satellite. Four satellite viruses are known (Ban et al., 1995):

- *Panicum mosaic virus satellite* associated with PMV (*Panicum mosaic virus, Panicovirus, Tombusviridae*);
- *Maize white line mosaic virus satellite* associated with MWLMV (*Maize white line mosaic virus*, not classified);
- *Tobacco mosaic virus satellite* (STMV) associated with tobamoviruses TMV (*Tobacco mosaic virus*) and TMGMV (*Tobacco mild green mosaic virus*);
- *Tobacco necrosis virus satellite* (STNV) associated with TNV (*Tobacco necrosis virus, Necrovirus*).

Each satellite virus is associated with a virus that ensures the functions of replication. In the STMV, the RNAs of the virus and the satellite present homologies of the 3' end; the rest of the molecule is different. In STNV, the analogy of 3' regions exists mostly in the secondary structures, which could explain the access of the helper virus and the satellite to a single replicase.

How can the origin of satellite viruses be explained? The comparison of proteins coded by the four satellite viruses shows no phylogenetic relationship. A hypothesis can be advanced about STMV. This satellite virus resembles a subgenomic RNA coding for a capsid having acquired by recombination a 3' end of a virus present in the same cell. This allows it to be recognized by the replicase of this virus, which then becomes its helper virus; the 5' end will be picked up from another virus. Even though this scenario is speculative, it is probable: STMV has a 3' end that resembles that of TMV (*Tobacco mosaic virus, Tobamovirus*) and a 5' end that resembles that of CMV (*Cucumber mosaic virus, Cucumovirus*). These two viruses, as well as others, can co-infect tobacco (Dodds, 1998).

Messenger satellite RNAs

Messenger satellite RNAs have genomes of 0.8 to 1.5 kb. They depend on their helper virus not only for replication, but also for encapsidation, since they use the capsid protein of their helper virus. There is little sequence homology between the satellite RNA and that of its assistant virus, except in the terminal structures. About 10 satellite RNAs are known, associated respectively with ArMV (*Arabis mosaic virus, Nepovirus*), GFLV (*Grapevine fanleaf virus, Nepovirus*), SLRSV (*Strawberry latent ringspot virus, Nepovirus*), TBRV (*Tomato black ring virus, Nepovirus*), BaMV (*Bamboo mosaic virus, Potexvirus*), PEMV (*Pea enation mosaic virus, Enamovirus*), etc. They only slightly modify symptoms.

The protein coded by the satellite has been detected *in vivo* for the satellite of TBRV: it is necessary for the replication of the satellite (Hemmer et al., 1993). This 48 kDa protein is highly basic and hydrophobic, which suggests a function of attachment to the RNA, demonstrated for the protein P20 coded by the satellite of BaMV (*Bamboo mosaic virus, Potexvirus*) (Tsai et al., 1999).

Non-messenger linear satellite RNAs

Non-messenger linear satellite RNAs are competitors of the viral RNA. These satellites have genomes of less than 700 nucleotides and have no known messenger function. Seven have been described, notably the following:

- satellite of CMV (*Cucumber mosaic virus, Cucumovirus*);
- satellite of PSV (*Peanut stunt virus, Cucumovirus*);
- satellite of TCV (*Turnip crinkle virus, Carmovirus*);
- small satellite of TNV (*Tobacco necrosis virus, Necrovirus*).

Even though their RNAs contain sometimes small potential ORFs, nothing confirms their functionality (Roossink et al., 1992). Consequently, their biological activity results mostly from their sequence and secondary structure.

The satellite RNA most thoroughly studied is that of CMV, which is 332 to 405 nucleotides long; 49% of the nucleotides are engaged in secondary structures (Fig. 7.1). Several natural variants have clearly differentiated biological properties depending on the host. Some variants attenuate symptoms in tobacco and intensify symptoms by producing a generalized necrosis in tomato. They have caused serious epidemics in Southern Europe. Other variants attenuate the symptoms and are studied for possible use in control of CMV, a major pathogen.

The sequence differences between these two groups of variants have been studied to attempt to specify the determining factor of necrosis. The region responsible has been mapped by mutations, and the study of its conformation shows that its secondary structures, especially a hairpin with a UUAU loop, are determinants for triggering necrosis (Rodriguez-Alvaro and Roossink, 1997). The capacity to replicate a given satellite varies greatly with the strain of the helper virus.

How are these satellites, which have a marked action on the host as well as the helper virus, replicated? The hypothesis of competition for the viral replicase, in which the satellite can prove more effective than the virus, has been confirmed by the *in vitro* replication of the CMV satellite by purified viral replicase (Hayes et al., 1992). The capacity of a CMV strain to help a satellite in its replication may disappear

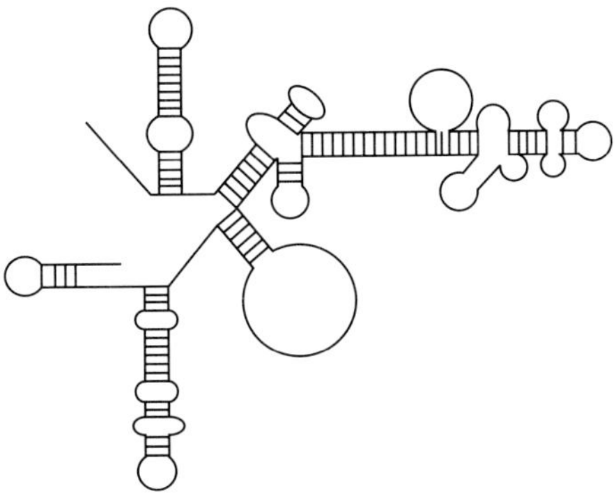

Figure 7.1 Structure of satellite of CMV (according to Rodriguez-Alvarado and Roossink, 1997).

following a single mutation in the helicase gene (Roossink et al., 1997). The TCV satellite, 356 nucleotides long, has a 3' region similar to that of its helper virus, the 6 terminal nucleotides are identical, and it can be replicated by the *in vitro* replication system of TCV (Stupina and Simon, 1997).

Non-messenger circular satellite RNAs

Non-messenger circular satellite RNAs have genomes of 220 to 350 nucleotides. They replicate by the rolling circle mechanism. Eight such RNAs are known, notably associated with the sobemoviruses (LTSV, *Lucerne transient streak virus*; SNMoV, *Solanum nodiflorum mottle virus*), nepoviruses (ArMV; TRSV, *Tobacco ringspot virus*) and a luteovirus (BYDV, *Barley yellow dwarf virus*) (Symons, 1997).

The infected tissues contain circular and linear RNAs, in which one or two remarkable structural domains can be identified, a hammerhead ribozyme or hairpin ribozyme in the TBRV satellite. The ribozyme allows the RNA to behave like an enzyme and self-cleave at a precise point on the RNA (Fig. 7.2).

The satellite associated with the TRSV has 69% of its 359 nucleotides engaged in secondary structures. It contains two ribozyme structures, one on the positive strand, the other on the negative strand. The ArMV satellite replicates by a rolling circle process that produces multimeric positive and negative strands (see further, Fig. 7.4A). The production of monomeric molecules is due to the self-cleaving ribozyme (Etscheid et al., 1995; Tanner, 1998).

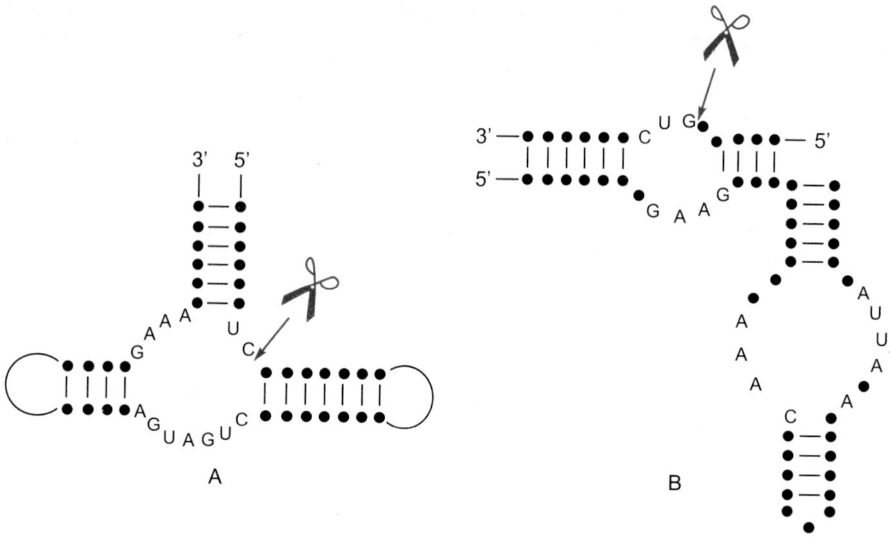

Figure 7.2 Self-cleaving RNA structures. The self-cleavage activity is linked to the nucleotides indicated. Although only the secondary structure is presented, ribozyme activity requires also tertiary folding. The arrow indicates the self-cleavage site. A: hammerhead ribozyme. This consensus sequence is found in the 12 satellite RNAs as well as in the viroids *Avsunviroidae*. B: hairpin ribozyme (according to Tanner, 1998).

The smallest satellite known (220 nucleotides) is associated with sobemovirus RYMV (*Rice yellow mottle virus*). Its rod structure resembles that of a viroid (Collins et al., 1998).

Satellites as parasitic agents

Like the viruses that help them, satellites form constantly evolving quasi-species, as observed in studies of the sequences of satellite populations, where nucleotide substitutions are frequent. The satellite is engaged in a relationship between three partners and undergoes selection pressures from the helper virus as well as from the plant. Hypotheses on the origin of satellites differ depending on whether the RNA of the satellite is or is not messenger. Satellites in which the RNA is messenger could have evolved from viral sequences by degenerative loss of functions. The small non-messenger satellites are rich in secondary structures, they resemble viroids and introns, and they could thus derive from small cellular RNAs (Kurath and Robaglia, 1995).

Viroids

Viroids were discovered by Diener in 1971 when he was searching for the causal agent of spindle tuber disease of potato. They are the smallest infectious pathogenic nucleic acids presently known. The symptoms greatly resemble those induced by viruses, with often a highly marked stunting. A viroid may propagate itself without apparent symptoms in one host but prove very serious for another, which makes it necessary to carry out epidemiological studies for the crops affected. The incidence on some crops may be very high; for example, coconut cadang-cadang viroid, one of the smallest known pathogenic RNAs (246 nucleotides), kills more than half a million coconut palms each year. Viroids have been described only in plants.

The means of transmission are vegetative propagation (in crops of potato, citrus, hop, chrysanthemum, coconut), mechanical means, the pollen and seeds, and insect vectors. Viroids are detected and identified on indicator hosts by polyacrylamide gel electrophoresis of plant extracts, hybridization with a probe, reverse transcription polymerase chain reaction (RT-PCR), or sequencing. These tests make it possible to eliminate the diseased individuals from propagation processes and thus to produce healthy crops (Singh and Dhar, 1998).

The structure of viroids is very simple in appearance: a small, circular, single-stranded RNA of 246 to 399 nucleotides. Viroids have no capsid and do not code for a protein. Viroid RNA has a series of regions comprising secondary structures separated by simple chain loops which gives it *in vitro* the compact conformation of a rod having two loops at the ends. The molecule is usually represented as in Fig. 7.3, which validates the terms upper chain and lower chain, right terminal loop and left terminal loop. There are linear forms that are also infectious.

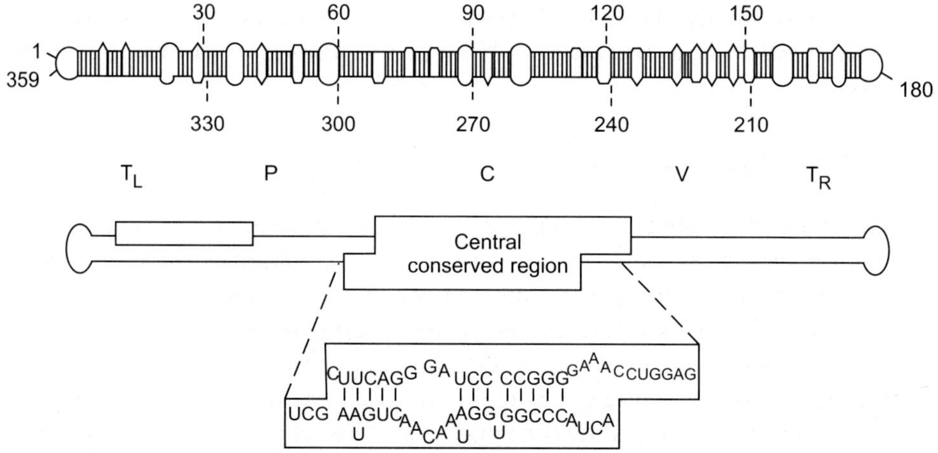

Figure 7.3 The circular RNA is represented in the form of a rod because of its numerous secondary structures. The regions T_L (left terminal), P (pathogenicity), C (central), V (variable), and T_R (right terminal) are shown.

Around 27 sequenced species are known (Table 7.1), and two families are distinguished (Flores et al., 1998):

- One small family (ASBVd, *Avocado sunblotch viroid*; PLMVd, *Peach latent mosaic viroid*; and CChMVd, *Chrysanthemum chlorotic mottle viroid*), the RNA of these viroids comprises a ribozyme structure that can cleave the RNA molecule by a self-cleaving reaction; these self-cleaving viroids constitute the family *Avsunviroidae*.
- A larger group (23 species, including the type species, PSTVd, *Potato spindle tuber viroid*) has a central conserved region (CCR), no ribozyme structure, and constitutes the family *Pospiviroidae*.

Properties of viroid RNA

Non-self-cleaving viroids (PSTVd group)

There are five regions in the RNA molecule (Keese and Symons, 1985):

- The central region is highly conserved throughout the group. It is made up of two series of conserved complementary nucleotides, surrounded in the higher chain by reverse sequences that may form a hairpin. The length of the sequence of this CCR can be used to classify it into three sub-families.
- The other regions are: P, domain associated with the host range and the expression of symptoms; V, the most variable domain; T_L, left terminal region, which contains the conserved sequence CCUC; and T_R, the right terminal region, which contains the conserved region CCUUC.

Table 7.1 Classification of viroids (Flores et al., 1998).

Family *Pospiviroidae*	
Genus *Pospiviroids*	PSTVd, *Potato spindle tuber viroid*
	MPVd, *Mexican papita viroid*
	TPMVd, *Tomato planta macho viroid*
	CSVd, *Chrysanthemum stunt viroid*
	CEVd, *Citrus exocortis viroid*
	TASVd, *Tomato apical stunt viroid*
	IrVd-I, *Iresine viroid I*
	CLVd, *Columnea latent viroid*
Genus *Hortuviroids*	HSVd, *Hop stunt viroid*
Genus *Cocadviroids*	CCCVd, *Coconut cadang-cadang viroid*
	CTiVd, *Coconut tinangaja viroid*
	HpLVd, *Hop latent viroid*
	CVd-IV, *Citrus IV viroid*
Genus *Apscaviroids*	ASSVd, *Apple scar skin viroid*
	CVd-IV, *Citrus IV viroid*
	ADFVd, *Apple dimple fruit viroid*
	GYSVd-1 and GYSVd-2, *Grapevine yellow speckle viroid 1 and 2*
	CBLVd, *Citrus bent leaf viroid*
	PBCVd, *Pear blister canker viroid*
	AGVd, *Australian grapevine viroid*
Genus *Coleviroids*	CbVd-1, -2, and -3, *Coleus blumei viroid 1, 2, and 3*
Family *Avsunviroidae*	
Genus *Avsunviroids*	ASBVd, *Avocado sunblotch viroid*
Genus *Pelamoviroids*	PLMVd, *Peach latent mosaic viroid*
	CChMVd, *Chrysanthemum chlorotic mottle viroid*

Viroids do not code for a protein; they depend entirely on cellular enzymes for their replication. Unlike viruses, they do not use the translation system, and their dependence of the host cell rely on the mobilization of the host transcriptional machinery to replicate their RNA. (Flores et al., 1998).

PSTVd has been observed in nuclei. The mechanism of replication by rolling circle is proved by the presence of monomeric or multimeric linear forms in extracts from infected plants. The infectious circular monomer (positive by convention) is copied into a multimeric negative strand, which serves in turn as template for the synthesis of multimeric positive strands. These are then cleaved into monomers that are circularized (Fig. 7.4A). Three enzymatic activities are necessary (Flores et al., 1997):

- A transcription activity effected by a cellular polymerase, RNA polymerase II, which is capable of producing positive chains *in vitro*: synthesis is inhibited by

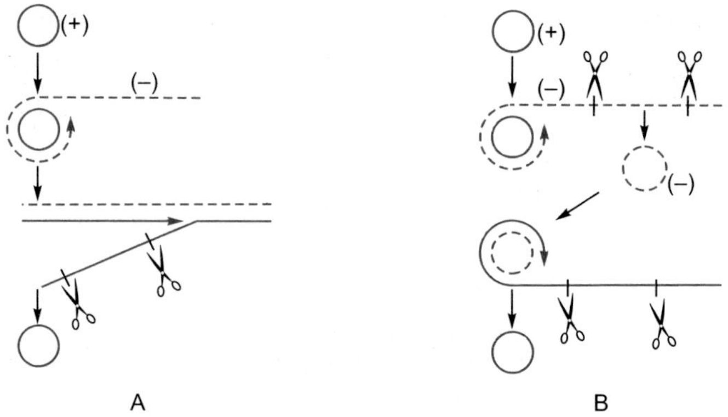

Figure 7.4 Rolling-circle RNA replication. A: PSTVd cycle. The multimeric (−) strand is copied in a multimeric (+) strand that is cleaved by a host enzyme, and monomers are circularized. B: ASBVd cycle. (+) strands and (−) strands are self-cleaved in monomers by hammerhead ribozyme structures.

α-amanitine at the same doses as for cellular messengers. Other enzymes can also be involved (Schindler and Mürhlbach, 1992).

- A ribonuclease activity to cleave the RNA.
- A ligase activity to circularize the monomers.

These two last activities have been observed *in vitro* (Tsagris et al., 1987). They involve probably host protein enzymes (Flores et al., 1997).

In vivo, the replication of PSTVd and CEVd (*citrus exocortis viroid*) is highly sensitive to temperature (× 100 or more between 18° and 30°C) and this fact could be linked to structural transitions necessary for the replication (Hu et al., 1996). *In vitro*, it is observed that the viroidal RNA can adopt metastable hairpin secondary structures. One involves the higher chain of the CCR and the adjacent reverse sequences; it resembles the origin of replication of the *Geminiviridae* and is a good candidate to be the origin of transcription. The other involves the lower chain. Tertiary structures have also been revealed that could explain certain aspects of replication: a loop at the level of the CCR greatly resembles cellular structures known to interfere with transcription factors or intranucleolar transport.

Self-cleaving viroids

Self-cleaving viroids have no CCR; they contain a hammerhead ribozyme sequence that occupies 30% of the molecule (see Fig. 7.2A). This structure is present in positive as well as negative strands and realizes self-cleaving.

Replication is insensitive to high doses of α-amanitine: it could be ensured by RNA polymerase I or more probably by a chloroplastic enzyme. ASBVd is located in the chloroplasts (Daros et al., 1994; Navarro et al., 1999). The rolling circle mechanism is different from that of PSTVd: the multimeric negative strand is cut and then joined again, and the negative monomer generates a second rolling circle to synthesize the positive strands (Fig. 7.4B). The hammerhead structures promote self-cleavage and a cell RNA ligase probably catalyses circularization (Flores et al., 1997).

Pathogenicity of viroids

Although some viroids produce serious diseases, others may multiply without symptoms. There is no direct relationship between symptoms and replication. Determinants of pathogenicity are likely to result from interaction between viroid RNA and host RNA or proteins. The principal means of study is the construction of hybrids between strains and the use of mutants that produce clearly differentiated symptoms.

The P domain is considered as the principal determinant of pathogenicity. Chimeric molecules constructed with highly and moderately pathogenic strains of CEVd produce symptoms directly related to the P domain they contain (Visvader and Symons, 1986). Other regions are also involved; when segments were exchanged between two similar species, it was found that the TL, P, V, and TR domains participate in various ways in the expression of symptoms observed: vein necrosis, dwarfism, and epinasty (Sano et al., 1992).

The effect of the thermodynamic stability and three-dimensional structure of the P region has been studied by comparing the structural and biological properties of 12 variants of PSTVd (Owens et al., 1996). The three-dimensional structure of the P domain, particularly the arrangement of three consecutive helices, is best correlated with the symptoms. It is observed that the most pathogenic strains often have a low thermodynamic stability, which favours the transition between the stable structure and the metastable structures necessary for replication. The geometry of the P domain is the main determinant of the RNA-protein relationship that is the source of symptoms and there is a direct correlation between the conformation of the molecule and its biological activity. The modalities according to which the viroidal RNA interacts with the cellular components remains mostly unknown. The phosphorylation of some proteins and the activation of a gene coding for a protein kinase have been demonstrated in relation with viroidal infection (Hammond and Zhao, 2000). The cell-to-cell movement of PSTVd has been studied by injecting fluorescent viroidal RNA into stomatal guard cells; the RNA remains within these cells deprived of plasmodesmata. In contrast, when injected into a mesophyll cell, it rapidly migrates through the plasmodesmata. The viroidal RNA is moreover capable of promoting the transport of a non-mobile RNA that is attached to it (Ding et al., 1997). It is highly probable that viroids borrow the plasmodesmata pathway and their movement is governed by a specific sequence or structural motif of the RNA.

The evolution of viroids

The origin of viroids remains hypothetical: they may represent a by-product of cell life or on the contrary be molecular fossils of RNA from precellular life. They evolve through general mechanisms of mutation, duplication, and recombination. Recombination is highly frequent not only between strains but also between species. It is favoured by the co-infection of a plant by several viroids; for example, tomato may harbour four different viroids (Singh et al., 1999). The study of sequences shows homologous sequences between non-related viroids. The interspecific recombination respects the functional domains; however, the upper chain and lower chain may have different origins (Kofalvi et al., 1997). Variability is also produced by mutations: more than 20 natural variants of PSTVd are known, presenting different pathogenicity, which vary mostly in the P and V domains. Viroids fit well with the concept of quasi-species, a replicating population of individuals that are very closely related but different (Gora-Sochaka et al., 1997). However, structural elements must be conserved, and the genetic divergence is tightly limited by constraints of replication (Ambros et al., 1999). *Avsunviroidae* viroids might have a common origin with the small satellite RNAs that have a ribozyme structure.

The source of viroids that multiply in cultivated plants is very probably wild plants, which often harbour viroids without manifesting symptoms. For example, MPVd (*Mexican papita viroid*), a possible ancestor for PSTVd and several viroids of this group by a series of adaptations to cultivated Solanaceae, was discovered in a wild *Solanaceae* of Mexico (Martinez-Soriano et al., 1996).

The Virus in the Agro-environment

8

CHAPTER

Virus Dissemination

To be transmitted or to disappear: a dilemma for plant viruses

Plant viruses use highly varied means to be disseminated. They are strict obligatory parasites that, in order to multiply, must find themselves in living cells. With a few rare exceptions, viral particles undergo rapid denaturation and loss of infectivity as soon as the host plant or host cell die. Therefore, in order to survive, the viral

Viruses cited

AbMV, *Abutilon mosaic virus, Begomovirus*; ACMV, *African cassava mosaic virus, Begomovirus*; ArMV, *Arabis mosaic virus, Nepovirus*; AYV, *Anthriscus yellows virus, Waikavirus*; BCMV, *Bean common mosaic virus, Potyvirus*; BCTV, *Beet curly top virus, Curtovirus*; BLCV, *Beet leaf curl virus, Nucleorhabdovirus*; BMYV, *Beet mild yellowing virus, Polerovirus*; BNYVV, *Beet necrotic yellow vein virus, Benyvirus*; BPYV, *Beet pseudoyellows virus, Crinivirus*; BSMV, *Barley stripe mosaic virus, Hordeivirus*; BWYV, *Beet western yellows virus, Polerovirus*; BYDV, *Barley yellow dwarf virus, Luteovirus*; CABMV, *Cowpea aphid-borne mosaic virus, Potyvirus*; CABYV, *Cucurbit aphid-borne yellows virus, Polerovirus*; CaMV, *Cauliflower mosaic virus, Caulimovirus*; CGMMV, *Cucumber green mottle mosaic virus, Tobamovirus*; CMV, *Cucumber mosaic virus, Cucumovirus*; CRSV, *Carnation ringspot virus, Danthovirus*; CSSV, *Cacao swollen shoot virus, Badnavirus*; CTV, *Citrus tristeza virus, Closterovirus*; CuNV, *Cucumber necrosis virus, Tombusvirus*; CYDV, *Cereal yellow dwarf virus, Polerovirus*; CymMV, *Cymbidium mosaic virus, Potexvirus*; CYSDV, *Cucurbit yellow stunting disorder virus, Crinivirus*; FBNYV, *Faba bean necrotic yellows virus, Nanovirus*; GFLV, *Grapevine fanleaf virus, Nepovirus*; GVA, *Grapevine virus A, Vitivirus*; INSV, *Impatiens necrotic spot virus, Tospovirus*; LBVaV, *Lettuce big-vein-associated virus, Varicosavirus*; LIYV, *Lettuce infectious yellows virus, Crinivirus*; LSMV, *Lettuce speckles mottle virus, Umbravirus*; MLBVV, *Mirafiori lettuce big-vein virus, Ophiovirus*; MNSV, *Melon necrotic spot virus, Carmovirus*; MRFV, *Maize rayado fino virus, Marafivirus*; ORSV, *Odontoglossum ringspot virus, Tobamovirus*; OYDV, *Onion yellow dwarf virus, Potyvirus*; PEBV, *Pea early-browning virus, Tobravirus*; PEMV, *Pea enation mosaic virus, Enamovirus*; PFBV, *Pelargonium flower break virus, Carmovirus*; PLRV, *Potato leafroll virus, Polerovirus*; PMMoV, *Pepper mild mottle virus, Tobamovirus*; PNRSV, *Prunus necrotic ringspot virus, Ilarvirus*; PSbMV, *Pea seed-borne mosaic virus, Potyvirus*; PVY, *Potato virus Y, Potyvirus*; PYDV, *Potato yellow dwarf virus, Nucleorhabdovirus*; PYFV, *Parsnip yellow fleck virus, Sequivirus*; RDV, *Rice dwarf virus, Phytoreovirus*; RpRSV, *Raspberry ringspot virus, Nepovirus*; RSV, *Rice stripe virus, Tenuivirus*; RTBV, *Rice tungro bacilliform virus, "Rice tungro bacilliform-like viruses"*; RTSV, *Rice tungro spherical virus, Waikavirus*; SBMV, *Southern bean mosaic virus, Sobemovirus*; SLCV, *Squash leaf curl virus, Begomovirus*; SMV, *Soybean mosaic virus, Potyvirus*; SqMV, *Squash mosaic virus, Comovirus*; SYVV, *Sowthistle yellow vein virus, Nucleorhabdovirus*; TBRV, *Tomato black ring virus, Nepovirus*; TBSV, *Tomato bushy stunt virus, Tombusvirus*; TMV, *Tobacco mosaic virus, Tobamovirus*; TNV, *Tobacco necrosis virus, Necrovirus*; ToMV, *Tomato mosaic virus, Tobamovirus*; TRSV, *Tobacco ringspot virus, Nepovirus*; TRV, *Tobacco rattle virus, Tobravirus*; TSV, *Tobacco streak virus, Ilarvirus*; TSWV, *Tomato spotted wilt virus, Tospovirus*; TYLCV, *Tomato yellow leaf curl virus, Begomovirus*; WDV, *Wheat dwarf virus, Mastrevirus, Geminiviridae*; WSMV, *Wheat streak mosaic virus, Tritimovirus*.

population must be able to maintain itself either by transmission to the progeny of the infected plant or by contamination of new plants of the same species or of other species. Viruses use a variety of transmission means (Fig. 8.1) corresponding to two strategies: an internal route and an external route.

The internal transmission route (or *vertical transmission*) makes it possible for viruses to perpetuate themselves in the living tissues of the infected plant by invading the reproductive organs. This mode of dissemination is most common in plants that are reproduced vegetatively: viruses are transmitted to most if not all the progeny resulting from cuttings, tubers, stolons, bulbs, cloves, or grafts taken from contaminated mother plants. In rare cases, viruses can be transmitted through seeds (Chapter 10). The internal transmission route allows the virus to perpetuate itself from generation to generation but does not enable the virus to infect new individuals of the same generation (Fig. 8.1). The virus remains in hosts that have a genetic background identical (vegetative propagation) or similar (seed transmission) to that of the initial infected plant, which reduces the constraints of adaptation. This mode of transmission surely contributed to the dissemination of viruses throughout the world through international seed trade.

The external transmission route (or *horizontal transmission*) involves a third agent (the vector) whose role is to take (acquire) the virus from an infected plant and to inoculate (transmit) it to a healthy plant. Thus there is, for short or longer periods, a stage in which the virus is found outside the plant cells. Vectors are found among rather diverse, mobile biological organisms: arthropods (insects, mites), nematodes, and fungi (Fig. 8.1). The interactions between a virus and its vector are generally complex and involve highly specific molecular recognition mechanisms.

The pecto-cellulose cell wall that surrounds plant cells forms a barrier that prevents the virus from crossing it on its own. This explains why plant viruses are not transmitted by rain or wind. The vector must thus be able to create small injuries that allow it to acquire and inoculate viral particles into living cells from which the virus can, eventually, initiate a systemic infection. This mode of dissemination allows the virus to invade hosts that are either genetically identical to the source plant if it spreads within a field of a single crop, or completely different hosts if it is transmitted to other plant species (weeds, other crops). Moreover, this mode of transmission allows spatial spread of viruses, sometimes over long distances, while their initial plant hosts remain immobile.

Finally, in rare cases, transmission by direct contact between two neighbouring plants or through humans activities during cultural operations can be encountered for highly stable viruses (*Tobamovirus*, *Potexvirus*, *Hordeivirus*). This mode of mechanical transmission has been adapted to carry out experimental inoculation of a large number of viruses; it is a technique presently commonly used in laboratories.

■ Vegetative propagation and grafting

Since virus diseases are generalized, all the organs used for vegetative propagation and coming from infected mother plants can transmit the disease to their progeny.

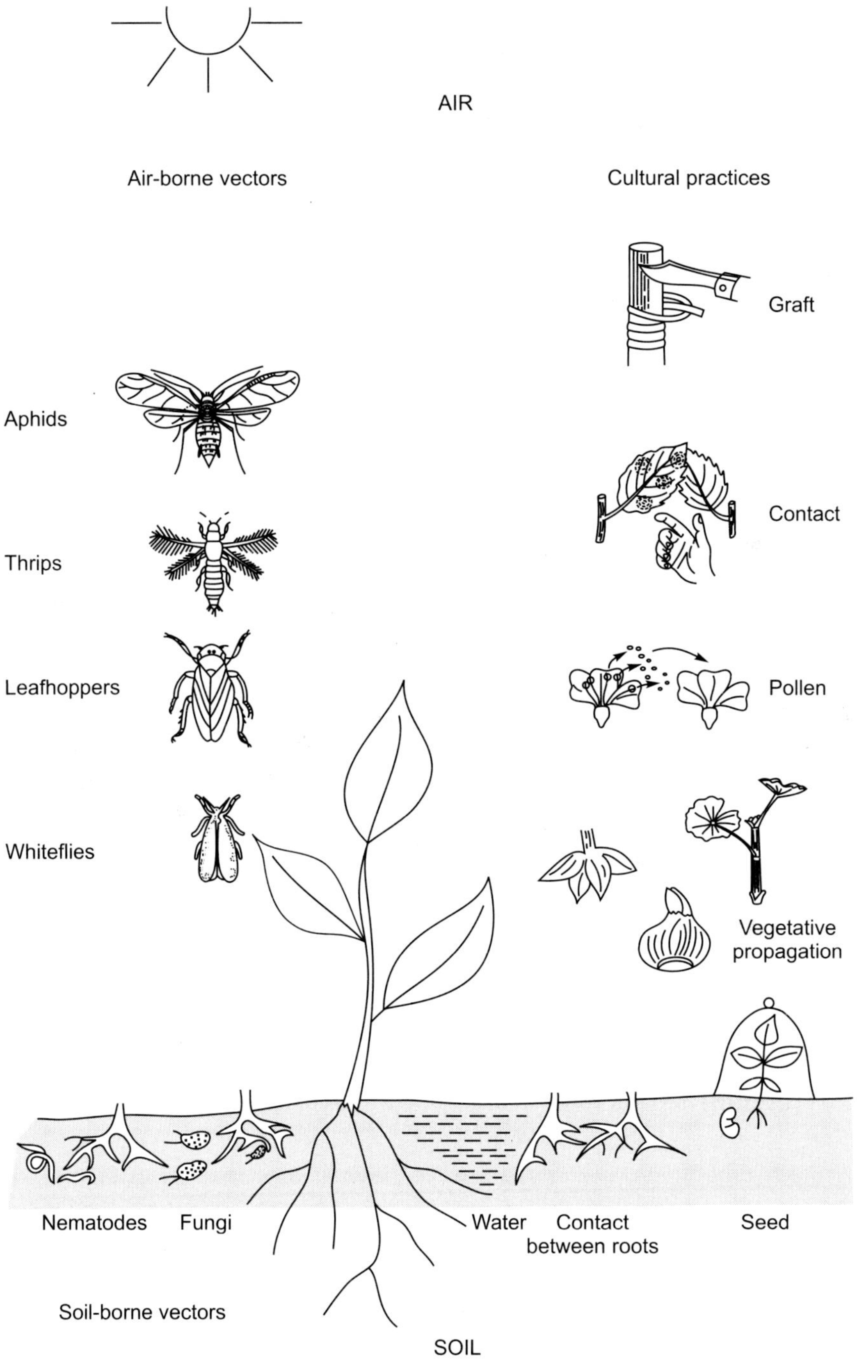

Figure 8.1 Different modes of plant virus dissemination: transmission through vegetative propagation material or through seed, contact transmission and transmission through air-borne or soil-borne vectors.

Examples can be found with tulip bulbs, potato tubers, and cuttings used for pelargonium or grapevine cultivation. Vegetative propagation is a reproductive practice essentially linked to human agricultural activities, although it is occasionally found in the wild. Today it is applied to a large number of fruit trees and ornamental or vegetable plants, as well as industrial crops in temperate and tropical regions.

When the viral infection is chronic, the virus is generally transmitted to all the progeny. On the other hand, in the case of late infection, the virus may migrate only to parts of the reproductive organs. In garlic, for example, contamination by OYDV (*Onion yellow dwarf virus, Potyvirus*) in May, one month before harvest, leads to contamination of only about 10% of the cloves, the others giving rise to healthy plants.

Vegetative propagation and *in vitro* propagation, its miniature form, are common practices today. When they are used on a large scale and without sanitary control, they could effectively serve for a massive dissemination of viruses. Such dissemination sometimes has disastrous economic consequences and entire cultivars or clones could be completely infected and thus become unproductive. Hence the term "degenerative diseases" attributed to viral disease in plants that are propagated vegetatively.

Grafting is commonly used for propagation of fruit and ornamental trees. If the graft has been taken from an infected mother plant, it will transmit the virus to the root-stock, and the entire plant thus becomes infected. In certain root stock-scion combinations, viral infection could lead to grafting incompatibilities that will prevent the development of the scion, as in the case of CTV (*Citrus tristeza virus, Closterovirus*). Transmission of viruses through grafting thus seems to be a particular example of vegetative transmission.

Almost all viruses can be transmitted experimentally through grafts, which is why this method is currently used to index woody plants in which diseases are difficult or impossible to transmit mechanically (Chapter 9). In certain cases, the roots of two neighbouring plants (most often trees) come into contact and anastomose to form a sort of natural graft. If one of the plants is infected, the virus may be transmitted to the other, which contributes to the progressive spread of the virus.

It is likely that most potato viruses were disseminated throughout the world by vegetative propagation, wherever this new crop was introduced. Today, globalization further increases this flow of international exchanges between countries and continents. One of its consequences is the introduction of viruses or vectors into geographic zones in which they were previously absent. To limit the consequences of these commercial exchanges, national and international rules have been established. In the European Union, a *phytosanitary passport* has been created that must guarantee, by proper indexing or suitable visual inspections, the absence of pathogens in plant multiplication material.

Some viruses are transmitted through the seeds

Seed-borne transmission involves only a small number of viruses. A precise inventory listed 108 viruses transmitted through seeds in at least one of their hosts, essentially potyviruses (20), nepoviruses (18), ilarviruses (8), and comoviruses (7), which represent around 10% of known viruses (Mink, 1993). In addition, seed-borne transmission is the rule for alphacryptoviruses and betacryptoviruses, since they have no other means of transmission from plant to plant. Seed transmission is also frequent for viroids. However, it remains that for most viruses, the progeny issued from seeds collected from an infected plant is healthy.

Most viruses infect mother tissues of the seed but are not transmitted to the plantlet

Tobacco mosaic virus (TMV, *Tobamovirus*) is not transmitted by seeds in tobacco: seeds collected from infected tobacco plants give rise to a population of healthy plantlets. However, if a sample of these seeds is washed, ground in buffer with a mortar and pestle and inoculated to healthy tobacco plants, the test plants will be infected. The seeds thus contain infectious TMV but the virus is not transmitted to the plantlets after germination. The localization of TMV has been studied by immunofluorescence either on seed sections or by seed dissection and testing of different tissues. TMV is not observed in tissues resulting from fertilization (embryo, endosperm); it is located in the tissues of maternal origin, i.e., in the residual layer of the nucellus and the teguments (Fig. 8.2).

Infection of the embryo is the key factor of transmission

The potyvirus PSbMV (*Pea seed-borne mosaic virus*), as its name indicates, is seed-transmitted in pea and lentil. Study of the distribution of PSbMV in seeds collected from infected peas shows that the virus is detected in most of the teguments and in a percentage of embryos corresponding, after germination, to the percentage of infected plantlets (Maury et al., 1987).

These two studies and many others show that seed-borne transmission of a virus depends mainly on the infection of the embryo. Nevertheless, there are two exceptions.

Exceptions to the rule

The embryo is infected but the virus is not seed-transmitted

During the maturation of the seed of certain bean cultivars, the sobemovirus SBMV (*Southern bean mosaic virus*) is totally or partly inactivated in the embryo (Uyemoto and Grogan, 1977). In the case of the potyvirus SMV (*Soybean mosaic virus*) on soybean cultivars Midwest and Merit, the infection rates of embryos in immature seeds, 94% and 58% respectively, drop at maturity to 66% and 0.8% (Bowers and Goodman,

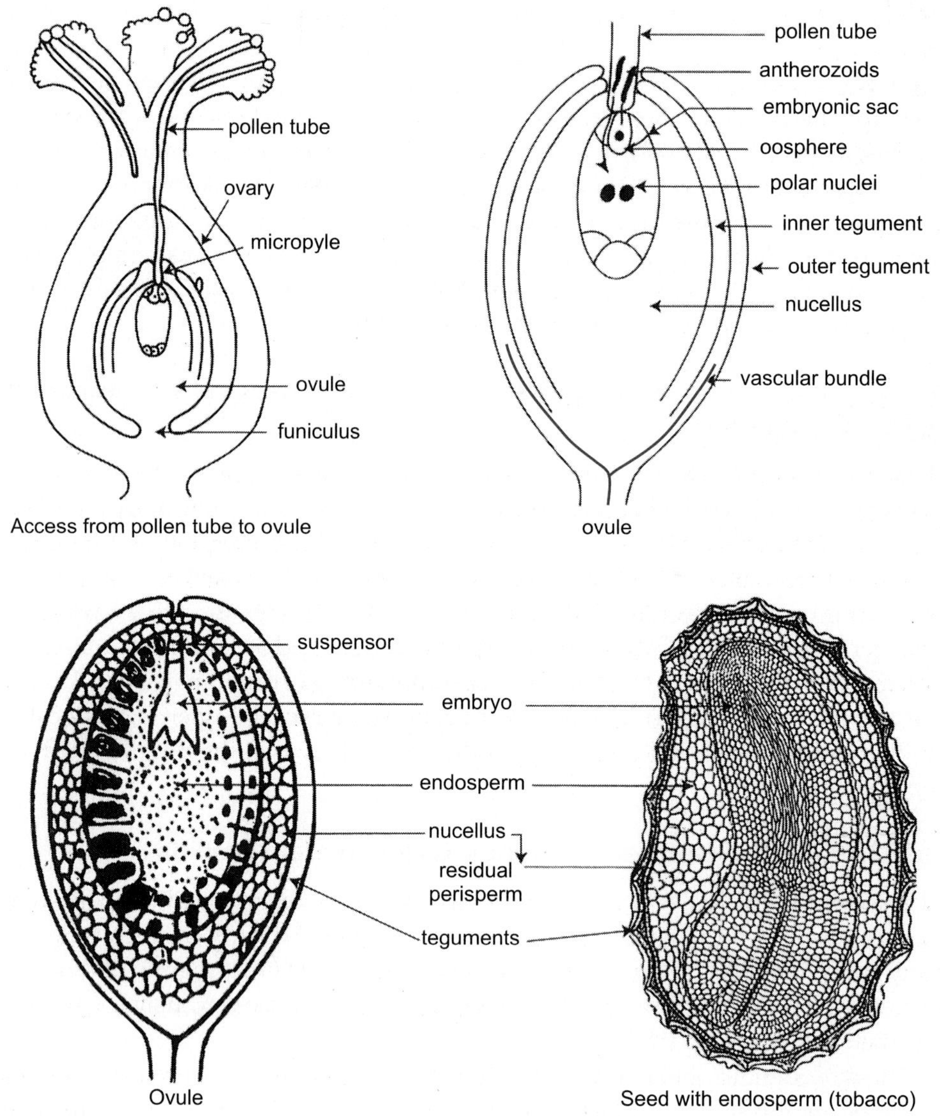

Figure 8.2 From ovule fertilization to seed: the major stages of seed formation and cell structures or tissues involved.

1979). Virus inactivation in the embryo could continue after the period of seed maturation during the storage period. For example, transmission of SqMV (*Squash mosaic virus*, *Comovirus*) by several melon seed lots evolved from 11-31% to 0% after two years of storage (Powell and Schlegel, 1970).

The embryo is not infected but the virus is seed-transmitted

The virus may be mechanically transmitted by wounds made to rootlets by the infected teguments, when they come out of the seed coat during germination. Several tobamoviruses are transmitted in this way: e.g., TMV, ToMV (*Tomato mosaic virus*) and PMMoV (*Pepper mild mottle virus*) in tomato and pepper, CGMMV (*Cucumber green mottle mosaic virus*) in cucurbits. Disinfection and dry heat treatment can eliminate the virus contaminating the outer layers of the seeds (Chapter 10).

An original way of seed transmission is the *vector-assisted seed transmission* (VAST) described for MNSV (*Melon necrotic spot virus, Carmovirus*) in melon. MNSV is found in internal and external teguments of the seed, and virions are released into the soil water or leachate as the seeds become soaked with water and germinate. MNSV may be then specifically and efficiently transmitted to seedling roots if zoospores of the vector, *Olpidium bornovanus*, are present in the soil. No seed transmission is observed if seeds are sown in sterile soil (Furuki, 1981; Campbell et al., 1996).

A necessary condition for the contamination of the plantlet from the teguments is that the virus conserve its infectivity there. This is true only for tobamoviruses and some other viruses, but in most cases the virus is rapidly inactivated in the seed coat during the seed maturation phase.

There are two possible ways for the infection of the embryo

In a maturing seed, the embryo cannot be infected from nearby maternal tissues. Indeed, these tissues have been infected like most of the tissues of the mother plant, but they are not linked to the embryo by plasmodesmata. Moreover, the embryo wall is very thick.

The embryo is thus infected at an earlier stage of development, either by infection of the pollen and transmission to the embryonic sac before fertilization (*indirect invasion of the embryo*) or by *direct invasion of the embryo* just after fertilization (Fig. 8.2). Most experiments report an indirect invasion but indications of direct invasion have also been presented. It is possible that both modes contribute to the final transmission rate (Johansen et al., 1994; Maule and Wang, 1996).

Indirect invasion of the embryo

Cross-pollination experiments were the first approach to show that gametes are infected, for example in bean infected by BCMV (*Bean common mosaic virus, Potyvirus*).

Two ultrastructural studies made it possible to observe the virus in pollen mother cells and in the pollen, in the mother cell of the megaspore, and in the embryonic sac.

In the case of BSMV (*Barley stripe mosaic virus, Hordeivirus*) transmitted by barley seeds, cytological changes were followed in the reproductive tissues during meiosis and embryogenesis in relation with the distribution of a seed-transmitted strain and a seed non-transmitted strain. It was observed that the presence of the virus of the seed-transmitted strain in the mother cells precedes the formation of a callose layer and the rupture of intercellular communications with the mother tissue cells just before meiosis. The virus then accumulates in the pollen and the embryonic sac at maturity. A non-transmitted isolate cannot be observed in the gametes (Carroll and Mayhew, 1976a and b).

In the case of TRSV (*Tobacco ringspot virus, Nepovirus*) transmitted by soybean seeds, the virus was observed in the embryonic sac and in the pollen. The high transmission rate is attributed to the capacity of the virus to infect the meristem and the mother cells of the megaspores. The pollen also is infected but its viability is greatly affected and cross-pollination experiments show that it plays a negligible role in transmission (Yang and Hamilton, 1974). TRSV is also transmitted by tobacco seeds, a host in which it induces a phenomenon of recovery (Valleau, 1941). The transmission rate is very high. In this case, the virus is observed not only in the meristematic zone but also in the cells of the initial ring (Roberts et al., 1970).

The correlation between the infection of gametes and the capacity of the virus to infect the meristems has also been studied on the Azuki bean–BCMV system: viral particles and pinwheel-type inclusions are observed in the meristematic zone, between 100 and 200 µm of the upper part of the crown, while CABMV (*Cowpea aphid-borne mosaic virus, Potyvirus*), not seed-borne in this host, is observed beyond the 200-300 µm zone (Tsuchizaki and Hibino, 1971). Infection of meristems seems to be the rule for *Alphacryptovirus* (Kassanis et al., 1978).

It emerges from these observations that for the virus to be transmitted by seeds, the mother plant must be inoculated before flowering, a condition necessary for the infection of the floral meristem and mother cells of gametes. Nevertheless, according to the experiment reported in the following section, if the floral meristem is not infected, the virus will still have a chance of reaching the embryo after fertilization.

Direct invasion of the embryo

An immunocytochemical study of the passage of PSbMV from vegetative tissues to the embryo in pea cultivars Vedette and Progreta raises certain arguments in favour of the direct invasion of the embryo after pollination (Wang and Maule, 1992). The cultivar Progreta does not transmit PSbMV through seeds. In contrast, in the cultivar Vedette, PSbMV is seed-borne with a high frequency but indirect invasion of the embryo does not seem to occur in this cultivar since cross-pollination experiments do not reveal pollen infection.

Observation of tissues of the future seed before pollination reveals the presence of the virus in the vascular bundles. Pollination seems to stimulate the transport of the virus toward the non-vascular tissues of the teguments in parallel with the mobilization of photo-assimilates toward the embryo. The virus thus progressively invades the entire tegument up to the micropyle, where it finds a tissue capable of leading it to the embryo: the *suspensor* (Fig. 8.2). This structure communicating with the embryo through plasmodesmata transports nutrients up to an early stage of embryo development but degenerates as soon the cotyledons develop. The virus thus would have no further opportunity for direct invasion. The mechanism of passage from mother tissues to the suspensor is nevertheless not clear since the cell walls at the point of contact, even though it is highly convoluted, does not seem to have plasmodesmata.

The property of seed-borne transmission varies with virus and host genotypes

For a virus infecting a plant species, seed-borne transmission is not an all-or-nothing phenomenon: indeed, it varies with the virus strain and the host cultivar. The influence of cultivar genotype has been seen in the experiment reported above in which the PSbMV-28 strain is transmitted at a frequency of 74% by the cultivar Vedette and is not transmitted through seeds of the cultivar Progreta (Wang and Maule, 1992).

The influence of the virus genotype is revealed by a variation of the seed transmission rates in a single cultivar, some strains not being transmitted at all (Johansen et al., 1996).

Viral determinants of seed-borne transmission control replication and movement

The existence of virus strains that may be seed-borne or not after infection of a same cultivar suggests that there are viral genes or sequences involved in the seed transmission process. When the viral genome has several components, the first approach is exchange of genomic components between transmitted and non-transmitted strains (pseudo-recombination). It is also possible to carry out true recombinations, when the genome has only one component, and to compare the behaviour of the chimera thus obtained (Chapter 11, box on identifying avirulence genes).

The property of seed-borne transmission has been found to be linked to RNA 1 of two nepoviruses, RpRSV (*Raspberry ringspot virus*) and TBRV (*Tomato black ring virus*), which excludes the role of the capsid, coded by RNA 2. Moreover, it is known that the capsid protein plays a determinant role in transmissibility through nematodes. Thus, the two major mechanisms of virus survival and dissemination are determined by different parts of the genome, which in nature must be subjected to different selection pressures (Hanada and Harrison, 1977).

Similarly, the property of seed-borne transmission has been found to be linked to RNA 1 in the CMV (*Cucumber mosaic virus, Cucumovirus*)-bean pathosystem (Hampton and Francki, 1992). In the BSMV-barley system, analysis of chimeras showed that the 5′ UTR part of the RNAγ, a repeat sequence of 369 bases in the γa gene, as well as the γb gene constitute the major determinants (Edwards, 1995). In PSbMV, the same approach revealed a role in seed-borne transmission of the 5′ UTR part, the HC-Pro gene, and sequences that are still undefined between the VPg gene and the 3′ end of the RNA (Johansen et al., 1996). Finally, the property of being seed-borne in PEBV (*Pea early-browning virus, Tobravirus*) depends on a gene coding for a

12 kDa protein on RNA 1 (Wang et al., 1997). A common characteristic of the major determinants of seed-borne transmission of BSMV, PSbMV, and PEBV is the presence of cysteine-rich motifs in the polypeptides concerned.

The frequency of seed-borne transmission is not correlated with the virus concentration in vegetative tissues. However, the major determinants of seed-borne transmission of BSMV, PSbMV, and PEBV involve the replication and movement functions: access to the meristem could depend on the synthesis in this very peculiar tissue of the same cellular cofactors present in the other tissues of the infected plant, cofactors that are necessary to replication and/or to movement of the viral strain.

Transmission by contact

Some highly stable viruses can be mechanically transmitted by direct or indirect contact from an infected plant to a nearby healthy plant, by the foliage or by the roots. This is the nearly exclusive natural mode of transmission of tobamoviruses and potexviruses. These particularly stable viruses, present in high concentrations in infected plants, are particularly damaging in high density crops such as greenhouse crops. In order to enable virus transmission, there must be an injury that damages the cells of the leaf epidermis or breaks the epidermal hairs to allow direct contact of the viral inoculum with living cells. Such injuries may occur simply when nearby plant leaves are rubbed together under the effect of wind, when workers walk through the fields, or under the effect of farm machinery (mowers).

At the root level, the growth of rootlets in the soil can also cause small injuries that may serve as an entry point for certain highly stable viruses. In fact, *Tobamovirus*, *Tombusvirus*, and *Dianthovirus* can persist in the soil for several months in plant debris or adsorbed to colloidal particles. An earlier infected crop can in this way be a source of contamination for the subsequent crop: this is the case, for example, with orchids repotted in used substrate contaminated by CymMV (*Cymbidium mosaic virus, Potexvirus*) or by ORSV (*Odontoglossum ringspot virus, Tobamovirus*).

Infectious virus particles of TMV, TBSV (*Tomato bushy stunt virus, Tombusvirus*), and CRSV (*Carnation ringspot virus, Dianthovirus*) were found, by means of specially adapted virus concentration techniques, in drainage water, rivers, or lakes. These viruses, which are particularly stable, are highly concentrated in the plant, have sometimes a wide host range, and keep their infectious capacity for a long time (Koenig, 1986; Koenig et al., 1988). The implication of the aqueous environment in the dissemination of these viruses is an important element, particularly in horticulture, where hydroponic cultivation practices are used with recycling of the nutrient solutions. It has been demonstrated, for example, that the carmovirus PFBV (*Pelargonium flower break virus*) is released in fertirrigation solutions and can contaminate the roots of healthy pelargoniums (Krczal et al., 1995).

Another mode of mechanical transmission may occur through the intermediary of pollen. During pollination, the virus can mechanically contaminate the mother tissues and infect the mother plant, as in the case of ilarviruses PNRSV (*Prunus necrotic ringspot virus*) and TSV (*Tobacco streak virus*) (Mink, 1993). For this to occur, there must be pollen contamination, transport of the pollen by pollinating insects to a healthy plant, and presence in the female flowers of pollen-feeding insects (*Thrips*) capable of creating injuries in the mother tissues that will allow mechanical transmission of the virus. In Rosaceae, dissemination of PNRSV by pollen, carried by the wind or by pollinating insects, contributes to the long distance transport of the virus. Several viroids are also transmitted by pollen.

Finally, farm workers can also disseminate viruses in intensive vegetable or ornamental crops, particularly in glasshouses and plastic tunnels, through cultural practices (cutting, pinching out, budding, harvesting). During these operations, a farmer's hands, clothes, or tools may become contaminated with highly stable viruses (tobamoviruses, carmoviruses), which are thus transmitted from plant to plant.

Transmission by vectors

Viruses can be transmitted from one plant to another through vectors, which effectively contribute to the survival and dissemination of viruses. Plant virus vectors have an astonishingly high diversity and specificity of transmission. Although the majority of virus vectors are insects of the piercing-sucking type, there are also other groups of organisms living in the air or soil, such as mites, nematodes, or fungi that may serve as efficient virus vectors (Fig. 8.1, Table 8.1). Among the insects, the most numerous vectors are found in the order Homoptera (aphids, whiteflies, leafhoppers), but there are also vector species belonging to the orders Thysanoptera (thrips), Hemiptera (bugs), or Coleoptera (beetles). Aphids constitute the principal group of vectors, in terms of the number of vector species (192 belonging to 13 genera) as well as the number of viruses transmitted (more than 275) (Matthews, 1991; Hull, 2002).

It was in Japan, at the end of the 19th century, that the involvement of an insect (the leafhopper *Inazuma dorsalis*) in the dissemination of a plant disease, rice dwarf, now known to be caused by the phytoreovirus RDV (*Rice dwarf virus*), was demonstrated for the first time (Bos, 1983). Doolittle (1916) proved the role of aphids as vectors of viruses by demonstrating the transmission of CMV by *Aphis gossypii*. The role of nematodes as vector of nepovirus GFLV (*Grapevine fanleaf virus*) was demonstrated later (Hewitt et al., 1958), as was the role of a soil fungus, *Olpidium brassicae*, in the transmission of the necrovirus TNV (*Tobacco necrosis virus*) (Teakle, 1962).

The strategies of virus transmission vary greatly: they generally involve highly specific molecular interactions for each virus-vector combination. This explains why, in general, a given virus is transmitted by only one vector species or taxonomically closely related species. Similarly, most often, all viruses of a genus (e.g., potyviruses) are transmitted by the same type of vectors (aphids in this case). Nevertheless, there are exceptions, and some viral genera (e.g., *Closterovirus*) may include viruses transmitted by different types of vectors (aphids, mealybugs) (Table 8.1).

Table 8.1 Major plant virus genera classified by vector and by mode of transmission (for definitions of modes of transmission, see Table 8.2; the viruses transmitted by fungi have particular modes of transmission, see box on acquisition and inoculation of viruses by fungi, later in this chapter). The numbers in parentheses correspond to the number of viral species whose vectors are known in each genus; the genera preceded by an asterisk include species transmitted by different vectors.

Vectors	Non-circulative viruses		Circulative viruses	
	Non-persistent	Semi-persistent	Non-propagative	Propagative
Aphids	*Alfamovirus* (1) **Carlavirus* (55) *Cucumovirus* (3) *Fabavirus* (2) *Potyvirus* (186) *Macluravirus* (2)	*Caulimovirus* (17) **Closterovirus* (10) *Sequivirus* (2) **Trichovirus* (1) **Waikavirus* (1)	*Enamovirus* (1) *Luteovirus* (13) *Polerovirus* (5) *Nanovirus* (5) *Umbravirus* (10)	**Cytorhabdovirus* (3) **Nucleorhabdovirus* (7) **Phytoreovirus* (1)
Beetles	**Machlomovirus* (1)		*Bromovirus* (6) **Carmovirus* (3) *Comovirus* (14) *Sobemovirus* (6) *Tymovirus* (21)	
Fungi	*Necrovirus* (3) *Tombusvirus* (1) **Carmovirus* (3)		*Bymovirus* (6) *Furovirus* (12) *Ophiovirus* (2) *Varicosavirus* (2)	
Leafhoppers		**Badnavirus* (1) **Waikavirus* (2)	*Curtovirus* (4) *Mastrevirus* (13)	**Cytorhabdovirus* (4) *Fijivirus* (6) *Marafivirus* (3) **Nucleorhabdovirus* (10) *Oryzavirus* (2) *Phytoreovirus* (5) **Phytoreovirus* (2) *Tenuivirus* (10)
Mealybugs		**Badnavirus* (3) **Closterovirus* (2) **Trichovirus* (1)		
Mites		**Trichovirus* (1)	*Rymovirus* (7)	
Nematodes		*Nepovirus* (39) *Tobravirus* (4)		
Thrips	**Machlomovirus* (1)			*Tospovirus* (5)
Whiteflies		*Crinivirus* (6) *Ipomovirus* (2) **Carlavirus* (1)	*Begomovirus* (41)	*Begomovirus* (1)

Aphids: insects designed for the transmission of plant viruses

Aphids, which represent the most important vector group, belong chiefly to the family *Aphidoideae* and the subfamily *Aphididae* (including the genera *Aphis, Macrosiphum, Myzus*, etc...) (Fig. 8.3). Numerous biological and morphological characteristics (structure of mouth parts, mode of feeding, searching behaviour for hosts, biological cycle alternating apterous and winged generations, reproduction rate) combine to give them these outstanding vector capacities.

Figure 8.3 Apterous adult aphid (a): *Macrosiphum euphorbiae*, vector of a large number of viruses transmitted non-persistently (such as PVY) or persistently (such as PLRV). Winged adult aphid (b): *Rhopalosiphum padi*, vector of BYDV.

Feeding behaviour and biology of aphids

Aphids have mouthparts that are particularly effective in transmitting viruses; they are formed of two pairs of flexible stylets supported by the labium. Two mandibular stylets flank two maxillary stylets delimiting an ascending alimentary canal, which conducts the nutrients towards the head and the digestive tube, and a descending salivary duct, coming from salivary glands and allowing saliva to be injected into the plant (Fig. 8.4). These salivary secretions, which surround the stylets during their penetration into the plant tissues, make it possible to follow the stylets intercellular trajectory towards the phloem in leaf sections.

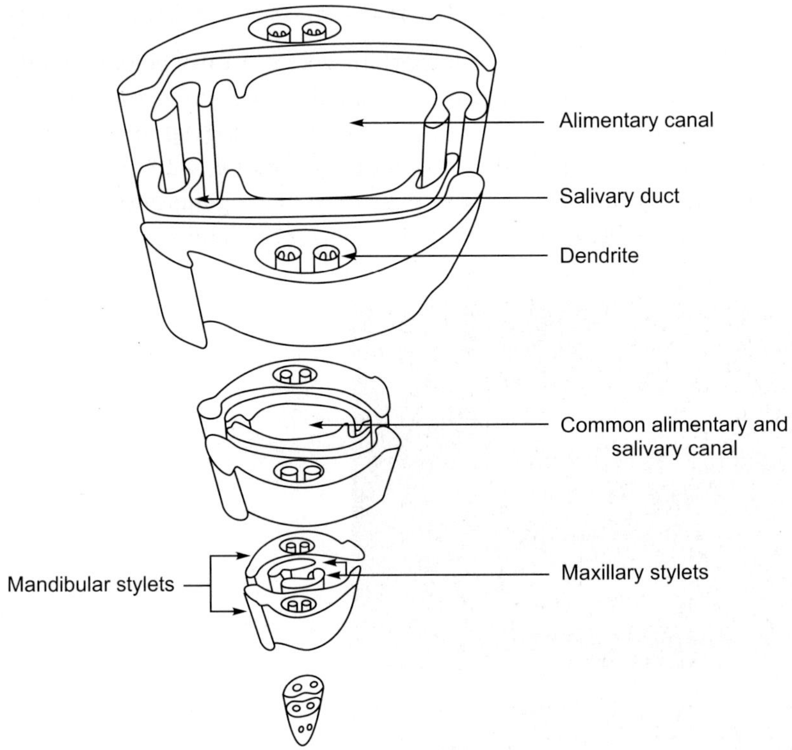

Figure 8.4 Perspective of the tip of an aphid stylet (modified from Taylor and Robertson, 1974).

Aphids have host ranges specific to each species or even biotype. They have a particular host search behaviour (Fig. 8.5). An aphid in flight cannot identify a favourable host by sight. It is attracted by certain colours (yellow) or by the contrast of colours of foliage and soil. When it lands on a plant, the aphid will make probes to "taste" the plant and determine whether it is suitable for its development. These probes are very brief (a few seconds), often repeated and superficial. This behaviour corresponds to optimal conditions for the acquisition or transmission of non-persistent viruses (Fig. 8.6a, Table 8.2). If the "tasted" plant is suitable, the aphid stops, and inserts its stylets all the way to the phloem, and feeds for a long time on the phloem sap. It is during these feeding probes that the aphid acquires or transmits viruses that multiply in the phloem, particularly persistent viruses (Fig. 8.6b, Table 8.2). Otherwise, the aphid flies on to another plant and the process is repeated (Fig. 8.5).

Finally, aphids have a very high potential for reproduction and aerial spread: their complex biological cycle (Cornuet, 1987) is an alternation of winged and apterous forms that allows the dissemination of the species as well as that of viruses. Most aphid reproduction occurs by parthenogenesis with very short generation times. This leads, during certain periods of the year, to population explosions that result in large flights sometimes observed even at high altitude.

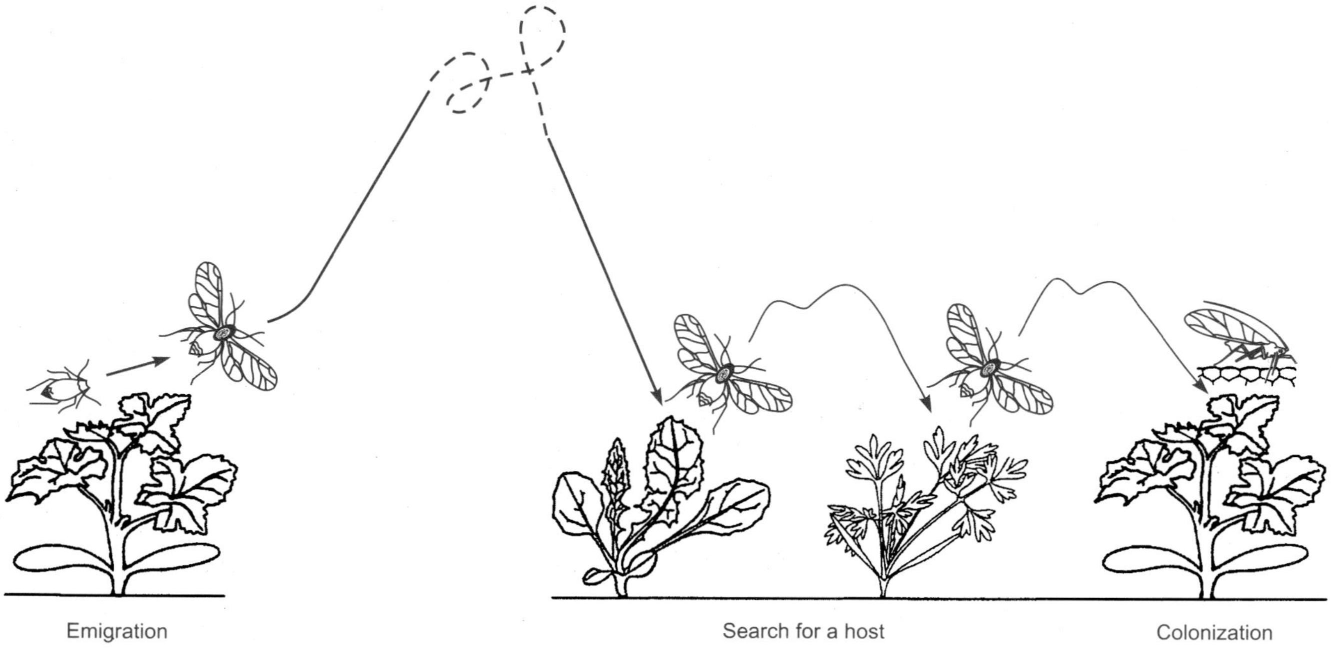

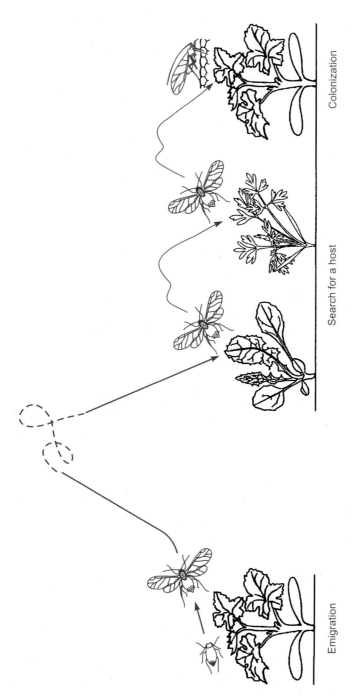

Figure 8.5 Flight and host search behaviour of a winged aphid.

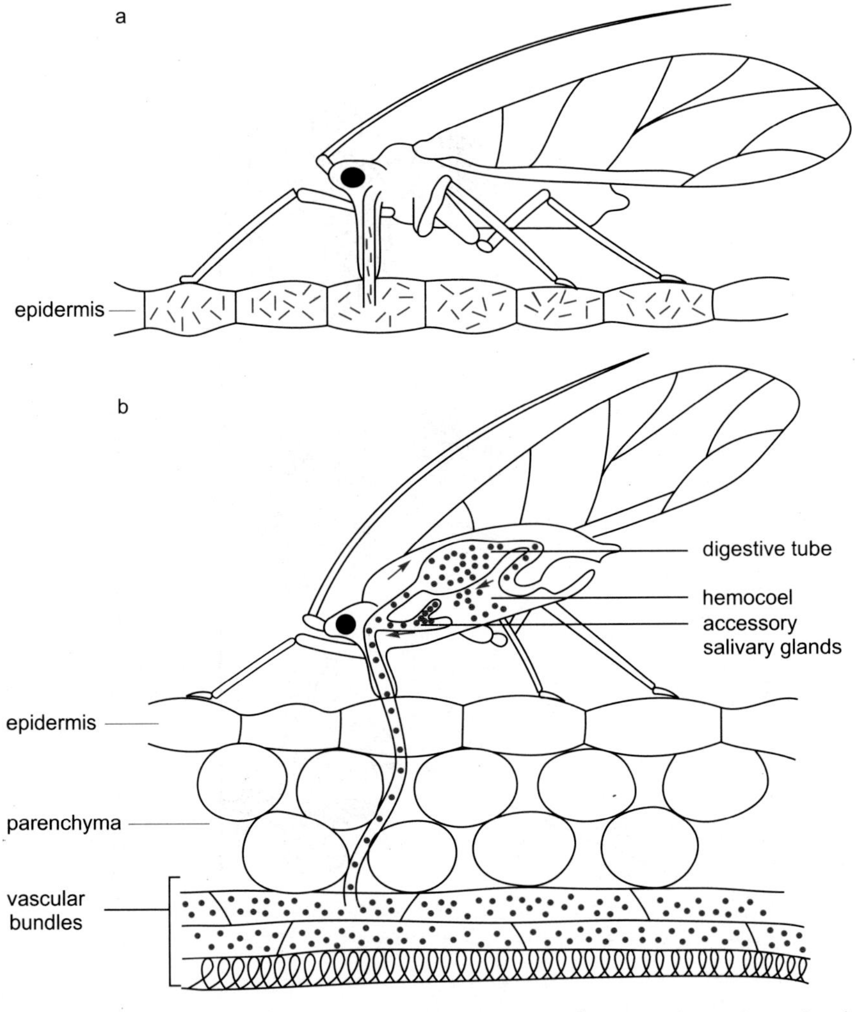

Figure 8.6 Virus acquisition by an aphid: (a) Acquisition of non-persistent (non-circulative) virus during a brief probe in the leaf epidermis. (b) Acquisition of persistent (circulative) virus during a feeding probe in the phloem: circulation of viral particles in the digestive tube and then in the hemocoel up to the accessory salivary glands.

Classification of the different modes of virus transmission by aphids

The relationships between aphids and plant viruses have been particularly well described and presently serve as a reference for all virus-vector interactions. However, the classification systems have evolved with the deepening of understanding of the biological and molecular basis of virus-vector interactions.

The different types of transmission were at first defined according to quantitative parameters describing the major steps of the virus transmission process. These steps are the following:

Table 8.2 Principal characteristics of the different modes of virus transmission by aphids (modified from Leclant, 1982).

	Non-circulative viruses (1)		Circulative viruses (1)	
	Non-persistent viruses (2) Stylet-borne viruses (4)	Semi-persistent viruses (3)	Persistent viruses (2)	
			Non-propagative	Propagative
Acquisition	very brief (seconds)	brief (minutes)	long (hours)	long (hours)
Latent period	no	no	yes	yes
Retention	short (few hours)	rather long (several hours to few days)	long (several days to lifelong)	long (lifelong)
Conservation after moulting	no	no	yes	yes
Transmission to progeny	no	no	no	no/yes
Specificity of transmission	low	narrow	narrow	narrow

(1) Harris, 1977; (2) Watson and Roberts, 1939; (3) Sylvester, 1956; (4) Kennedy et al., 1962

— *Acquisition period* corresponds to the time required for the aphid to acquire a virus from an infected plant.
— *Inoculation period* corresponds similarly to the time required for the aphid to transmit the virus to a healthy plant.
— *Latent period* corresponds to the interval between the time an aphid acquires a virus and the time it can transmit the virus.
— *Retention period* is the period during which an aphid retains its ability to transmit a virus; it is then called viruliferous or infectious.

These criteria were used by Watson and Roberts (1939) and then by Sylvester (1956) to establish the first classification distinguishing viruses transmitted according to the *non-persistent*, *semi-persistent*, and *persistent* modes (Table 8.2). A major advantage of these definitions is that they characterize a virus by the transmission capacity of its vector as a function of time.

Other nomenclatures were then proposed, including that of Harris (1977), which takes into account the location of the virus in the aphid (Table 8.2). Harris distinguished two types of association between viruses and their vectors. If the viral particles that will be transmitted remain external, it is a *non-circulative* virus. On the other hand, if the virus is internalized, i.e., when the virus goes into the body of the insect before being inoculated, it is a *circulative* virus (Fig. 8.6b). Besides this, circulative viruses can be *propagative* if there is multiplication of the virus in the vector (see box).

Different modes of virus transmission by aphids

Non-circulative, non-persistent viruses (Fig. 8.6a): The transmission of these viruses by aphids is quick and is enhanced by a fasting period of 15 min to 2 h before the acquisition feeding period. The transmission is much more efficient with a short probing time (a few seconds to a few minutes) and the aphid rapidly loses its viruliferous capacity. These viruses can generally be inoculated experimentally by mechanical means and multiply in all plant tissues.

Several hypotheses have been developed to explain this type of transmission. It was at first thought that there was a passive contamination of the external parts of the stylets (stylet-borne viruses, defined by Kennedy et al., 1962, who compared the aphid to a "flying needle"). A second theory suggested an active transmission mechanism: the virus is acquired during a probe and then regurgitated through the alimentary canal when the insect makes another probe. This is the ingestion-egestion principle: Harris (1977) compared the aphid to a "flying syringe". In this case, the virus is retained on specific receptor sites on the alimentary canal or on the anterior intestine. Later, Martin et al. (1997), using the technique of electropenetrography (EPG), suggested that transmission occurs by ingestion-salivation, the virus being retained on the distal part of the stylet, where the alimentary and salivary ducts are fused (Fig. 8.4).

Non-circulative, semi-persistent viruses: These viruses have transmission parameters intermediate between those of non-persistent and persistent viruses. The aphid may remain viruliferous one or a few days and can thus infect several plants in succession.

Circulative, non-propagative viruses (Fig. 8.6b): These viruses are most often linked to the phloem tissues and cannot be transmitted mechanically. To acquire or inoculate them, the aphid must make deep feeding probes in order to reach the phloem, and hence the relatively long periods of acquisition and transmission (from several minutes to a few hours). The virus goes through the aphid body, without multiplying, before being transmitted, which explains the latent period (Fig. 8.6b). The aphid remains viruliferous for several days, sometimes even for its whole lifetime.

Circulative, propagative viruses: These viruses multiply in the vector, which remains viruliferous all its life. They can invade all the tissues of the vector but more generally they replicate in only some types of cells.

Hebrard et al. (1999) proposed to extend the classification of Harris to other groups of vectors by distinguishing (1) viruses with non-circulative transmission that are attached to the mouth parts cuticle or to the anterior digestive tract of their vectors, (2) viruses with circulative transmission that enter the vector internal cavity by selectively crossing barriers formed by cell membranes of the digestive tube and salivary glands, but which do not replicate in the vector, and (3) viruses with propagative transmission that are internalized and replicate in the vector as well as in the plant.

There are many other efficient air-borne vectors

Whiteflies (Fig. 8.7) are vectors of viruses that have a significant economic impact in field crops in warm regions (in Mediterranean or tropical climates) or in protected

Figure 8.7 Whiteflies: (1) *Trialeurodes vaporariorum* (adults and larvae), vector of *Beet pseudo yellows virus* (BPYV); (b) mating of *Bemisia tabaci*, vector of a large number of viruses.

crops in more temperate regions. Three whitefly species have been reported as virus-vectors, but one of them, *Bemisia tabaci*, is particularly efficient in transmitting certain begomoviruses, criniviruses, carlaviruses, and ipomoviruses. Some of these viruses, such as TYLCV (*Tomato yellow leaf curl virus, Begomovirus*) or CYSDV (*Cucurbit yellow stunting disorder virus, Crinivirus*), have in recent years emerged in many new regions of the world, causing considerable damage. The spread of these viruses seems to be linked to significant increases in the vector populations as well as a shift in the dominant biotype of *Bemisia* (Polston and Anderson, 1997; Wisler et al., 1998).

Leafhoppers and *delphacids* (*planthoppers*) (Fig. 8.8), well known for their capacity as vectors of phytoplasma, also transmit viruses belonging to several families or genera (including *Rhabdoviridae, Geminiviridae, Reoviridae, Tenuivirus*). These viruses are transmitted on the circulative mode and are sometimes propagative. In the case of tenuivirus RSV (*Rice stripe virus*), the virus is also transmitted to the progeny of the vector *Laodelphax striatellus*. RSV reduces the fecundity of the vector but not its longevity. It has been transmitted by transovarial passage to 40 successive generations over a period of 6 years (Shinkai, 1962).

Mealybugs of the genus *Planococcus* are not very mobile but they contribute to the transmission of badnaviruses such as CSSV (*Cacao swollen shoot virus*) or one of the vitiviruses associated with leafroll disease in grapevine, GVA (*Grapevine virus A*).

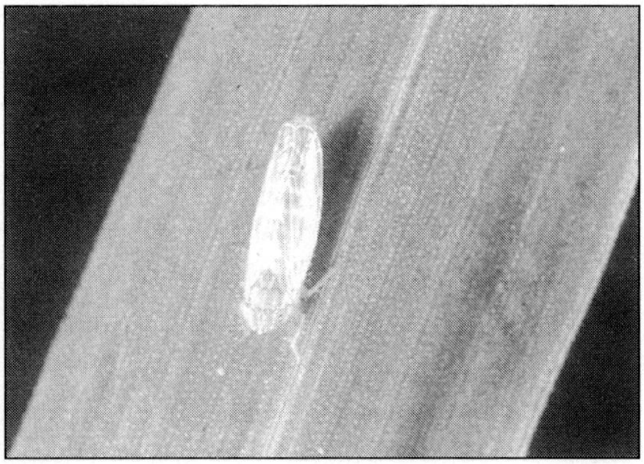

Figure 8.8 Adult leafhopper *Psammotettix alienus*, vector of *Wheat dwarf virus* (WDV).

In the order Thysanoptera, family *Thripidae*, the *thrips* include different species of the genera *Thrips, Frankliniella,* and *Scirtothrips*, which are vectors of tospoviruses (Fig. 8.9). The mouthparts of thrips are transformed into stylets: the massive mandible perforates and the retracting maxilla form an aspiration tube with a tip shaped like a sharp bevel. To feed, the thrips (larvae and adults) perforate the leaf epidermis and after the cell is lysed by the saliva injected, they suck in the cell contents, rather than the sap, using their pharyngeal pump. The species *Frankliniella occidentalis*, a particularly efficient vector recently introduced in Europe, caused severe epidemics of TSWV (*Tomato spotted wilt virus, Tospovirus*) and INSV (*Impatiens necrotic spot virus, Tospovirus*) in the 1990s in vegetable and ornamental crops (Moury et al., 1998). Another species that is also highly polyphagous, *Thrips palmi*, has increased its distribution area in tropical countries to an alarming extent. It is particularly noteworthy that, in thrips, only the two first larval stages can acquire tospovirus particles, the virus being transmitted subsequently by larvae and adults.

Coleoptera, insects with crushing mouthparts, include a certain number of vectors, such as phytophagous beetles that transmit viruses specifically through their alimentary regurgitants. The viruses transmitted are essentially isometric, stable, and concentrated in the plant (*Bromovirus, Comovirus, Sobemovirus,* and

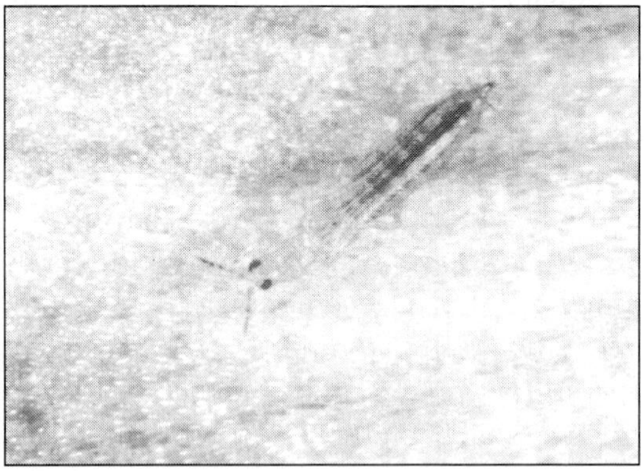

Figure 8.9 Thrips *Frankliniella occidentalis*, vector of *Tomato spotted wilt virus* (TSWV).

Tymovirus). Finally, a *bug*, *Piesma quadratum* (Hemiptera), transmits a nucleorhabdovirus, BLCV (*Beet leaf curl virus*).

Alongside the class of Insects so well represented as vectors of plant viruses, the class of Arachnids contains a single order comprising virus vectors, eriophyid *mites*, also of the piercing-sucking type. Mites transmit trichoviruses and rymoviruses. Although they lack wings, mites can be disseminated quite effectively by wind because of their small size. *Aceria tulipae* is a vector of WSMV (*Wheat streak mosaic virus, Tritimovirus*); the viral particles acquired by the larvae are transmitted by all the other stages, and dense accumulations of the virus are observed in the digestive parts and the salivary glands.

Nematodes and fungi are soil-borne vectors

Nematodes are parasites that feed on the absorbing hairs or the root cortical cells by piercing them with their long stylets to extract the cellular content (Fig. 8.10). Nematodes that are vectors of viruses belong to two families: the *Longidoridae* comprising the genera *Xiphinema* and *Longidorus*, which transmit *Nepovirus* including GFLV and ArMV (*Arabis mosaic virus*), and the *Trichodoridae* comprising the genera *Trichodorus* and *Paratrichodorus*, which transmit the tobraviruses TRV (*Tobacco rattle virus*) and PEBV.

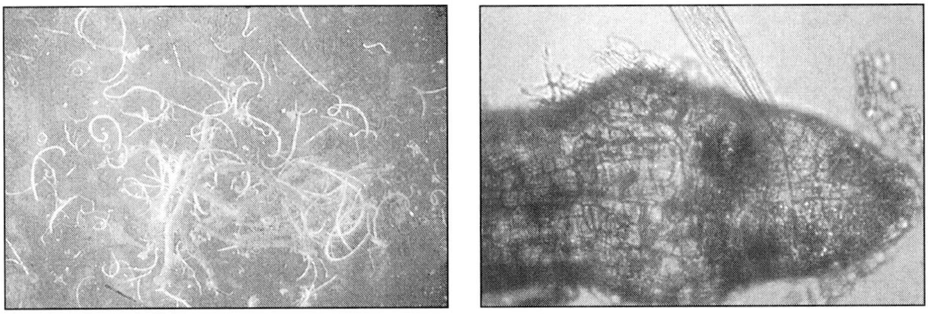

Figure 8.10 Nematodes *Xiphinema index* isolated from a colony grown in controlled conditions (at left). These nematodes are responsible for the transmission of grapevine fanleaf disease (GFLV). Feeding probe on a root (at right).

The viruses transmitted by nematodes are non-circulative and can be considered semi-persistent or persistent, because although latency is not observed between acquisition and inoculation, the retention period of the virus in the vector is quite long: some weeks or even months for adults, mostly in the absence of feeding (Gray, 1996). The virus can thus be inoculated successively to several different plants. The viral particles are adsorbed into specific sites on the cuticle of the anterior alimentary tract (Fig. 8.11). Large quantities of viral particles are observed by electron microscopy to be attached on the oesophagus. The virus is gradually released from the retention site during the absorption of nutrients; this dissociation occurs under

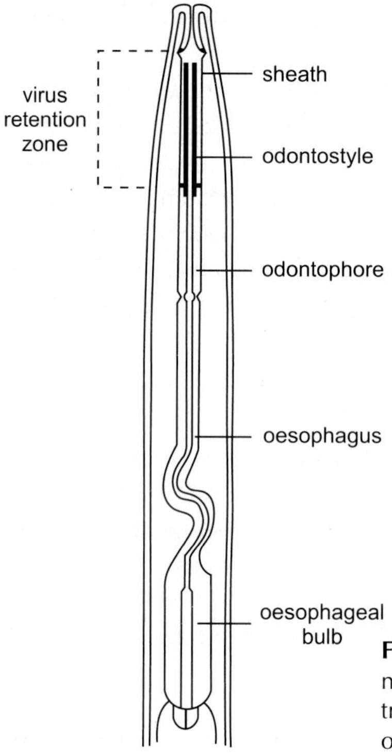

Figure 8.11 Mouth parts and digestive apparatus of a nematode of the genus *Longidorus*. The nepoviruses transmitted by these nematodes are adsorbed specifically on the odontostyle.

the action of oesophageal secretions inducing a change in pH (Taylor and Brown, 1997).

Fungi that are capable of transmitting viruses are all obligatory parasites; they belong to Chytridiomycota (*Olpidium brassicae* and *O. bornovanus*) and to plasmodiophorids (*Polymyxa graminis, P. betae,* and *Spongospora subterranea*). The systematic position of these genera within the fungi is presently disputed: Chytridiomycota remain in the fungi, while plasmodiophorids might be classified as protists rather than fungi. The vector activity is ensured by mobile uniflagellate (*Olpidium*) or biflagellate (*Polymyxa* and *Spongospora*) zoospores. They have conservation forms made up of resting spores (*Olpidium*) or cystosores (*Polymyxa* and *Spongospora*) (Campbell, 1996) (Fig. 8.12). About 20 plant viruses are transmitted by fungi. They belong to different genera: *Necrovirus, Carmovirus, Tombusvirus, Ophiovirus,* and *Varicosavirus,* the vectors of which are *O. brassicae* or *O. bornovanus*; *Benyvirus, Bymovirus, Furovirus, Pecluvirus,* and *Pomovirus,* which are transmitted by *P. graminis, P. betae,* or *S. subterranea.* The virus-vector relationships are specific and are not easily integrated into the groups defined for viruses transmitted by Insects (see box).

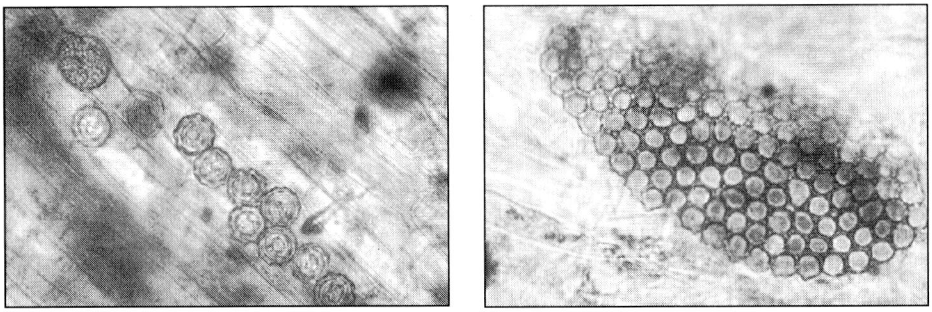

Figure 8.12 Resting spores of the fungus *Olpidium brassicae* (left), vector of *Tobacco necrosis virus* (TNV) and *Mirafiori lettuce big-vein virus* (MLBVV). Cystosore of the fungus *Polymyxa betae* (right), vector of *Beet necrotic yellow vein virus* (BNYVV).

Acquisition and inoculation of viruses by fungi, original modes of transmission

There are two types of virus–fungal vector relationships depending on the mode of acquisition of the virus and the presence or absence of virus in the resting spores (Campbell, 1996).

The first, unique among plant viruses, is illustrated by TNV. The zoospores of *Olpidium brassicae* probably acquire virions *in vitro*, i.e., outside the host plant. The zoospores released by the resting spores or by the zoosporangia in the soil and in aqueous media come into contact with virus particles coming from crop debris or roots of infected plants. This acquisition can also be made to occur experimentally by mixing a suspension of zoospores with an extract from an infected plant or with a purified viral preparation. The particles bind to the surface of the zoospore membrane or to the flagellum, and thus remain located on the outside of the spore. Inoculation of the virus occurs when the zoospore penetrates the root cortical cells.

For LBVaV (*Lettuce big-vein-associated virus, Varicosavirus*) and MLBVV (*Mirafiori lettuce big-vein virus, Ophiovirus*) transmitted by *O. brassicae* or BNYVV (*Beet necrotic yellow vein virus, Benyvirus*) transmitted by *Polymyxa betae,* the virus is acquired *in vivo,* i.e., inside the host, during the phase of intracellular parasitism of the fungus. The zoospores released by the cystosores or by the zoosporangia produced in the epidermal cells of roots of infected plants will be viruliferous. The virus can be conserved for several years in resting spores present in the soil or in plant debris (Fig. 8.12). It is interesting to note, as is the case for aphids, that a single species, *O. brassicae,* can transmit several viruses in very different ways.

Specific molecular interactions between viruses and vectors

Even the most stable viruses are not transmitted passively by vectors. This immediately suggests the existence of active and specific interactions between viral particles and vectors to achieve transmission. The existence of viral strains that have lost their capacity to be transmitted by vectors and the development of molecular biology have made it possible, over the past few years, to identify the viral proteins and sometimes the functional domains involved in virus transmission. On the other hand, little is still known on the vector receptors that are involved in interactions with viruses.

For most plant viruses, the capsid protein is the principal—even the only—structural protein. It is found in direct contact with the external environment. It is thus not surprising that it plays an important role in virus-vector recognition, but other viral proteins may also be involved. Some examples will show the different strategies developed by plant viruses, over the course of evolution, to ensure their transmission. It is interesting to observe that viruses that have very different genetic organization can use the same types of molecular interactions with their vectors.

■ The capsid: a key protein for transmission

The CMV capsid is the only determinant of transmission by aphids

Cucumber mosaic virus (CMV) is efficiently transmitted by numerous aphid species in the non-persistent mode. Several experimental approaches have made it possible to demonstrate that the capsid protein is the only protein implicated in recognition by the vector. Pirone and Megahed (1966) first showed that CMV could be transmitted by aphids after *in vitro* acquisition from a purified viral preparation through a parafilm® membrane, but not from an RNA preparation. Pseudo-recombination experiments between RNA of two strains of CMV, one transmitted by aphids and the other not, revealed that the determinant of transmissibility was carried by RNA 3, which codes for two proteins, including the capsid protein (Mossop and Francki, 1977). But experiments by Gera et al. (1979) and Chen and Francki (1990) of *in vitro* reconstitution of viral particles from RNA and capsid proteins of transmissible or non-transmissible cucumovirus strains definitively proved the role of the capsid protein in transmission of CMV by aphids (Fig. 8.13). During these experiments it was even possible to encapsidate the RNA of TMV (a virus that is not transmitted by aphids) in the capsid protein of a cucumovirus, and to observe the transmission by aphids of the particles thus reconstituted.

Comparison of sequences of CMV strains that are non-transmissible and transmissible by aphids, as well as reverse genetics experiments, revealed the importance of three amino acids located in the central part of the capsid protein in restoring transmissibility by the aphid *Aphis gossypii*. Surprisingly, two supplementary mutations are necessary to restore transmission by the aphid *Myzus persicae* (Perry et al., 1998). Four of these five amino acids are located in an internal position in the protein and are not exposed to the surface of virions; they are implicated in the stability of viral particles, which could be different in the mouthparts of the two aphid species (Perry et al., 1998).

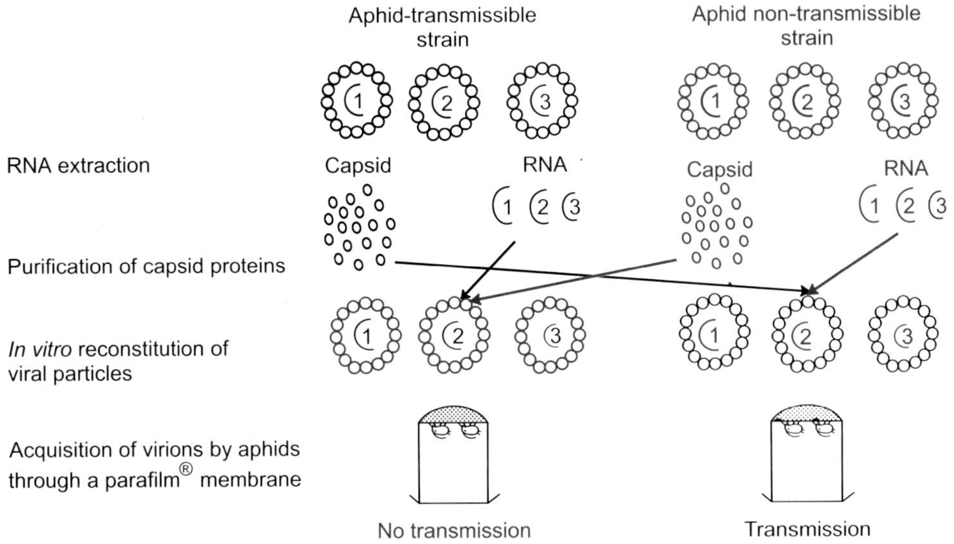

Figure 8.13 *In vitro* reconstitution experiments of cucumovirus particles, demonstrating the essential role of the capsid protein for aphid transmission (based on studies by Gera et al., 1979 and Chen and Francki, 1990).

Other viruses depend exclusively on the capsid for their transmission by fungi, nematodes or whiteflies

This is the case with the tombusvirus CuNV (*Cucumber necrosis virus*) transmitted by *O. bornovanus*. The role of the capsid in the adsorption of viral particles on the zoospores *in vitro* has been demonstrated through the creation of a chimera of TBSV, a virus not transmitted by *O. bornovanus*, in which the capsid protein was replaced by that of CuNV. The chimera virus is effectively transmitted by *O. bornovanus* (McLean et al., 1993). In the *O. bornovanus*-CuNV interaction, it seems that the virus-vector recognition signal is of the ligand-receptor type. Virus particle attachment to the zoospore would also involve a conformational change (resembling the "swollen state" that can be produced *in vitro*). The binding of CuNV particles to zoospores would be in this way similar to that described for poliovirus-animal cells interaction (Rochon et al., 2004).

Gene exchange experiments have shown that GFLV transmission by the nematode *Xiphinema index* depends only on the capsid protein (Andret-Link et al., 2004). For the crinivirus LIYV (*Lettuce infectious yellows virus*) transmitted by *B. tabaci* in the semi-persistent mode, the situation is slightly different. The viral particles are made up of a major capsid protein of 28 kDa, and one minor capsid protein of 52 kDa, which encapsidates one of the distal extremities of virions (Tian et al., 1999). LIYV can be acquired *in vitro* by *B. tabaci* from purified viral preparations through a parafilm® membrane; experiments of immuno-neutralization using antiserums against the two types of capsid showed that only the antiserum directed against the minor capsid inhibited the transmission (Tian et al., 1999). These results suggest that this protein intervenes in the interaction of LIYV with its vector.

The capsid can also be the sole determinant of transmission in certain circulative viruses

Geminiviridae are circulative viruses transmitted principally by two groups of vectors: whiteflies (*Begomovirus*) and leafhoppers (*Curtovirus* and *Mastrevirus*). The capacity of the whitefly *B. tabaci* to acquire *in vitro* the begomovirus SLCV (*Squash leaf curl virus*) from a purified virus preparation through a membrane indicates the importance of the capsid, the only structural protein of the virion, in the interaction with the vector (Cohen et al., 1983). This was confirmed by experiments of capsid protein exchanges between a curtovirus, BCTV (*Beet curly top virus*), and a begomovirus, ACMV (*African cassava mosaic virus*). A chimera virus made of the ACMV genome in which the capsid protein gene was replaced by that of BCTV was transmissible, after injection, by the leafhopper *Circulifer tenellus* (Briddon et al., 1990). Similarly, the replacement of the capsid protein gene of begomovirus AbMV (*Abutilon mosaic virus*), which is not transmissible by *B. tabaci*, by that of a transmissible virus restores the transmission (Hofer et al., 1997). A point mutation on the capsid protein of TYLCV leads to a loss of transmissibility by *B. tabaci* (Noris et al., 1998). A mutation on a neighbouring amino acid (Asp^{131}) of the capsid protein of another begomovirus has the same effect (Kheyr-Pour et al., 2000). This amino acid is present at this position in all the begomoviruses transmissible by *Bemisia*: it could belong to a domain particularly important for virus-vector interaction.

Geminiviridae, like *Luteoviridae*, must complete a cycle within the body of their vectors before being transmitted. In the case of *Begomovirus*, the passage from the intestine towards the hemocoel does not seem to be specific. It occurs in non-vector species or for non-transmissible strains of the virus (Hofer et al., 1997). On the other hand, the passage from the hemolymph towards the salivary glands is highly specific and occurs only when there is transmission (Briddon et al., 1990). In the case of curtoviruses and mastreviruses, the specificity is found at the intestine-hemocoel passage.

■ A helper component may cooperate with the capsid for transmission

Discovery of the helper component of potyviruses

The classic experiments of Govier and Kassanis (1974) and those of Govier et al. (1977) made it possible to understand the original mechanism of transmission of a potyvirus, *Potato virus Y* (PVY), by aphids. If aphids are used to acquire PVY through a parafilm® membrane from purified viral preparations, no transmission is observed. If an extract from infected plants (not containing infectious viral particles, such as a supernatant obtained after ultracentrifugation) is added to purified virus preparation, there is transmission (Fig. 8.14). On the other hand, a healthy plant extract has no effect. Govier and Kassanis (1974) deduced from this observation that, for PVY to be transmitted, a soluble substance present in infected plant but not in healthy plants is required. This component was later identified as a protein coded by the viral genome that was called "helper component" (HC) (Govier et al., 1977; Pirone and Blanc, 1996).

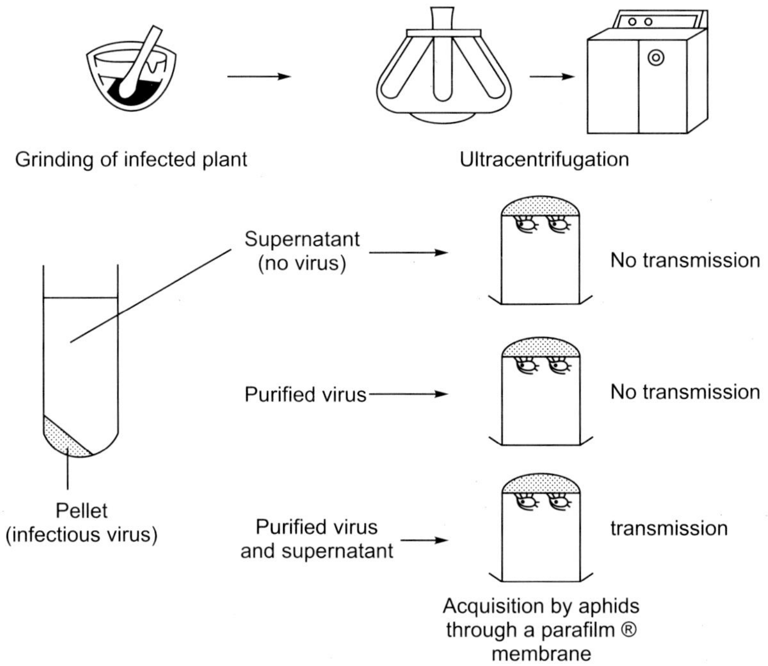

Figure 8.14 A helper component is required for aphid transmission of potyviruses (based on the studies of Govier et al., 1977).

Helper component of potyviruses: a multifunctional protein

The helper component protein is coded by a gene located in the 5′ part of the viral RNA. It has a molecular weight of 53 to 58 kDa depending on the potyvirus and its active form is a dimer. This dimerization could occur by means of a divalent metallic ion linking two molecules at a cysteine-rich region called the *zinc finger* (Thornbury et al., 1985; Robaglia et al., 1989). The helper component is a multifunctional protein: it also has a protease activity in its C-terminal part (hence its name HC-Pro) and is implicated in virus movement and replication, symptomatology, and the suppression of gene silencing established by the host (Maia et al., 1996; Revers et al., 1999). In the infected plant, it can accumulate in the form of amorphous inclusion bodies. The helper component is effective for transmission only when it is provided to the aphid before or at the same time as the virus.

The capsid protein and helper component have functional domains that are important for transmission

There are quite a large number of potyvirus strains that have lost their capacity to be transmitted by aphids. *In vitro* acquisition experiments have shown that this could be due to a loss of the activity of the helper component or to a modification of the properties of the virion itself. Comparison of sequences of capsid proteins and of the helper component of transmissible and non-transmissible strains has made it possible to identify domains important for the transmission process.

A triplet of conserved amino acids DAG (aspartic acid-alanine-glycine) is present in the N-terminal part of the capsid protein of most potyviruses transmissible by aphids, in a region that is highly variable (Harrison and Robinson, 1988). Numerous natural mutations (DTG, DAN...) on this triplet lead to the loss of transmissibility. Experiments of directed mutagenesis on the infectious cDNA of various potyviruses have confirmed the essential role of the triplet DAG for transmission by aphids, but also of the neighbouring amino acids context (Atreya et al., 1995).

Two conserved motifs implicated in transmission have been identified in the helper component: a KITC motif (lysine-isoleucine-threonine-cysteine) or KLSC motif (lysine-leucine-serine-cysteine) in the N-terminal part, and a PTK motif (proline-threonine-cysteine) in the C-terminal part (Granier et al., 1993; Atreya and Pirone, 1993; Huet et al., 1994). Two types of natural mutations leading to the loss of transmissibility are known (EITC or ELSC and PAK).

The hypothesis most often evoked to explain the role of the helper component of potyviruses is that of a "double-sided adhesive tape" mechanism, which would allow fixation of viral particles on the cuticle of the tip of aphid stylets. This mechanism must be reversible to ensure the ultimate release of virions at the moment of inoculation. Recent experiments of directed mutagenesis associated with studies of protein-protein interactions *in vitro* have elegantly confirmed the validity of this hypothesis. Indeed, the DAG domain of the N-terminal part of the capsid protein of potyviruses interacts with the helper component (Blanĉ et al., 1997) and more specifically with the PTK domain of this protein (Peng et al., 1998). The KITC (or KLSC) motif of the helper component could thus interact with the cuticle of the aphid (Wang et al., 1996) (Fig. 8.15).

Experiments of acquisition of radioactive labelled purified virus in the presence or absence of the helper component have shown that potyviruses, to be transmitted, must attach to the distal part of the stylet, confirming the hypothesis of transmission by ingestion-salivation (Martin et al., 1997). Moreover, an aphid can transmit a potyvirus after having acquired just about 50 viral particles (Pirone and Thornburry, 1988). This reveals the remarkable efficiency of the helper component strategy, and most of all the importance of transmission by vectors as a bottleneck reducing the variability of viral populations (Chapter 13).

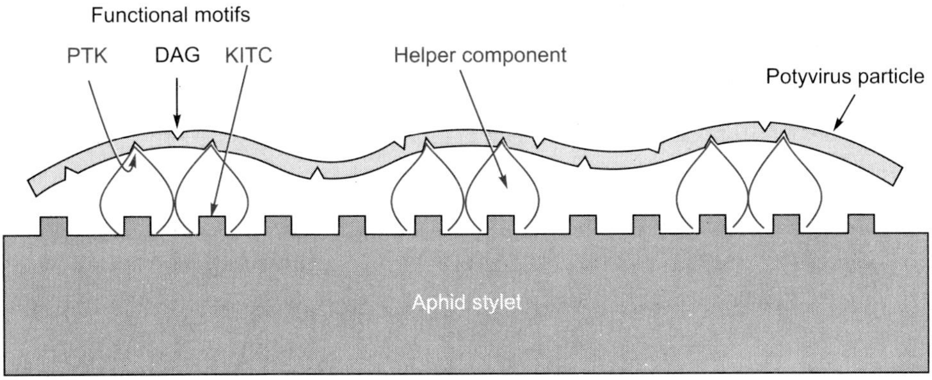

Figure 8.15 A tentative model for interaction between potyvirus, helper component, and aphid stylet (Huet, personal communication).

Other viruses use the helper component strategy

The caulimovirus CaMV (*Cauliflower mosaic virus*) also uses a helper factor for its transmission by aphids. It was long thought that the product of ORF II, the 18 kDa (P2), was the only non-structural protein needed for transmission (Pirone and Blanc, 1996). Nevertheless, one experimental result was disturbing: a purified preparation of the aphid transmission factor mediated the transmission of a strain that did not produce it, from extracts of infected plants but not from purified viral preparations. The explanation proposed was that the viral particles underwent a degradation during the purification steps, making them non-transmissible. Leh et al. (1999) showed that the cause of this phenomenon was entirely different. In fact, a second non-structural viral protein (P3, product of ORF III) is necessary for transmission of CaMV by aphids: this protein was present in infected plant extracts but not in purified viral preparations. Experiments of protein-protein interactions on membrane made it possible to confirm *in vitro* the affinity between P2 and P3. The situation of caulimoviruses seems thus to be quite original. They require two helper proteins to achieve their transmission by aphids. Deletion experiments on the two genes involved have shown that the N-terminal of P3 interacts with the C-terminal of P2, probably by the intermediary of α-helices (Leh et al., 1999). Interestingly, in infected cells, mainly P2-P3 or P3-virus complexes were found by immunogold labelling. Therefore, it is hypothesised that transmissible P2-P3-virus complexes are produced in aphids during acquisition on infected plants (Drucker et al., 2001).

Franz et al. (1999) indicated, for the first time, the probable existence of a helper factor required for transmission of the nanovirus FBNYV (*Faba bean necrotic yellows virus*), a virus transmitted by aphids according to the circulative mode.

Recently, the involvement of the HC-Pro in mite transmission was demonstrated for WSMV (*Wheat streak mosaic virus, Tritimovirus*), indicating that the helper component strategy is shared by viruses belonging to different *Potyviridae* genera and having different vectors (Stenger et al., 2005).

Readthrough protein: a second structural protein involved in transmission

The transmission of *Luteoviridae* by aphids is a complex phenomenon that has been particularly well studied in BYDV (*Barley yellow dwarf virus*, *Luteovirus*), PLRV (*Potato leafroll virus*, *Polerovirus*), BWYV (*Beet western yellows virus*, *Polerovirus*), CABYV (*Cucurbit aphid-borne yellows virus*, *Polerovirus*), and PEMV (*Pea enation mosaic virus*, *Enamovirus*). These circulative viruses undergo a cycle in their vector before being transmitted. The transmission mechanisms require successive passages through several biological barriers, which all present a certain degree of specificity.

Structural proteins of *Luteoviridae*

The *Luteoviridae* are made up of a major capsid protein (around 24 kDa) and a minor component, the minor capsid protein or readthrough protein (around 75 kDa). This protein is synthesized by means of a readthrough mechanism with an amber stop codon at the 3′ end of the capsid protein gene. This fusion protein has the capsid protein at its N-terminal end and the readthrough domain at the C-terminal end. It is found in small numbers in plants and in virions.

Mutants of PEMV having lost the readthrough protein are not transmitted by aphids (Demler et al., 1996). The essential role of this readthrough protein has been confirmed by mutagenesis in the case of BWYV (Brault et al., 1995) and BYDV (Chay et al., 1996). The relative efficacy of transmission of natural isolates of PLRV seems to be linked to mutations in the C-terminal part of the readthrough protein, but the capsid protein also contains domains important for virus-vector interaction (Jolly and Mayo, 1994).

Virions take a complex trajectory in the vector

Circulative viruses undergo a cycle in their vector and accumulate in the salivary glands. When an aphid reaches the phloem of an infected plant to feed, it ingests the virus, which will follow the alimentary tract up to the intestine. The virus is thus faced with a first barrier: the intestinal epithelium. Once this barrier is crossed (acquisition step), the virus reaches the hemocoel that fills the general cavity of the insect. The virus particles that do not cross this barrier continue their transit and are eliminated in the aphid honeydew. The viral particles that reach the hemocoel spread through the hemolymph up to the salivary glands, where they accumulate. They are then released into the salivary canal and injected with the saliva into a plant, where they eventually initiate a new infection (Fig. 8.16).

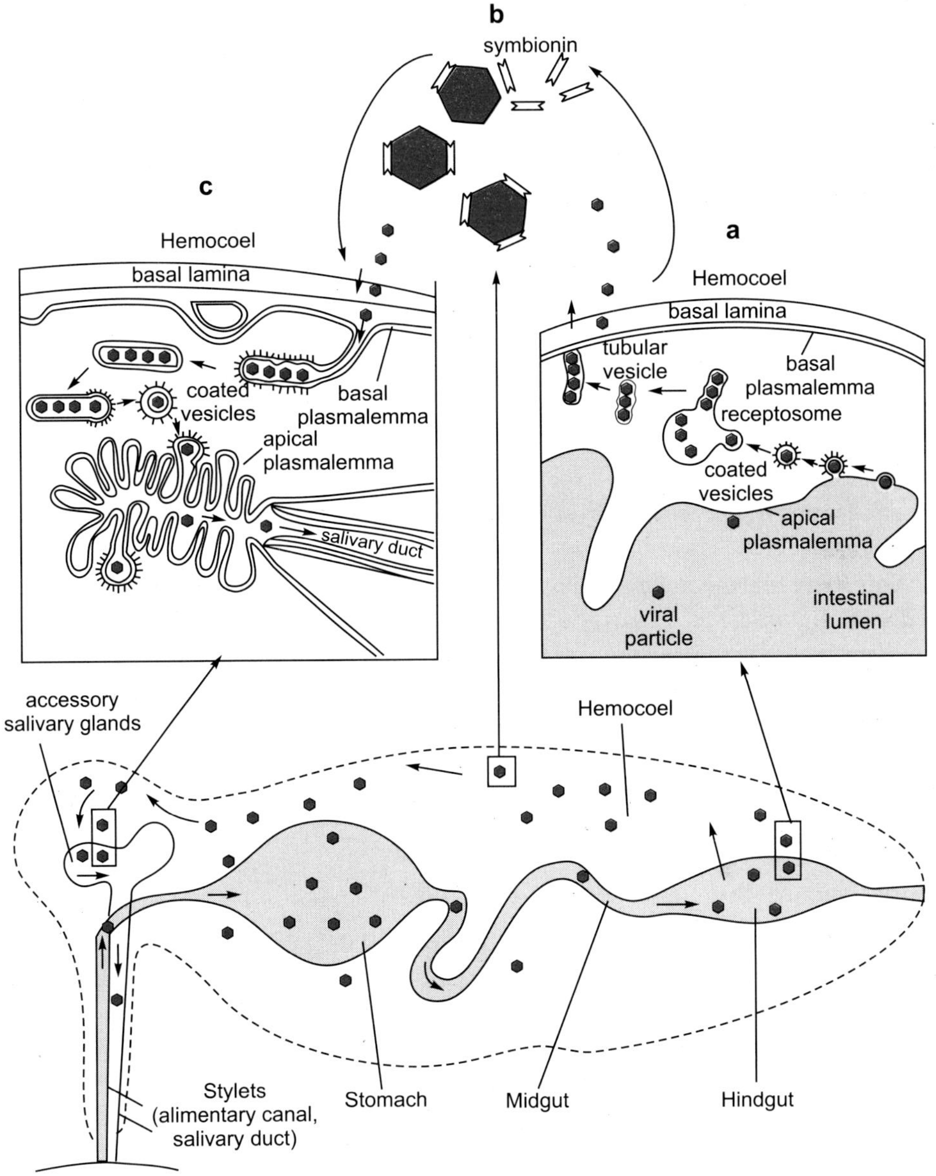

Figure 8.16 Transit of a luteovirus in the body of its aphid vector. When an aphid feeds on an infected plant, it ingests the virus, which follows the alimentary tract up to the intestine (midgut and hindgut). After having crossed the intestinal epithelium, the virus reaches the hemocoel or general cavity (a). There, it may interact with symbionin that is though to stabilize virions (b). The viral particles reach the accessory salivary glands, where they accumulate. Then, the virions are released in the salivary duct (c) and injected, with the saliva, into a plant, where they eventually initiate a new infection (Gildow, 1982, 1985; Gildow and Gray, 1993).

Crossing the intestinal wall

The virus can cross the intestinal wall to reach the hemocoel at various sites, depending on the virus. It requires a phase of recognition with receptors on the epithelium of the digestive tube (Fig. 8.16). In the case of BYDV (Gildow, 1985), the passage from the intestine to the hemocoel occur in the epithelium of the hindgut, while in PLRV this passage occurs in the midgut (Garret et al., 1993). Surprisingly, for CABYV, it was shown that the virus could cross the epithelial barrier at both sites: midgut and hindgut (Reinbold et al., 2003). The transport across the epithelial cells of the intestine is active: it is realized by an endocytosis-exocytosis mechanism. In the first place there is interaction between domains of the surface of the virus particles and receptors of the apical plasmalemma of epithelial cells, then formation of coated cytoplasmic vesicles. These coated vesicles ensure the transport of the virus into the cytoplasm. They fuse with larger cytoplasmic vesicles, the receptosomes. Tubular vesicles disengage from the receptosomes to allow the crossing of the basal plasmalemma to release the virus into the hemocoel (Fig. 8.16) (Gildow, 1985; Garret et al., 1993).

This first barrier can be specific: in *Metopolophium dirhodum*, CYDV-RPV (*Cereal yellow dwarf virus, Polerovirus*) does not cross the intestinal epithelium; the virus is not recognized by the receptors of the intestine and continues its path to be excreted in the honeydew, like many other viruses. Nevertheless, this barrier is not always selective, and certain isolates of BYDV are present in the hemocoel of aphid species that do not transmit them. The readthrough domain of the minor capsid protein has been recently involved in regulating the virus-vector specificity and the gut tropism for CABYV and BWYV (Brault et al., 2005)

Migration in the hemocoel of the aphid

Viral particles will diffuse in the hemolymph until they come into contact with the salivary glands. But this environment is not favourable to the virus, since it is there that various aphid defence mechanisms may operate. Van den Heuvel et al. (1994) have shown that the PLRV particles have a high affinity for symbionin, a protein synthesized in large quantity by endosymbiotic bacteria of the genus *Buchnera*, present in all aphids. It seems that this protein may even be indispensable for the PLRV transmission: if *M. persicae* are fed on a medium containing an antibiotic in order to eliminate the symbiotic bacteria, their vector capacity is greatly reduced (Van den Heuvel et al., 1994).

Symbionin is a "chaperone" protein homologous with the GroEL protein of *Escherichia coli*. In interacting with the virus, symbionin would have a protective effect by maintaining the integrity of the viral particles (Fig. 8.16). The symbionin-virus interaction was observed even in virus–non-vector aphid combinations, which indicates that they do not intervene in the specificity of vection. The behaviour of a series of BWYV deletion mutants was used to show that symbionin interacts with the

N-terminal part of the readthrough domain but not with the capsid protein (Van den Heuvel et al., 1997).

Recently, a similar interaction was observed between TYLCV and symbionin in *B. tabaci*; it could thus be a quite general phenomenon for circulative viruses (Morin et al., 1999).

Accumulation of viral particles in the salivary glands

Accessory salivary glands play an important selective role. This was demonstrated by Harris and Bath (1972) and Harris et al. (1975), who observed in them virus particles of a transmissible PEMV strain but not those of a non-transmissible strain. Combining electron microscopy with immunogold labelling, Gildow (1982, 1985) showed similarly that in *Sitobion avenae* the BYDV-MAV, transmitted by this aphid, is present in the hemocoel and in the accessory salivary glands, while CYDV-RPV, not transmissible by *S. avenae*, is present in the hemocoel and only in the basal lamina of the accessory salivary glands cells (Fig. 8.16).

To be transmitted, the *Luteoviridae* must therefore be first recognized by external receptors of the basal lamina. A second specific barrier is the basal plasmalemma, where coated vesicles form, containing the viral particles that will cross the cell by endocytosis. The virions are then excreted in the salivary duct by exocytosis. This system of active membrane transport keeps viruses from being in direct contact with the cell cytoplasm. Once the particle is in the saliva flow, it is injected into the phloem of the plant on which the aphid feeds (Fig. 8.16) (Gildow and Gray, 1993).

The crossing of cells of the accessory glands is thus an important step for the transmission specificity of *Luteoviridae*. Studies carried out on BYDV (Gildow, 1982, 1985; Gildow and Gray, 1993) show that the transport of viral particles can be regulated specifically and independently at three levels: attachment to the basal lamina (filter and concentration), transport through the basal lamina, and finally attachment to the basal plasmalemma with formations of cytoplasmic vesicles.

Readthrough proteins also intervene in certain transmissions by fungi

Other viruses such as *Benyvirus*, *Furovirus* and *Pomovirus*, which are transmitted by plasmodiophorids, have also adopted a strategy involving a readthrough protein of the capsid. Interestingly, the readthrough protein is located at one extremity of the viral particles. In BNYVV, for example, a readthrough protein of 75 kDa coded by RNA 2 is necessary for the transmission of viruses by *P. betae* as well as for the assembly of virions (Tamada et al., 1996; Rochon et al., 2004).

Propagative viruses

Insect viruses or plant viruses? Members of several virus families (*Reoviridae*, *Rhabdoviridae*, *Bunyaviridae*...) can multiply in their plant hosts as well as in their vectors (aphids, leafhoppers, thrips). These families also include members that do not infect plants but only animals (invertebrates or vertebrates). The question arises whether these viruses could be insect viruses that have evolved to acquire the capacity to multiply in plants as well (Gray and Banerjee, 1999; Hebrard et al., 1999). Indeed, several viruses of aphids or leafhoppers are known to be transmitted horizontally from insect to insect through the intermediary of plants on which the insects feed, although these viruses cannot multiply in plants (Gildow, 1999). In the case of propagative viruses, it thus seems difficult to decide whether it is the plant that transmits the virus to the insect or the other way around (Hebrard et al., 1999).

It is evident that all the essential viral functions allowing infection (replication, movement) also condition the transmission of propagative viruses, to a certain extent. Nevertheless, there are also proteins that play a particular role in the recognition between the virus and its vector. In the case of RDV, which has been particularly well studied, the minor capsid protein P2, implicated in virus adsorption into cells of the vector insect *Nephotettix cincticeps*, is necessary for transmission by leafhoppers, but not for multiplication in the host plant (Omura and Yan, 1999). The glycoprotein spikes protruding from the virion surface of PYDV (*Potato yellow dwarf virus*, *Nucleorhabdovirus*) appear to be involved in the recognition of insect cell surface receptors, because blocking the G protein by antibodies or its enzymatic removal reduces infectivity for the leafhopper vector cells (Gaedigk et al., 1986).

In the vector, the virus is sometimes observed to be transmitted to the progeny. In RSV transmitted by the leafhopper *L. striatellus*, transmission of the virus has been observed over 40 successive generations (Shinkai, 1962). Some TYLCV strains present very particular properties that have not yet been demonstrated in other *Geminiviridae*: they can multiply in the vector, be transmitted to the progeny, and be sexually transmitted. In this case, it seems highly likely that proteins other than the capsid intervene in the complex interactions of TYLCV with its vector *B. tabaci* (Czosnek and Ghanim, 1999).

Virus interactions for transmission

Mixed infections of a single host by viruses of the same genus or different genera are very frequent in nature. Viral replication sometimes occurs in the same cells, which can lead to various molecular interactions that may affect relationships with vectors. If the viruses are transmitted by the same vectors with the same efficiency, these interactions will undoubtedly have no major consequences for vection. On the other hand, if the viruses have different vector specificities, or if one of the viruses is not transmissible, interactions between these viruses could considerably modify the virus-vector relationships. There are two potential types of interactions between viruses related to transmission: hetero-encapsidation and hetero-assistance.

Multiplication of certain plant viruses in their vectors

Various observations have shown that certain plant viruses do multiply in their insect vectors: repeated transovarian transmission of a virus to the progeny or the presence of aggregates of viral particles in certain insect cells. The multiplication of nucleorhabdovirus SYVV (*Sowthistle yellow vein virus*) in its aphid vector was demonstrated by series of successive injections of extract from a viruliferous aphid into healthy aphids. The concentration of the initial injection was estimated at 105 viral particles, and each transmission represented a dilution of 1/70. The aphids were always infectious after the fifth series of injections, and it became evident that the virus multiplied inside the vector (Sylvester and Richardson, 1969). A similar approach was used to demonstrate PYDV multiplication in its leafhopper vector.

Multiplication inside the vector is quite frequent in the case of leafhopper-borne viruses (*Rhabdoviridae, Tenuivirus, Phytoreovirus*). The vector of marafivirus MRFV (*Maize rayado fino virus*) is the leafhopper *Dalbulus maides*. MRFV induces severe symptoms in maize that could lead eventually to the death of infected plants. In the vector, multiplication of the virus begins 10 d after acquisition and reaches a maximum after 20 d, but viral infection of numerous organs of *D. maides* affects neither the fertility nor the longevity of the leafhoppers (Gamez and Leon, 1986).

It has long been investigated whether TSWV replicates in its thrips vector *F. occidentalis*. The proof was brought about by the observation of a non-structural protein (NSs), produced only in the course of infection, in the salivary glands of thrips larvae and adults (Ullman et al., 1993). It was also shown that TSWV multiplies preferentially in two types of organs: the digestive tube and the salivary glands of thrips vectors (Nagata et al., 1999). The acquisition of TSWV by thrips vectors is restricted to a short well-defined period—first and early second larval stages—although transmission will occur during the lifetime of the thrips. At these early developmental stages, there is a temporary association between midgut, visceral muscles, and salivary glands, and TSWV can thus readily migrate from the intestine lumen to the salivary glands. At later developmental and at the adult stages, healthy thrips cannot acquire and transmit TSWV even though they feed on infected plants. Indeed, the thrips undergo major morphological changes and the loss of tight contact between midgut and salivary glands lead to a strong flow of virus particles, after crossing the midgut epithelium, towards hemocoel and the malpighian tubules that prevents the virus from reaching the salivary glands (Moritz et al., 2004).

Hetero-encapsidation is an exchange of structural proteins

Hetero-encapsidation involves viruses in which the efficiency or specificity of transmission depends on structural proteins of the virion. The first biological demonstration of hetero-encapsidation was provided by the classic experiments of Rochow (1970) on BYDV. These experiments showed that a BYDV strain normally transmitted by *Macrosiphum avenae* could also be transmitted by *Rhopalosiphum padi*, if it was in mixed infection with a strain specifically transmitted by this aphid species. The hypothesis advanced to explain this phenomenon was that the capsid protein of a strain could totally encapsidate (trans-encapsidation) or partly encapsidate (phenotypic mixing) the RNA of another strain. In a more detailed study using four BYDV strains, monoclonal antibodies, and specific probes, Wen and Lister (1991) demonstrated that there was trans-encapsidation when the strains are genetically distant, while there is phenotypic mixing (particles contain capsid proteins of both strains) when the strains are closely related. Even though it is not yet demonstrated, it is likely that in *Luteoviridae* hetero-encapsidation concerns the major capsid proteins as well as the minor capsid (readthrough) proteins.

Other virus genera, particularly *Potyvirus*, can present hetero-encapsidations (Fig. 8.17). In this case, the phenotypic mixing is very easily demonstrated by immuno-electron microscopy experiments. Hetero-encapsidation can thus allow transmission by aphids of a strain in which the capsid protein is deficient for transmission (Bourdin and Lecoq, 1991). The same phenomenon was observed in transgenic plants expressing the functional capsid protein of a potyvirus (Chapter 12).

Hetero-encapsidation can intervene between viruses belonging to different genera or even families. Poleroviruses assist aphid transmission of umbraviruses, viruses that do not code for a capsid protein. For example, LSMV (*Lettuce speckles mottle virus, Umbravirus*) is transmitted by *Myzus persicae* only from plants that are also infected by BWYV (Falk and Duffus, 1981). Similarly, numerous satellite RNAs are encapsidated by the capsid protein of their helper viruses, and thus transmitted by hetero-encapsidation. It can also be observed that there are stable associations between different viruses or between virus and satellites that survive by means of hetero-encapsidation, constituting new pathogenic entities (Falk and Duffus, 1981). PEMV represents an ultimate stage of polerovirus-umbravirus coevolution: the RNA 1 corresponds to genomic RNA of a polerovirus that has lost its movement protein function, a movement function that is complemented by RNA 2, which is similar to the genomic RNA of an umbravirus that has no coat protein gene (Demler et al., 1996).

Hetero-assistance is a functional aid between two viruses

Hetero-assistance is observed only in viruses that have adopted the "helper component" strategy. The first example of this phenomenon was presented by Kassanis and Govier's experiments (1971). The C strain of PVY (PVY-C) is not transmissible by aphids. On the other hand, aphids that have made probes on a plant infected by a transmissible strain of PVY, once transferred to a plant infected by PVY-C, acquire and effectively transmit PVY-C. The aphids in this case acquired from a PVY-infected plant a functional helper component that then mediated transmission of PVY-C. The phenomenon of hetero-assistance thus does not necessarily imply that the two viruses concerned multiply in the same cell, or even in the same plant. It is also based on the fact that certain helper components do not present a high specificity.

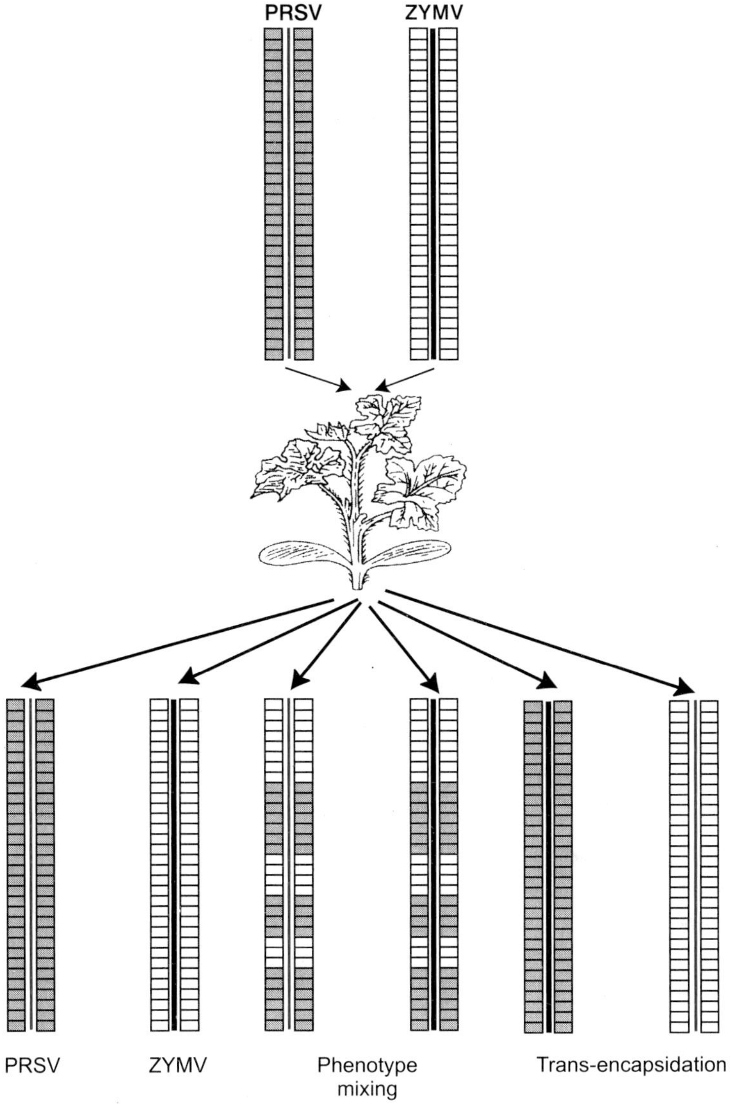

Figure 8.17 Infection of a melon plant by two potyviruses (ZYMV and PRSV) leads to the production of viral particles identical to the parents, or containing the RNA of one strain and the capsid proteins of both strains (phenotype mixing or chimera particles), or the RNA of one strain and the capsid protein of the other strain (trans-encapsidation).

Experiments of mixed infections by two strains of ZYMV that are not transmissible by aphids, one deficient for the capsid protein and thus transmissible only by hetero-encapsidation and the other deficient for the helper component and thus transmissible by hetero-assistance, revealed that the mechanism of hetero-assistance was more effective than that of hetero-encapsidation (Desbiez et al., 1999).

Hetero-assistance can also involve viruses belonging to two different genera: the helper factor of *Waikavirus* allows transmission of different viruses. AYV (*Anthriscus yellows virus, Waikavirus*) helper factor mediates transmission by the aphid *Cavariella aegopodi* of the sequivirus PYFV (*Parsnip yellow fleck virus*) and the helper factor of RTSV (*Rice tungro spherical virus, Waikavirus*) allows transmission of RTBV (*Rice tungro bacilliform virus*) by the leafhopper *Nephotettix virescens* (Pirone and Blanc, 1996).

Hetero-encapsidation or hetero-assistance can have important epidemiological consequences that have been debated (Falk and Duffus, 1981; Pirone and Blanc, 1996; Hammond et al., 1999). In particular, they allow transmission and thus the survival of non-transmissible virus isolates or virus species (*Umbravirus, Sequivirus*) as well as the transmission of transmissible viruses in the absence of their specific vectors (*Luteoviridae*). These phenomena also permit maintenance of some molecular variability within viral quasi-species, in not eliminating too quickly genotypes that are poorly or not transmissible (Pirone and Blanc, 1996).

The epidemiology of viral diseases

Epidemiology describes the movement of a viral disease within a population of healthy host plants. Epidemic processes must be analysed carefully in order to design or choose appropriate control methods against viruses. Several factors can determine the dissemination of a virus in a given space:

— specific properties of the virus (stability of viral particles, host range, virus concentration in the host, symptom severity ...);
— availability of external virus sources;
— abundance and activity of vectors;
— type of virus-vector relationships;
— susceptibility of the cultivated crop.

These parameters, intrinsic to the plant-virus-vector system, can also vary as a function of environmental conditions (climatology, nature of soils, topography, cultural practices) (Thresh, 1978; Harrison, 1981).

■ Some important factors

The abundance of virus sources within a crop or at its immediate proximity will most often directly determine the earliness of epidemics. The sources can be found within the crop itself in the case of viruses transmitted by vegetative propagules or through the seed. The infected plants are thus generally distributed at random in the field. Sources from outside the crop can be highly varied (see Figs. 10.4 and 10.7), such as weeds, volunteers, alternative hosts, or nearby infected crops.

The activity of vectors is the second important factor that regulates virus dissemination. This is related to the reproductive activity that determines the number of vectors, to the vector activity itself (i.e., local and long distance movements from plant to plant), and the transmission efficiency. The biology of the vector, its potential for dissemination, and the virus-vector and vector–host plant relationships are other factors that can influence the development of epidemics. The evaluation of vector populations is an important parameter for epidemiology studies. In the case of soil-borne vectors (nematodes), vector density can be estimated directly by counting individuals in a soil sample (Cotton, 1979). In the case of air-borne vectors, particularly aphids, different types of traps are used (Moericke traps or yellow traps, suction traps, fishing-line sticky traps). The counts of aphids thus collected must then be correlated with the number of aphids effectively visiting the plants. The total number of aphids must then be weighted by the vector capacity for each species or biotype present in the sample. Indeed, it has been shown that transmission capacity can vary considerably from one species to another (Labonne et al., 1982; Raccah, 1986).

Significant *differences in susceptibility* may exist within a single plant species. As the plant develops, its response to viral infection may evolve. In squash, for example, it is observed that adult plants are less easily infected by CMV than young plantlets. In addition, generally, the earlier the infection, the greater the incidence on host development and thus the impact on the final yield.

The development of epidemics is also closely linked to climatic conditions. A harsh winter may destroy a number of reservoir plants or virus vectors and reduce the inoculum available at the beginning of the cropping season. For example, a simple relationship can be established between the intensity of epidemics of polerovirus BMYV (*Beet mild yellowing virus*) and the frequency of frost days during the previous winter (Dewar and Smith, 1999). Dry periods will be unfavourable to the dissemination of viruses transmitted by fungi. The immediate environment of the crop, for example, the presence of wind-breaks, can have a significant influence on the spatial dissemination of aphid-borne viruses (Quiot et al., 1979).

The different stages of the development of a plant virus disease epidemic

Evaluation of the number of infected plants in the field

The temporal evolution of an epidemic in a field can be represented by the cumulative percentage of infected plants as a function of time. The number of diseased plants can be evaluated on the whole crop or on a sample of individuals. Regular visual surveys for symptomatic plants is a simple method of data collection, but it often remains imprecise because of the frequent presence of mixed virus infections and the convergence of symptoms caused by different viruses. A more accurate approach is to collect samples that will be analysed in the laboratory, eventually for the presence of several different viruses, using suitable diagnostic methods (such as ELISA, Chapter 9). The precision of these data will depend on the number of plants observed or tested and the frequency of observations. Moreover, there is an incubation period between the time a plant is inoculated by a virus and the time the virus can be detected visually or serologically. It is important to take this delay into account when correlations are to be established with vector populations (Thresh, 1983; Labonne and Quiot, 1988).

There are two steps in the development of epidemics of a viral disease

Primary contaminations

Primary contaminations can occur within the crop, for example, when viruses are transmitted by seeds or when the soil harbours the infectious vector, e.g., nematodes (Plate II.1) or fungi (Plate V.5). They can also come from outside the crop (weeds, nearby crops) through the intervention of vectors (Plate IV.2). If only external primary contaminations contribute to the development of the epidemic, and if they occur regularly, the representative curve of the epidemic is of the "simple interest" (Van der Planck, 1963) or "monocyclic" type (Zadoks and Schein, 1979). It corresponds to an equation of the following type:

$$dy/dt = r\,(1 - y) \quad \text{or} \quad y = 1 - (1 - y_0)e^{-rt}$$

where y is the proportion of infected plants, t is the time, r is a parameter measuring the epidemic progression rate, and y_0 is the initial contamination rate.

Secondary contaminations

When contaminated plants in the crop become sources of virus, they lead to secondary contaminations. Most often, these sources quickly become much more effective than the external sources. The increase in the number of plants infected becomes exponential. The representative curve of the epidemic is of the "compound interest" (Van der Planck, 1963) or "multicyclic" type (Zadoks and Schein, 1979). It has a sigmoid shape and responds to an equation of the logistic form:

$$dy/dt = ry\,(1 - y) \quad \text{or} \quad \log(y/1 - y) = rt + \log(y_0/1 - y_0)$$

where y is the proportion of infected plants, t is the time, r is a parameter measuring the epidemic progression rate, and y_0 is the initial contamination rate.

Other transformations can be used to describe plant virus epidemics (Madden and Campbell, 1986; Nutter, 1997) and to better adjust to the experimental data.

The study of the spatial development of epidemics is also important. It can prove the presence of clusters or foci of diseased plants or reveal contamination gradients from sources within or outside the field. In this case, all the plants must be regularly observed and appropriate spatial analysis should be conducted.

Virus spread

The different steps of virus spread in a field are illustrated by three examples, which are developed in the boxes that follow.

Modelling and prediction of epidemics

The analysis of epidemic development curves or of the evolution of the spatial distribution of diseased plants in a field contributes to the overall description of the pathosystem affecting a crop. These observations do not allow a precise analysis of the incidence of different factors acting on epidemics. This is why models have been developed taking into account each of numerous parameters intervening in the development of epidemics, including virus sources, vector efficiency, virus-vector relationships, and plant susceptibility (Ruesink and Irwin, 1986; Ferriss and Berger, 1993; Holt and Chancellor, 1997). Each relationship between the different elements of the system must thus be formulated and validated by experimental data.

The advantage of such models is two-fold: simulation and prediction. The simulation of a given situation makes it possible, for example, to test beforehand the advantage of a cultural practice, a partial resistance, or the use of healthy seeds to slow down viral epidemics. This analysis can help reduce the number of field experiments, which could be limited to validation experiments. Prediction should help the farmer, with a thorough understanding, to apply adapted control methods or to choose the optimal crop or cultivar type on the basis of risks. There are at present operational systems of risk prediction at the regional level (Nelson et al., 1994; Dewar and Smith, 1999; Fabre et al., 2004), but unfortunately there are still no tools by which producers can estimate the epidemic risks that will be encountered in each of their parcels.

A virus transmitted by aphids in the non-persistent mode: *Cucumber mosaic virus* (CMV) in melon (Fig. 8.18)

Stages of the epidemic

In melon, *Cucumber mosaic virus* (CMV) is not transmitted by seeds. If the plants are produced in a nursery protected from aphids, all the plants at the planting stage are healthy. But sometimes there are large numbers of weeds around the fields. Some of them (such as chickweed) can transmit CMV through seed, while others survive through the winter (such as shepherd's purse, common groundsel or woody nightshade), and may be virus reservoirs (CMV infected weeds are represented in red in the figure) as well as overwintering hosts for the aphids (Quiot et al., 1983).

The first contaminations occur in spring, shortly after planting, during the first aphid flights. Aphids come from outside the field. Some do not carry CMV (1) but may acquire it by visiting infected weeds on which they make probes. If the weed is a non-preferred host for the aphid, it will fly away and then contaminate a young melon plant. Other aphids (2) may have acquired CMV on nearby winter crops (such as lettuce or spinach) and will directly infect melons. The first diseased plants in the field are the primary infection foci.

The aphid flights, particularly of species that do not colonize melon ("visitor" aphids), will then allow very rapid dissemination of CMV throughout the crop. This corresponds to the secondary development of the epidemic. A large number of weeds will also be infected: some of them will become new reservoirs that will allow the virus to survive during the next winter and contaminate subsequent crops.

Development of the epidemic over time and space

The efficiency of aphid transmission according to the non-persistent mode, the abundance of reservoir plants, and large aphid populations (especially in spring) contribute to a rapid development of epidemics. Epidemic development curves have an overall S shape. The diseased plants are often distributed in large patches, sometimes on the border of the fields. These patches rapidly extend and join each other, which leads to the contamination of the entire crop a few weeks later.

Which control methods should be used?

Limit the sources of viruses and/or vectors: carefully weed the field borders, avoid planting a new crop next to an old one, where infected plants are often numerous.

Prevent and/or slow down contaminations: carefully protect the nurseries or young crops (if possible under nets or non-woven covers), use plastic mulches that efficiently repel aphids. Insecticide treatments generally do not act quickly enough to prevent dissemination of CMV by aphids.

Use virus-resistant varieties; the use of aphid-resistant varieties could slightly slow down the development of epidemics.

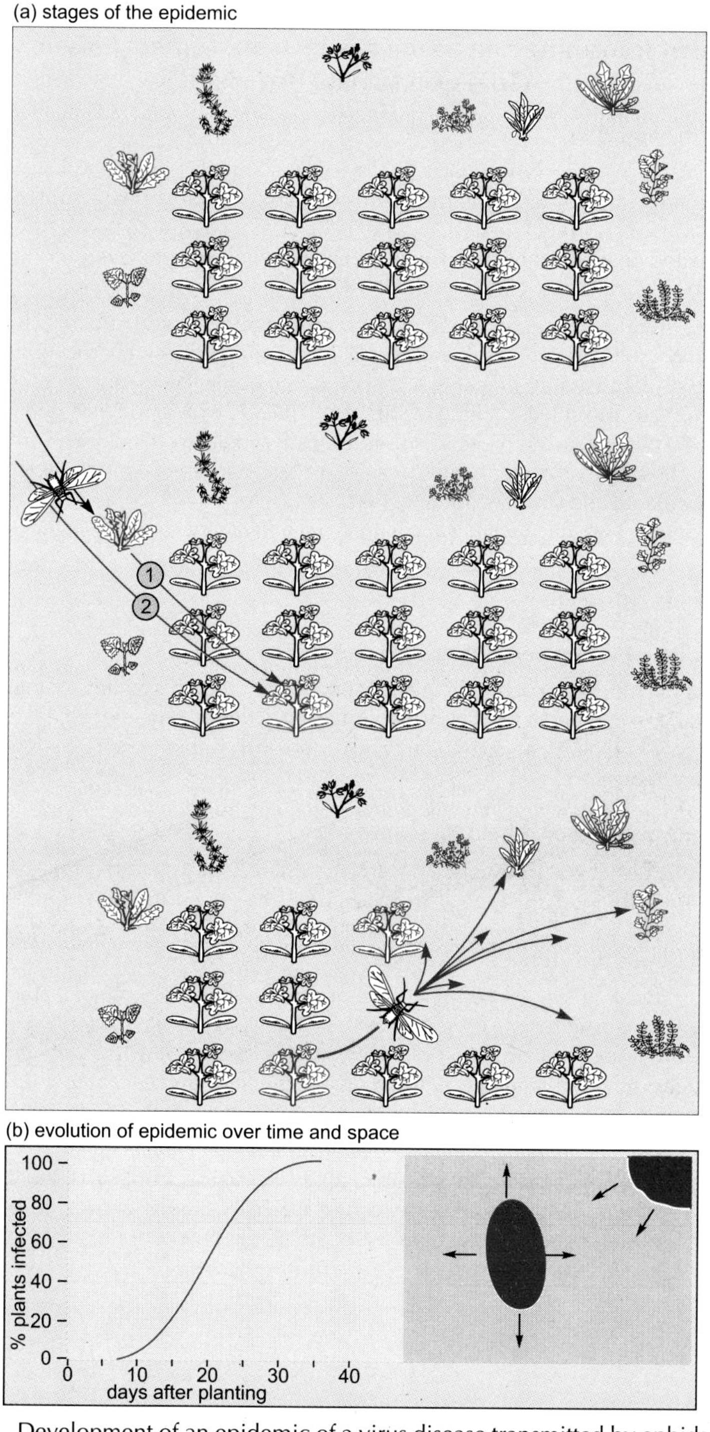

Figure 8.18 Development of an epidemic of a virus disease transmitted by aphids in the non-persistent mode: *Cucumber mosaic virus* (CMV) in melon.

A virus transmitted mechanically by man: *Tomato mosaic virus* (ToMV) in tomato (Fig. 8.19)

Stages of the epidemic

Tomato mosaic virus (ToMV) (Plates I.3 and III.1) is highly stable and can survive for several months in plant debris in the soil or contaminate greenhouse structures. It can also be transmitted through seeds in tomato. If the farmer has used a poor quality seed lot, some infected plants (represented in red) randomly distributed through the field or greenhouse can be observed (Broadbent, 1976).

During the frequent farming operations (tying, pinching of tips, staking, harvesting), workers may contaminate their hands, tools, and even clothes with ToMV by touching the diseased plant (simply breaking a few of the glandular hairs covering the stems or leaves of tomato can release a little of the plant "juice", which contains numerous viral particles). ToMV, unlike many other viruses, is not immediately denatured. Farm workers, by just handling a nearby plant, will cause again micro-injuries through which the virus will be inoculated and the infection will start.

ToMV thus disseminates rapidly from one row to the next. There is often a particular distribution of infected plants along the row, which follows the path of the farmers as they tend to the tomato plants.

Development of the epidemic over time and space

Because of the frequency of handling and traffic in the crop, the stability of the virus, and the efficiency of the mechanical transmission, a rapid development of epidemics is often observed. After being propagated along a row, the virus is then generalized to the entire field or greenhouse.

Which control methods should be used?

Limit the sources of the virus: use virus-free certified seed lots; avoid repeating tomato crops on the same parcel, especially when there has been a ToMV epidemic in the preceding crops and if there is root debris in the soil; before planting, carefully disinfect the greenhouse structures, stakes, and tools.

Prevent and/or slow the contamination: limit the handling of contaminated plants or handle them only after having finished with the healthy plants (for example, pass through a tunnel containing contaminated plants only after having finished operations in the healthy tunnels); avoid smoking in the crops (cigarette tobacco can contain infectious TMV particles); frequently disinfect hands and tools (e.g., with 3% or 10% trisodium phosphate respectively).

Use resistant varieties as far as possible. One biological control method using an attenuated strain of TMV and cross-protection was successfully applied in greenhouse crops, but it is no longer practical because of the development of resistant varieties.

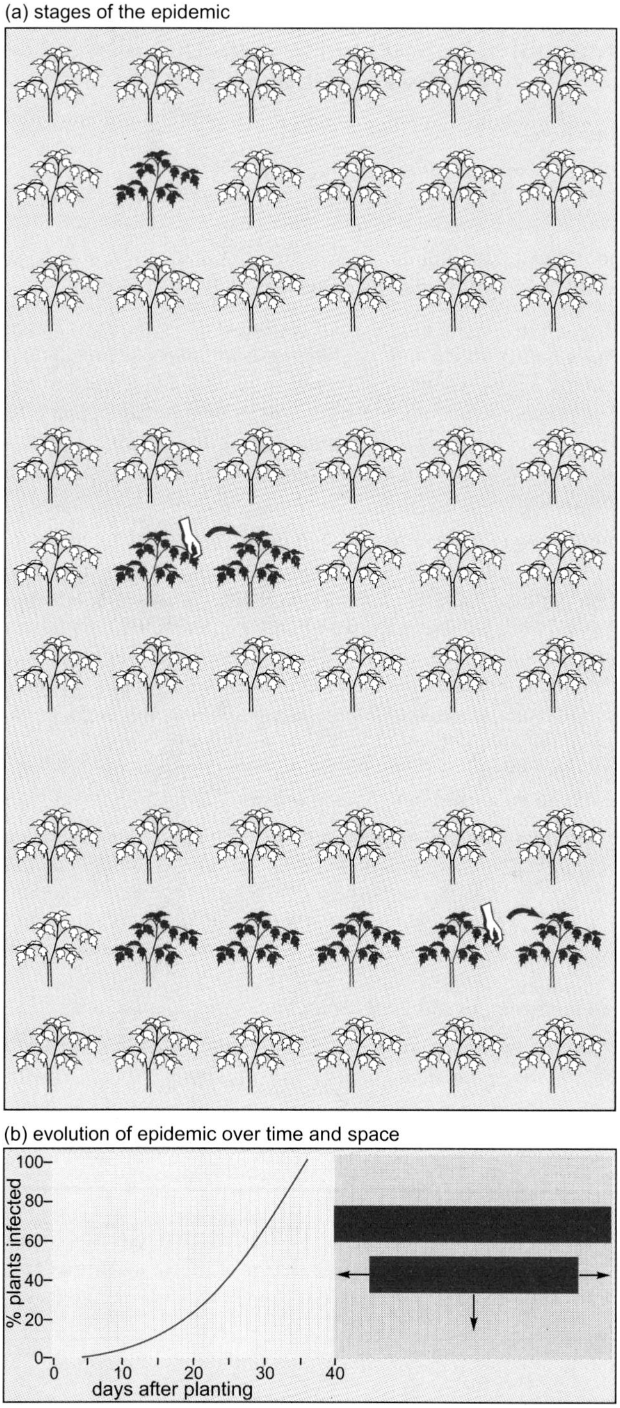

Figure 8.19 Development of an epidemic of a virus disease transmitted mechanically by man: *Tomato mosaic virus* (ToMV) in tomato.

A virus transmitted by a soil fungus: *Mirafiori lettuce big-vein virus* (MLBVV) in lettuce (Fig. 8.20)

Stages of the epidemic

Mirafiori lettuce big-vein virus (MLBVV) is not seed-borne. If the plantlets were produced in sterile soil, they are all healthy at the time of planting. On the other hand, the field or greenhouse soil may contain viruliferous resting spores of the fungus vector *Olpidium brassicae*. These spores are highly resistant, can survive several years in the soil, and represent a very efficient form of virus conservation (Falk, 1997).

After planting, probably under the effect of substances produced by the roots, the resting spores germinate and produce one or perhaps several zoospores. The zoospore has a flagellum that allows it to "swim" and move in the thin film of water that surrounds the colloidal soil particles. It reaches the root system of a lettuce plant and penetrates an epidermal cell of a young root. The zoospore will thus transmit MLBVV to the plant (represented in red) and then evolve into a zoosporangium, a sort of sac filled with a large number of zoospores.

If there is enough water in the soil, the zoosporangium will release a large number of zoospores a few days later. These can then contaminate other roots of the same plant (1) and thus increase the inoculum potential. They can also penetrate the roots of a neighbouring plant and transmit the disease. If the conditions remain favourable, a new zoosporangium will form (2), which will allow the big-vein disease gradually to spread. If, on the other hand, conditions become unfavourable (drought, senescent plants), the zoospore evolves into a resting spore that will be released (3) in the soil after decomposition of the root debris.

Development of the epidemic in time and space

The reduced mobility of zoospores makes the disease spread slowly into small patches in the fields, especially if there were originally few resting spores in the soil (A). On the other hand, if the density of resting spores is high, the disease can quickly affect the entire crop (B). This is the type of rapid development that is most often observed in hydroponic cultivation in greenhouses.

Which control methods should be used?

Limit virus sources: avoid parcels that have been planted with lettuce for several successive crops (where the density of infectious resting spores can be high), follow regular rotations, disinfect soils.

Prevent and/or slow down epidemics: choose well-drained fields (the fungal zoospores need water to move) and in the case of hydroponic cultivation use specific authorized fungicides or active ingredients (surfactants).

Use resistant varieties. Big-vein disease develops its characteristic symptoms in winter and early spring. Just because a variety has no symptoms in a summer crop, that does not mean that it is resistant to the disease or that the vector did not multiply.

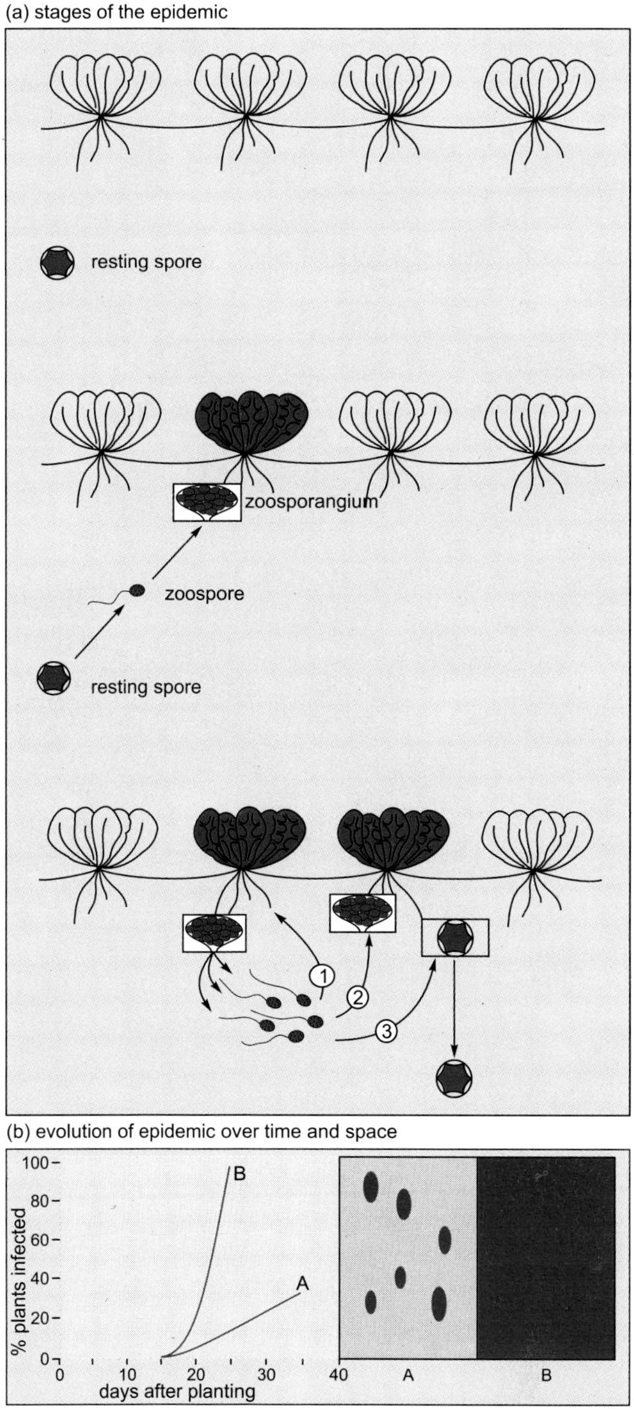

Figure 8.20 Development of an epidemic of a virus disease transmitted by a fungus: *Mirafiori lettuce big-vein virus* (MLBVV) in lettuce.

9

CHAPTER

Diagnostic Methods

For most studies in plant virology, it is essential to be able to establish a proper diagnosis, that is, to know how to recognize a disease and identify precisely the causal virus or viruses. To characterize the aetiology of a new viral disease, it is necessary to identify the virus, reveal its presence in an infected plant, and verify that it is actually the agent responsible for the symptoms observed with respect to the rules of Koch's postulate. Highly efficient, quick, and economical diagnostic techniques are also needed to study the epidemiology of a viral disease, to evaluate the efficiency of control methods, or to guarantee that plants or seeds are virus-free.

Diagnostics involves two complementary components, *identification* and *detection*, corresponding to two types of questions that can be illustrated by the following examples. A farmer who observes diseased squash plants in a field will ask, "What is the virus infecting these plants?" The techniques to be used must be broad enough to lead to the *identification* and eventually the characterization of the virus or viruses responsible for the disease, among the thirty or so viruses known to infect squash. A seed supplier knows that LMV (*Lettuce mosaic virus, Potyvirus*) is seed-borne in lettuce. He wonders about the sanitary status of a seed lot that he wants to import: "Is this seed lot contaminated by LMV? To what extent?" This is a matter of *detection* to track down a single virus, LMV, giving eventually a quantitative response.

The approach that will lead to a diagnosis depends thus on the problem posed and on a judicious choice of visual and/or laboratory tests. There is no key for virus

Viruses cited

BCMNV (*Bean common mosaic necrosis virus, Potyvirus*); BNYVV (*Beet necrotic yellow vein virus, Benyvirus*); BYDV (*Barley yellow dwarf virus, Luteovirus*); BWYV (*Beet western yellows virus, Polerovirus*); CMV (*Cucumber mosaic virus, Cucumovirus*); CYSDV (*Cucurbit yellow stunting disorder virus, Crinivirus*); DMV (*Dahlia mosaic virus, Caulimovirus*); EMDV (*Eggplant mottled dwarf virus, Nucleorhabdovirus*); JGMV (*Johnsongrass mosaic virus, Potyvirus*); LMV (*Lettuce mosaic virus, Potyvirus*); LSLV (*Lily symptomless virus, Carlavirus*); MLBVV (*Mirafiori lettuce big-vein virus, Ophiovirus*); MRMV (*Melon rugose mosaic virus, Tymovirus*); PEMV (*Pea enation mosaic virus, Enamovirus*); PLCV (*Pelargonium leaf curl virus, Tombusvirus*); PLRV (*Potato leafroll virus, Polerovirus*); PPV (*Plum pox virus, Potyvirus*); PRSV (*Papaya ringspot virus, Potyvirus*); PSbMV (*Pea seed-borne mosaic virus, Potyvirus*); PVX (*Potato virus X, Potexvirus*); PVY (*Potato virus Y, Potyvirus*); TBV (*Tulip breaking virus, Potyvirus*); TEV *Tobacco etch virus, Potyvirus*); TMV (*Tobacco mosaic virus, Tobamovirus*); TNV (*Tobacco necrosis virus, Necrovirus*); TRV (*Tobacco rattle virus, Tobravirus*); TSWV (*Tomato spotted wilt virus, Tospovirus*); TYMV (*Turnip yellow mosaic virus, Tymovirus*); WMV (*Watermelon mosaic virus, Potyvirus*); ZYMV (*Zucchini yellow mosaic virus, Potyvirus*).

identification by a simple dichotomous choice, as can be done with phytopathogenic fungi, and there is no universal reagent that can identify all viruses in a single proof. An entire arsenal of methods is available to the practitioner. The most appropriate are selected and used on a case-by-case basis, depending on the precision required (specificity, sensitivity), the equipment available in the laboratory, and of course the funds available for the analysis.

The observation of symptoms is generally insufficient to establish a reliable and precise diagnosis of plant viruses, but it sometimes proves highly useful in guiding it. For a precise identification, the practitioner often uses various methods based on biological, morphological, biochemical, or immunological properties of viruses. The means of detection available are generally more varied and efficient for viruses that are well characterized, particularly at the molecular level. Some of these tests, such as serological tests, have been used for a long time and are the subject of constant technological innovation. Others developed more recently, such as molecular tests, are based on new knowledge acquired on virus genomes.

Symptoms observed on the plant

A diversity of symptoms

Plant viruses usually cause more or less significant and characteristic changes in appearance on all or part of a plant. They result in anomalies of pigmentation (mosaics, flower breakings, yellows) or growth (stunting), necroses, or deformations that are described with precision by pathologists. The observation of symptoms on the entire plant and the analysis of the circumstances of their appearance and their evolution constitute the first steps of diagnosis.

Symptoms and virus names

It is interesting to note that symptomatology is the source of description of viral diseases and even of virus names. Conventionally, the name of a virus is chosen on the basis of the symptoms that it causes on the plant on which it was first identified. The term *mosaic*, which describes the presence of leaf areas with different intensities of coloration, is part of the name of many viruses belonging to a wide range of genera. Examples are "tobacco mosaic virus", "cucumber mosaic virus", "apple mosaic virus", and "sugarcane mosaic virus", each belonging to a different *Family*. Likewise, there are many "necrosis viruses", "mottle viruses", or "yellows viruses", according to the symptoms they cause.

Mosaics

An abnormal distribution of chlorophyllian pigments in the leaves, often associated with an alteration of the structure of chloroplasts, results in the juxtaposition of variously coloured areas on a leaf that resemble a mosaic. This irregular coloration is particularly clearly visible on young leaves and sometimes accompanied by leaf deformations (Plates I.2, I.3, I.5, I.6, V.1).

Various types of mosaics are observed. In the case of typical light green or yellow (aucuba) mosaics, the discoloured cells are grouped in islands regularly distributed over the entire surface of the leaf; this is the classic case of TMV (*Tobacco mosaic virus, Tobamovirus*) in tobacco. If the areas are diffuse and less contrasted, it is called *mottle*. Sometimes the spots are in simple or concentric rings (*ringspot*) (Plate I.5). When the coloration is more intense along the veins it is called *vein banding* (Plate I.4) and if the coloration is lighter it is called *vein clearing*.

Flower breaking

Typical examples of flower breaking can be seen in tulips infected by TBV (*Tulip breaking virus, Potyvirus*) (Plate I.1). Light stripes are due to the absence of vacuolar anthocyanins in certain zones of the petals, while dark stripes are due to an accumulation of these pigments.

Yellows

Yellows often characterize infections by viruses multiplying in the vascular tissues. Yellows are generally associated with a thickening and rolling of leaves (Plate II.3); they are particularly visible in older leaves (Plate II.2). Yellowing observed in barley infected by BYDV (*Barley yellow dwarf virus, Luteovirus*) is due to the accumulation of viral particles in the phloem tissues that impede normal sieve flow. This creates a state of induced deficiency, which causes atrophy of roots and stunting of the plant (Plates II.1, V.3, V.4).

Necroses

The death of infected cells leads to a brown coloration on the leaves and the formation of necrotic spots (Plates II.4, II.6); these spots may remain localized or combine to form large dry patches, as can be observed in tulip infected by TNV (*Tobacco necrosis virus, Necrovirus*). Certain viruses, such as TSWV (*Tomato spotted wilt virus, Tospovirus*), induce in numerous species (lettuce, chrysanthemum) necroses that could lead to the death of the plant (Plate II.5). Necroses can also be observed on fruits, tubers, as in the case of potato infected by PVY^{NTN}, a particularly severe isolate of PVY (*Potato virus Y, Potyvirus*) (Plate III.4), or seeds, such as those collected from peas infected by PSbMV (*Pea seed-borne mosaic virus, Potyvirus*) (Plate III.6) or beans infected by BCMNV (*Bean common mosaic necrosis virus, Potyvirus*) (Plate III.5).

Growth reduction and deformations

Virus multiplication is accompanied by metabolic alterations that generally lead to a reduction in the vigour of infected plants and a delay in their development. Dwarfing may be characteristic of a disease, as in the case of BYDV (Plate V.3). Generally, this slowing of growth is accompanied by smaller leaves and short internodes, as well as smaller or fewer flowers (Plate I.1), fruits, or seeds than those produced by a healthy plant. All these effects combine to result in often highly significant yield losses.

More or less spectacular malformations can sometimes be observed on leaves or fruits (Plate VI.1); these may consist of blisters or rugosity affecting the leaf blade (Plate II.7). CMV (*Cucumber mosaic virus, Cucumovirus*) is responsible for deformation of leaves, which take on a characteristic filiform shape (shoe-stringing), in numerous plant species that it infects (tomato, tobacco, or dahlia, for example). Other viruses such as PEMV (*Pea enation mosaic virus, Enamovirus*) cause outgrowths, called "enations", that appear on the veins on the underside of leaves (Plate IV.1). In the case of sugarbeet rhizomania due to BNYVV (*Beet necrotic yellow vein virus, Benyvirus*), there is unexpected growth of an abundant root system (hairy roots).

Certain viruses cause no symptoms on the plants they infect. They are latent and their hosts are called "healthy carriers". They may not appear detrimental to the plant and are often spread unnoticed in crops or inadvertently propagated in planting material. But these viruses can in fact have severe synergistic effects with other viruses infecting the same plant. This is true for LSLV (*Lily symptomless virus, Carlavirus*), which, with CMV, causes necrotic spot disease on bulbs.

Can symptomatology provide a reliable diagnosis?

The symptoms of a viral disease may vary as a function of numerous parameters, including the nature of the virus strain or the host cultivar, the time of infection, and environmental conditions. It is therefore quite difficult to establish a universal typology of symptoms caused by a virus in a given plant species. The observation of symptoms is thus insufficient to establish a reliable diagnosis.

On tobacco, various isolates of PVY may be responsible for slight mottling, deforming mosaic, or serious necrosis. Similarly, melons infected by certain strains of ZYMV (*Zucchini yellow mosaic virus, Potyvirus*) develop either severe mosaics and leaf deformations or a sudden wilt followed by lethal necrosis, depending on whether the melon cultivar possesses the dominant gene *Fn*.

Symptom expression depends also on the physiological stage at which the plant is infected: a young plantlet generally develops more pronounced symptoms than an older plant. This characteristic has given rise to a cultural practice: the topping of potato plants during certified seed production favours the appearance of marked symptoms on the young shoots and a more efficient elimination of infected plants.

Environmental conditions (temperature, light intensity, nutrition) play a major role in the intensity and development of symptoms. Infected plants may, in particular environmental conditions, present attenuated symptoms or not show symptoms at all. Certain viral diseases are seasonal. In the case of pelargonium infected by PLCV (*Pelargonium leaf curl virus, Tombusvirus*), plants show severe symptoms (asteroid spots, leaf holes, necroses) (Plate II.7) only between November and March, while during the rest of the year they look healthy although they are infected. In lettuce, the big vein disease caused by MLBVV (*Mirafiori lettuce big-vein virus, Ophiovirus*) is expressed only in the winter and spring crops, while the yellows caused by BWYV (*Beet western yellows virus, Polerovirus*) are expressed mainly in summer crops. Unlike these phenomena of seasonally apparent viruses, viruses that are called "latent" never cause symptoms on the plants they infect.

Another difficulty encountered in the establishment of a viral diagnosis through symptomatology is the possible confusion with other pathological or physiological disorders. The symptoms of mosaic are, generally, quite specific to a viral infection; only some rare cases of phytotoxicity (particularly linked to hormone or herbicide treatments) or mite attack (tarsonems) also induce such symptoms. This is not true of yellows or necroses, which could be caused by numerous biotic or abiotic factors (e.g., phytopathogenic bacteria or fungi, nutrient deficiencies, phytotoxicity). Apart from these possible confusions, different viruses may cause very similar symptoms in a single host, making a distinction impossible.

Finally, infections by several viruses (mixed infections) are common in nature: the symptoms observed may therefore be quite different from symptoms caused by each virus individually. In particular, there is sometimes an accentuation of symptoms due to synergy between the viruses of the complex; this is often observed in mixed infections between potyviruses and CMV or PVX (*Potato virus X, Potexvirus*).

Guiding the diagnosis

In order to establish a visual diagnosis, the plant health professional must either have a perfect personal knowledge of the major viruses infecting a crop in a given area or have access to databases that can help in the process. Among available databases are monographs on diseases affecting specific crops: some are in the form of abundantly illustrated books or CD-ROMs (e.g., Blancard et al., 1994, 2004, *Crop Compendia* and *The Crop Protection Compendium*, published by the American Phytopathological Society and CABI Publishing, respectively). Another approach, still experimental, is to use computers and artificial intelligence software to develop expert systems that can be used by farmers themselves.

In addition to the detailed description of symptoms on different plant organs, these diagnostic tools exploit other important information including the analysis of distribution of diseased plants in the field as well as the history of cultural practices and problems that may have occurred to the crop (particularly the possible build-up of vector population).

Symptoms observed at the cellular level

Beyond the analysis of macroscopic symptoms at the leaf or whole plant levels, it is possible to observe symptoms induced by viruses at the cellular level. In 1903,

Ivanowski observed particular structures, called *X bodies*, in cells of tobacco infected by TMV. Inclusion bodies were then observed in cells of numerous plants infected by viruses; they may be amorphous or crystalline and represent either clusters of viral particles or accumulations of other proteins coded by the virus.

The use of viral inclusion bodies in diagnosis has attracted greater attention in recent years. Fragments of the epidermis or sections taken freehand can be used, after simple staining, to observe inclusion bodies sometimes specific to a given genus (*Potyvirus, Carlavirus, Comovirus, Tobamovirus*) (Table 9.1, box below, and Plate VI.1b) (Christie and Edwardson, 1986). This technique, which requires some experience but no sophisticated equipment (a simple optical microscope will suffice), constitutes in many cases a quick and economical diagnostic method.

The symptoms induced by a virus at the cellular level can be observed with even greater precision by the observation of ultrathin sections with an electron microscope. The samples undergo a complex treatment of fixation, inclusion, and staining. The alterations observed are most often specific to the genus of the virus (Table 9.1). This type of analysis is very useful for the characterization of a new virus, but cannot be considered part of a routine diagnosis because of its cost and the time required to carry it out.

Table 9.1 Examples of inclusion bodies associated with viruses belonging to different genera (Christie and Edwardson, 1986).

Genus	Staining	Type of inclusion body	Type of tissue
Closterovirus	Azure A	Striate bodies	Phloem
Cucumovirus	Azure A	Virus aggregates	Epidermis, mesophyll
Potyvirus	Orange-green	Inclusions: -cylindrical, cytoplasmic* -amorphous, cytoplasmic** -crystalline, nuclear**	Epidermis, mesophyll
Tobamovirus	Azure A and orange-green	Virus crystal stocks, X bodies	Epidermis, mesophyll

*All viruses. **Some viruses only.

Apart from inclusion bodies, viruses can induce cytopathological modifications that are sometimes characteristic of a virus family, genus, or species (Figs. 9.1, 9.2). These effects chiefly pertain to the cellular organelles and the membrane system. Among the most frequent cytological manifestations are hypertrophy of mitochondria caused by TRV (*Tobacco rattle virus, Tobravirus*), the aggregation of chloroplasts linked to infection by TYMV (*Turnip yellow mosaic virus, Tymovirus*), the accumulation of lipids or starch induced by PLRV (*Potato leaf roll virus, Polerovirus*), the proliferation of endoplasmic reticulum (various rhabdoviruses), or the formation of cytoplasmic vesicles (CMV).

Cellular inclusion bodies associated with infection by a potyvirus

The case of potyviruses is particularly interesting. Apart from the capsid protein, the genomic RNA of *Potyvirus* codes for at least 8 non-structural proteins. Some of these proteins, whose function is now known, accumulate in infected cells in the form of characteristic inclusion bodies (Fig. 9.1). Three types of inclusion bodies are distinguished:

— Cylindrical cytoplasmic inclusion bodies (CI) are specific to the family *Potyviridae*; they constitute a criterion of identification. They are made up of polymers of a viral protein of 70 kDa, a protein with helicase function implicated also in the transport of the virus from cell to cell. They are observed in cells of plants infected by all potyviruses, in the characteristic form of pinwheels in cross-section or in the form of bundles or laminated aggregates in longitudinal section (Pl.VI.1c). A classification of *Potyvirus* into four subgroups has been proposed (Edwardson et al., 1984); it takes into account the type of association and the form of these inclusion bodies.

— Nuclear inclusion bodies (NI) are observed only with certain potyviruses, including TEV (*Tobacco etch virus*) or PPV (*Plum pox virus*). These inclusion bodies are seen as regularly shaped plates in the nucleoplasm or sometimes in the nucleolus. They are crystalline, fibrous, or granular. In the case of TEV, two proteins have been identified in these inclusions; protein NIa of 49 kDa, with proteolytic activity, and protein NIb of 58 kDa, which is the polymerase.

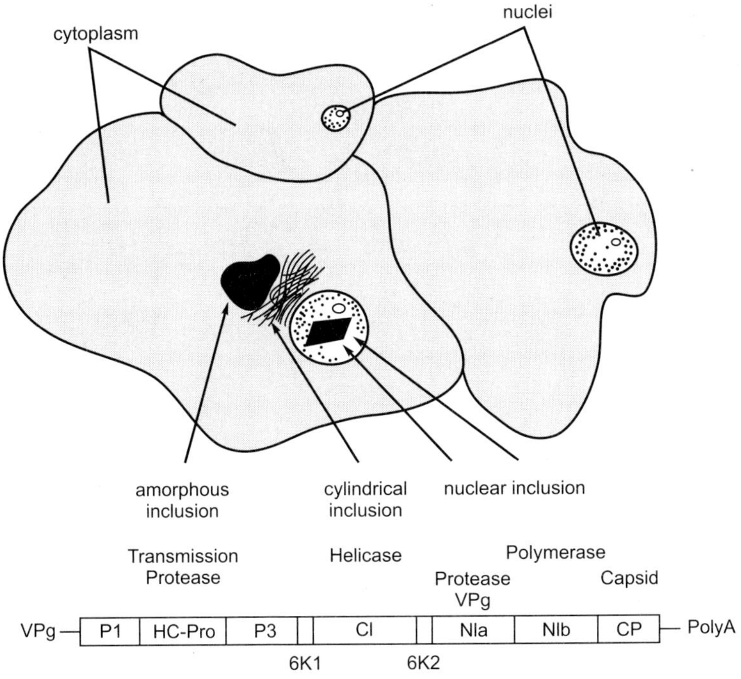

Figure 9.1 Different types of cellular inclusion bodies caused by potyviruses.

— Amorphous inclusion bodies (AI) of irregular form are observed in the cytoplasm of cells infected by some potyviruses, such as PVY. They are granular and are constituted of the helper component (HC-Pro) of 56 kDa, a protein that plays an essential role in the transmission of the virus by aphids.

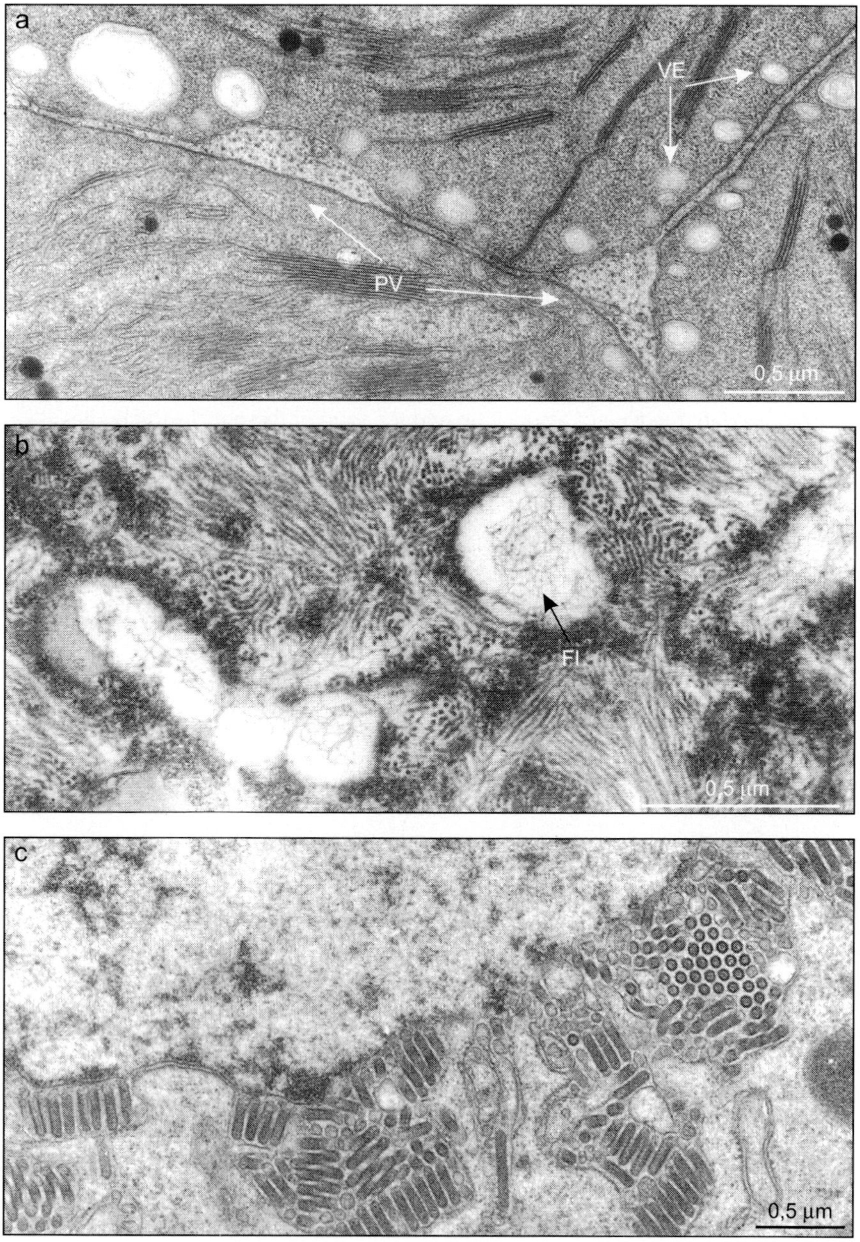

Figure 9.2 Characteristic cytopathological modifications induced by various virus genera. (a) Aggregation of chloroplasts in a melon cell infected by a tymovirus (MRMV). Note the presence of "virus pockets" (PV) between the chloroplasts and numerous vesicles (VE) in the chloroplasts. (b) Accumulation of viral particles (seen in cross-section and longitudinal section) in phloem cells of a cucumber infected by a crinivirus, CYSDV. Note the presence of vesicles containing fibrils (FI). (c) Numerous particles of a nucleorhabdovirus (EMDV) in the nuclear membrane of an infected tobacco.

Diagnosis through biological means

Each virus is characterized by a host range, that is, by a set of plant species that it can infect, under natural or experimental conditions. The indicator plants or differential hosts can be cultivated or wild species. Some of them develop particular reactions (local lesions, systemic symptoms) that can be characteristic of a given virus. Experimental transmission of a virus to test plants or biological indexing is thus based only on the infectious properties of the virus. This approach is still very important for detecting the presence of a virus in an infected plant, especially when it is an unknown virus for which there is no other diagnostic method.

Mechanical inoculation

Mechanical inoculation to herbaceous hosts is the most frequently used method (Plate VII.1b and 1c; Fig. 9.3). It is noteworthy that certain species, such as *Chenopodium quinoa* and *C. amaranticolor*, or *Nicotiana benthamiana* and *N. clevelandii*, are sensitive to more than 100 different viruses and are nearly universal test plants. To ensure a high sensitivity and reproducibility of symptoms, the experimental inoculations are better carried out on young plants that are subsequently kept (incubated) in controlled conditions (temperature about 20°C and extra light intensity). Certain plants such as cucumber can be used at the cotyledonary stage while others such as *Chenopodium* or *Nicotiana* are used at the stage of 4-6 developed leaves, well before floral induction, which could modify the plant reactions.

The crude extract obtained by grinding a piece of infected plant with a mortar and pestle in an appropriate buffer (generally a phosphate buffer) is deposited on the leaf

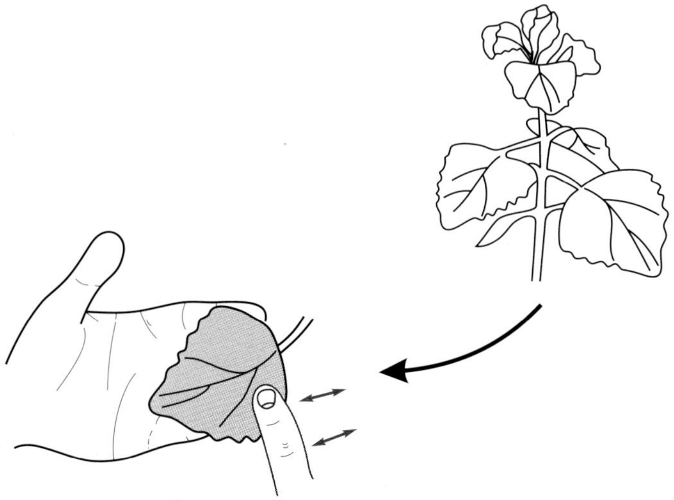

Figure 9.3 Mechanical virus inoculation of a herbaceous indicator plant.

surface of a test plant. The mechanical inoculation is done by rubbing with a spatula or a sponge, or simply with the finger sheathed in a rubber fingerstall or glove (Fig. 9.3). The virus easily penetrates through the slight injuries caused by rubbing on the epidermis or in the leaf glandular hairs. In order to facilitate this penetration, an abrasive such as carborundum® can be used. It is recommended that the leaves be washed immediately after inoculation. The plants are then placed in a greenhouse or in a climate-controlled chamber in conditions that favour the multiplication of the virus and the manifestation of symptoms (with sufficient light and a temperature of 20-25°C).

■ Inoculation by vectors

Some viruses cannot be transmitted mechanically, particularly those restricted to the phloem (*Luteovirus, Polerovirus, Nanovirus, Tenuivirus, Marafivirus, Reovirus*, and certain members of *Closteroviridae* and *Geminiviridae*). In this case, indicator plants can be inoculated by vectors. This requires that colonies of healthy vectors (aphids, whiteflies, leafhoppers, etc.) be maintained in the laboratory. The inoculation can be carried out as shown in Fig. 9.4.

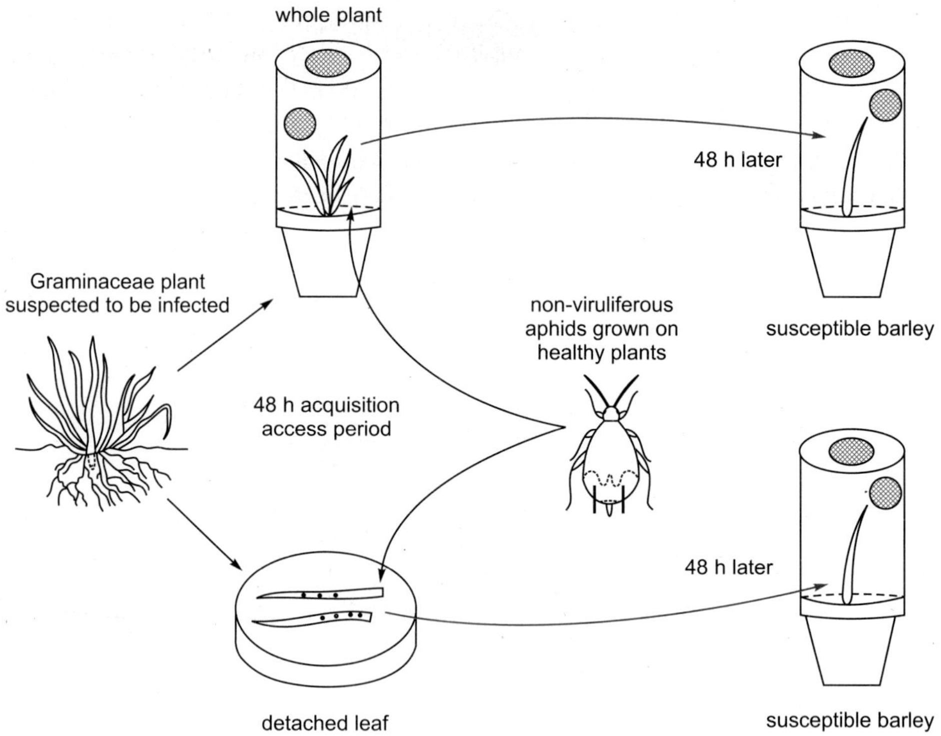

Figure 9.4 Experimental transmission of BYDV by the aphid *Rhopalosiphum padi* (from M. Beuve).

Grafting

Grafting (Fig. 9.5) on indicator plants is another type of biological indexing still widely used for the detection of viruses or infectious agents of unknown or poorly known aetiology, particularly on fruit species or grapevine. The ligneous indicators used are selected for their good and rapid symptom expression. This sometimes occurs only after several months to a year, such as in *Vitis vinifera* cv. Pinot noir or Cabernet franc for the detection of closteroviruses responsible for grapevine leafroll disease.

Principal types of reaction of differential hosts

The analysis of symptoms developed by test plants inoculated under controlled laboratory conditions has long been used for diagnosis in plant virology. The

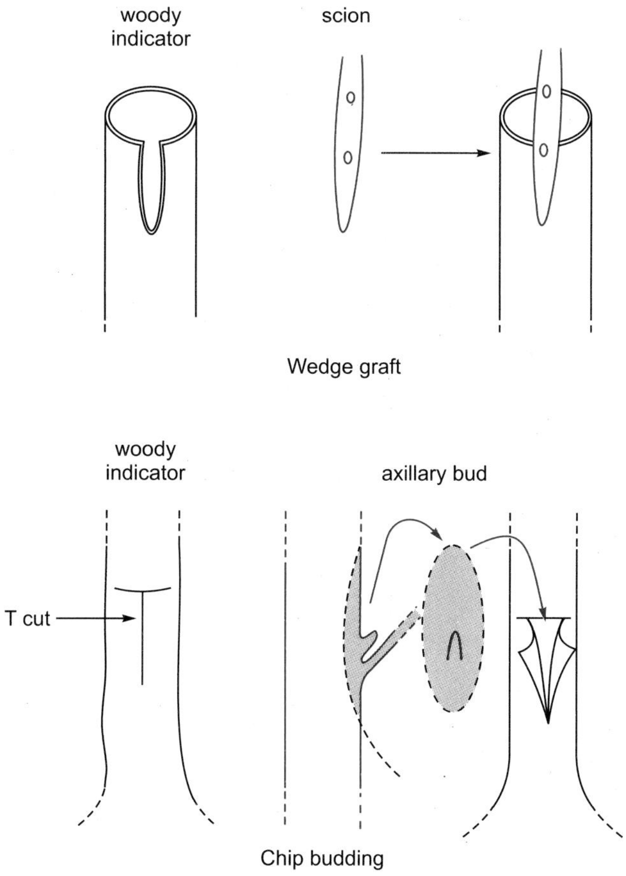

Figure 9.5 Virus detection by graft transmission: grafting a scion or a bud from a supposed infected plant on to susceptible woody indicator plant.

reactions observed on a susceptible test plant are either systemic—the virus invades the plant partly or totally from the inoculated leaf—or localized and restricted to the inoculated leaf.

Local reactions, chlorotic or necrotic lesions of variable size depending on the virus or on the strain, are rapidly observed, in 2 to 7 days after inoculation. Systemic reactions are observed generally only after 7 to 10 days, sometimes later.

Four major viruses cause mosaics in cucurbits in Mediterranean regions: CMV, WMV (*Watermelon mosaic virus, Potyvirus*), ZYMV, and PRSV (*Papaya ringspot virus, Potyvirus*). These viruses can easily be distinguished by the reaction of four differential hosts, particularly according to whether they cause or do not cause local lesions (Fig. 9.6). In contrast, when these viruses are found in mixed infection (which is common in the field), the diagnosis becomes difficult because certain viruses can be masked by the reaction of indicator plants to other viruses. In this example (Fig. 9.6), if the plant to be analysed is infected by the four viruses together, the presence of CMV and WMV could be suspected because of the reactions of *Vigna* and *Lavatera*, but the presence of ZYMV and PRSV will go unnoticed.

The number of local lesions depends on the initial concentration of infectious viral particles in the inoculum (Fig. 9.7). Virologists have long used this method to evaluate virus concentrations in infected plants, in systems such as TMV–*N. tabacum* cv. Xanthi or CMV–*Vigna sinensis*, by referring to the linear relationship between number of local lesions and the concentration of infectious viral particles in the extract, within certain concentrations.

Biological indexing (mechanical inoculation, grafting, transmission by vector) is particularly valued for its wide range of detection (viruses, phytoplasma, viroids), sensitivity, and capacity to detect unknown viruses. This method is also the only one at present that can be used to characterize the pathogenicity of a specific virus strain (pathotype, severity of symptoms). One of its major constraints is the need for (and cost of) regular production and maintenance of indicator plants in insect-proof greenhouses. Moreover, biological indexing sometimes takes quite a long time to yield results and results may be difficult to interpret in the case of simultaneous infection by several viruses.

Serological diagnostic methods

Antigen-antibody reaction: the basis of serological methods

Serological techniques use the specific interaction of two types of proteins: antigens, proteins of viral origin, and antibodies, proteins specific to these antigens that are elaborated by an animal (generally a rabbit) in response to the injection of the antigen.

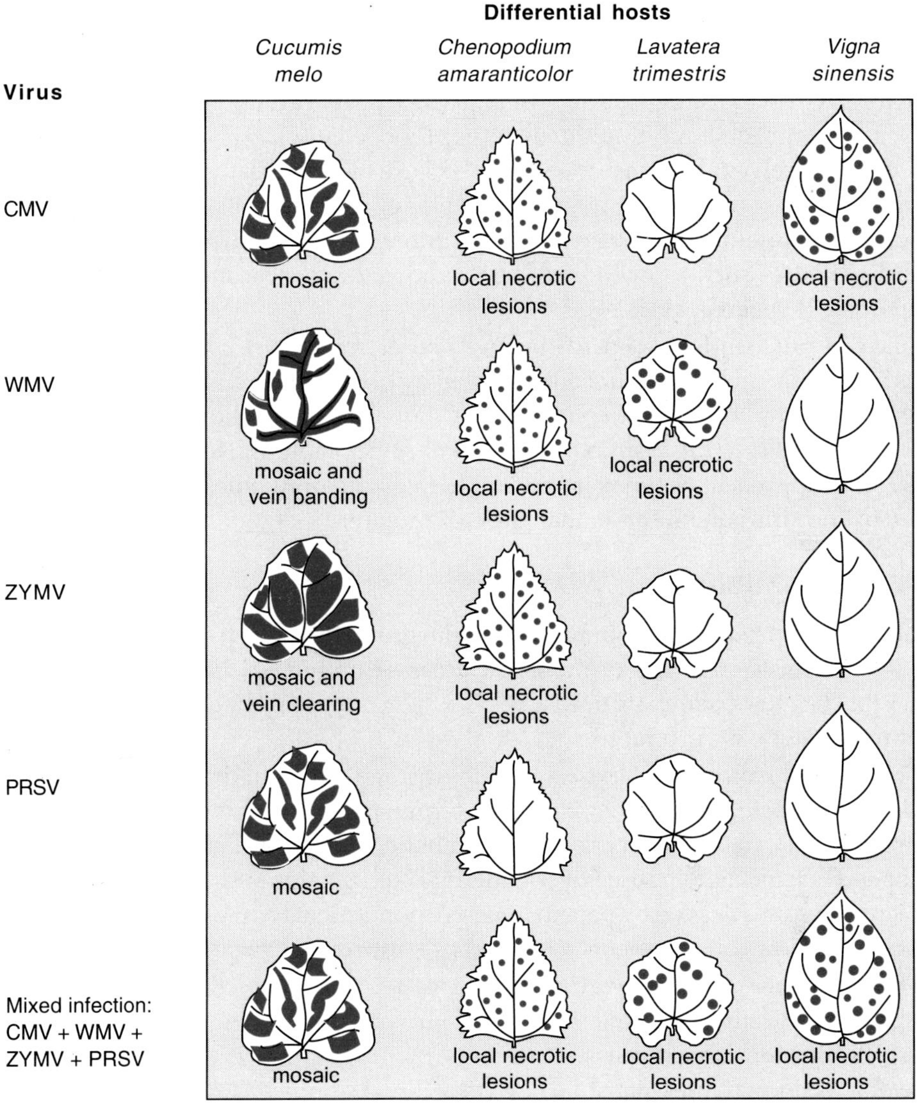

Figure 9.6 A simple host range makes it possible to identify a virus by means of specific reactions (CMV/*Vigna*, for example) or differential reactions. In case of mixed infection, the interpretation is more difficult (see text). ▲

Figure 9.7 Local lesions caused by TMV in *Nicotiana glutinosa*. ▶

Viral antigen

An antigen is a molecule that can trigger an immune response in the animal. Viral proteins have several antigenic determinants (epitopes) varying in their property of inducing the production of antibodies (immunogenicity). For a long time, viral particles themselves were used as antigens because they are relatively easy to obtain, for many viruses, as purified preparations free of contaminants of cellular origin. Other viral proteins could also serve as antigens, particularly the proteins of inclusion bodies such as cylindrical, amorphous, or nuclear inclusion bodies of potyviruses (Hiebert et al., 1984).

Since the nucleotide sequences of many viruses are known, it is now possible to mass produce a viral protein uncontaminated by plant proteins. The corresponding viral gene is simply cloned in a plasmid that allows the production of the protein in the bacterium *Escherichia coli* (Nikolaeva et al., 1995; Hourani and Abou-Jawdah, 2003). This approach also makes it possible to obtain antibodies against proteins coded by the viral genome other than the coat protein.

Polyclonal antibodies

Antibodies are proteins of the immunoglobulin group (Ig), which are found in blood serum. The basic structure of the five classes of immunoglobulins is the same (Fig. 9.8). The most commonly used are IgG, which represent about three quarters of the immunoglobulins produced.

The molecule comprises two heavy chains (H) and two light chains (L) connected by disulphide bridges (Fig. 9.8). The NH_2-terminal regions of the heavy and light chains (in red) are highly variable and constitute recognition sites specific to epitopes. Each IgG is composed of two identical recognition sites of an epitope: it is bivalent. It may be cleaved by papain into two monovalent *Fab* (antibody fragments) and one crystallizable fragment *Fc*, or by pepsin into one bivalent *F(ab')*2 and *Fc*.

The protocols of immunization vary greatly. The most commonly applied comprise an intramuscular injection of the virus or purified viral proteins (0.1 to 1 mg); this first immunization is followed, one month later, by several intramuscular booster injections at intervals of 1 to 2 weeks. Blood samples are taken regularly to evaluate the level of specific antibodies and the absence of antibodies reacting with the normal proteins of the host. The antibodies thus obtained are called "polyclonal" because the serum contains a wide variety of antibodies, each of them is directed against one of the antigenic determinants present on the viral particles or proteins used as immunogens.

The production of polyclonal antibodies is generally carried out in a rabbit (Fig. 9.9). The injection of a purified virus preparation or of viral proteins induces the production by activated B lymphocytes of a complex combination of antibodies that are collected in the serum. The evolution of this combination has been studied during immunization with JGMV (*Johnsongrass mosaic virus, Potyvirus*). The potyvirus

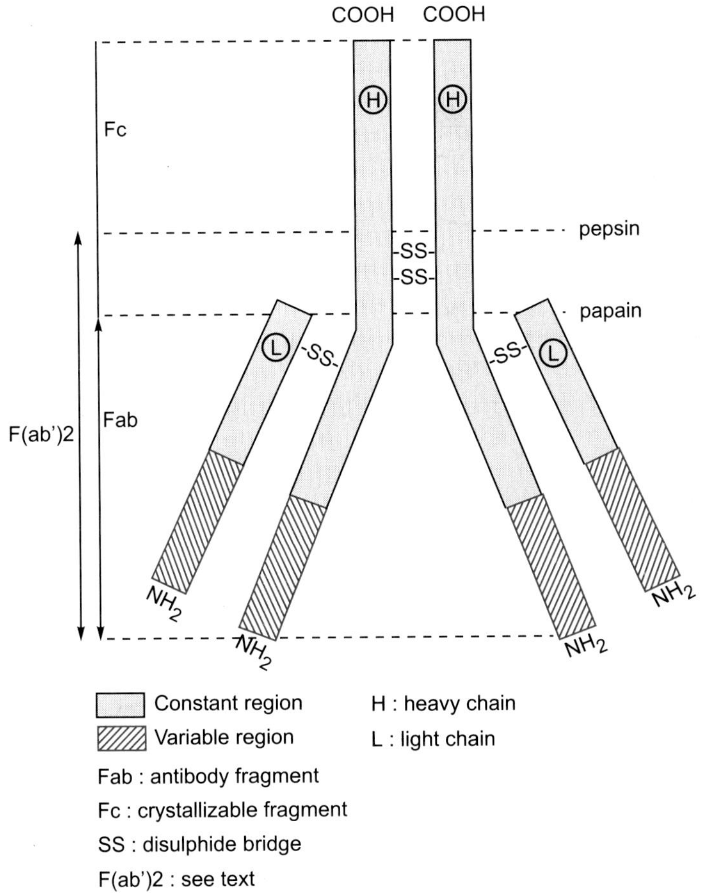

Figure 9.8 Structure of an immunoglobulin G (IgG).

particle (see Fig. 1.8) is formed by the assembly of circa 2000 capsid subunits around the viral RNA; the N- and C-terminal ends of each capsid protein are exposed on the surface of the viral particle (and can be hydrolysed *in vivo* or *in vitro* by proteases without the form of the particle being modified). Antiserum preparations collected at different times during the immunization process are tested with a set of 295 overlapping octapeptides, covering the complete sequence of the capsid protein. The interaction of an antibody with an octapeptide is evaluated at 405 nm by an ELISA-type detection method (see ELISA section below). Figure 9.10 represents the reactions of antibodies obtained (1) after two injections, (2) after five injections, and (3) after multiple booster injections (Shukla et al., 1989). Whenever *intact* particles are used as immunogens, the antibodies produced after two or five injections (Fig. 9.10(1) and (2)) are almost exclusively directed against the N-terminal part (67 amino acids) of the capsid protein: *this N-terminal region is immunodominant*. The antibodies directed

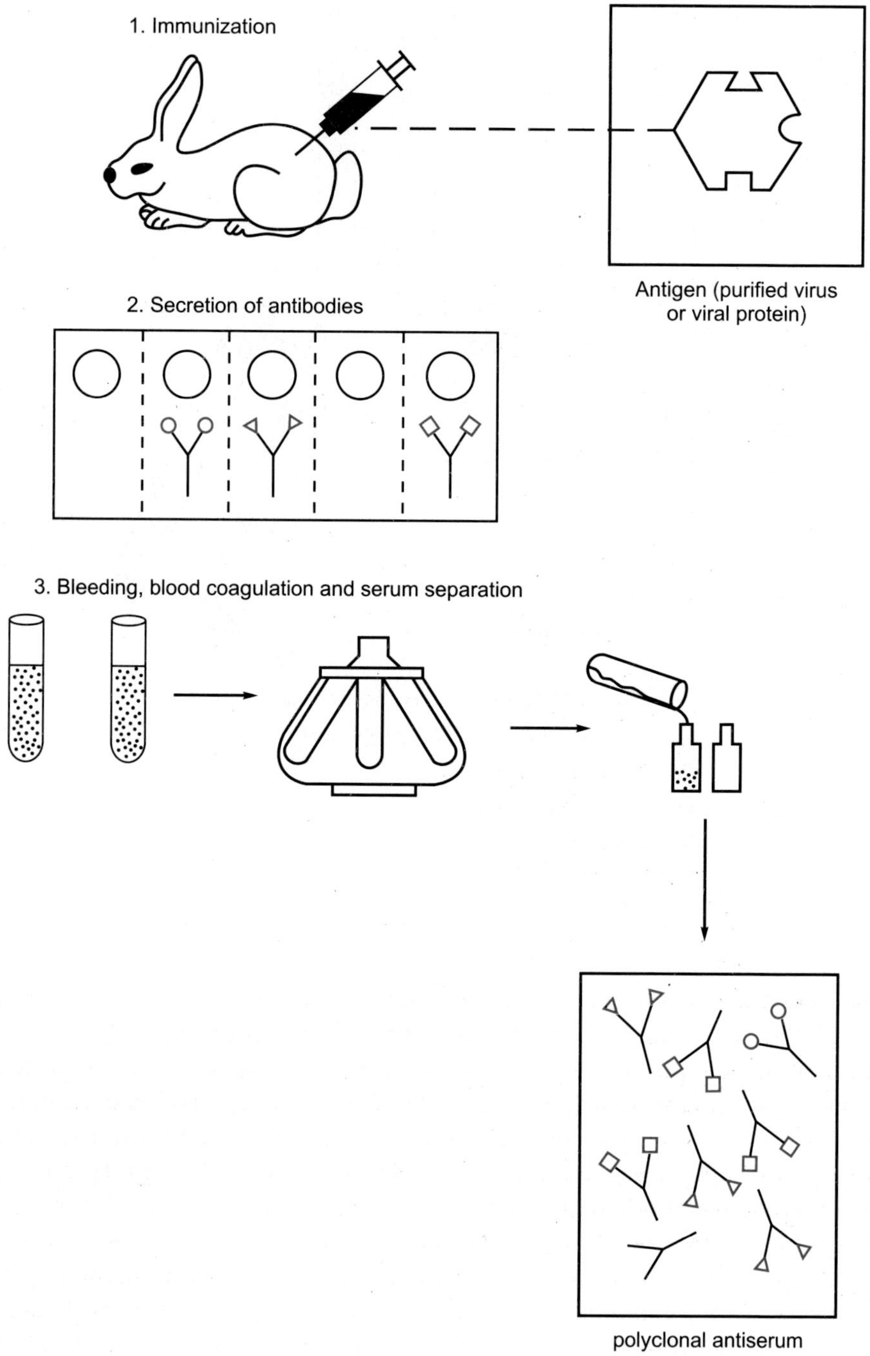

Figure 9.9 Production of a polyclonal antiserum. It contains a wide variety of immunoglobulins corresponding to numerous epitopes present on the virion or viral protein.

against the central part (218 amino acids) and the C-terminal part (18 amino acids) of the capsid protein (Fig. 9.10(3)) appear only after several booster injections (Shukla et al., 1989).

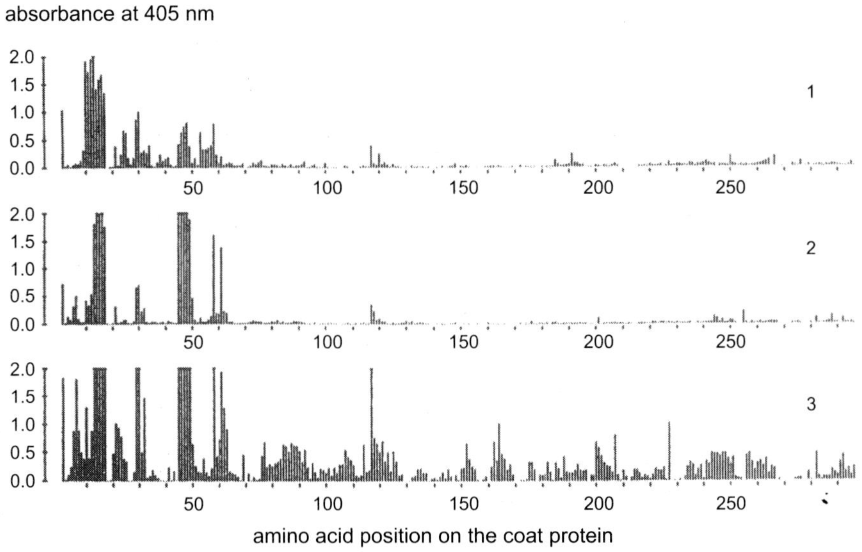

Figure 9.10 Identification of epitopes of the capsid protein of JGMV recognized by antibodies produced by a rabbit following an immunization protocol with freshly prepared intact particles of JGMV. The protocol comprises (1) two injections, (2) five injections, and (3) numerous booster injections (Shukla et al., 1989).

Monoclonal antibodies

The specificity of the antigen-antibody reaction makes serological methods very good tools for the identification of viruses. Nevertheless, the polyclonal antibodies are reactants produced in limited volume; their performance varies from one bleeding to another (Fig. 9.10) and from one rabbit to another. To overcome this disadvantage, monoclonal antibodies (Mabs) can be produced, each reagent containing only one type of antibody, which reacts with only *one epitope* of a viral protein (Fig. 9.11). If the epitope corresponds to a *variable* part of the protein, this method makes it possible to distinguish virus strains in which this epitope is present or not. If, on the contrary, the epitope corresponds to a *conserved* motif of the protein, the Mab will allow a very wide detection of all virus strains. In this way, it has been possible to produce a monoclonal antibody that reacts with practically all potyviruses (Jordan, 1992).

Another major advantage of Mabs is that the hybridoma cells thus obtained are theoretically eternal cell lines that make it possible to produce the same antibody constantly in a volume adaptable to the demand. Monoclonal antibodies, singly or in

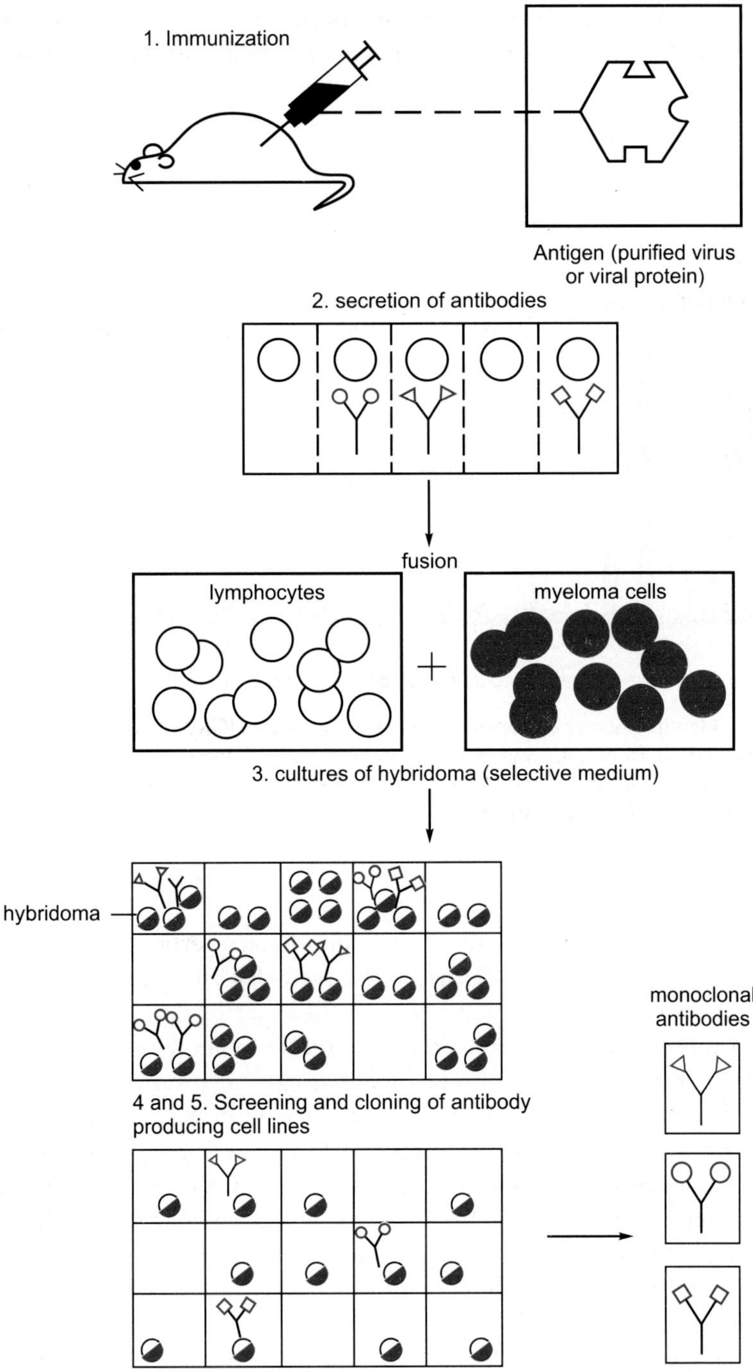

Figure 9.11 Production of monoclonal antibodies. Each hybridoma line produces only one type of antibody.

a defined combination, are reagents particularly well adapted for an international *standardization* of serological protocols.

Monoclonal antibodies are produced in mice or rats, animals for which there are eternal lines of cancerous cells (myeloma) (Fig. 9.11). The hybridoma cells obtained after fusion between the spleen cells of the immunized animal (lymphocytes) and the cells of the myeloma are multiplied *in vitro* on a selective medium. The culture supernatants containing the antibodies secreted by the hybridoma cells are then tested. This screening, which is generally done by ELISA (see section on the ELISA technique, below) with respect to the antigen that is to be detected, makes it possible to choose the hybridoma cells presenting the best reactivity. The hybridoma cells selected are then cloned, by limit dilution, in order to obtain cell lines that secrete a single type of antibody having the required specificity. These cell lines can be conserved in liquid nitrogen.

Recently, a new method of producing single type antibodies similar to monoclonal antibodies has been developed, using molecular biology instead of the conventional immunization approach. The technique, known as "phage display", uses libraries of DNA sequences coding for functional fragments of antibody molecules, obtained by polymerase chain reaction (PCR) from a number of animal species (McCafferty et al., 1990). The antibody repertoire of a vertebrate is estimated to contain more than 10^8 different antibodies. The antibody fragments contain the heavy and light chain variable domains (see Fig. 9.8), the regions of hypervariable sequences that confer specificity to antibodies. The recombinant antibody proteins are displayed on the surface of a bacteriophage and screened against the target viral protein. This technique has been used to produce antibodies to a range of plant viruses including PVY (Boonham and Barker, 1998) and BNYVV (Griep et al., 1999).

Viral epitopes

The identification of epitopes on the capsid protein (or epitope mapping) is done by testing Mabs with a series of octapeptides representing the entire polypeptide chain of the protein (as for JGMV, Fig. 9.10). This makes it possible to detect only *continuous epitopes* that correspond in general to sequences of 5 to 15 amino acids.

There are also *discontinuous epitopes* formed for example by two folds of an amino acid chain (Fig. 9.12). Discontinuous epitopes have been identified on the surface of TMV particles using Mabs produced against virions (Van Regenmortel, 1986) and testing TMV coat protein mutants. For example, Mab 20 recognized substitutions at positions 66 or 140 on the amino acid sequence of the capsid protein but did not recognize a substitution at position 65. This Mab reacts specifically with a discontinuous epitope comprising residues 66, 67, and 140-143. Moreover, 80% of Mabs prepared against the virion do not recognize a series of peptides covering the complete sequence of the capsid protein. These results suggest that Mabs obtained against a native protein recognize essentially epitopes that depend closely on the conformation of this protein.

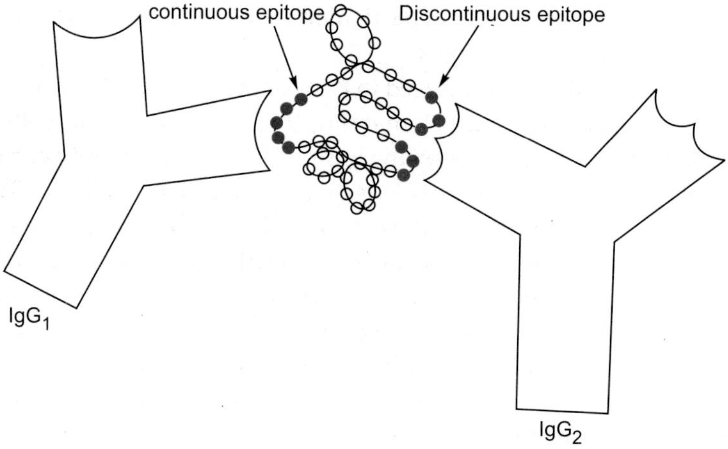

Figure 9.12 Continuous/discontinuous structure of epitopes.

The existence of *neotopes*, or epitopes specific to the quaternary structure of the virion, has been observed in all the virus genera. A demonstration is provided by "cross-absorbing" an antiserum with dissociated capsid subunits: the cross-absorbed antiserum no longer recognizes the dissociated capsid protein but still interacts with intact virus particles. This has been confirmed in the case of TMV: out of a total of 18 Mabs prepared against the virus, 8 did not react with the capsid subunits (Altschuh et al., 1985). The conformation of neotopes is strongly modified when the viral particle is directly adsorbed on a plastic surface (microtitre ELISA plate); the virus thus becomes equivalent in antigen terms to dissociated capsid subunits (Altschuh et al., 1985).

Finally, certain epitopes of the capsid protein are not exposed to the surface of viral particles, and these are *cryptotopes*. Out of a total of 30 Mabs prepared against the capsid protein of TMV, 5 did not react with the viral particles and are thus specific to cryptotopes (Al Moudallal et al., 1985).

The determination of conformational epitopes is facilitated by the use of "biosensors" that analyse the specific molecular interactions as soon as they occur. One of the reagents (the antibody) is immobilized on the surface and the other (the antigen) is introduced in a solution that flows on this surface. The principle of the detection is based on a change in the refraction index near the surface. The increase in the refraction index for the proteins is proportional to the mass present in the volume detected; the changes in the refraction index are monitored continuously and recorded. This technology has been used for the serological study of the capsid protein of TMV (Dubs et al., 1992) and the cylindrical inclusion protein of PVY (Boudazin et al., 1994).

The affinity of an antibody for an epitope

The intermolecular attractive forces implicated in the interaction between an antigen and an antibody are *hydrogen bonds*, *electrostatic forces* due to the attraction of ionic groups of opposite charges, *Van der Waal's forces* created by the interaction between these different electronic clouds, and *hydrophobic bonds* produced by the association of non-polar and hydrophobic groups from which water molecules are excluded (the hydrophobic bonds can contribute to the force of the antigen-antibody bond) (Fig. 9.13). These forces require close contact between the different reactive groups. Their global intensity measures the *affinity* of an antibody for an epitope.

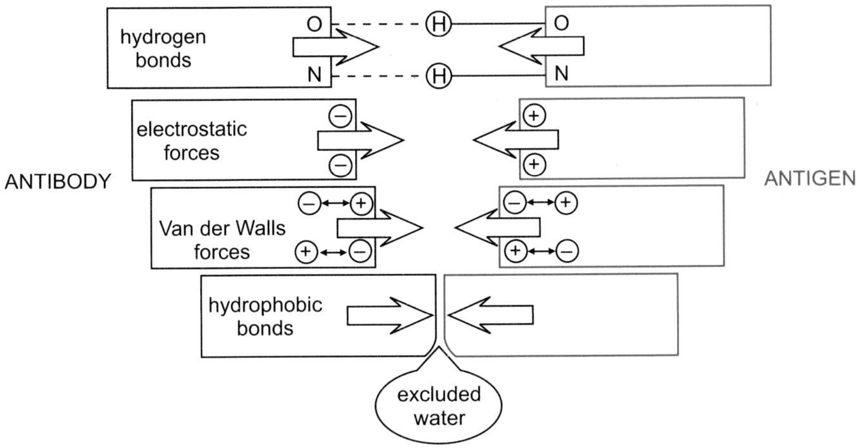

Figure 9.13 Various forces at work in the affinity of the antigen-antibody complex. The optimal distance between the reactive groups varies with the type of bond.

In an antiserum, the affinities of different antibodies are highly variable. The average value, which can be measured, constitutes the *avidity* of this serum (Van Regenmortel, 1986).

Immunoprecipitation and immunodiffusion tests

The first serological tests used in plant virology were immunoprecipitation and immunodiffusion tests.

The bivalence of antibodies and the presence of numerous identical capsid proteins in a viral particle are the source of the formation of large insoluble virus-antibody aggregates when an extract of infected leaf is mixed with a specific antiserum. This reaction, known as immunoprecipitation, occurs only when the relative concentrations of the virus and antibodies are within a range known as equivalence.

The precipitation reaction can be conducted in a liquid medium on a glass slide, in test tubes (ringtest), or in a Petri dish under a film of oil: four drops of dilute antiserum are deposited in the dish, then some drops of plant extract clarified by a slight centrifugation are added, and the precipitate forms in a few minutes (Fig. 9.14). The antigen-antibody precipitate can be seen with the naked eye or under a light microscope on a dark background at low magnification (×10-100). This quick but not very sensitive method was widely used for the detection of certain viruses of potato and bulbs in the 1950s. Today it is mostly of historical interest. The use of antibodies adsorbed on latex particles has made it possible to better see the precipitates and bring the sensitivity to around 10-100 ng/ml.

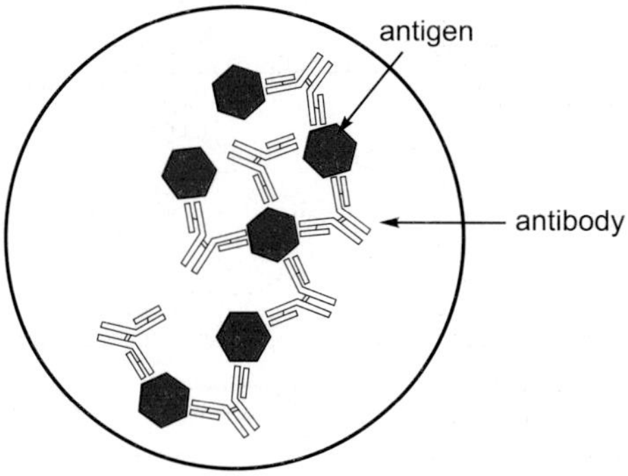

Figure 9.14 Formation of a flocculate in a microprecipitation test on a slide in liquid medium. If there is excess of antigen or antibody, the precipitate will not be visible.

In techniques of immunodiffusion, the reaction is achieved in an agar gel medium in which the reagents diffuse from wells in which they have been deposited. The double immunodiffusion test is most commonly used: the specific serum (antibodies) is deposited in the central well, the different antigens at the periphery. When there is equivalence between the antigens and the antibodies, a white precipitation line forms. The number, shape, and type of connection of precipitation lines allows for an accurate study of serological relationships between virus strains (Fig. 9.15). The spur in Fig. 9.15 shows partial identity between antigens A and B. The continuous precipitation lines shows complete identity between A and C. The absence of reaction with D and E indicates that there is no serological reaction. The addition of an anionic detergent such as sodium dodecyl sulphate (SDS) to the gel, at concentrations compatible with the serological reaction, disorganizes the structure of filamentous viruses such as potyviruses, which do not migrate in the gel because of their size. In these conditions they can thus be detected efficiently (Purcifull and Hiebert, 1978).

The sensitivity of immunodiffusion tests may vary from 1 to 20 μg/ml of virus depending on the quality of the antiserum and the type of medium. This is generally

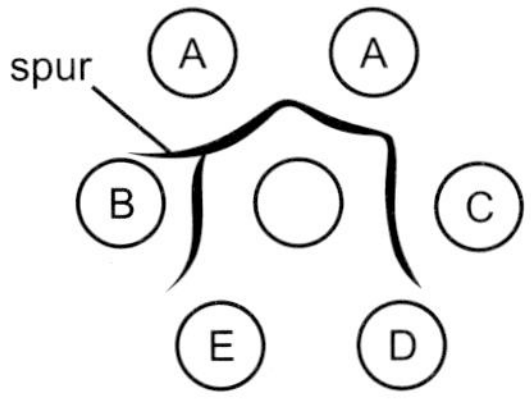

Figure 9.15 Immunodiffusion reactions in an agar gel. The central well contains the antiserum. The peripheral wells contain crude extracts of plants infected by different viral strains (A, B, C, and D). Well E contains a crude extract of a healthy plant. The absence of reaction with well E indicates that the antiserum contains no antibodies against the "normal" proteins of the host. For the interpretation of precipitation lines, see text.

sufficient for the detection of numerous viruses (potyviruses, cucumoviruses, comoviruses). This technique does not allow the detection of viruses that are in low concentration in plant extracts (such as luteoviruses). Moreover, immunodiffusion tests require fairly large quantities of crude antiserum, which makes them quite costly.

The ELISA technique

Labelling antibodies with an enzyme increases the sensitivity of detection

The *enzyme-linked immuno-sorbent assay* (ELISA) has revolutionized the field of plant virus diagnosis. The principle consists not in direct observation of the antigen-antibody precipitate but in revealing this interaction by labelling immunoglobulins with enzymes generating a coloured reaction. The conjugation with alkaline phosphatase, most often used for the detection of phytoviruses, is effected with glutaraldehyde (Avraméas, 1969).

There are several ELISA protocols that use polystyrene microtitre plates as a support (Plate VII.2a) but the most commonly used is the *double antibody sandwich* or DAS-ELISA (Clark and Adams, 1977). A DAS-ELISA test comprises several stages (Fig. 9.16). The antigen is placed between two layers of specific antibodies. The affinity between the IgG adsorbed to the walls of the well and the virus is so strong that it allows a concentration of the virus on the antibodies layer. To reveal the presence of the virus thus retained, the same IgG are used, but conjugated with an enzyme: alkaline phosphatase, for example. After several washes to eliminate all the antibodies not linked to the antigens, the activity of the alkaline phosphatase is revealed by the transformation of a soluble substrate, colourless *p*-nitrophenylphosphate, into yellow *p*-nitrophenol. The intensity of the coloration is measured with a spectrophotometer at a wavelength of 405 nm; the absorbance may, in a certain range of concentrations, be proportional to the logarithm of the concentration.

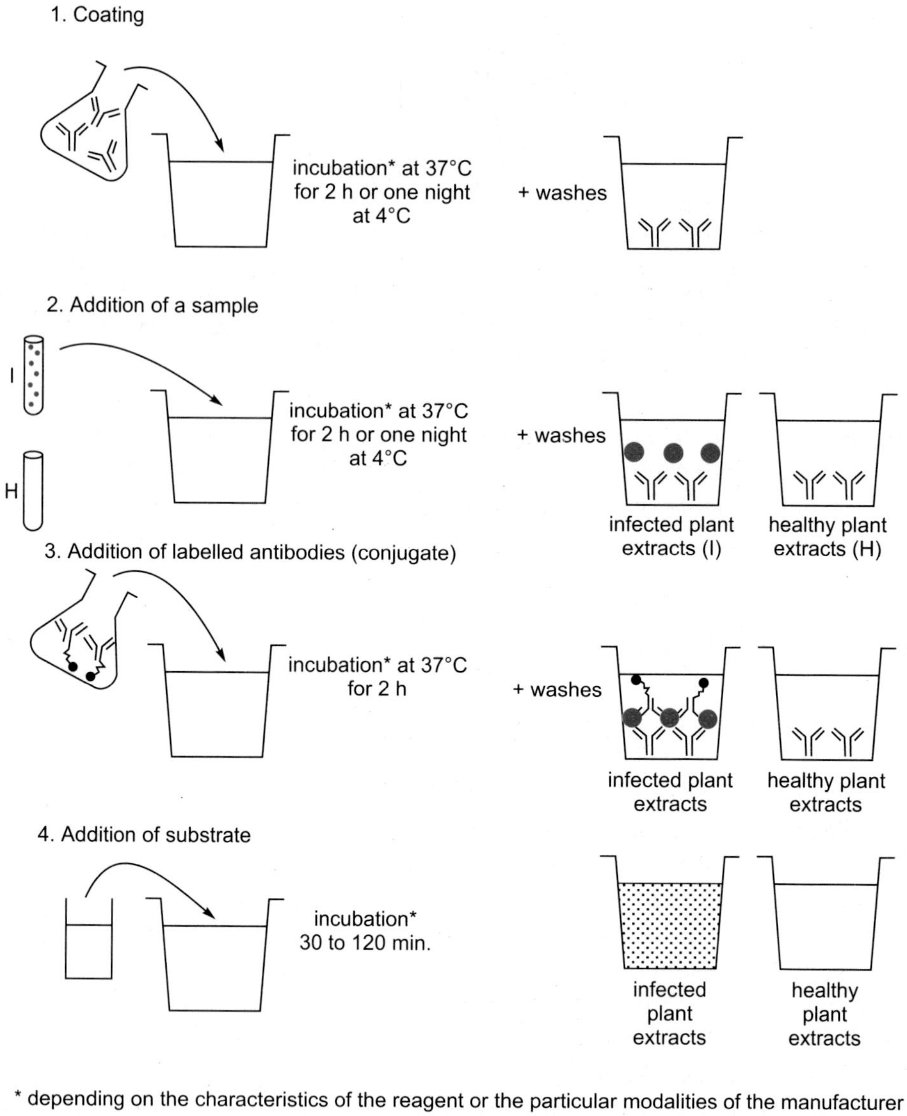

Figure 9.16 Stages of a DAS-ELISA test.

The sensitivity of the DAS-ELISA test is between 1 and 10 ng/ml, depending on the virus considered. It is nearly 1000 times that of an immunodiffusion test. To further improve this sensitivity and broaden the spectrum of strains detected, the conjugation step to the enzyme with glutaraldehyde, which alters a significant proportion of antibodies, can be avoided. The detection antibodies can be labelled without damage with biotin or used unlabelled. In the latter case, the virus-trapping antibodies must be modified (see box below and Fig. 9.17) or antibodies from a different animal must be used. Indirect detection protocols comprising three layers of

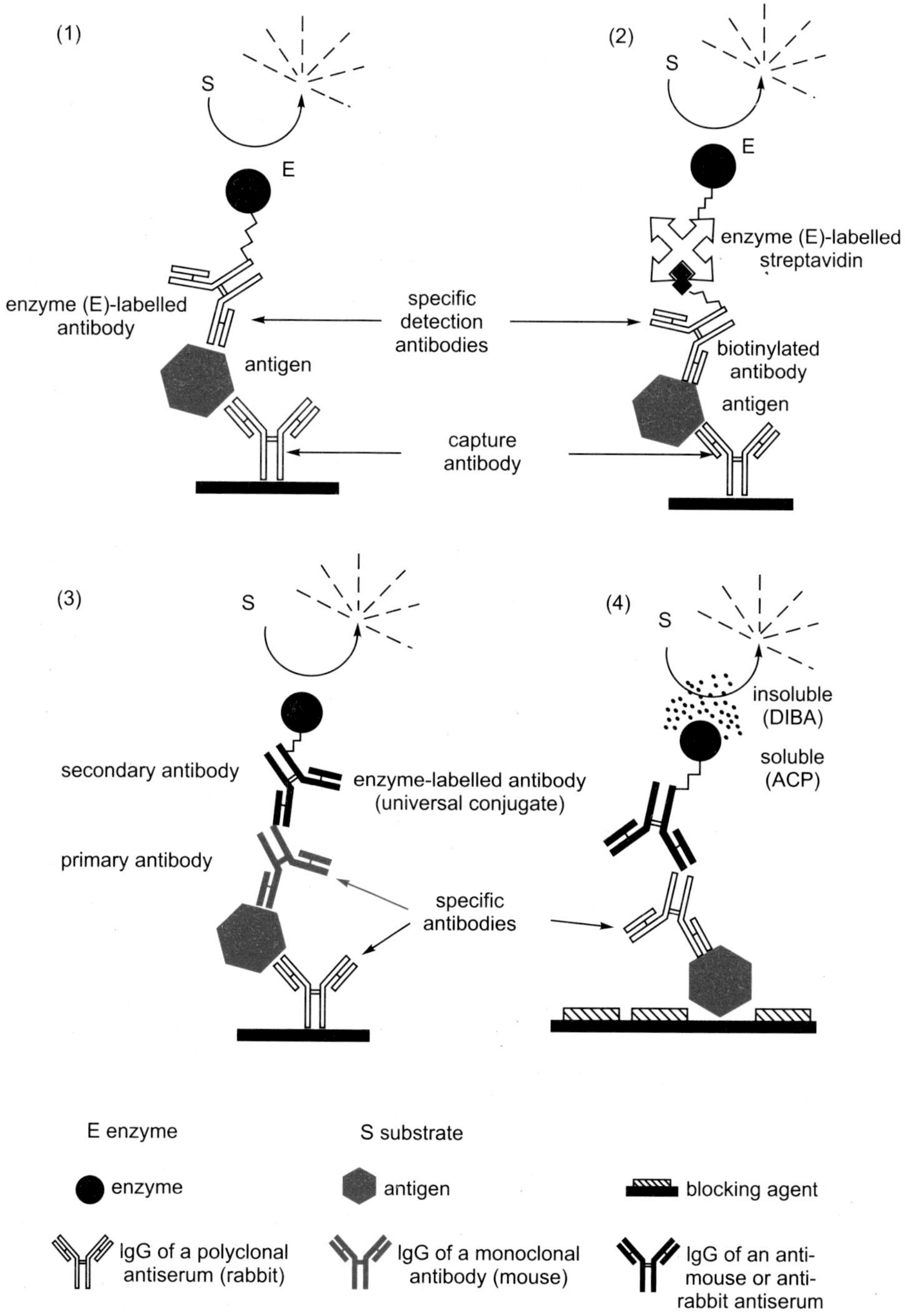

Figure 9.17 Comparison of protocols for conventional DAS-ELISA (1), streptavidin-biotin DAS-ELISA (2), TAS-ELISA with a monoclonal antibody (3), and DIBA on a membrane or ACP-ELISA (4).

antibodies (*triple antibody sandwich* or TAS-ELISA) (Fig. 9.17) generally yield a sensitivity less than 1 ng/ml, reaching sometimes 10 pg/ml, and allow for the detection of a larger number of virus strains (Devergne et al., 1981; Koenig, 1981).

In practice, two factors are predominant in the successful use of the ELISA technique: the serum quality and a judicious choice of the positive threshold.

Serum quality has a decisive influence on the sensitivity and specificity of the test

The use of a serum with high avidity favours the sensitivity of the test; the absence of antibodies against the plant proteins is critical for the specificity of the response. The qualitative choice of polyclonal antibodies can be critical in achieving certain objectives. For example, a potyvirus transmitted by seeds in pea, PSbMV, has its capsid protein intact in the embryo and truncated of its N- and C-terminal parts in the teguments. If the objective of the test is to determine the number of seeds that have come from infected plants (the number of infected teguments), a serum obtained after multiple boosters is chosen (Fig. 9.10(3)). If, in the same seed extracts, the objective is to detect selectively the virus within the embryos (i.e. that will be transmitted to the plantlet), a serum that recognizes only the N-terminal part of the capsid protein has to be chosen (Masmoudi et al., 1994) (Fig. 9.10(1) or (2)) (Plate VII.2a).

Monoclonal antibodies (Halk and de Boer, 1985) are particularly useful in increasing the specificity of the ELISA test. Nevertheless, extreme specificity may have disadvantages when a monoclonal antibody is used for routine tests; there is a risk that certain virus strains having mutations in the region coding for the corresponding epitope may escape detection. This risk is greatly reduced by the use of a defined mix of monoclonal antibodies reacting with different epitopes.

The choice of positive threshold: a compromise between sensitivity and specificity

The sensitivity and specificity of virus detection by ELISA also depend on the choice of a *positive threshold.*

If healthy plant extracts are deposited in all the wells of a microtitre plate, the ELISA test generally gives a "background" and the numbers of wells by absorbance (405 nm) are distributed according to a normal law. If a virus with a final concentration of around 1 μg/ml is added to each of the healthy plant extracts, all the samples containing the virus will react with absorbance clearly higher than those of the healthy extracts, and the distribution curve is displaced towards higher absorbance (405 nm). But if the virus is added to the healthy extracts at a final concentration as low as 10 ng/ml, the curve is only slightly displaced towards the higher absorbance. It intersects with the curve of healthy extracts. There is thus no threshold that separates a healthy sample from a sample containing 10 ng/ml of the virus with a probability of 1.

If all the samples to be identified as infected contain a high concentration of the virus, the choice of a positive threshold is not difficult (for example, 3 times the mean absorbance values of healthy samples: the specificity of the test is excellent, the sensitivity is low). But when the ELISA test is to be used to detect very low virus concentrations, the sensitivity of detection must be increased at the cost of specificity. The positive threshold chosen becomes a *compromise between sensitivity of detection and specificity*. It can be defined as a *probability that a healthy sample will go over the positive threshold*, on the basis of absorbance obtained on the microtitre plate for healthy samples. For example, if a threshold is chosen such that a healthy sample has a probability of 0.005 of exceeding it, and if 10 healthy samples are deposited per microtitre plate, the threshold is calculated as the value of $x + 3.25s$, x being the mean of 10 "healthy" absorbance levels, s the standard deviation, and 3.25 a coefficient given by the Student Fischer table for $(10 - 1) = 9$ degrees of freedom. The sound basis of the choice is confirmed or not confirmed by the simultaneous use of another type of test, for example a biological test. Different ways of determining the positive threshold have been compared (Sutula et al., 1986).

DIBA and TIBA, two variants of ELISA on a membrane

Certain variants of the ELISA test do not use the advantage of providing a result in the numerical form (absorbance values at 405 nm) but propose only a visualization of the coloured reaction. The test uses nylon or nitrocellulose membrane as a support.

The principle of the *dot-blot* or *dot immuno binding assay* (DIBA) (Plate VII.2b) consists in depositing the plant extract to be tested (from 2 to 5 μl) on the membrane and using an indirect reaction. The substrate of the enzyme has to produce an *insoluble* coloured degradation product in order to see the reaction on the membrane (Fig. 9.17). This simple test, which is quick (taking less than 3 h) and easy to carry out, can be used routinely (Clauzel et al., 1994). The *antigen-coated plate* (ACP) ELISA protocol is very similar to DIBA (Fig. 9.17), except that, as for other ELISA protocols, the enzyme substrate and its degradation product are soluble.

The *tissue-blot immuno assay* (TBIA) involves a direct application of fresh sections of infected tissue (stem, leaf, or bulb) on the membrane to visualize the *in situ* distribution in the tissues of numerous viruses (potyviruses, cucumoviruses, luteoviruses, cloteroviruses, etc.) with a high sensitivity (Plate VII.2c) (Makkouk et al., 1993). The possibility of using low-cost paper for blotting samples makes this technique useful for laboratories with limited resources (Makkouk and Kumari, 2002).

In conclusion, the ELISA test and its variants, because they have significantly increased sensitivity, have greatly improved diagnostic methods in plant virology (see box). The commercial availability of high quality reagents for the major plant viruses has made it possible to extend the use of these tests not only to research laboratories, but also to a large number of public or private establishments.

Towards plant virus diagnosis in the fields

There has been great progress in recent years in the development of rapid diagnostic kits that can be used in the fields. The kits do not require specific training or equipment, are ready to use, and can provide a visual response within a few minutes. These immunological tests use the principle of the lateral flow (Danks and Barker, 2000; Ward et al., 2004). Commercial kits are available for some important plant viruses, either as a single- or two-step dipstick assay or as a single-step lateral flow plastic device.

ELISA: reasons for a success

The spectacular development of the ELISA technique over the past few years is due to several factors. Some are related to the protocol itself: it is simple and easily standardized. The equipment needed to carry out the test (oven, photometer) is relatively cheap. The technique is highly sensitive and perfectly adapted to the treatment of numerous samples (routine tests) and to simultaneous search for several different viruses. It thus perfectly answers the need to track down and identify viruses. Some stages of the ELISA test can be automated (reagent deposits and washes). High quality reagents that allow the detection of most of the major plant viruses are marketed by several international companies (e.g., Sediag, Bioreba, Agdia, Adgen).

The ELISA technique has greatly benefited from the considerable improvements in sample preparation procedures. The systems now available vary widely (roll-press, ball-press, pneumatic press, extraction in bags) and make it possible to rapidly obtain extracts from numerous samples of all types of plant organs (leaves, stems, bulbs, tubers, bark, seeds). Apart from the method of preparing plant extracts, the optimization of the tests also relies on the definition for each virus–host plant combination of specific parameters such as the timing for indexing (which must correspond to the period in which the viral concentration in the infected plant is maximal), the sampling protocol, and the extraction buffer that should be adapted for plants that contain substances (in particular tannins) incompatible with a successful ELISA test (Albouy et al., 1980).

Moreover, and this is undoubtedly one of the major advantages of the technique, the results of ELISA tests are numerical, so they can easily be presented, analysed and archived through simple computer software. Once the positive reaction thresholds are defined, the results are very easy to interpret.

Finally, the ELISA technique, because of its sensitivity, can be used in *group testing*: on the basis of a small number of tests, the percentage of infected individuals in a population can be determined with a controlled precision. For example, a large sample (3000 leaves) is divided into 60 equal groups (60 × 50 leaves). The percentage of leaves infected is deduced by a statistical relation of the number of groups negative according to the ELISA test with a determined interval of confidence (Maury et al., 1985). This method has interesting applications in epidemiological studies.

Contribution of electron microscopy

Direct observation of virus particles may guide the diagnosis

Since the early development of electron microscopes, the possibility of observing various types of virus particles directly in crude extracts from infected plants has been exploited for the identification of plant viruses (Brandes and Wetter, 1959). The principle of the dip method is simple: a drop of crude plant extract is deposited on a formwar membrane placed on an electron microscope grid, and then a drop of contrastant (generally a heavy metal salt such as ammonium molybdate or uranyl acetate) is added. During the observation, the viral particles appear light on a dark background (negative staining).

Two principal morphological groups are distinguished: viruses that have isometric particles and those that have elongated particles, either rigid or flexuous rods (Chapter 1, Fig. 1.1). The size and shape of viral particles are characteristic of a viral genus (Chapter 14). These criteria are not sufficient, however, to distinguish two viruses belonging to the same genus, or even in certain cases to distinguish two viruses belonging to genera of similar morphology. Among viruses of isometric symmetry, except for *Caulimovirus*, whose particles of 50 nm diameter are easily recognizable, numerous genera such as *Carmovirus, Cucumovirus, Tombusvirus*, or *Nepovirus* have a size close to 30 nm in diameter and cannot be easily differentiated. Similarly, for viruses of elongated shape, the rigid particles of *Tobravirus* and *Tobamovirus*, respectively of 46-215 nm and 300 nm length, are clearly distinguished from the flexuous particles of genera *Potexvirus, Carlavirus, Potyvirus*, and *Closterovirus*, the modal length of which varies from 420 nm to more than 2000 nm. However, it is more difficult to distinguish between viruses belonging to these last genera, especially when there are broken particles in the preparations. Other genera have a particular morphology that makes them easy to recognize: for example, the bacilliform particles of *Alfamovirus* and *Rhabdoviridae*, the geminate particles of *Geminiviridae*, or the pleomorphic particles of *Tospovirus* (see Fig.1.1).

Apart from these direct observations in crude plant extracts or in purified virus preparations, the analysis of infected tissue in ultrathin sections provides information on the location of virus particles within the cell, the different types of inclusion bodies, and the cellular alterations that serve as useful additional information for diagnosis (see Fig. 9.2 and box on cellular inclusion bodies).

Combination of serology and electron microscopy

The techniques of immuno-electron microscopy (IEM) combine the observation of viral particles and visualization of the antigen-antibody reaction. They constitute one of the most precise means for virus identification. The simplest and fastest technique consists in the examination of a mixture of crude plant extracts and specific antiserum, or the addition of antibodies to a grid on which there are already virus particles deposited. The antibodies fixed on the antigenic motifs of the capsid form a sort of decoration (or coating) around the viral particle that is very easily observed (Fig. 9.18).

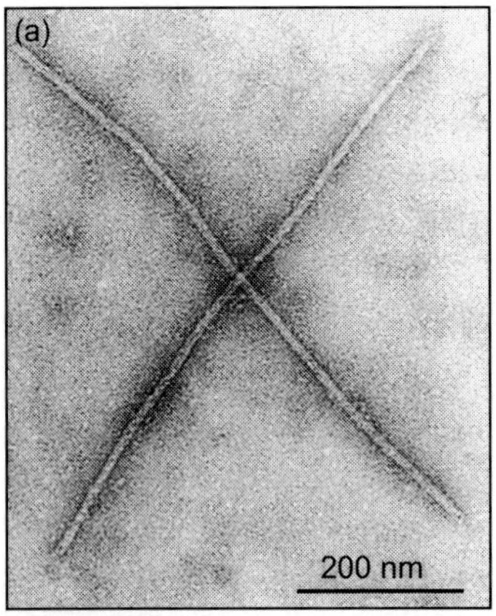

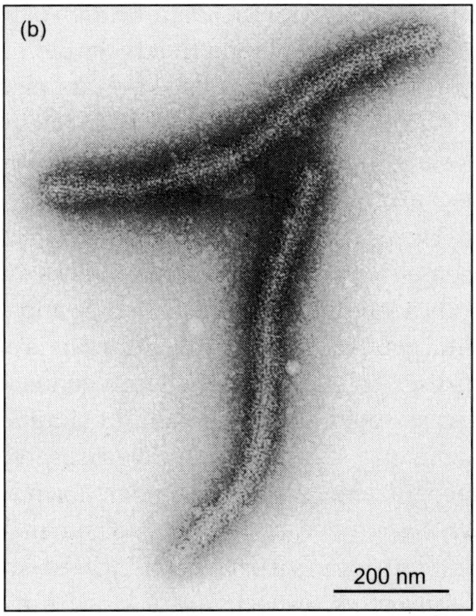

Figure 9.18 The decoration technique allows visualization of the antigen-antibody reaction. Extracts from plants infected by two potyviruses, ZYMV (a) and PPV (b), have been mixed with a PPV specific antiserum. The particles of ZYMV (a) have the normal appearance of a potyvirus, while the particles of PPV (b) are "coated" by antibodies; they are decorated.

In the Derrick method (Derrick, 1973) (Fig. 9.19), the grid is first activated with antibodies, which allows the adsorption of a greater number of viral particles. The viral particles are then decorated. This technique not only results in an increase in sensitivity (100 to 1000 times that of a direct observation), useful for detecting viruses that are in low concentration in the plant extracts, but also improves the quality of observation since the virus is selectively trapped and a large amount of cellular debris is eliminated by successive rinses (Milne and Luisoni, 1975; Milne, 1988). Other improvements can be made using antibodies labelled with colloidal gold particles for the decoration.

The IEM techniques can be used only in laboratories that have an electron microscope. They are less suitable for treating a series of samples, but on the other hand they are of great interest in the detection of viruses that exist in low concentration and in the identification of different viruses occurring as a complex in a single sample.

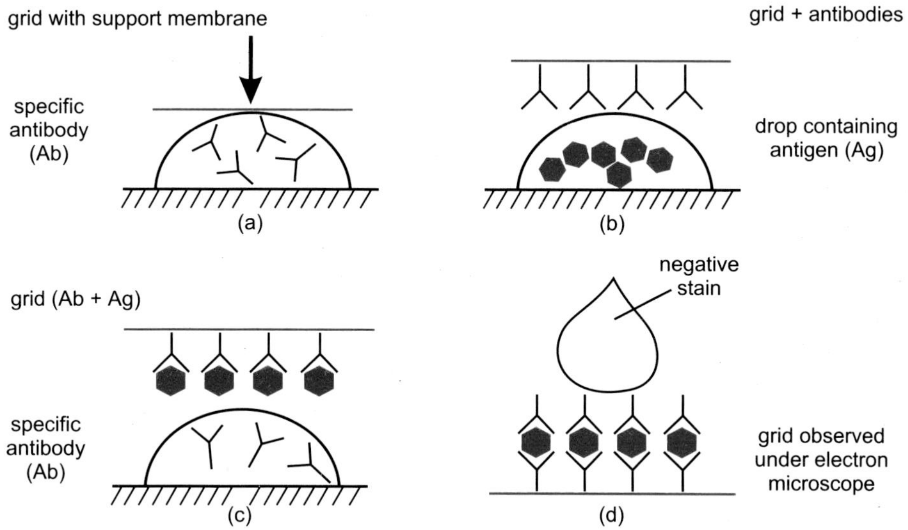

Figure 9.19 Principle of the Derrick method: (a, b) trapping and (c, d) decoration of virus particles.

Detection of viral nucleic acids

The knowledge of RNA or DNA sequences of plant virus genomes opens up an avenue to the development of new techniques of great interest for diagnosis. In contrast with serological techniques, which generally detect the virus capsid protein, techniques based on the sequence of nucleic acids allow in theory the detection of any region of the viral genome. Two techniques are now widely used for diagnosis: molecular hybridization and PCR.

■ Molecular hybridization procedures

The principle of hybridization is based on the formation of a molecule of double-stranded nucleic acid (duplex or hybrid) obtained by hybridization between a target molecule represented by a fragment of the viral genome sequence and the probe constituted by a complementary nucleic acid (either cDNA or cRNA) of that sequence. The hybridization is revealed by labelling the probe either with a radioactive isotope (^{32}P) or a non-radioactive reporter such as biotin, acetylaminofluorene (AAF), fluorocytosine, or digoxigenin.

The test itself is quick and relatively easy to apply (Fig. 9.20). The samples to be tested (only 4-50 µl) are deposited on a nitrocellulose or nylon membrane (dot-blot test). After baking at 80°C or fixation with UV, the membranes can be conserved for several weeks at 4°C or even at room temperature before being analysed, which is very useful for field work. The hybridization takes place in the presence of a labelled

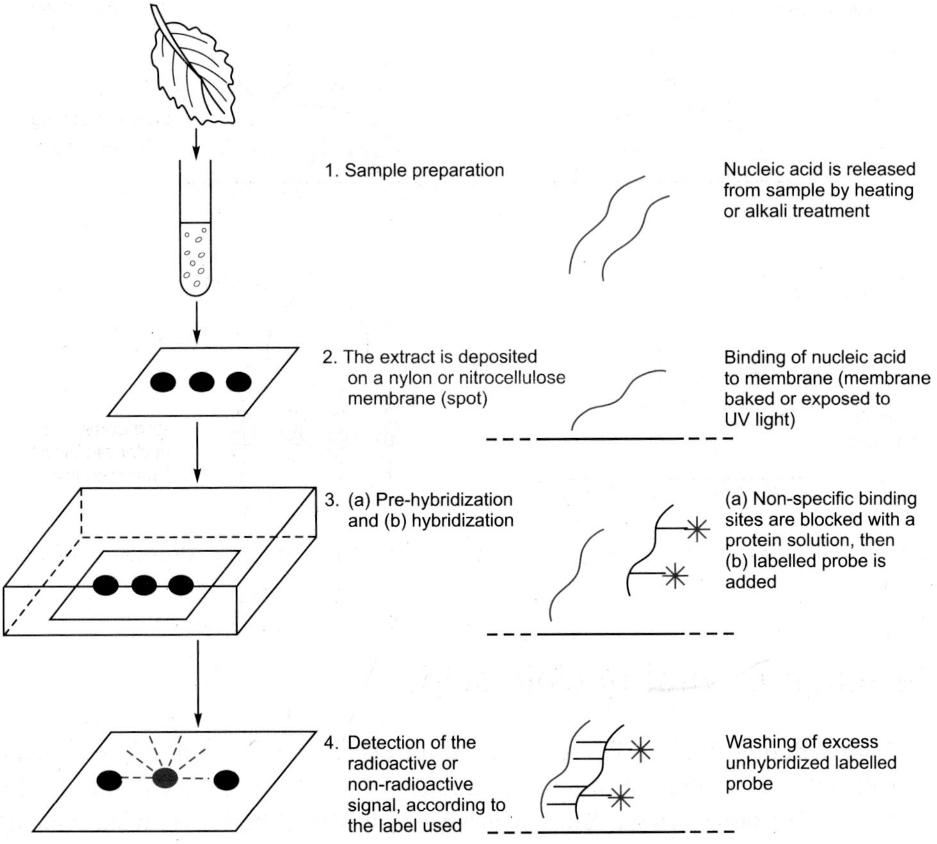

Figure 9.20 Steps in molecular hybridization with a labelled probe.

probe and, after washing and drying, the membranes can be revealed by autoradiography (radioactive probes), immuno-enzymatic reaction, or chemioluminescence (non-radioactive probes) (Fig. 9.21).

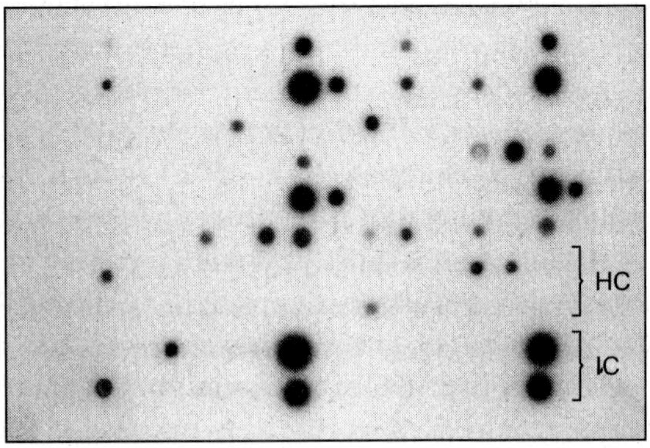

Figure 9.21 Molecular hybridization: detection of DMV (*Dahlia mosaic virus, Caulimovirus*) with a ^{32}P radioactive probe. HC, healthy control; IC, infected control.

The typical sensitivity of a test with a ^{32}P probe is from 1 pg of viral RNA for cDNA probes to 0.1 pg for cRNA probes. This is higher than the sensitivity of a DAS-ELISA test. Nevertheless, radioactive labelling requires particular precautions and limits the use of these probes, which can be manipulated only in authorized laboratories. Tests using non-radioactive probes are thus preferable despite their often lower sensitivity and risks of toxicity during the preparation of probes (Wang et al., 1988).

Amplification of nucleic acid sequences

The first techniques of enzymatic amplification of nucleic acids were developed in 1983 (Seika et al., 1983). They were adapted during the 1990s for the detection of plant viruses. They constitute another major breakthrough in the field of diagnosis (Henson and French, 1993).

The technique of enzymatic amplification *in vitro* can be used to specifically copy exponentially part or all of a viral nucleic acid found in a mixture by means of recognition and hybridization of short complementary sequences (primers) and the action of a heat-stable DNA polymerase. This makes it possible to detect the viral sequence thus amplified as well as characterize it precisely by sequencing (Fig. 9.22).

Numerous methodological adjustments can be used to apply PCR to a wide range of virus-plant combinations. In the case of viruses with an RNA genome, it is necessary to include, before the amplification reaction itself, a step of reverse transcription (RT) during which the RNA is copied into cDNA by a reverse transcriptase (RT-PCR). Depending on the protocols adopted, these two steps (transcription and amplification) can be conducted separately or simultaneously in a single tube.

The preparation of the sample is a critical step because the extracts of certain plants (such as pelargonium or grapevine) may contain inhibitors of enzymatic reactions. Consequently, the extraction or purification protocol of nucleic acids must be adapted.

The sensitivity and specificity of the PCR or RT-PCR tests depend on the choice of primers, temperatures of hybridization and elongation, and the number of cycles. For example, if a specific test for a virus species or virus strain is to be carried out, the primers are chosen in a variable region of the genome. If, on the other hand, several viruses of the same genus or family need to be detected, the primers will be defined in a conserved region within the genus or the family. In practice, in light of the extreme sensitivity of this technique, the impact of the molecular variability of the virus must not be underestimated.

Sometimes, a single or a few mismatches between a primer sequence and a viral nucleic acid target sequence can lead to a negative reaction in PCR or RT-PCR. One way to overcome this difficulty is to use degenerate primers, which consist of a mixture of closely related primers with only a few nucleotide differences at certain positions. Another way to improve primers annealing to the target sequence is to

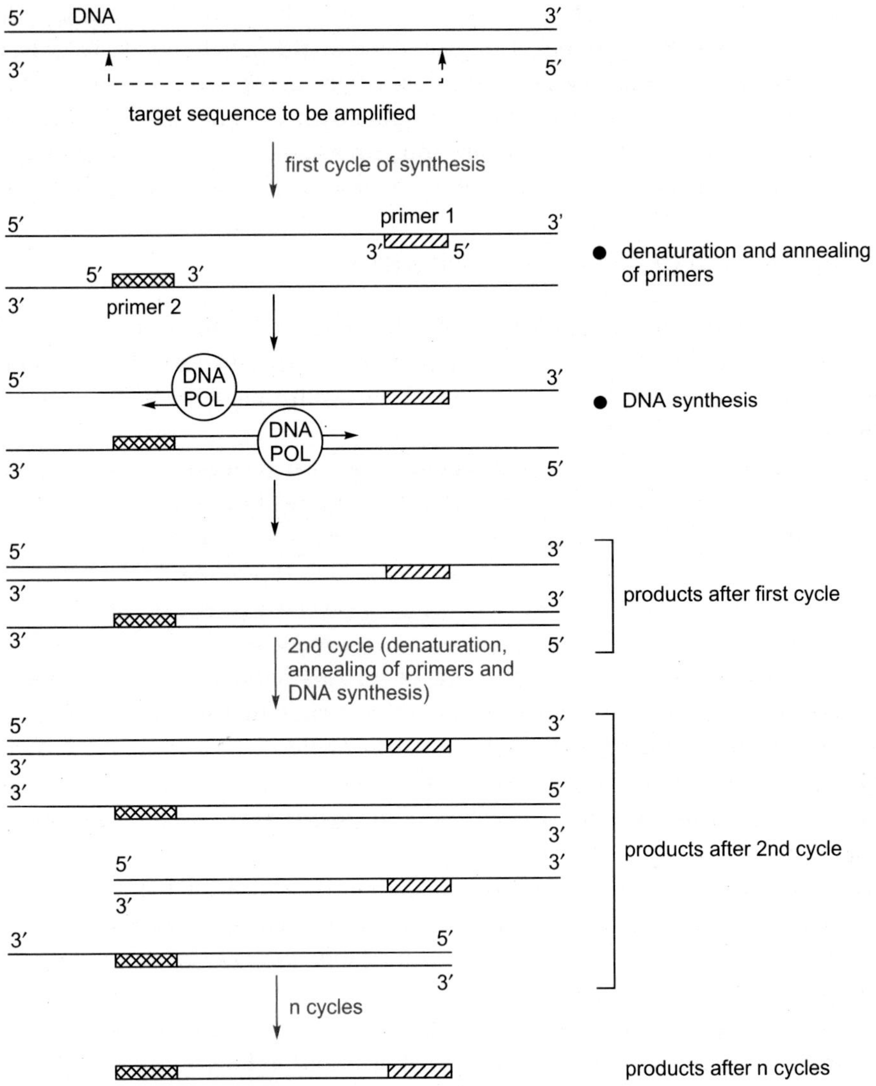

Figure 9.22 The polymerase chain reaction.

change the annealing temperature within the PCR cycle. An increase in annealing temperature will provide a more specific detection, while a decrease will broaden the potential of detection.

Amplification products are currently visualized by observation of bands separated by gel electrophoresis and stained with ethidium bromide (Fig. 9.23). However, the refinement of techniques for detection of the amplification products (amplicons) by colorimetry can lead to not only an increase in sensitivity but also a significant simplification of the analysis when dealing with numerous samples. In the variant called PCR-ELISA, digoxigenin incorporated in the form of DIG-dUTP in

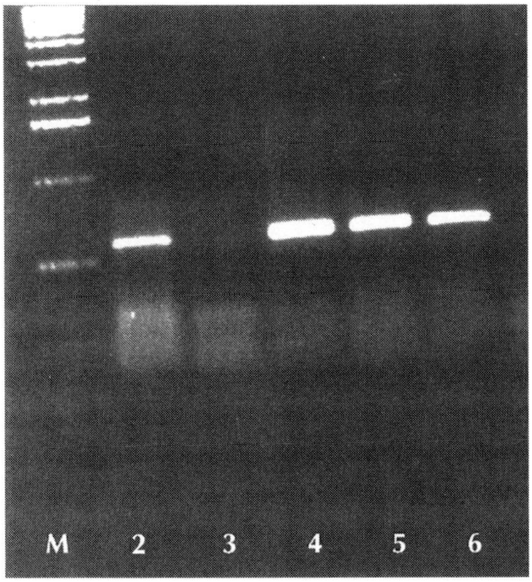

Figure 9.23 RT-PCR test for amplifying a fragment of the capsid protein gene of PFBV (*Pelargonium flower break virus, Carmovirus*) (500 bp). M, molecular ladder; 2, positive control (viral RNA); 3, healthy control; 4, 5, 6, samples from infected pelargonium.

the amplification products is revealed by anti-digoxigenin antibodies conjugated with an enzyme producing a coloured reaction. If a primer is itself labelled with biotin, the reaction can occur in microtitre plates in which the wells have been activated with avidin.

PCR tests are constantly being modified and improved to broaden their field of application: for example, a mixture of enzymes optimized to realize reverse transcription and amplification simultaneously, colorimetric tests (Candresse et al., 1998). In one variant, the virus is trapped using specific antibodies before the RT-PCR. This technique, called immunocapture (IC-) RT-PCR, can be applied for routine virus detection from crude plant extracts (Wetzel et al., 1992). Other variants (multiplex PCR) are designed to use specific primers of several viruses or virus strains in the same reaction so as to detect these viruses simultaneously, which obviously reduces the cost of detection considerably (Jacobi et al., 1998). However, in this case, the amplified fragments should have significant differences in size for each virus, in order to be differentiated during the electrophoresis of the PCR products. The development and diffusion of these techniques are closely dependent on their cost per sample. They are particularly adapted to test plants with high added value (for instance nuclear stock for certification scheme), to conduct molecular epidemiology studies, or to detect viruses for which no serological detection kits are available.

Diagnosis by RT-PCR provides a notable increase in sensitivity; it thus creates particular experimental constraints to prevent contaminations and false positives. Special organization of the laboratory (separating the activities of sample preparation from those of amplification) and a rigorous cleanliness of laboratory benches and premises must be respected.

The possibility of quantifying target DNA (or RNA) has been considerably simplified by the recent development of real-time PCR. In this method, the amplified products are detected by a built-in fluorometer as they accumulate, using either non-specific DNA-binding dyes or fluorescent probes that are specific to the target DNA. Advantages of this method are the possibility of detecting several targets (different viruses or virus strains) during the same test by using probes with different fluorescent reporter dyes, the capacity to compare virus concentrations in different plant genotypes, and a significant reduction of the time required for the test.

Carrying out a PCR test

The relatively simple principle of PCR is based on the capacity of a replication enzyme, DNA polymerase, to synthesize a DNA fragment using as primers short nucleotide sequences of around 20 bases. These two primers, complementary to each of the two strands of the DNA respectively, make it possible to amplify the DNA fragment between the annealing zones (Fig. 9.22). At each cycle, the sequence defined by the two initial primers is thus duplicated. However, a chain reaction takes place, because this sequence is duplicated not only from the initial molecule but also from newly synthesized strands. It is the repetition of several cycles of denaturation-hybridization-extension (around 20 to 40 cycles) that leads to an exponential accumulation of the DNA.

The denaturation of products at each cycle occurs by exposure to a temperature of 95°C. The discovery of *Taq polymerase*, which resists these temperatures, was at the origin of the development of PCR tests. This enzyme is extracted from the thermophilic bacterium *Thermus aquaticus*.

In practice, the amplification occurs in a microtube or in a microtitre plate in which all the necessary reagents are added. The protocol comprises three successive phases, carried out in different thermal conditions:

- Denaturation, from 15 sec. to 2 min. at 95°C, which allows the separation of the complementary chains of double-stranded DNA.
- Hybridization, from 1 to 2 min., at a defined temperature between 37°C and 60°C, depending on the composition (AT/GC ratio) and the primer length; during this step the primers hybridize with their complementary sequences, flanking the region to be amplified. The specificity of the PCR is based on the quality of this pairing linked to the size of primers and their sequences.
- Extension, from 30 sec. to 25 min. at 72°C, which leads to the synthesis of a sequence complementary to the template strand in the presence of the polymerase and precursor nucleotides, the enzyme copying the DNA starting from the primers.

Thermocyclers are used to carry out these three steps automatically.

Once the DNA fragment is amplified, it must be characterized. After a gel electrophoresis that separates the DNA molecules according to size, a fluorescent substance, ethidium bromide, that intercalates between the two DNA strands makes it possible to visualize it and estimate the molecular weight (and thus the length) of the amplified fragment. The amplified products can ultimately be characterized by the study of restriction fragment length polymorphism (RFLP); a more refined characterization by direct sequencing is also commonly carried out.

Towards a judicious use of diagnostic methods

For several years, increasingly sensitive and specific diagnostic methods have been developed. These techniques are perfectly adapted to detection, confirming (or not confirming) a hypothesis posed beforehand on the possible presence of a given virus in an infected plant sample. In other words, they can efficiently detect *only* the target virus. That could, in the case of mixed infections, lead to partial or even wrong diagnoses in revealing the presence of only a single virus among a complex, a virus that may not be the one responsible for the symptoms observed.

At present, there is a need for methods or reagents that are less specific and can detect in a single test a large number of different strains or viruses, including emerging viruses that appear regularly in various parts of the world. Biological methods and electron microscopy offer many advantages in this regard but can generally be used only in research laboratories. Serological and molecular methods could undoubtedly meet these needs through the development of adapted kits. For example, in the case of potyviruses, there already exists a monoclonal antibody allowing the detection of most potyviruses by ELISA, and there are RT-PCR primers chosen in highly conserved parts of the genome that can amplify the coat protein gene of all *Potyviridae* (Jordan, 1992; Gibbs and Mackenzie, 1997).

Highly sensitive diagnostic techniques are often difficult to use routinely because they demand great care in laboratory organization, choice of material, and the carrying out of manipulations. Any risk of contamination of samples and appearance of false positives must be avoided. This is why the techniques must be used within a framework of voluntary or compulsory quality control protocols and good laboratory practices. Private companies or plant clinic laboratories need to conform to national or international regulation (particularly for international trade and the establishment of phytosanitary passports). This implies strict following of officially established protocols.

The evolution of diagnostic methods towards simplification of protocols and much faster response will certainly continue in the years to come. Kits are already available to detect the presence of a virus in just a few minutes (Ward et al., 2004). Other highly effective kits, based on genome amplification, are being developed or tested (Martin, 1998). But we should keep in mind that more conventional and less demanding diagnostic methods can still be improved and they remain very useful in resolving problems posed by virus diagnosis, particularly in developing countries (Makkouk and Kumari, 2002).

10
CHAPTER

Control of Plant Viral Diseases: Prophylactic Measures

There is at present no chemical treatment against viruses that can be applied directly in the field. This is why the diseases they cause are still called "incurable", since once a plant is systemically infected by a virus it will remain infected all its life. Advances in the control of viral diseases can result from the use of measures deduced from a thorough knowledge of viral biology (Jones, 2004). The application of rational preventive methods, adapted to each type of crop and to each virus, makes it possible today to control viral diseases quite effectively and in many cases to limit the damage they cause to an economically acceptable level.

Methods of virus control comprise three major components.

Viruses cited

ArMV, *Arabis mosaic virus, Nepovirus*; BBSV, *Broad bean stain virus, Comovirus*; BBTMV, *Broad bean true mosaic virus, Comovirus*; BCMNV, *Bean common mosaic necrosis virus, Potyvirus*; BCMV, *Bean common mosaic virus, Potyvirus*; BSMV, *Barley stripe mosaic virus, Hordeivirus*; BYDV, *Barley yellow dwarf virus, Luteovirus*; BYMV, *Bean yellow mosaic virus, Potyvirus*; BYSV, *Beet yellow stunt virus, Closterovirus*; CABMV, *Cowpea aphid-borne mosaic virus, Potyvirus*; CeMV, *Celery mosaic virus, Potyvirus*; CGMMV, *Cucumber green mottle mosaic virus, Tobamovirus*; CLCuV, *Cotton leaf curl virus, Begomovirus*; CMV, *Cucumber mosaic virus, Cucumovirus*; CTV, *Citrus tristeza virus, Closterovirus*; CymMV, *Cymbidium mosaic virus, Potexvirus*; DMV, *Dahlia mosaic virus, Caulimovirus*; GFLV, *Grapevine fanleaf virus, Nepovirus*; HPDV, *High plain disease virus*; IPCV, *Indian peanut clump virus, Pecluvirus*; LMV, *Lettuce mosaic virus, Potyvirus*; MDMV, *Maize dwarf mosaic virus, Potyvirus*; MNSV, *Melon necrotic spot virus, Carmovirus*; ORSV, *Odontoglossum ringspot virus, Tobamovirus*; PCV, *Peanut clump virus, Pecluvirus*; PEBV, *Pea early-browning virus, Tobravirus*; PeMoV, *Peanut mottle virus, Potyvirus*; PLRV, *Potato leafroll virus, Polerovirus*; PMMoV, *Pepper mild mottle virus, Tobamovirus*; PPV, *Plum pox virus, Potyvirus*; PRSV, *Papaya ringspot virus, Potyvirus*; PSbMV, *Pea seed-borne mosaic virus, Potyvirus*; PVA, *Potato virus A, Potyvirus*; PVM, *Potato virus M, Carlavirus*; PVS, *Potato virus S, Carlavirus*; PVX, *Potato virus X, Potexvirus*; PVY, *Potato virus Y, Potyvirus*; SBWMV, *Soil-borne wheat mosaic virus, Furovirus*; SMV, *Soybean mosaic virus, Potyvirus*; SMYEV, *Strawberry mild yellow edge virus, Potexvirus*; SqMV, *Squash mosaic virus, Comovirus*; SYVV, *Sowthistle yellow vein virus, Nucleorhabdovirus*; TBV, *Tulip breaking virus, Potyvirus*; TMV, *Tobacco mosaic virus, Tobamovirus*; ToMV, *Tomato mosaic virus, Tobamovirus*; TRSV, *Tobacco ringspot virus, Nepovirus*; TSWV, *Tomato spotted wilt virus, Tospovirus*; TYLCV, *Tomato yellow leaf curl virus, Begomovirus*; ULCV, *Urdbean leaf crinckle virus*; WMV, *Watermelon mosaic virus, Potyvirus*; WSMV, *Wheat streak mosaic virus, Tritimovirus*; ZYMV, *Zucchini yellow mosaic virus, Potyvirus.*

First of all, healthy seeds must be used in a healthy environment. This requires the implementation of selection programmes that guarantee the farmer access to virus-free seeds or propagules. Also, the sources of viruses in and around the fields must be eliminated.

Second, some methods can be used that disturb the activity or efficiency of vectors. These most often involve adaptation of particular cultural or phytosanitary practices on a case-by-case basis.

Finally, plants that are more resistant to viral infection can be produced. This objective can be achieved by mild-strain cross-protection, which is a form of biological control of phytopathogenic viruses, as well as by selection of resistant varieties using conventional breeding methods (Chapter 11). The recent development of genetic engineering opens up new perspectives to obtain transgenic resistant plants (Chapter 12).

Control of plant viral diseases does not involve only one plant, one field, or even one farm. It must be addressed at the regional level and thus requires a collective discipline.

Virus-free seeds and vegetative propagules

Sources of viruses that initiate an epidemic can sometimes be present within the crop, as early as the planting stage. This situation very frequently occurs in plants that are vegetatively propagated: one infected mother plant will produce, through cuttings, rhizomes, tubers, bulbs, or grafts, a progeny that will be most often totally infected. For plants propagated by seed, this is less common, since most plant viruses are not transmitted through seeds (Chapter 8). It is therefore very important to limit virus transmission through planting material and seeds in order to prevent the development of early epidemics in individual fields but also to avoid global virus dissemination through international seed trade.

It is thus essential for the farmer to have access to planting material or seeds that are guaranteed to be healthy through suitable controls. The principle of sanitary selection consists in choosing or obtaining virus-free plants (nuclear or foundation stocks) and then propagating them in such a way that they are protected from virus infections. In this way, seed lots (in the wide sense, including true seeds and vegetative propagules) can be commercialized with a contamination rate that is guaranteed to be below a defined or regulatory threshold.

Sanitary selection has proved effective for about 50 years and has been particularly beneficial in keeping healthy a large number of annual and perennial, temperate and tropical cultivated species. Its implementation is based on producing healthy stock from infected plants (through thermotherapy or meristem tip culture), rapid propagation in controlled conditions (*in vitro* culture), and of course the application of cultural practices that limit the development of viruses in seed production fields. To achieve these objectives, sensitive and reliable diagnostic methods must be available to monitor the health status of seed crops quickly and at any given time (Chapter 9).

Production of virus-free stocks of vegetatively propagated plants

In a sanitary selection programme of a plant species that is vegetatively propagated, it is crucial to start with a virus-free plant, constituting the initial material, which must be protected from subsequent viral contamination during the entire propagation process. This virus-free plant may be simply collected from the field, during the inspection of a crop or varietal collection, but its sanitary status must be rigorously controlled by appropriate diagnostic methods. Virus-free plants can also be obtained in laboratory from infected plants using curative methods that allow a healthy clone to be regenerated. The two techniques of thermotherapy and meristem tip culture have made it possible to regenerate a wide range of plant species that were initially totally virus-infected, while conserving the original agronomic properties of the varieties concerned.

Thermotherapy

The use of high temperature, either by exposure to hot air or by immersion in hot water, was developed after the studies of Kunkel in 1935, which were designed to eliminate various diseases caused by phytoplasma in peach. In 1949, Kassanis achieved the recovery of potato tubers infected by PLRV (*Potato leafroll virus, Polerovirus*) by hot air treatment at 36°C for 20 days. Since then, thermotherapy has been used to produce healthy stocks of numerous woody species (rose, ornamental shrubs, fruit trees) and herbaceous species (strawberry) (Mink et al., 1997).

The objective of the method, which is still highly empirical, is to reach temperatures that are lethal for the virus but not for the host plant. The efficiency of thermotherapy treatments seems to be linked to a significant reduction or complete stop of virus multiplication at temperatures close to 40°C (Dawson, 1976). These effects could be linked, among other things, to destabilization of virus particles or inactivation of viral replicase (Mink et al., 1997).

Many viruses have been eliminated by hot air treatments of whole plants or plant fragments in adapted containers (the plant may be kept for a week to several months between 35°C and 40°C, depending on the plant species and the virus). Treatments in hot water or humid chambers (from 5 min. to a few hours, between 35°C and 50°C) are highly effective against phytoplasma but are much more rarely used against viruses (Nyland and Goheen, 1969). There is nevertheless a wide variability in the efficiency of heat treatments in a single plant species, depending on the virus under consideration. In potato, for example, PVY (*Potato virus Y, Potyvirus*) or PLRV are quite easily eliminated, unlike PVS (*Potato virus S, Carlavirus*) or PVX (*Potato virus X, Potexvirus*) (Mink et al., 1997).

Meristem tip culture has been used successfully to produce virus-free stocks of a large number of cultivated plant species

A brief history

Organogenesis in plants is ensured by meristems which are constituted of clumps of undifferentiated cells that have the capacity to divide actively. The possibility of regenerating from this type of explant plantlets that are genetically identical to the original plant has had a highly significant impact on agriculture. Studies of virus distribution in tobacco shoot bud meristems (Limasset and Cornuet, 1949) showed progressive reduction in the number of viral particles towards the apical parts, the meristem-dome being free of them. This absence of viruses in the meristematic cells of certain infected plants inspired Morel and Martin (1952, 1955) to isolate and cultivate apical meristems of dahlia and potato *in vitro* in an attempt to obtain healthy plants. They obtained healthy stocks from dahlia infected by DMV (*Dahlia mosaic virus, Caulimovirus*) and from the entire French collection of potato varieties infected by PVY (see box and Fig. 10.1).

Meristem tip culture: the principle

Meristem tip culture involves *in vitro* culture of a very small fragment of the apical crown of caulinary meristems (measuring a few tens of millimetres) (Fig. 10.1). The still undifferentiated meristem cells are thus capable of regenerating, on an adapted nutrient medium, a plantlet that conserves all the genetic characteristics of the mother plant (Plate VIII.1). Nevertheless, the agronomic properties of the clones thus obtained need to be verified ultimately in the field because some somaclonal variability may appear during the different stages of *in vitro* culture.

Virus eradication is not always total and the plantlets obtained are not all virus-free. The success rates depend on the virus that was meant to be eliminated and its capacity to invade the meristem cells, the species or variety from which virus-free stock was to be generated, and most of all the size of explant taken (0.2 to 0.5 mm). The size adopted for the sampling is a result of a compromise between the chances of regeneration of the explant (meristem-dome with one or two leaf primordia) and the chances of obtaining virus-free explants. In carnation, for example, the percentage of healthy plants obtained from meristems of 0.1 to 1 mm varies from 67% to 0% (Stone, 1968).

This important discovery has since been widely used throughout the world to eliminate numerous viruses in more than 40 species that are vegetatively propagated (Faccioli and Marani, 1998). These include ornamentals, vegetables, industrial crops, or fruit-trees of temperate regions (e.g., carnation, *Pelargonium*, garlic, potato, strawberry, apple, plum, grapevine) as well as of tropical regions (e.g., orchids, cassava, sweet potato, sugarcane, pineapple, banana).

Meristem tip culture is thus the starting point for any *in vitro* micropropagation with the aim of propagating only plants free of viral infection. It is presently practised by numerous public laboratories and private firms in the field of horticulture.

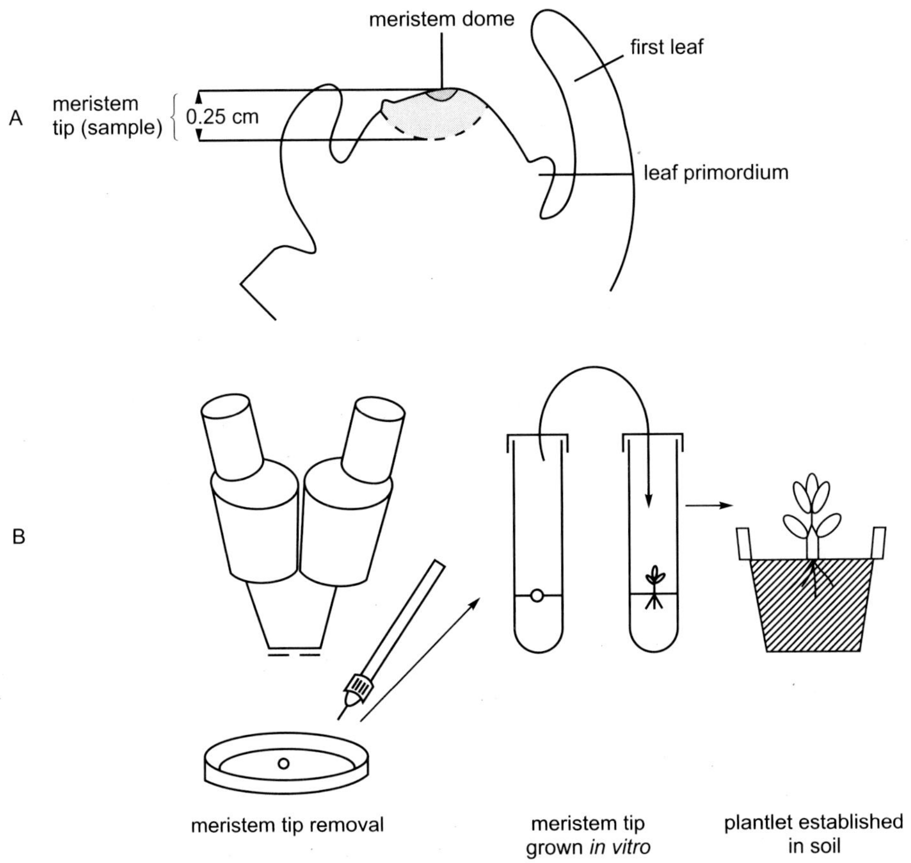

Figure 10.1 A. Longitudinal section of a shoot bud meristem. On either side of the dome, there are leaf primordia; the part to be sampled for the meristem tip culture is shown in grey. B. Different stages of meristem tip culture for virus elimination. At left: dissection of a part of the meristem crown using a microscope under laminary flow hood. Young leaves and the leaf primordia are removed, and the tools are sterilized each time so as not to risk contaminating the meristem. At centre: the meristem is removed and placed on the surface of the culture medium. Each plant species requires development of a particular culture medium, which generally contains mineral salts (according to the formula of Murashige and Skoog, 1962), a carbon source (most often sucrose), vitamins, and hormones (cytokinin and then auxin). At right, after several subcultures, plantlets are established in soil and diagnostic tests can then be conducted to ensure that the plant is indeed virus-free (modified from Cornuet, 1987).

Obviously, the efficiency of the method in producing virus-free plants must be validated by rigorous virus detection controls.

Optimizing meristem tip culture

Are meristems always virus-free? Sensitive methods of detection have made it possible to detect certain viruses in the meristem clusters. In orchids, for example, ORSV (*Odontoglossum ringspot virus*, *Tobamovirus*) is detected just within the apical dome and is thus not eliminated during meristem tip culture, while CymMV (*Cymbidium mosaic virus, Potexvirus*) is absent from the meristem. The absence of a virus in the meristematic tissues could be due to the lack of cofactors necessary for viral multiplication or absence of cell-to-cell migration. On the other hand, it is surprising to observe that certain viruses that are present in the meristem tip could be eliminated during *in vitro* culture; hormonal conditions (cytokinins), accumulation of polyphenols, and metabolic competition are among the factors that could interfere with and block viral replication (Faccioli and Mariani, 1998).

For plants infected by viruses that are difficult to eliminate, thermotherapy can be combined with meristem tip culture, either by treating the mother plant before the meristem is taken, or by carrying out thermotherapy of the plantlet *in vitro*. In strawberry, for example, 82% of plants were free of SMYEV (*Strawberry mild yellow edge virus, Potexvirus*) after a treatment of mother plants for 6 weeks at 36°C, as against just 25% from untreated plants (Mullin et al., 1974).

Chemotherapy can also be combined with *in vitro* culture. Some inhibitors of viral replication have been identified, but these substances also alter the host metabolism and are often phytotoxic. These molecules are analogues of bases involved in the synthesis of nucleotides or nucleosides such as 8-azaguanine or 2-thiouracil, antibiotics, or antimetabolites. Ribavirin, a synthetic analogue of guanosine, better known under the commercial name Virazole®, is an antiviral substance with a very wide spectrum of action, since it is also used in human virology (influenza, hepatitis, HIV, Lhasa fever). In plants, this substance incorporated in an *in vitro* culture medium slows or blocks viral multiplication and increases the rate of production of healthy stocks. For example, ribavirin, at doses of 10 to 100 mg/l, makes it possible to produce healthy stocks of various plant species including fruit trees (Deogratias et al., 1989), orchids (Albouy et al., 1988; Toussaint et al., 1993), and bulb plants (Blom-Barnhoorn et al., 1985).

Micrografting of meristem tips *in vitro*

In ligneous plants and fruit trees in particular, meristem tip culture is very difficult. Most often, the meristem is incapable of undergoing all the stages that lead to the formation of a plantlet. The obtaining of virus-free ligneous plants thus relies on the development of a specific technique, grafting meristem tips on a healthy root-stock (grown from seed or through *in vitro* micropropagation) (Plate VIII.2). This method, initially developed by Holmes (1956) for chrysanthemum, made it possible to eliminate several diseases caused by viruses, viroids, or phytoplasma in citrus (Navarro et al., 1975), diverse *Prunus* and *Malus* species (Navarro et al., 1982; Deogratias et al., 1986), and grapevine.

Various modifications were proposed in order to obtain more rapid growth of the graft in species of the genus *Prunus*. Moreover, pre-treatment of mother plants in a thermotherapy chamber increases the efficiency of virus elimination (Deogratias et al., 1989). These successive improvements have made it possible to obtain a success rate close to 80%.

I.1. Colour 'break' symptoms on tulip due to TBV (*Tulip breaking virus, Potyvirus*)

I.2. Mosaic on tobacco due to an "aucuba" strain of TMV (*Tobacco mosaic virus, Tobamovirus*).

I.3. Mosaic on tomato caused by ToMV (*Tomato mosaic virus, Tobamovirus*).

I.4. Vein banding on melon leaf infected by SqMV (*Squash mosaic virus, Comovirus*).

I.5. Ring spot on papaya leaf infected by PRSV (*Papaya ringspot virus, Potyvirus*).

I.6. Distribution of symptoms on leaves of tobacco in which a basal leaf has been inoculated with PVY (*Potato virus Y, Potyvirus*). Very young leaves (17, 18) have more marked symptoms on the apical part; young leaves (14, 15, 16) have homogeneous symptoms of mosaic; leaves in source-sink transition (12, 13) have symptoms limited to the basal part; other leaves (11) have no symptoms. Note: leaves 15, 16, 17 and 18 are enlarged (× 1.2) in relation to leaf 14.

PLATE **II**

II.1. Court-noue of grapevine caused by GFLV (*Grapevine fanleaf virus, Nepovirus*); GFLV is disseminated by the nematode *Xiphinema index*.

4

II.2. Yellowing of old leaves of melon caused by CABYV (*Cucurbit aphid-borne yellows virus, Polerovirus*).

II.3. Curling and yellowing of tomato leaves caused by TYLCV (*Tomato yellow leaf curl virus, Begomovirus*).

II.4. Necrosis on *Phalaenopsis* caused by the CyMV + ORSV complex (*Cymbidium mosaic virus, Potexvirus; Odontoglossum ringspot virus, Tobamovirus*).

II.5. Necrosis on *Gloxinia* caused by TSWV (*Tomato spotted wilt virus, Tospovirus*).

II.6. Necrotic spots on cucumber caused by MNSV (*Melon necrotic spot virus, Carmovirus*).

II.7. Curling of a pelargonium leaf caused by PLCV (*Pelargonium leaf curl virus, Tombusvirus*).

7

III.1. Heterogeneous coloration of tomato fruits caused by ToMV (*Tomato mosaic virus, Tobamovirus*).

III.2. Ring and curled patches on tomato fruits caused by TSWV (*Tomato spotted wilt virus, Tospovirus*).

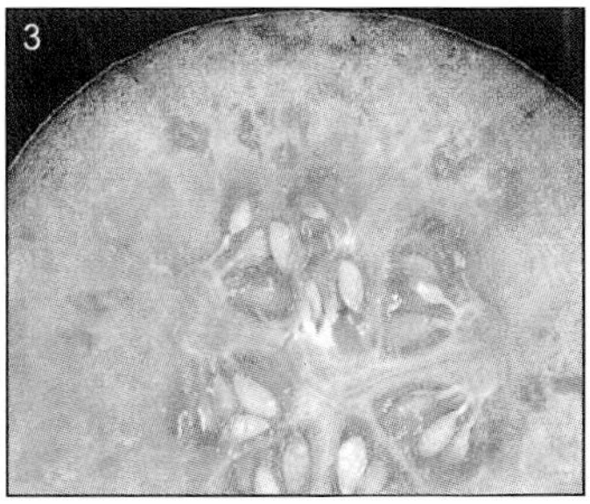

III.3. Internal necrosis of melon caused by ZYMV (*Zucchini yellow mosaic virus, Potyvirus*).

III.4. Necrosis on potato tuber infected by PVY-NTN (necrotic strain of *Potato virus Y, Potyvirus*).

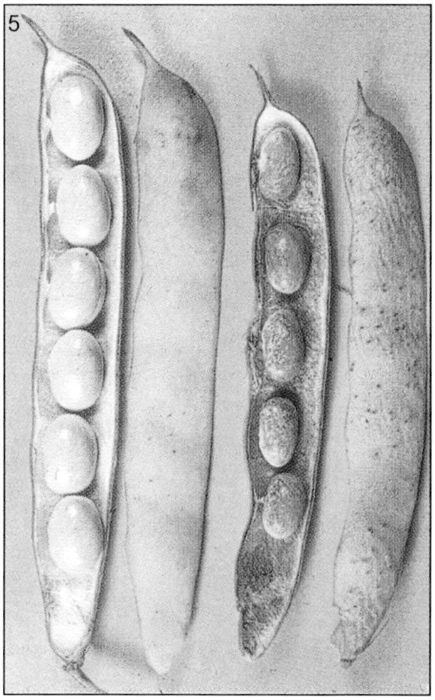

III.5. Necrosis on bean pod and seeds infected by BCMNV (*Bean common mosaic necrosis virus, Potyvirus*) (right), healthy pod and seeds (left).

III.6. Necrosis on teguments of pea seeds infected by PSbMV (*Pea seed-borne mosaic virus, Potyvirus*).

6

PLATE **IV**

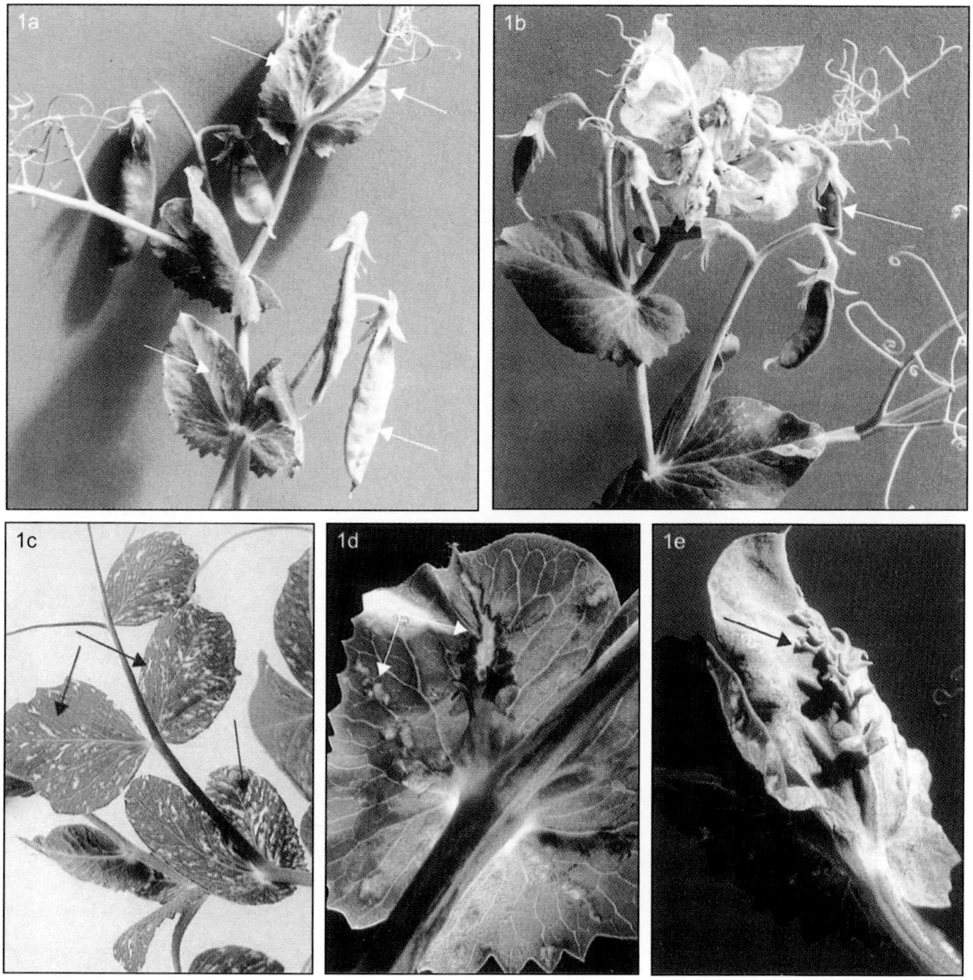

IV.1. Variability of symptoms caused on peas by PEMV (*Pea enation mosaic virus, Enamovirus*); normal internodes and generally green colour (1a) or rosette and apical yellowing (1b) with a typical symptom, enations (arrows): on leaflets and stipules seen from the upper side (1c); on pods (1a, 1b); on stipules seen from the lower side (1a, d, e).

IV.2. Foci of PEMV in a pea field. The virus is disseminated by the aphids *Acyrthosiphon pisum* or *Myzus persicae* from reservoirs which are wild legume plants.

V.1. Symptoms of mosaic in wheat (*Triticum aestivum*) caused by SBWMV (*Soil-borne wheat mosaic virus, Furovirus*).

V.2. Varietal trials for testing susceptibility of durum wheat (*Triticum durum*) to SBWMV.

V.3. Symptoms on oat (*Avena sativa*) of yellowing and dwarfing due to BYDV-PAV (*Barley yellow dwarf virus-PAV, Luteovirus*).

V.4. Symptoms of dwarfing on barley (*Hordeum sativum*) caused by WDV (*Wheat dwarf virus, Mastrevirus*).

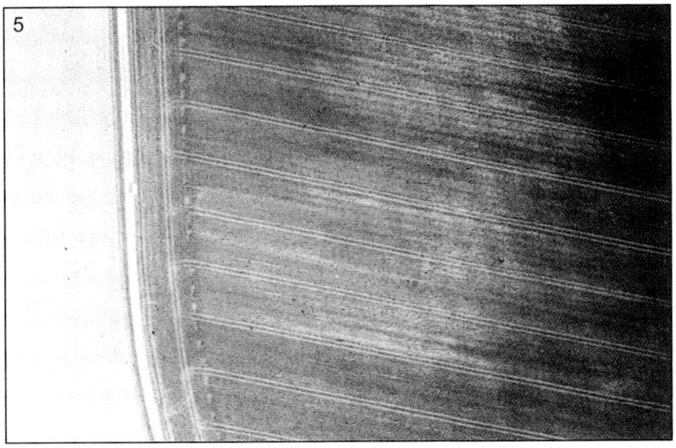

V.5. Aerial photograph (400 m) showing foci of wheat spindle streak mosaic in a field of wheat (*Triticum aestivum*). WSSMV (*Wheat spindle streak mosaic virus*) is a *Bymovirus* spread by a soil fungus, *Polymyxa graminis*.

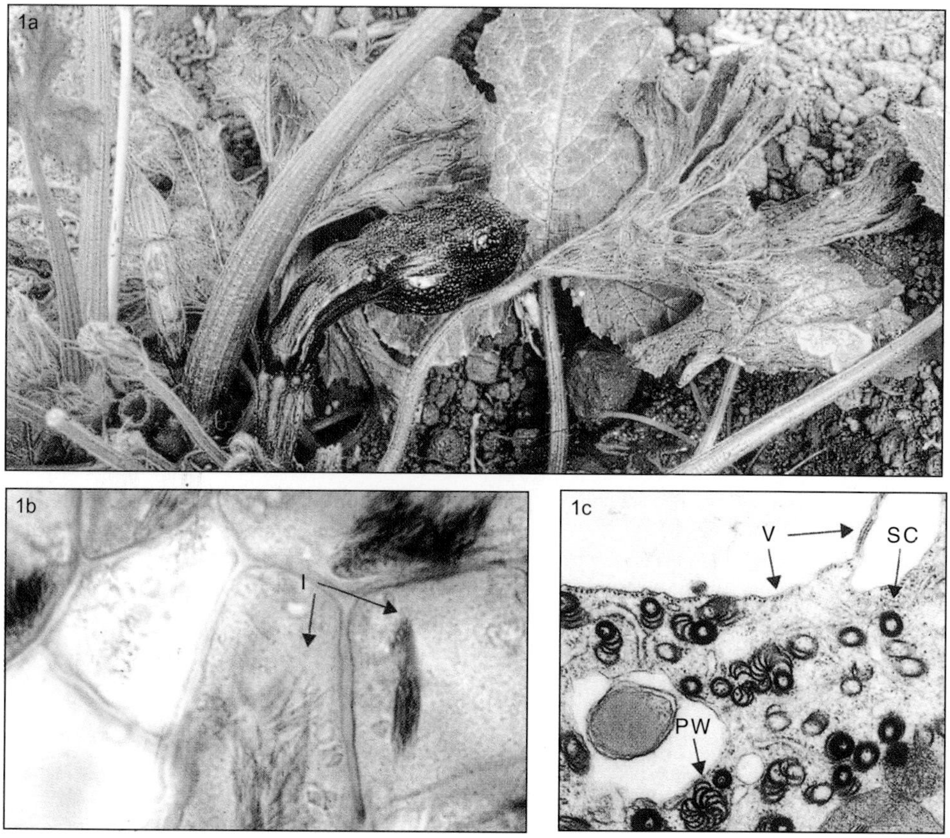

VI.1. Symptoms caused by a potyvirus, *Zucchini yellow mosaic virus* (ZYMV) at different levels of observation.

1a: macroscopic symptoms of mosaic and deformation on leaves and fruits.

1b: cellular inclusions (I) characteristic of *Potyviridae* observed under light microscope after orange-green coloration.

1c: cylindrical inclusions, some of pinwheel type (PW) and others of scroll type (SC), observed in ultrathin sections examined in electron microscopy. There are also viral particles (V) seen in transversal section along the tonoplast of the vacuole or aligned in the cytoplasmic bridges.

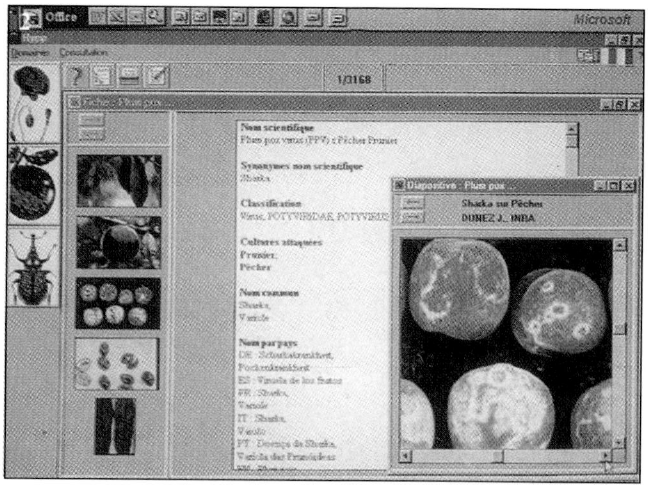

VI.2. HYPP: Diagnostic aid profile for sharka disease (*Plum pox virus, Potyvirus*).

VII.1. Detection of viruses by biological tests.

1a: indexing of curling and vein mosaic grapevine diseases by grafting on the woody indicator *Vitis riparia* "Gloire"; at right, healthy control.

1b: detection on *Chenopodium amaranticolor* of numerous viruses known to induce local lesions (arrow);

1c: detection of *Potato virus Y* (PVY) by mechanical inoculation of *Solanum demissum* A6.

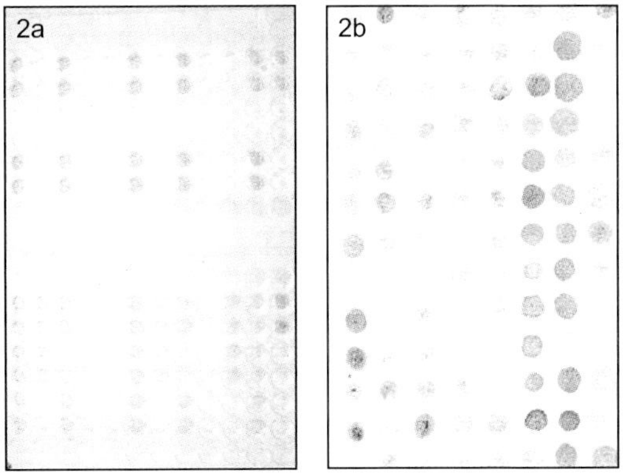

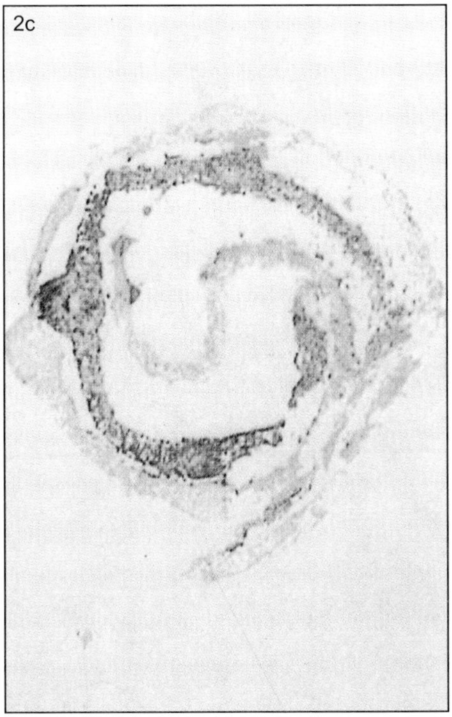

VII.2. Detection of viruses by serological tests.

2a: DAS-ELISA, detection of PSbMV (*Pea seed-borne mosaic virus, Potyvirus*) in pea seeds with two different serums. The serum used in the upper microtitre plate detects the virus in the embryo but not in the teguments, while the serum used in the lower plate detects both.

2b: Dot Blot (DIBA), detection of PFBV (*Pelargonium flower break virus, Carmovirus*) in pelargonium leaf extracts.

2c: Immuno-print (TBIA), detection of TSWV (*Tomato spotted wilt virus, Tospovirus*) in a section of a rolled up leaf of chrysanthemum.

PLATE **VIII**

VIII.1. *In vitro* culture of a garlic plantlet from meristem in order to obtain a virus-free plant.

VIII.2. Micrograft of apricot meristem on healthy stock grown from seed.

VIII.3. Transgenic *Nicotiana tabacum* cv. Xanthi expressing the CP gene (minicapsid construct) of LMV (*Lettuce mosaic virus, Potyvirus*): high level of resistance to PVY (*Potato virus Y, Potyvirus*) (left).

VIII.4. Transgenic *Nicotiana benthamiana* expressing the CP gene of LMV: strong attenuation of symptoms induced by PepMoV (*Pepper mottle virus, Potyvirus*) (right).

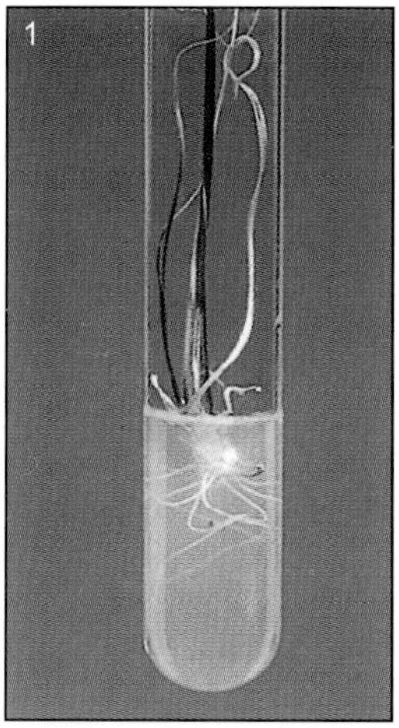

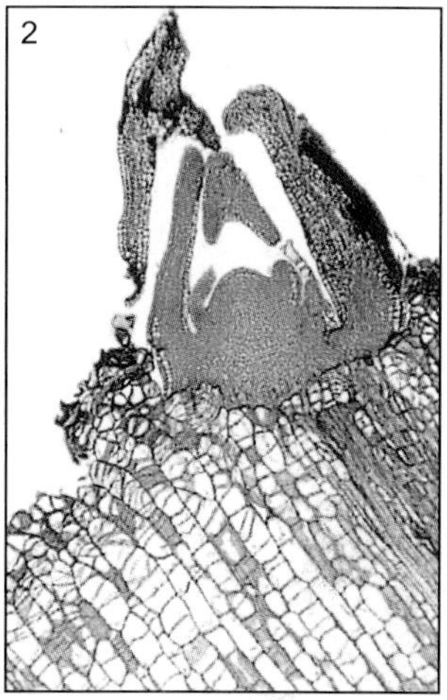

VIII.5. Transgenic squash expressing the CP genes of CMV (*Cucumber mosaic virus, Cucumovirus*), ZYMV (*Zucchini yellow mosaic virus, Potyvirus*) and WMV (*Watermelon mosaic virus, Potyvirus*): high level of resistance to infection by CMV + WMV + ZYMV (top right). Controls: susceptible squashes inoculated respectively by WMV, ZYMV and CMV.

VIII.6. Transgenic tomato expressing the CP gene of CMV: high level of resistance to infection by CMV (right). Control: non-transformed tomato (left).

VIII.7. Transgenic papaya (*Carica papaya*) expressing the CP gene of PRSV (*Papaya ringspot virus, Potyvirus*): very good resistance to PRSV in natural conditions (on right). Control: non-transformed papaya (on left) (experiment by Professor Yeh in Taiwan).

Meristem tip culture is generalized to most herbaceous crops but also to ligneous plants by means of micrografting (see box). This method remains remarkably efficient in obtaining healthy plant material. It can then be amplified by traditional or *in vitro* propagation and constitute the basis of a repository of healthy plants. In fact, healthy *in vitro* plantlets can be conserved for several years at low temperatures and in some species meristems can be stored by cryoconservation at –196°C.

The health of the regenerated material must be carefully monitored

It is important to ensure, at the end of thermotherapy or meristem tip culture treatments, that the plant material obtained is effectively healthy, and that viruses present in the mother plants have been thoroughly eliminated. Similarly, along the successive generations required to supply a quantity of seeds or planting material corresponding to the needs of producers, the products must be regularly tested to detect possible recontamination. Any diagnostic method can be used, from simple symptom observations in the propagation field to the most sophisticated molecular methods (Chapter 9). The most sensitive techniques, and often the most costly (biological, immunochemical, or molecular tests), make it possible to eliminate infected individuals rapidly from the first generations. Routine large-scale analyses, for monitoring lots that are being propagated as well as for controlling commercial seed lots, preferably use field observations and diagnostic techniques that can easily be automated, such as ELISA.

When the symptoms are sufficiently characteristic of the disease and recognizable in the fields, infected plants can be eliminated to maintain a satisfactory sanitary status. This is the case, for example, with mass selection practised for ornamental bulbs, as for tulips against TBV (*Tulip breaking virus, Potyvirus*).

Even though the practice of grafting on indicator plants is long and laborious, it remains essential for controlling the health of grapevine or fruit trees (Martelli and Walter, 1998). A notable improvement has recently been made in the practice of herbaceous cutting (see box). Similarly, in strawberry, the control procedure recommended against about 20 viruses and phytoplasma uses grafting on clones of *Fragaria vesca* and *F. virginiana* that clearly express typical symptoms.

■ Certification schemes

Technical rules define production conditions for certified plants and seeds

Certification schemes involve not only viral diseases but also bacterial and fungal diseases; the overall objective is the production of healthy seeds and plants. This objective requires rigorous organization and control of the different phases of propagation. Production programmes for healthy plants have been elaborated in

Biological grapevine indexing by grafting on indicator varieties

Single-bud cuttings or buds isolated from the sample to be analysed are grafted on cuttings of different varieties of grapevine chosen because they express characteristic symptoms when they are infected by certain viruses (Plate VII.1a). The plants are then observed periodically for 2 to 3 years in the nursery. Traditional indexing with dormant stem is relatively long and costly and can be replaced by a much faster protocol of indexing by grafting on herbaceous cuttings (Walter et al., 1990).

Indicator varieties used for tracking and identifying the principal viruses in grapevine are the following (Walter, 1998):

Indicator varieties	Viral disease
• *Vitis rupestris* St Georges	Decline, mottle, Rupestris stem pitting, asteroid mosaic
• *V. vinifera* Cabernet franc, Pinot noir or other red types of vine	Leafroll
• Kober 5BB	Kober stem grooving
• LN33	Corky bark, enations, LN33 stem grooving
• *V. riparia* Gloire de Montpellier	Vein mosaic
• *V. rupestris* × *V. berlandieri* 110R	Vein necrosis

many countries for the main species propagated vegetatively (e.g., potato, grapevine, fruit trees, pelargonium).

Certified plants and seeds are produced according to technical regulations elaborated by competent authorities (such as ministries of agriculture, research and technical institutes, plant protection services, seed producers' and farmers' associations), which may differ from one country to another. However, these schemes and virus monitoring protocols are being streamlined at the European level. The regulation must evolve and be capable of integrating new techniques and adapting to different types of enterprises. Apart from phytosanitary aspects, the norms should consider genetic, physiological, and morphological aspects that guarantee a quality product. Once they are streamlined, the certification is official and the enterprises can be accepted for production of one type of seed or plant. In France, conformity with norms is regularly controlled by SOC (an official control establishment). The material produced, if it respects the allowed sanitary tolerances, is commercially sold under the official label "certified plants".

In vitro micropropagation is naturally included within the framework of a certification scheme. Any technique that shortens the propagation time of a healthy clone and bypasses field cultivation is potentially beneficial since it reduces the risks of recontamination. Micropropagation, which carries on in a few cubic metres of laboratory space the production that would otherwise occupy several hectares, has revolutionized the production of many horticultural species but requires numerous sanitary and physiological controls. The choice of clone, the verification of the absence of virus, and the guarantee of maintaining varietal conformity are factors that are particularly important for the success of this strategy.

Certification schemes often require significant modifications in production systems. Their establishment leads to technical constraints that considerably increase the cost of plants and seeds. Two examples, the production of virus-free potato seeds and the clonal selection of grapevine, illustrate the different steps involved.

Production of certified potato seeds in France

Production of certified potato planting material was initiated in the 1920s by farmers from Brittany who organized themselves to create a seed production network of good sanitary quality. This was undoubtedly one of the first attempts to establish a sanitary selection scheme in France. Initially, the principle consisted of detecting healthy plants (F0), then multiplying them into "families" of the first year (F1), then of the second year (F2), and so on for 7 to 8 generations. This gave rise to the system called "genealogical" sanitary selection. The presence of a single diseased plant in an F1 or F2 made it necessary to eliminate the whole family. In subsequent generations, only a small percentage of virus-infected plants were tolerated.

Today, the combination of thermotherapy and meristem tip culture has made it possible to regenerate most varieties that are otherwise chronically infected (Merlet et al., 1996). The "family" method, based on propagation of a tuber from a healthy plant, now represents only a small part of the production. It has been supplanted by the cuttings method, in which the first generation is obtained by *in vitro* cutting techniques. Finally, the vitroplants method using microtubers produced *in vitro* so far accounts for only a small proportion of production. Setting aside these differences in the manner in which the first generations are obtained (Fig. 10.2), the later steps of the propagation always follow the same principle as that of genealogical selection, and the maximum contamination rates tolerated at the end of the cycle are the same.

The successive stages of propagation in the field are subject to various constraints and many controls. The first recommendation is that the first stages of the propagation be carried out in areas unfavourable to the multiplication of aphids that are vectors of the major potato viruses (cool and windy regions) and far from zones where potatoes for consumption are produced. In France, these are essentially coastal areas (Brittany, Normandy, Northern region) or mountain areas (Massif Central). Fields are surveyed early and regularly to detect and eliminate diseased plants, examining not only their leaves but also the young tubers. This roguing is done essentially on the basis of symptoms but may also involve ELISA tests. Specific cultural practices can be used to limit the risks of virus dissemination: elimination of volunteers or weeds (which could be virus reservoirs), application of insecticides (to limit aphid pullulation and the dissemination of PLRV) or mineral oils to prevent the dissemination of PVA, PVM, PVS, and PVY. Early killing of the haulms by chemical or mechanical means in the fields keeps the plants free of late contaminations that could pass undetected. Finally, to prevent the dissemination of viruses transmitted

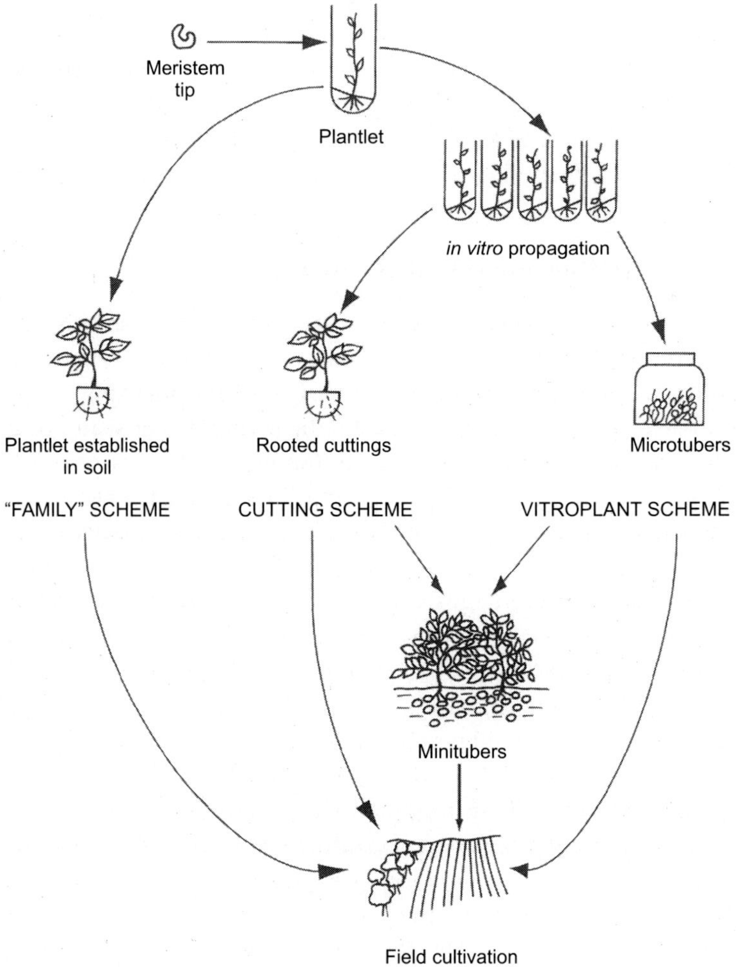

Figure 10.2 Different sanitary selection schemes for potato in France (modified from Merlet et al., 1996).

by contact (PVX and PVS), it is recommended that a path be left clear through the crops for farm machinery.

All these propagations are carried out either in production units approved by SOC (for the first generations) or by farmers specialized in seed production (for later generations). The virus contamination rate (VCR) must be determined so that lots can be classified into different categories: super elite (VCR < 1%), elite (VCR < 2%), class A certified plants (VCR < 5%), and class B certified plants (5% < VCR < 10%). Plants are classified in pre-culture, after lifting of the dormancy, according to a protocol defined by SOC. In France, the use of certified potato seeds is mandatory, even for private gardens.

Grapevine is also subject to rigorous sanitary selection

Clonal selection and pomological (genetic) selection of grapevine in France is designed to select and propagate clones that are best adapted to quality and production constraints (Walter, 1998). Clonal selection of grapevine began in France in 1946 in order to limit the diffusion of GFLV (*Grapevine fanleaf virus, Nepovirus*) (Plate II.1) (Vuittenez and Dalmasso, 1978; Huglin et al., 1980).

Selection is carried out in several steps (Fig. 10.3). The approval of a clone necessarily includes biological indexing on the highly sensitive grapevine varieties; other analyses (ELISA) are complementary and if they are positive, the clones are eliminated. The approved clones are then propagated. At present, the French regulatory certification scheme takes into account fan-leaf and leaf-roll for all cultivars and, in addition, mottle for root-stocks. It also takes into account the results of analysis for the rugose wood disease complex: Rupestris stem pitting, Kober stem grooving, and corky bark.

This approach to quality production is based most importantly on the competence and dependability of the propagating institutions. Official certification can only

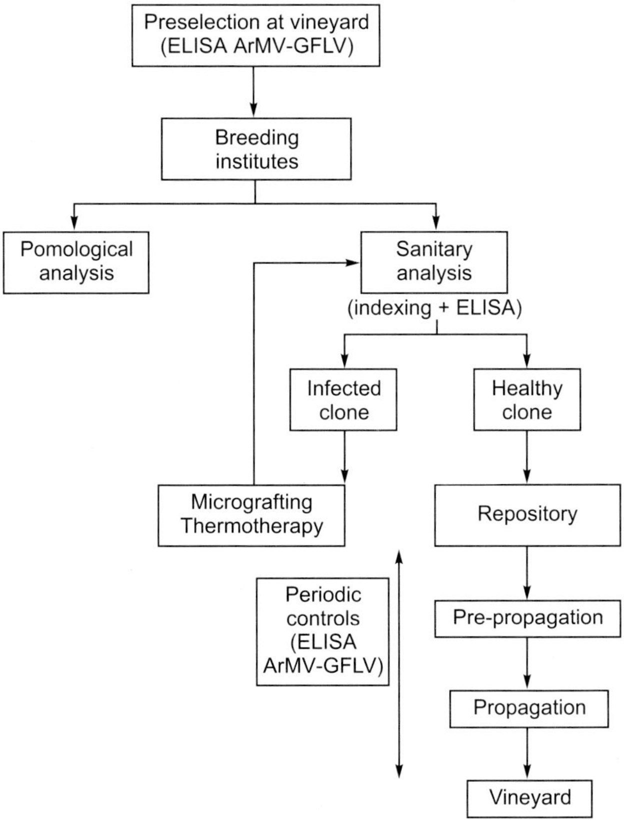

Figure 10.3 Organization of sanitary selection of grapevine in France (Walter, 1998).

attest that norms established and recorded in the technical regulations have been followed. For other crops, where there is as yet no official certification (as for certain horticultural products), the producers have themselves voluntarily set up specific protocols guaranteeing optimal sanitary quality of their plants.

Virus seed transmission

About 100 viruses are reported to have been transmitted by seeds in at least one of their hosts (Mink, 1993) but seed transmission is economically significant only for about 30 of them (Table 10.1).

Table 10.1 Major viruses for which seed transmission has economic significance.

In legume crops			
Comovirus	BBSV	*Broad bean stain virus*	broad bean
	BBTMV	*Broad bean true mosaic virus*	broad bean
Cucumovirus	CMV	*Cucumber mosaic virus*	peanut, bean, cowpea, lupin
Nepovirus	TRSV	*Tobacco ringspot virus*	soybean
Pecluvirus	IPCV	*Indian peanut clump virus*	peanut
	PCV	*Peanut clump virus*	peanut
Potyvirus	BCMV	*Bean common mosaic virus*	peanut, bean, cowpea
	BCMNV	*Bean common mosaic necrosis virus*	bean
	BYMV	*Bean yellow mosaic virus*	lupin
	CABMV	*Cowpea aphid borne mosaic virus*	cowpea
	PeMoV	*Peanut mottle virus*	peanut
	PSbMV	*Pea seed-borne mosaic virus*	lentil, pea
	SMV	*Soybean mosaic virus*	soybean
Tobravirus	PEBV	*Pea early browning virus*	pea
Not classified	ULCV	*Urdbean leaf crinckle virus*	*Vigna mungo* and *V. radiata*
In vegetable crops			
Carmovirus	MNSV	*Melon necrotic spot virus*	melon
Comovirus	SqMV	*Squash mosaic virus*	squash, melon
Potyvirus	LMV	*Lettuce mosaic virus*	lettuce
Tobamovirus	CGMMV	*Cucumber green mottle mosaic virus*	cucurbits
	PMMoV	*Pepper mild mottle virus*	pepper
	TMV	*Tobacco mosaic virus*	pepper, tomato
	ToMV	*Tomato mosaic virus*	pepper, tomato
In cereal crops			
Furovirus	SBWMV	*Soil-borne wheat mosaic virus*	wheat (UK), rye (Poland)
Hordeivirus	BSMV	*Barley stripe mosaic virus*	barley
Pecluvirus	IPCV	*Indian peanut clump virus*	wheat
Potyvirus	MDMV	*Maize dwarf mosaic virus*	maize
Not classified	HPDV	*High plain disease virus*	maize

Importance of seed quality control for local and international exchange

For certain viruses that have no reservoir host between two crops (such as SMV (*Soybean mosaic virus, Potyvirus*) in soybean or PSbMV (*Pea seed-borne mosaic virus, Potyvirus*) in pea), selection pressure for transmission through seed is constant; the disease depends on the transmission of the virus through the seed. A method of producing or selecting healthy seeds has a determining preventive effect for these viruses. For viruses with a very broad host range, such as CMV (*Cucumber mosaic virus, Cucumovirus*), transmission through seed may be believed to be less important; however, it has been shown that in peanut, CMV introduction through seed constitutes also the primary inoculum responsible for epidemics (Reddy, 1998). Thus, in most cases, quality control of seeds is a very important prophylactic measure (see box on quality control of seeds).

The production of entirely virus-free seeds is most often technically impossible; nevertheless, for certain vegetable crops, seed-producing plants can be protected from viral contamination if they are grown under insect-proof tunnels. As for the elimination of viruses from infected seeds, positive results have been obtained by heat treatment only when transmission occurs through contaminated teguments (in the case of tobamoviruses infecting cucurbits, pepper, and tomato, see box on seed disinfection).

Quality control of seeds for seed-borne viruses

Sampling: The quantity of seeds tested in the laboratory is minimal with respect to the size of the seed lot it represents (e.g., 20 t for leguminous seeds). To obtain reproducible results in the assessment of seed quality, it is essential that the successive samples (primary samples, sample submitted for analysis, and working samples) be representative of the seed lot being tested. Mechanized operations that make it possible to reach a high level of representativity are described in the international rules for seed testing (1999).

Evaluation of virus transmission rate for a seed lot: The virus transmission rate, often low, must be determined on a working sample comprising several thousands of seeds. For example, lettuce seeds produced for export to the United States are tested for LMV (*Lettuce mosaic virus, Potyvirus*) on a sample of 30,000 seeds. These 30,000 seeds are distributed into 60 groups of 500 each. Each group is subjected to ELISA. The number of positive seeds is related statistically to the number of positive groups, with a confidence interval (Maury et al., 1985). This protocol applies only to seed transmission through the embryo. The correlation between the serological seed testing and the actual seed transmission rate requires that (1) only the virus present in the embryo is detected and (2) the virus is still infectious in the embryo (Maury *et al.*, 1998).

Quality control with respect to a threshold of tolerance: It is possible to determine whether the seed lot is infected above or below the threshold of tolerance (even without knowing the transmission rate) with a different statistical approach and a smaller number of serological tests (Masmoudi et al., 1994).

Seed transmission also poses the problem of international seed trade and risk of introduction of a disease into a country in which it did not exist. Quality control of a seed lot with respect to a particular virus (see discussion of sampling in box on seed quality control) could guarantee that the percentage of infected seeds in a lot is lower than a threshold but does not in any case guarantee the complete absence of the virus and consequently that the virus will not be introduced with this particular seed lot; in this case, the decision to import a commercial seed lot is a political one. On the other hand, if a virus of the producing country is already present in an importing region, the seed lots are accepted as long as their contamination rate is less than a threshold of tolerance beyond which the virus will cause significant losses to the crop itself. This threshold is determined on the basis of epidemiological studies.

Disinfection of seeds contaminated by tobamoviruses

Tomato and pepper seeds collected from plants infected by tobamoviruses (e.g., ToMV, *Tomato mosaic virus* or PMMoV, *Pepper mild mottle virus*) have healthy embryos and infected teguments; moreover, they are contaminated on the outside by these viruses, which are known for their exceptional stability (Plate III.1). Viruses within the teguments or on the outside constitute a potential source of contamination of the plantlet through microinjuries that occur during germination (Chapter 8). In these very particular cases, seed disinfection significantly reduces or even eliminates virus seed-transmission. Seeds are disinfected during extraction by incubation for 24 h in a solution of 0.2% HCl and 3 g/l of pectinase and then, after drying, by a treatment with dry heat (80°C) for 24 h. For each species or variety, it must be verified that such treatments do not reduce the seed germination rate.

Determination of tolerance threshold is based on epidemiological studies

In the case of SMV, intensive epidemiological studies have made it possible to indicate the key role of the initial seed infection rate in the intensity of subsequent virus epidemics. Moreover, vector density is critical in the epidemic if it is manifested early in the season. A secondary infection before flowering increases the chances of dissemination, exerts a depressive effect that is more accentuated on the plant, and allows the passage of the virus to the next generation. These studies have made it possible to develop a method that can be used in most soybean-producing regions in the world to predict the impact of the virus on the yield (Ruesink and Irwin, 1986). The model uses parameters such as the aphid population, planting date, and varietal characteristics; it calculates the yield reduction at harvest as a function of the percentage of seeds infected.

According to the model, regions that have regularly low densities of aphid vectors early in the season can tolerate seed infection rates of 1%, which corresponds to around 1 source plant of the virus per square metre (or 10,000 sources per hectare). On the other hand, in regions with high vector aphid densities, significant losses can be predicted if more than one seed out of 10,000 is infected. This model thus can be used to define the tolerance threshold, 1% or 0.01% depending on the agro-climatic characteristics of the region.

In many cases in which no model has been worked out, the thresholds are determined in a more empirical fashion. For example, the incidence of initial input of LMV on the production of lettuce has been the subject of studies in various agro-climatic contexts. In regions with high aphid populations, an initial inoculum of 0.5% can cause a total loss of the yield. The losses are significant if the percentage of virus transmission through seed is greater than 0.1% (Dinant and Lot, 1992). These data are useful indications for seed producers in different countries but they are only recommendations.

More interesting is the process of threshold adjustment, practised in California by lettuce producers (Grogan, 1980). After several years of collective observations, it appeared that the threshold of 0.1% was insufficient and that a threshold of 0.02% was preferable. This demand resulted from a test called "0 infected seeds in a representative sample of 30,000 seeds" (with a confidence level of 0.95). The threshold adjustment was complemented by a legal prohibition of sale or import of lettuce seed lots in California that did not satisfy the test. This is one of the rare cases of official seed certification for seed-borne viruses.

Genetic resources must be completely virus-free

Transmission of viruses by seeds also poses the problem of the risk of introducing new viruses, or new viral strains, through germplasm exchange. Genetic resources (gene banks) are key elements in any plant breeding programme and constitute a rich variability of genotypes originating from all over the world. Theoretically, in regions in which sources of resistance have been found, the selection pressure for emergence of virulent strains is high. When the virus is seed-borne, the diversity of genetic resources may be associated with a diversity of the virus. Then, the risk is that virulent seed-borne strains will be imported along with the genetic resources imported for breeding programmes. Germplasm exchange involves generally only a small number of seeds, and the establishments that are in charge of genetic resource maintenance most often apply quarantine protocols and provide seed samples taken only from healthy plants after cultivation in a confined environment.

Preventing and reducing virus dissemination

The development of a viral epidemic in a field results from a complex set of interactions that involve three major components: plants (as sources of virus or as a crop in which the disease may spread), vectors (sometimes only transient partners that are, however, essential for the survival of the virus), and of course the viral populations themselves (Chapter 8). Because of the great diversity of these three components, there are a very large number of agricultural practices that can interfere more or less effectively in plant-vector-virus interactions (Lecoq, 1995, 1996; Jones, 2004).

It is impossible to act directly against viruses, but one can reduce the abundance of virus sources and the efficiency of virus dissemination by vectors. These interventions can be planned only with a thorough knowledge of each pathosystem (virus host range, type of virus-vector relationships, virus prevalence, and dynamics of vector populations). Because of the diversity of biological situations encountered, cultural control methods and prophylactic measures vary greatly and are often specific to each virus-host combination, or even to the particular type of crop (protected crops or open fields).

Elimination of virus sources in the environment

Weeds: abundant sources of viruses

The importance of weeds as a source of viruses has been established since the beginning of plant virology. In 1925, Doolittle and Walker proved the role of various perennial weed species as reservoirs of CMV and proposed weed elimination as a method of control. Since then, the importance of weeds as alternative hosts for viruses, as well as reservoirs of vectors, has been abundantly illustrated (Duffus, 1971; Thresh, 1981). The total destruction of weeds is not possible for practical and obvious environmental reasons. It would be, for example, not so easy to eliminate virus reservoirs in neighbouring fields belonging to another farmer and moreover the spontaneous flora harbours numerous auxiliary insects that are highly useful to agriculture.

Careful weeding around nurseries and within and around fields is nevertheless a generally recommended measure, but its efficiency varies greatly from one ecosystem to another (Quiot et al., 1982). Intensive use of selective herbicides can lead to changes in weed populations: in the Salinas Valley (California), it contributed to an increase in the population of sow-thistle, host of BYSV (*Beet yellow stunt virus, Closterovirus*) and SYVV (*Sowthistle yellow vein virus, Nucleorhabdovirus*), which seemed correlated to an increase in epidemics of these viruses in lettuce crops (Duffus, 1971).

Other possible sources of viruses include private flower or kitchen gardens. Sources of TSWV (*Tomato spotted wilt virus, Tospovirus*) in greenhouse crops of tomato (Plate III.2) are often flowers cultivated alongside, and sources of ZYMV (*Zucchini yellow mosaic virus, Potyvirus*) (Plates III.3, VI) and WMV (*Watermelon mosaic virus, Potyvirus*) affecting melon or watermelon crops in the Imperial Valley (California) are cucurbits cultivated in gardens in residential areas (Perring et al., 1992).

Contamination may come from nearby crops

The crops themselves may ensure the persistence of the inoculum. It is important, therefore, to create a break in the infection cycle. For species with a short vegetative period, such as vegetable or ornamental crops, the producers often plant successive crops during the year. Sometimes, heavily infected fields at the end of their production cycle harbouring high vector populations border young plantings that will quickly become contaminated. Such situations, which are frequent because of the boom in protected crops, are best avoided, and early crops should be eliminated as soon as they are no longer productive. In the same way, seed productions of biennial plants (sugarbeet, carrot) should be kept distant from commercial crops and volunteers should be carefully eliminated (potato, onion, sugarbeet) (Zitter and Simons, 1980).

In regions with intensive agriculture and when producers are well organized, "crop-free" periods can be imposed to break the viral cycle. This community effort has proved highly effective against viruses that have a limited host range. Examples are CeMV (*Celery mosaic virus, Potyvirus*) in celery in California and Florida, WSMV (*Wheat streak mosaic virus, Tritimovirus*) in cereal crops in Alberta, and CLCuV (*Cotton leaf curl virus, Begomovirus*) in cotton in the irrigated plain of Gezira in Sudan (Broadbent, 1964). More recently, the establishment of a "crop-free" period (mid-June to mid-July) in the Arava valley in Israel considerably reduced attacks by viruses transmitted by aphids or whiteflies in vegetable crops (Ucko et al., 1998). In regions with more diversified agriculture, changing the planting date may be sufficient to avoid peaks in vector populations (Zitter and Simons, 1980).

Eradication, a radical method effective in perennial crops

Eradication, i.e., the elimination of infected plants as soon as they are detected in a crop, has long been the only advice given to farmers faced with viral attacks. This practice, consisting of uprooting and burning, was often recommended only as a stopgap measure. Eradication is still used in seed production of vegetatively propagated crops, such as potato or ornamental bulbs.

This method of control has proved highly effective in the case of perennial crops. It requires observation and testing campaigns in orchards and a regulatory arsenal that allows competent authorities to enforce the uprooting of diseased plants and grant financial compensation to the producers. A rigorous eradication campaign resulted in the elimination of PPV some years after its introduction in Switzerland. At present, this method limits epidemics of PPV in France. Eradication campaigns have also successfully limited the incidence of CTV (*Citrus tristeza virus, Closterovirus*) in citrus orchards in California and Israel (Thresh, 1988).

■ Disturbing the efficiency of vectors

Phytosanitary treatments against air-borne vectors

The efficiency of chemical control against air-borne vectors is variable and the strategies recommended must take into account the virus-vector relationships, the type of crop, and its environment.

In the case of viruses transmitted non-persistently by aphids, contact insecticides generally do not take effect quickly enough to prevent transmission and thus do not protect the crop efficiently. Sometimes it is even observed that, in causing a transitory increase in aphid activity, a treatment may favour the dissemination of these viruses. However, pyrethroid insecticides could, in particular conditions, have some efficacy in slowing down virus spread (Asjes, 1985).

On the other hand, in the case of epidemics caused by viruses transmitted persistently, some recommended insecticide products are effective by acting on vector populations. The active ingredients are derived from diverse chemical families (organophosphates, organohalogens, synthetic pyrethroids) and act by contact or ingestion. They can be applied by incorporation in the soil, seed coating (see box on seed coating), or spraying on leaves. Because of the toxicity of some of these products to humans, plants, or auxiliary insects, as well as the appearance of resistance, it is desirable to limit the frequency of treatments. For this purpose one must define the optimal conditions for their use: seasonal sensitivity of the plant, planting density, presence of viruliferous insects, and climatic conditions during treatment.

Identifying and eliminating virus sources in protected and open field crops

Preventive virus control consists in identification of the potential virus sources near crops and their elimination to prevent risk of dissemination.

In protected crops (Fig. 10.4), sources of insect-borne viruses are most often found outside; viruliferous insects can be kept out by use of woven or unwoven insect-proof nets to cover the openings or the use of UV-blocking plastics. Sources of viruses transmitted by fungi or nematodes can be soil, substrates, or nutrient solutions. Soil, substrates, and nutrient solutions should, therefore, be disinfected before any new crop if they are to be recycled. Certain viruses transmitted mechanically (tobamoviruses, carmoviruses) are highly stable. They can contaminate pots or greenhouse structures or remain infectious in crop residues. In this case, rules of hygiene must be strictly followed, especially careful disinfection and elimination of all crop residues (Broadbent, 1963).

In open fields (Fig. 10.5), for viruses transmitted by insects or mites, risks of infection come most often from sources outside the field (weeds, nearby crops). These sources can be numerous in the case of ubiquitous viruses, such as potyviruses, luteoviruses, or cucumoviruses. In the case of viruses transmitted by soil-borne vectors (nematodes, fungi), sources of contamination are often the vectors themselves, which may sometimes remain viruliferous in the soil for several months or even several years. These vectors may be introduced into the field by farm tools or irrigation water or come from weeds growing nearby (taking into account the low mobility of the vectors). Among the preventive methods to be used, careful weeding should eliminate virus reservoirs (often symptomless) along the field borders, old plants that are no longer productive should be destroyed, and appropriate measures should be taken against the vectors.

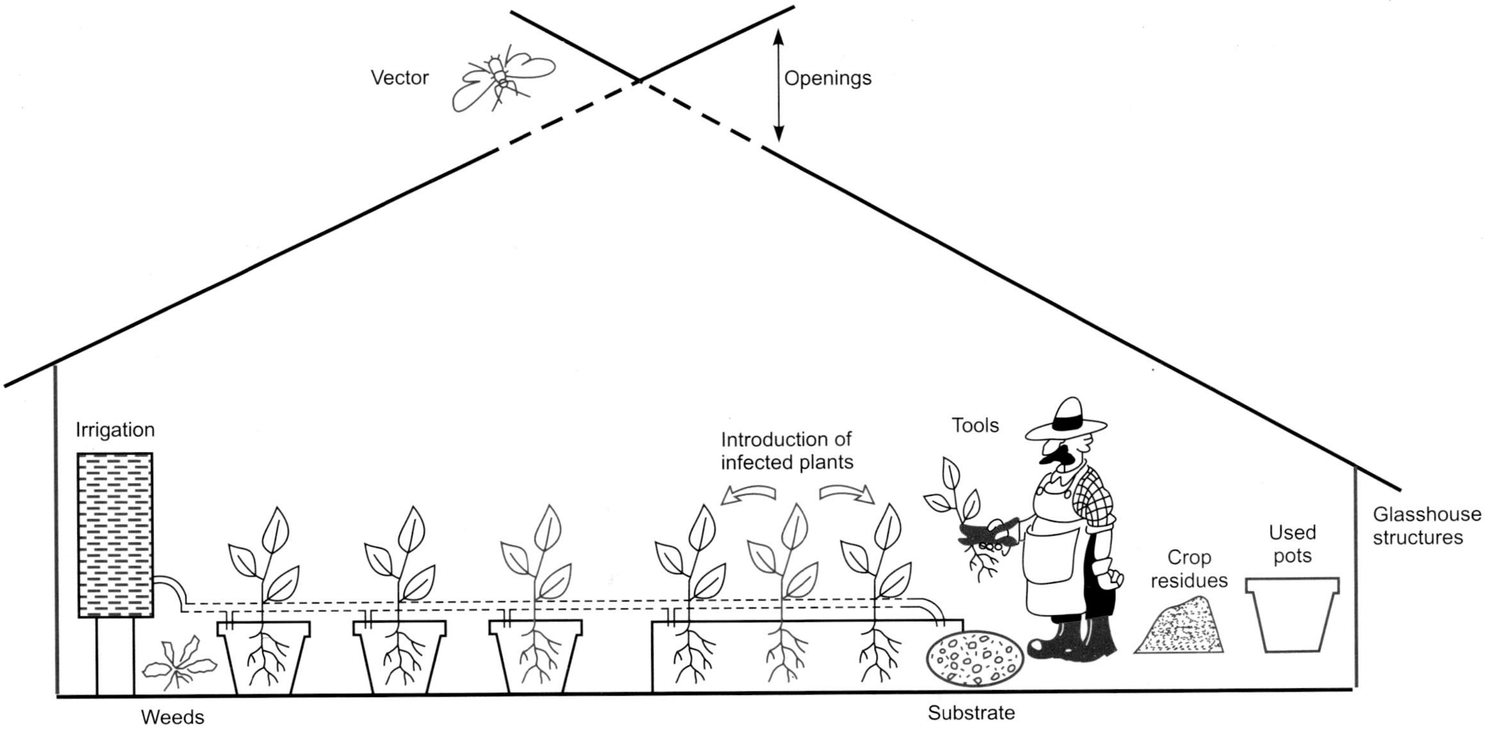

Figure 10.4 Potential virus sources (in red) are highly diverse in protected crops.

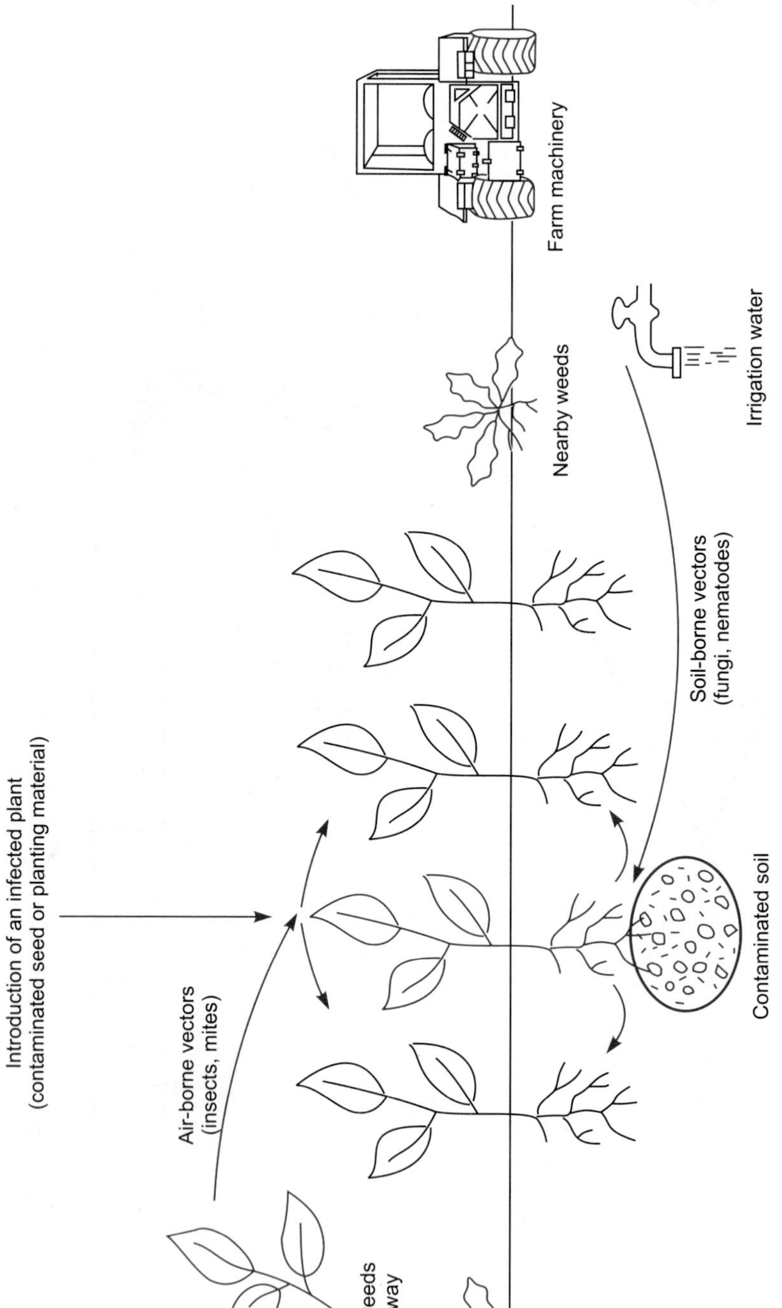

Figure 10.5 Principal virus sources (in red) for open field crops.

Cereal seed coating, a technique in full development

The main objective of seed coating is to protect autumn cereal seedlings against BYDV (*Barley yellow dwarf virus, Luteovirus*) (Plate V.3). Sowings have long been done as late as possible to prevent the last aphid flights. This solution, which often compromises the yields, is not always feasible or effective against aphids. The spraying of insecticides protects only the surface of mature leaves and requires a thorough knowledge of the level of activity of aphids. Several new insecticide molecules (of the nitromethylene group) applied to the seeds present systemic migration and show high aphicide efficiency over long periods. Imidacloprid, the most widely used of these molecules (70% of barley seeds and 50% of wheat seeds are presently treated in France), partly suppresses BYDV primary infections and completely stops the secondary dissemination of the virus. Several questions have been raised about these molecules, which risk slowing the development of alternative control strategies and which are presently too costly to be used in developing countries. Will their systematic use reduce the frequency of reservoirs? What is the risk of appearance of resistance in the major vector species? Finally, because of their persistence and very broad spectrum, will these molecules have unintended effects on beneficial insects when they are applied on spring cereal crops or on maize?

Mineral oils can be used to prevent the dissemination of viruses transmitted non-persistently by aphids. They have no direct toxic effect on aphids but could interfere with virus-aphid stylet interactions. They are difficult to use because crops must be sprayed repeatedly in order for the leaves to remain covered with a protective film on the upper and lower sides. This process nevertheless resulted in three-fold reduction of PVY contamination in potato seed production (Merlet et al., 1996). It is also used to protect ornamental bulbs in the Netherlands and vegetable crops in certain regions in Israel and the United States (Asjes, 1974; Zitter and Simons, 1980; Raccah, 1986).

Phytosanitary treatments against soil-borne vectors

Chemical control can also be used against soil-borne vectors (nematodes, fungi), which contribute sometimes to the conservation of viruses in the soil from one crop to the next (*Nepovirus, Bymovirus, Varicosavirus, Ophiovirus*).

The application of nematicides is a measure generally considered satisfactory in controlling nematode vectors and preventing the diseases they transmit. The length of nematode life cycle and their low reproduction rates make these treatments generally sufficient to protect annual crops (Vuittenez and Dalmasso, 1978). In contrast, for perennial crops, nematicides must be combined with other measures such as leaving a fallow period or planting bait plants. This is recommended at the time of planting mother vines in order to keep them protected for as long as possible against contamination by nepoviruses responsible for grapevine fanleaf, GFLV and ArMV (*Arabis mosaic virus*) (Plate II.1).

Chemical nematicides are classified as fumigants and non-fumigants. Fumigants, the best known of which are methyl bromide and D-D (dichloropropane-dichloropropene), are recommended for many crops (potato, strawberry, sugarbeet, grapevine). Their efficiency varies with the nature of the soil (disinfection being more difficult in heavy and deep soils) and the depth at which vectors are found. These products are often highly toxic and must be applied cautiously and by accredited agencies.

Methyl bromide is also effective in destroying resting spores of fungi that are vectors of viruses. In any case, the soil can quickly be recontaminated. Zoospores can also be controlled by treating the water with zinc salts or nutrient solutions with surfactants (Tomlinson, 1988). The imminent ban on the use of methyl bromide in agriculture, owing to its harmfulness to the environment, and the absence of an effective substitute, may cause serious problems in years to come for control of soil-borne vectors.

Disinfection of tools to control mechanically transmitted viruses

The farmer may transmit certain viruses from plant to plant during cultural operations (pruning or removing leaves, axillary buds, or cuttings). This occurs frequently in vegetable and ornamental crops grown in greenhouses, for highly stable viruses (tobamoviruses, carmoviruses, potexviruses) as well as sometimes for more labile viruses (potyviruses). Disinfection of tools with, for example, 3% trisodic phosphate after working on each plant prevents this type of dissemination (see box on identifying and eliminating viruses, and Fig. 10.4).

Plastics used in agriculture may disturb activity of air-borne vectors

In the case of ornamental and vegetable crops, various types of plastic film can be used to disturb the activity of air-borne vectors. Woven or non-woven polythene or polypropylene covers (Fig. 10.6) form a physical barrier that prevents vectors from landing on crops or entering a protected area (nurseries and greenhouses). Plastic mulches (Fig. 10.7), particularly transparent ones, have a double effect: they allow soil heating and thus advance the growth of plants but they also reflect light, which repels aphids. Thus, a delay of several weeks can be observed in the development of viral epidemics, but the repellent effect decreases to the extent that the growing plants gradually cover the plastic (Lecoq, 1992). Antignus et al. (1996) demonstrated that the use of UV-absorbing polypropylene films for construction of plastic tunnels is highly effective in protecting plants against pests (aphids, thrips, whiteflies) and the viruses they transmit. The elimination of UV rays from the light spectrum seems to interfere with the capacity of insects to orient themselves under the tunnel. This method has proved particularly effective in limiting infections by TYLCV (*Tomato yellow leaf curl virus, Begomovirus*) transmitted by *Bemisia tabaci* in protected tomato crops.

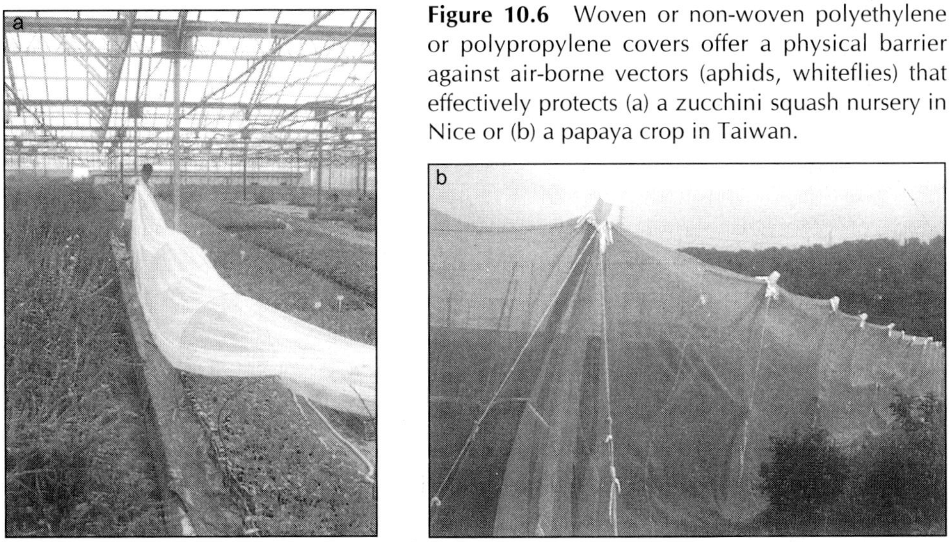

Figure 10.6 Woven or non-woven polyethylene or polypropylene covers offer a physical barrier against air-borne vectors (aphids, whiteflies) that effectively protects (a) a zucchini squash nursery in Nice or (b) a papaya crop in Taiwan.

Figure 10.7 Plastic mulches reflect light and have a marked repellent effect against aphids. They can be used to delay the development of viral epidemics for a few weeks.

This list of different methods available to limit or delay the development of viral epidemics is by no means exhaustive. It reveals the great diversity and sometimes extreme specificity of possible levels of intervention. Control can be effective only if these means are applied knowledgeably, that is, with a thorough understanding of the biology of the virus to be controlled, sources of contamination, and agronomic constraints of the crop. Very often, it is the combination of different methods (oil treatments with pyrethroid-based insecticides, plastic mulches with careful weeding) that leads to effective crop protection (Jones, 2004).

■ Mild-strain cross-protection

Cross-protection: the principle

Cross-protection is the art of using viruses to control viruses. In 1929, McKinney demonstrated the phenomenon of cross-protection: a tobacco plant infected with a strain of TMV (*Tobacco mosaic virus, Tobamovirus*) could not be infected by a second strain of TMV that induced different symptoms. It very quickly became apparent that this phenomenon, observed with most plant viruses, could have applications in plant protection. The principle of mild-strain cross-protection thus consists in inoculating a virus strain producing mild symptoms into a plant in order to protect it against severe strains of the same virus (Fig. 10.8). Cross-protection has mostly been used for commercial purposes to control viruses of vegetable or fruit crops, undoubtedly because viruses could cause the most serious damage in these plants, particularly in terms of fruit quality (Fuchs et al., 1997; Lecoq, 1998) (see box on examples of cross-protection). Taking into account constraints linked to its implementation, cross-protection should be considered a transient control method to control particularly severe viruses while other solutions are pursued, such as breeding for resistant cultivars (Chapter 11).

Isolation of virus strains causing mild symptoms

Mild strains used for cross-protection have so far been obtained essentially empirically. They may be natural variants isolated from plants presenting mild symptoms in the field (CTV) or in the greenhouse (ZYMV) or mutants obtained by random mutagenesis after nitrous acid treatment (TMV, PRSV). Certain mild strains have been obtained by maintaining infected plants at either high or low temperature (Lecoq, 1998). Since the molecular bases of virus pathogenicity are now better understood and mutations involved in the attenuation of symptoms have been identified, a more rational approach can be considered for obtaining mild strains by directed mutagenesis on infectious cDNA (Gal-On and Raccah, 2000).

A mild strain that is to be used in practice must meet certain criteria:

— It must cause mild symptoms and not significantly affect the yield of the plant to be protected (Fig. 10.9) or that of other susceptible crops.
— It must be genetically stable (Chapter 13) and not evolve towards a more severe form.
— It must have a broad spectrum of protection and be effective against the largest number of severe strains possible.
— It must be easy to multiply, conserve, and inoculate.
— It is preferably not efficiently transmitted by vectors to prevent unintentional dissemination.

Limitations of cross-protection

Cross-protection, like any method of biological control, presents some risks. The major problems encountered so far have been some rare cases of synergism with other viruses (mild TMV strain and CMV in tomato) (Fulton, 1986). Moreover, there is sometimes a slight aggravation of symptoms caused by mild strains in senescent plants. This phenomenon seems more closely linked to the physiological state of aged plants than to an evolution of mild strains (Lecoq et al., 1991). However, reversion of mild strains towards more revere forms cannot be ruled out.

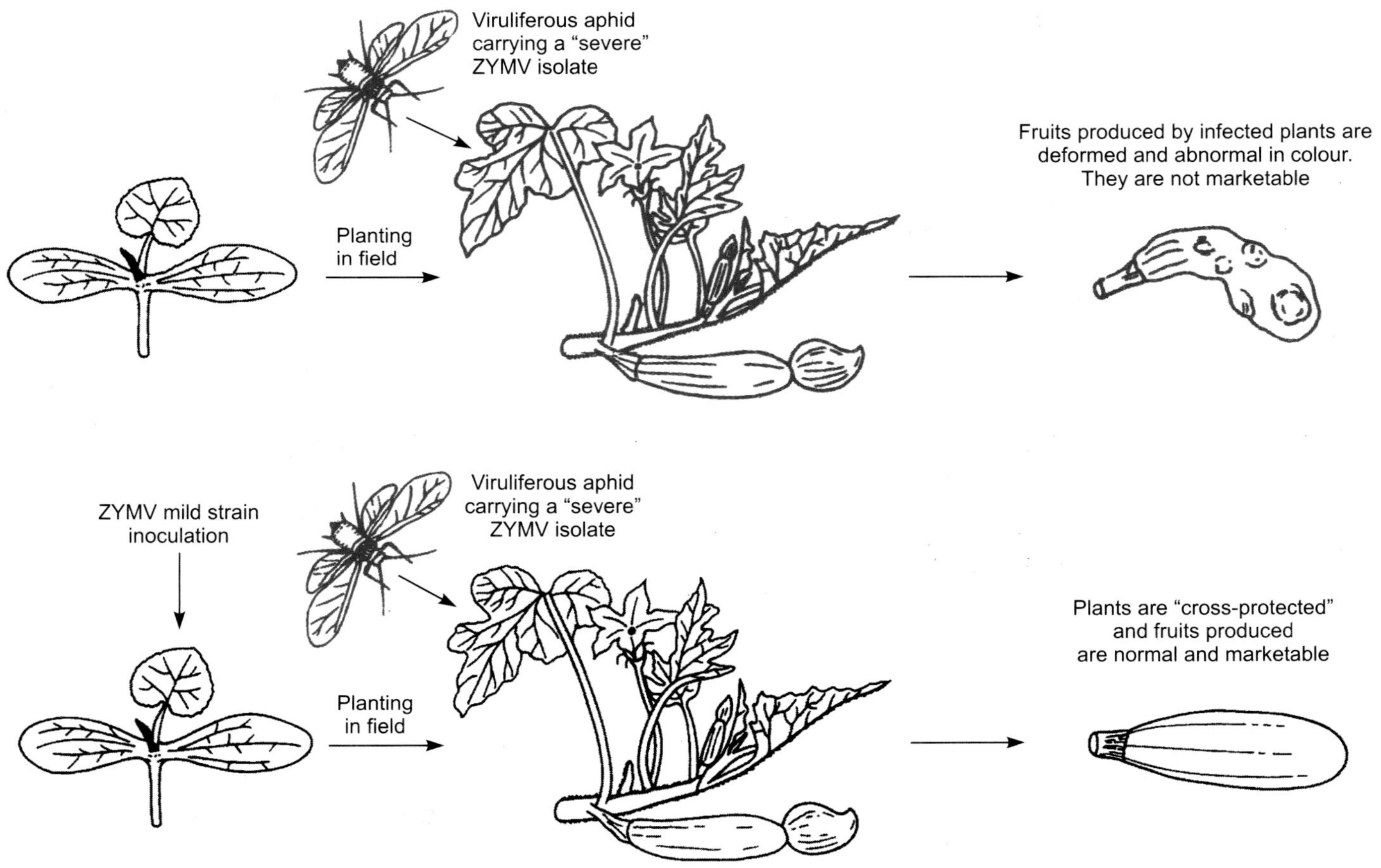

Figure 10.8 Principle of mild-strain cross-protection illustrated by the case of ZYMV (*Zucchini yellow mosaic virus, Potyvirus*) in zucchini squash.

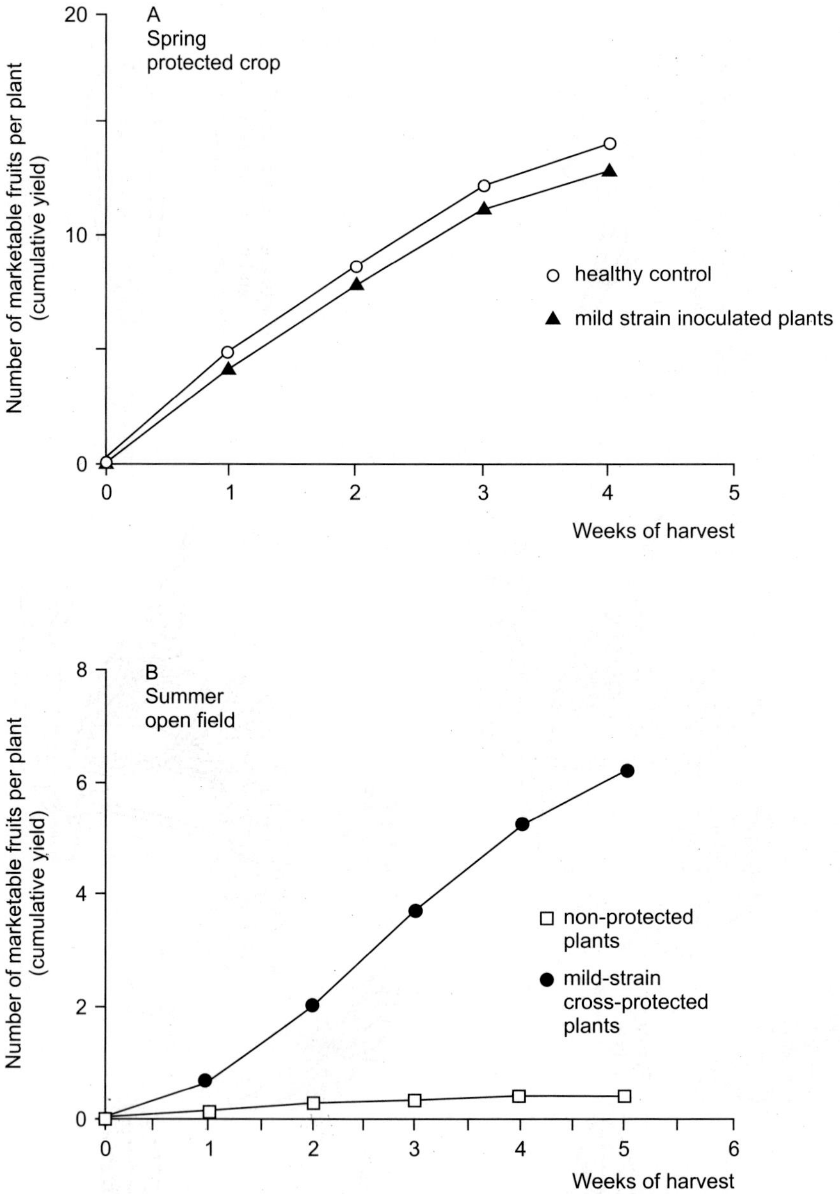

Figure 10.9 Effect of the mild ZYMV-WK strain on yield of squash, in the absence of severe strain (A, spring, protected crop) and in conditions of natural ZYMV epidemic (B, summer, open field). It is observed in A that the mild strain has practically no effect on yield as compared to a healthy control (○). On the other hand, when there is an epidemic of severe strains, the mild strain confers effective cross-protection (●).

Examples of commercial applications of cross-protection

ToMV-tomato pathosystem (Plate III.1): The mild strain ToMV MII.16 was obtained by nitrous acid random mutagenesis followed by single local lesions transfers. This strain was used between 1970 and 1985 in many European countries and in Canada, Japan, and New Zealand to inoculate protected tomato crops. In France alone, more than 13 million tomato plants were cross-protected every year. The development of resistant varieties of tomato (Chapter 11) gradually led to a decline in the use of cross-protection. The mild strain inoculum was produced by technical institutes that regularly conducted quality control tests before marketing. The inoculation was directly carried out by producers.

ZYMV-cucurbit pathosystem (Plate VI): The mild strain ZYMV-WK is a natural variant of an aphid non-transmissible strain; it is thus also non-transmissible. It practically does not affect the yield and confers cross-protection against most severe ZYMV strains (Fig. 10.9). It is today used successfully in many regions of the world (Europe, Israel, Taiwan, Hawaii, Tonga).

PRSV-papaya pathosystem (Plate I.5). The mild strain PRSV-HA5-1 was obtained from a Hawaiian strain by nitrous acid random mutagenesis followed by single local lesion transfers. In the field, this strain proved highly effective in Hawaii in protecting against local severe strains but less effective in other regions of the world (Taiwan, Thailand), probably because of the molecular variability of PRSV in these regions.

CTV-*Citrus* pathosystem: The situation is quite different for CTV and *Citrus*: in this case, a mild strain must be isolated for each rootstock-scion combination. Several natural mild strains have been isolated in Brazil and at present all the *Citrus* crops of that country are effectively cross-protected, which has allowed regeneration of an orchard of several million trees. Other promising assays were conducted in the United States, South Africa, and Reunion island.

(Fuchs et al., 1997; Lecoq, 1998)

One of the major limitations to cross-protection is its specificity. In effect, the protection is rapid and effective to the extent that the mild and severe strains are genetically rather similar. In the case of PRSV, for example, a mild strain originating from Hawaii effectively protects against severe strains from Hawaii, moderately against severe strains from Taiwan, and not at all against severe strains from Thailand. Similarly, the mild strain ZYMV confers no protection against greatly divergent strains from Reunion island (Fuchs et al., 1997; Lecoq, 1998).

Mechanisms at work

Diverse mechanisms of competition and regulation have been proposed to explain cross-protection (Ponz and Bruening, 1986; Sherwood, 1987):

- The capsid protein of the mild strain encapsidates the RNA of the severe strain from the time of its inoculation and prevents its replication.
- The RNA of the mild strain hybridizes with the RNA of the severe strain.

— The mild strain prevents the movement of the severe strain within the plant.
— The mild strain uses cellular co-factors or sites needed by the severe strain to establish an infection.

It is likely that one or several of these mechanisms intervene simultaneously or successively to establish cross-protection, and that different mechanisms are at work in different virus genera. More recently, the implication of silencing (Chapter 5) has been mentioned (Ratcliff et al., 1999). It is interesting to note that several of these mechanisms have also been proposed to explain pathogen-derived resistance in transgenic plants.

In the case of TMV, the importance of the capsid protein and of the corresponding RNA sequence was elegantly demonstrated by Culver (1996) using PVX as the vector of expression. Two constructions containing the capsid gene of TMV were used, one allowing synthesis of the TMV capsid protein, the other not. Both constructions conferred some protection against TMV but the construction allowing expression of the capsid protein was more effective. This confirms the implication of several different molecular mechanisms in the implementation of cross-protection.

11

CHAPTER

Controlling Plant Viral Diseases: Breeding for Resistant Varieties

The use of virus-resistant varieties often seems to be the most simple, effective, and economical way to control viral diseases. In theory, farmers just have to purchase seeds of a variety resistant to one or several viruses to prevent the risk of epidemics in their crops. Unfortunately, the situation is not as simple as it seems at first. The major cultivated species do not have resistances to all the economically important viruses. Moreover, the resistances that are identified sometimes provide only partial protection, or even transient protection when they are overcome by adapted viral strains. However, a diversity of virus resistances, having various genetic determinisms or modes of action, are already present in various cultivated species (see box below).

Search for and characterization of virus resistances

Genetic resources

The search for virus resistance genes begins within the botanical species to which the cultivated plant belongs. This approach consists in studying the behaviour of a

Viruses cited

BCMV, *Bean common mosaic virus*, *Potyvirus*; BWYV, *Beet western yellows virus*, *Polerovirus*; BYMV, *Bean yellow mosaic virus*, *Potyvirus*; ClYVV, *Clover yellow vein virus*, *Potyvirus*; CMV, *Cucumber mosaic virus*, *Cucumovirus*; CPMV, *Cowpea mosaic virus*, *Comovirus*; CPSMV, *Cowpea severe mosaic virus*, *Comovirus*; LMV, *Lettuce mosaic virus*, *Potyvirus*; MNSV, *Melon necrotic spot virus*, *Carmovirus*; PLRV, *Potato leafroll virus*, *Polerovirus*; PMMoV, *Pepper mild mottle virus*, *Tobamovirus*; PRSV, *Papaya ringspot virus*, *Potyvirus*; PSbMV, *Pea seed-borne mosaic virus*, *Potyvirus*; PVX, *Potato virus X*, *Potexvirus*; PVY, *Potato virus Y*, *Potyvirus*; TMV, *Tobacco mosaic virus*, *Tobamovirus*; ToMV, *Tomato mosaic virus*, *Tobamovirus*; TSWV, *Tomato spotted wilt virus*, *Tospovirus*; TuMV, *Turnip mosaic virus*, *Potyvirus*; TYLCV, *Tomato yellow leaf curl virus*, *Begomovirus*; WMV, *Watermelon mosaic virus*, *Potyvirus*; ZYMV, *Zucchini yellow mosaic virus*, *Potyvirus*.

Examples of resistance to viruses present in commercial varieties of some cultivated species

Through the efforts of plant breeders from research institutes or private seed companies, virus resistances are increasingly frequent in commercial varieties of various cultivated species. However, these new varieties are generally not adopted by farmers unless they also have agronomic characteristics that are as good as those of the susceptible varieties. Moreover, some of these resistances have already been overcome by adapted virus pathotypes.

Species	***Some virus resistances present in commercial varieties***
Bean	BCMV
Cucumber	CMV, PRSV, WMV, ZYMV
Lettuce	BWYV, LMV, TuMV
Melon	CMV, MNSV, resistance to virus transmission by *Aphis gossypii*
Pea	BYMV, PSbMV
Pepper	CMV, PMMoV, PVY, TMV
Potato	PLRV, PVX, PVY
Squash	CMV, ZYMV
Tomato	TMV, ToMV, TYLCV, TSWV

collection of varieties (accessions) of diverse origins after inoculation of the virus concerned. These collections are called "genetic resources" and are kept in "gene banks"; they include modern cultivated varieties and old local varieties or land races, originating from various regions throughout the world, as well as wild types of the species (Fig. 11.1). The finding that plants have geographic centres of origin in which the greatest genetic diversity can be found has actively favoured the search for sources of resistance. These centres of origin as well as the centres of secondary diversification are in fact the sites where the host plant and the virus may have co-evolved, and where selection pressure acts on each side in favour of the domination of the partner (Leppikk, 1970).

Figure 11.1 The diversity within genetic resources of the genus *Cucurbita* is revealed by a remarkable polymorphism of the fruits as well as widely varying susceptibilities to viruses.

When resistance cannot be found within the species, the behaviour of related species must be evaluated. These related species may prove rich in resistance genes but they are generally more difficult to use in breeding programs because they need to achieve inter-specific crosses with the varieties of agronomic interest. In the case of tomato (*Lycopersicum esculentum*), most of the virus resistance genes have been found in the related species, *L. peruvianum, L. hirsutum*, or *L. pimpinellifolium* (Laterrot, 1989). Genetic resources are generally conserved in national or international research centres, botanical gardens, or private companies. Access to genetic resources is today a strategic, economic and political issue on the global scale. The international convention in Rio, in 1992, and subsequent international conferences attempted to establish the bases of a more equitable sharing of genetic wealth by recognizing farmers' rights over their local land races. Specific institutions are in charge of coordinating these activities in different countries, as for instance the Bureau des Ressources Génétiques in France and the National Plant Germplasm System in the United States.

Choosing a virus strain from a collection of isolates

The choice of a virus strain to be used for evaluating the behaviour of genetic resources is of major importance. The first requirement is to have an access to a collection of isolates of diverse geographic origin that are representative of the known variability of the virus. This constitutes in a way a "gene bank" of the virus. In general, a strain representative of the most commonly found isolates in a region or country is selected. This could also be:

— a strain inducing clear-cut symptoms ("aucuba" strains inducing white or yellow mosaics) allowing easy sorting of susceptible and resistant plants;

— a strain not transmissible by vectors, to prevent risk of unintentional dissemination of the virus in experimental greenhouses or in the environment;

— a strain that has a specific virulence, when the aim is to find resistance to a given pathotype (see box).

When a virus is endemic in a region, the behaviour of genetic resources can be studied in the field under natural epidemic conditions (Plate V.2). Nevertheless, in such a case, neither the quantity nor the quality of the inoculum can be controlled, and these may vary greatly from one site or one year to another. It is thus preferable to use artificial inoculations that would allow the use of standardized protocols (uniform plant stage at inoculation, inoculum concentration, plant incubation conditions). The most commonly used method is the mechanical inoculation at the plantlet stage. This allows observation of a large number of plants over a limited surface area. After inoculation, plants may be incubated in a greenhouse or preferably in a climatic-controlled cabinet to prevent variations due to the season or to the environment. However, to study certain resistance types, or for viruses that cannot be transmitted mechanically, inoculations must be done with viruliferous vectors, which is a much more time-consuming and costly process. When the virus is

The two components of virus pathogenicity are virulence and aggressiveness

In many scientific papers and handbooks, "virulence" is used as a general term to describe the symptoms intensity or the severity of a disease caused by a virus in a susceptible plant. However, by analogy with the definitions of Vanderplanck (1975) for plant-fungus interactions, we prefer to use "virulence" to describe one of the two major components, with aggressiveness, of virus pathogenicity.

Virulence is the capacity of a strain to infect a variety having a resistance trait to this virus. A strain is virulent if it is capable of multiplying or causing a disease in a variety that has a resistance (this is a qualitative property), otherwise it is said to be avirulent. Strains of the virus can thus be combined into *pathotypes* as a function of their properties of virulence or avirulence with respect to one or several resistances.

Aggressiveness concerns the intensity of symptoms in a variety in which the infection may develop. For example, a strain may be more or less aggressive depending on whether it induces mild or severe symptoms of mosaic (this is a quantitative property).

not mechanically transmissible and vector transmission is not possible, *Agrobacterium*-mediated inoculation may be applied if an infectious cDNA clone of the virus and adequate containment conditions are available (Grimsley et al., 1987).

Analysis of the genetic determinants of resistance

When a resistance factor is identified in a population, it is necessary to fix this character by successive self-pollinations of resistant plants or by obtaining haploid plants and then doubling their chromosome numbers by using colchicine.

The inheritance of a resistance character is studied by using the classical methods of mendelian genetics. Various crosses are carried out, and from the proportion of susceptible or resistant plants in each generation, the genetic basis of the resistance can be established (Fig. 11.2, Table 11.1).

When complex inheritance is involved, it may be preferable to study the heredity using homozygous progenies such as doubled haploid lines or recombinant inbred lines (RIL). These progenies make it possible to carry out quantitative resistance tests requiring repetitions on several plants of the same genotype or tests with different strains.

Resistance genes exist in plants in three principal types of arrangements:

- Isolated genes with or without a range of distinct alleles, each conferring specific resistance. A single allele is present in a pure line.
- Gene clusters; groups of closely linked homologous genes with different specificities.
- Resistance genes to viruses, bacteria, or fungi, less closely linked, located in specific genomic regions called "major resistance complexes".

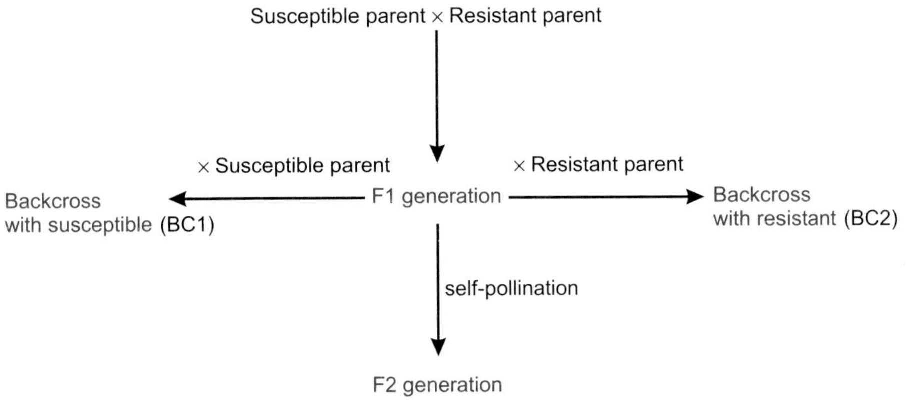

Figure 11.2 Obtaining of different generations necessary for studying the genetic determinism of a resistance character.

Table 11.1 Crosses used to study the inheritance of a virus resistance character and theoretical relative proportions of susceptible and resistant plants, in cases where resistance is controlled by a single dominant or a single recessive gene.

Population	One dominant gene		One recessive gene	
	Susceptible plants (%)	**Resistant plants (%)**	**Susceptible plants (%)**	**Resistant plants (%)**
Susceptible parent	100	0	100	0
Resistant parent	0	100	0	100
F1 generation	0	100	100	0
F2 generation	25	75	75	25
Backcross with susceptible (BC1)	50	50	100	0
Backcross with resistant (BC2)	0	100	50	50

For each of these arrangements, hypotheses on the mechanisms of evolution are proposed by Hammond-Kosack and Jones (1997).

Depending on the number and the efficiency of the genes controlling resistance, the breeder will choose different breeding strategies to introduce the resistance into the varieties of commercial interest (Scully and Federer, 1993). The most commonly used is the successive backcross method. The susceptible variety presenting useful agronomic characters is crossed with the resistant variety, often of poor quality. In each generation, resistant plants with better agronomic quality are conserved. These plants are again crossed with the susceptible variety, and the process is repeated until a plant is obtained that has the resistance and the desirable agronomic traits. In the case of recessive resistances, it is often necessary to complete a generation of self-pollination between every two backcrosses. A breeding programme for introducing a virus resistance generally lasts for 10 to 15 years, depending on the resistance inheritance.

In certain cases, a genetic linkage is observed between a resistance and an unfavourable agronomic character. The crosses must then be multiplied to attempt to break this linkage, which could slow down the breeding program and sometimes jeopardize it.

The development of genetic maps for numerous plant species allows location on the genome of genes and quantitative trait loci (QTL) involved in resistance to viruses. It is also possible to reveal resistance gene clusters or genes with pleiotropic effects. Ultimately, marker-assisted selection will become a routine tool that accelerates and simplifies breeding programmes for resistance to viruses.

Diversity of resistance mechanisms

■ Resistances may occur at any stage of the virus infection cycle

A resistance factor is any property of a plant that can disturb the virus biological cycle: at the cell, organ, individual plant or population (field) levels. The resistance type can be determined by comparing the virus behaviour in a susceptible variety and in the resistant accession. Diverse methods can be used: vector or mechanical inoculation, symptom assessment (type, localization, intensity, time of appearance), estimation of virus localization (by immunoprints) or multiplication rate (by quantitative detection methods such as DAS-ELISA or real-time RT-PCR (see Chapter 9)).

In the example of an aphid-borne virus, resistances can be broadly differentiated phenotypically according to the level at which they operate (Fig. 11.3).

(1) Resistance to virus inoculation by aphids

Long considered purely theoretical, this type of resistance was proved to exist in several cultivated species (Jones, 1998). In this case, the plant is susceptible to the virus but resistant to its transmission by its vector or one of its vectors. For example, resistance to transmission of *Cucumber mosaic virus* (CMV) and potyviruses by the aphid *Aphis gossypii* was identified in melon (Lecoq et al., 1979). This resistance is associated with a resistance of the plant to the aphid itself. Controlled by a dominant gene, the *Vat* gene (standing for *virus aphid transmission*), this resistance is now present in commercial hybrids. Resistances to the transmission of viruses by other vectors (fungi, nematodes, mites) have also been described (Jones, 1987).

(2) Tendency to escape infection

This "partial" resistance may be characterized as a lower probability of infection becoming established than in susceptible plants, using the same inoculum level. Some pepper genotypes have a "tendency" to escape infection by CMV (Palloix et al.,

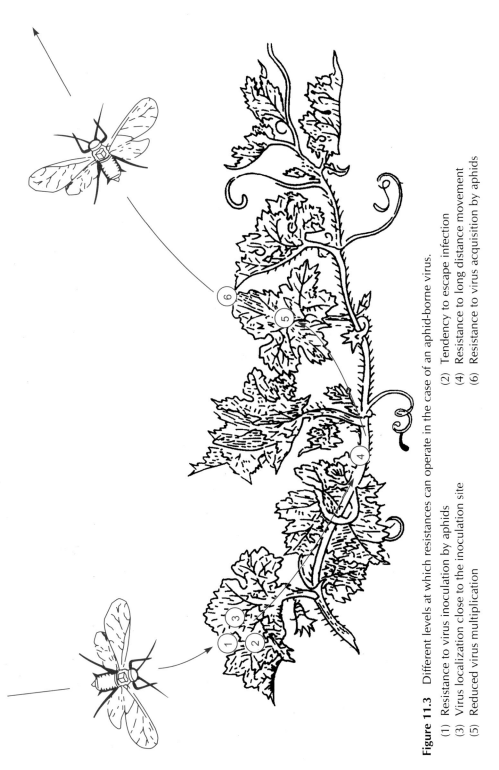

Figure 11.3 Different levels at which resistances can operate in the case of an aphid-borne virus.

(1) Resistance to virus inoculation by aphids
(2) Tendency to escape infection
(3) Virus localization close to the inoculation site
(4) Resistance to long distance movement
(5) Reduced virus multiplication
(6) Resistance to virus acquisition by aphids

1997, Caranta et al., 1997). Mature plant resistance is a form of escaping infection that is expressed at the adult stage but not at an earlier stage.

(3) Virus localization close to the inoculation site

Several different mechanisms may lead to virus "sequestration" in a cell or in a few cells close to the inoculation site. Sometimes replication does not occur (immunity) or is hardly detectable (extreme resistance) in inoculated cells or leaves because of the lack of some factor necessary for virus pathogenesis (Köhm et al., 1993; Legnani et al., 1995) (see Chapter 6). Resistance to cell-to-cell movement of the virus corresponds to an inhibition of movement functions of the virus: the virus may multiply in the inoculated cells but it is not able to move outside these cells or may move only towards a few neighbouring cells (see Chapter 4). A resistance of this type to PVY (*Potato virus Y, Potyvirus*) has been described in *L. hirsutum*, a wild species closely related to tomato (Legnani et al., 1995). Finally, virus localization is sometimes accompanied by hypersensitivity and formation of local necrotic lesions (I gene of resistance to BCMV, *Bean common mosaic virus, Potyvirus*, in bean) (Chapter 6) or chlorotic lesions (recessive oligogenic resistance to common strains of CMV in melon). These types of resistance are the one that are the most commonly used by breeders because they are very easy to select for (Fraser, 1985).

(4) Resistance to long distance movement of the virus within the plant

The virus multiplies in inoculated organs, but it either becomes systemic at a slower rate than in the susceptible variety or invades only part of the plant (Murphy and Kyle, 1995). To reveal this type of resistance, specific tests must sometimes be devised. In pepper, for example, the following test is used: plantlets at the 4-5 leaf stage are topped and then CMV is inoculated on the youngest remaining leaf. The dynamics of symptom appearance on the axillary branches can then be observed; in susceptible plants all the axillary branches have mosaics, while in the resistant accession, only one or a few branches show symptoms, and at a later stage (Lecoq et al., 1982; Palloix et al., 1997).

(5) Reduced virus multiplication

In this case, the virus spreads to the whole plant but it reaches a lower concentration than in susceptible plants; this reduction of the concentration is not always associated with a reduction of the symptom intensity or with a reduced impact on yield, which limits its agronomic interest. Resistances to the multiplication of CMV and WMV (*Watermelon mosaic virus, Potyvirus*) have been identified in melon (Lecoq et al., 1982; Gray et al., 1986; Dias-Pendon et al., 2005). The development of semi-quantitative methods of virus detection allows easier selection for this type of resistance (Chapter 9).

(6) Resistance to virus acquisition by aphids

This type of resistance, which concerns the last stage of the viral cycle in a plant, has not yet been used *per se* in breeding programmes. It could be linked, for example, with a lower concentration of viral proteins involved in transmission (Chapter 8) and could help to slow down the progression rates of epidemics in the field.

Many other types of resistance to viruses will undoubtedly be identified in the future, as our understanding of virus-plant interactions improves. Moreover, certain varieties may allow viral multiplication without inducing a significant effect on the commercial yield of the plant. Such plants are regarded as tolerant to the virus. This *tolerance* can, as in the case of squash and WMV (Adlerz et al., 1985), be expressed as an absence of mosaic symptoms on fruits, which are thus not depreciated. The use of tolerant varieties represents nevertheless a major risk for neighbouring crops. These varieties may indeed be efficient sources of viruses that could spread to other susceptible varieties or species.

A single phenotype of resistance (such as localization of virus close to the inoculation site) can result from totally different molecular mechanisms. Thus, for each resistance, it is important to develop specific tests that ultimately allow combining different mechanisms in a single genotype.

■ Resistance genes: two models

If two isogenic lines are compared, one being susceptible to a virus, the other differing only by having the resistance gene, the explanation at the molecular level of the resistance may involve two major types of plant-virus interactions. There might be the loss of a host function involved in a specific interaction with a viral factor (*negative model*). The resistance gene is then a mutated form of the gene coding for the susceptibility factor. There could also be specific interactions elicited by a viral protein (*positive model*): in this case, the resistance gene codes for an inhibitor or for the inducer of an inhibitor of a viral function (Fraser, 1990).

Resistance associated with the loss of a susceptibility factor is recessive

Plant viruses involve many host proteins to achieve major steps of their infection cycle (Chapters 2 and 3). In the *negative model*, resistance is linked to an absence of recognition between the product of the resistance gene (a non-functional cofactor) and a factor coded by a specific virus gene. If the resistant line is crossed with a susceptible line, the F1 hybrid still synthesizes some susceptibility factor and thus is not resistant. Therefore, the resistance gene is recessive. Virulence may result from the aptitude of a viral mutant to cooperate with the product of the resistance gene (Fig. 11.4, see box below).

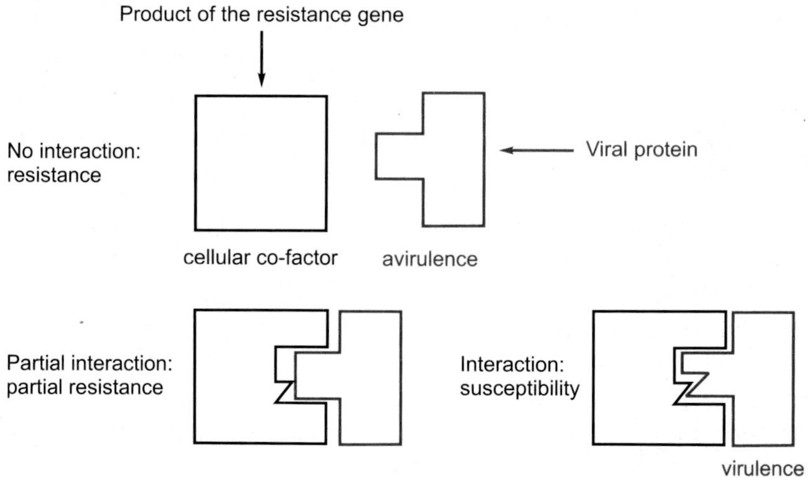

Figure 11.4 Simplified diagram of the negative model of resistance to a plant virus. In red, the viral factor concerned.

Resistance to PSbMV governed by recessive genes in pea (see box and Fig. 11.5) is clearly of monogenic type: *sbm1* alone confers a total resistance with respect to pathotype P1, a resistance that expresses itself also in protoplasts. The pathotype P4 is virulent with respect to *sbm1*. Through recombination by exchange of segments between the cDNA-P1 and the cDNA-P4 (see Fig. 11.7), it was demonstrated that the virus gene coding for the VPg protein was responsible for virulence of P4 with respect to *sbm1* (Keller et al., 1998). One interpretation of these results is that, in susceptible pea plants, a specific interaction between the PSbMV VPg protein and the product of the functional allele *sbm1* is necessary for the multiplication of the virus. It has been shown that *sbm1* gene codes for a eukaryotic translation factor (eIF4E), a protein known to interact with the VPg in the picorna-like virus superfamily (Gao et al., 2004; see Chapter 6). However, the role of a putative VPg-eIF4E interaction in potyvirus infection cycle is not yet known: it could concern translation, replication, or cell-to-cell movement. In any case, in avirulent PSbMV/resistant pea combinations the VPg and eIF4E would not interact properly and thus prevent virus multiplication or proper cell-to-cell trafficking (Gao et al., 2004). Previously, eIF4E and its isoform eIF(iso)4E were shown to be coded by recessive resistance genes to a series of potyviruses in pepper, lettuce, and tomato (Ruffel et al., 2002; Nicaise et al., 2003; Caranta et al., 2003). A few point mutations generally close to the eIF4E cap-binding domain are enough to confer resistance, and few mutation differences provide different resistance specificities to alleles at a single locus (Caranta et al., 2003; Gao et al., 2004).

Resistance linked to the production of an inhibitor is dominant

In the positive model, resistance is determined by a specific recognition between the product of the resistance gene and a viral factor (Fig. 11.6). If the resistant line is

Resistance of pea to PSbMV (*Pea seed-borne mosaic virus, Potyvirus*) is controlled by recessive genes

From lines originating from Ethiopia, the major centre of diversification of the genus *Pisum*, research on resistance has made it possible to identify three recessive genes, *sbm1*, *sbm3*, and *sbm4*, which confer a total resistance to PSbMV pathotypes P1, P2, and P4 respectively. In certain lines, these three genes are grouped (into a cluster) on chromosome VI with other recessive genes for resistance to potyviruses infecting pea (BYMV, *Bean yellow mosaic virus*; ClYVV, *Clover yellow vein virus*). In other lines, there is a second cluster of recessive genes for resistance to potyviruses infecting pea on chromosome II; in this second cluster, the gene *sbm2* also confers resistance to pathotype P2 of PSbMV (Fig. 11.5) (Alconero et al., 1986; Provvidenti and Alconero, 1988; Provvidenti, 1990). However, in more recent work it has been shown that *sbm1* codes for the initiation factor eIF4E, and that *sbm4* should be considered as an allele of *sbm1* having a different specificity (Gao et al., 2004).

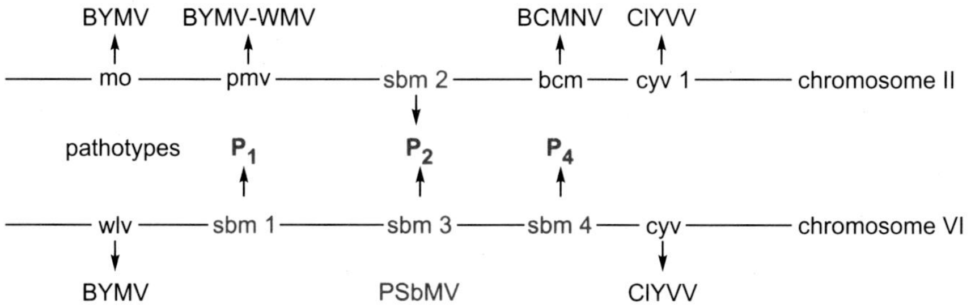

Figure 11.5 "Clusters" of genes for resistance to *Pea seed-borne mosaic virus* (PSbMV) and other potyviruses in pea.

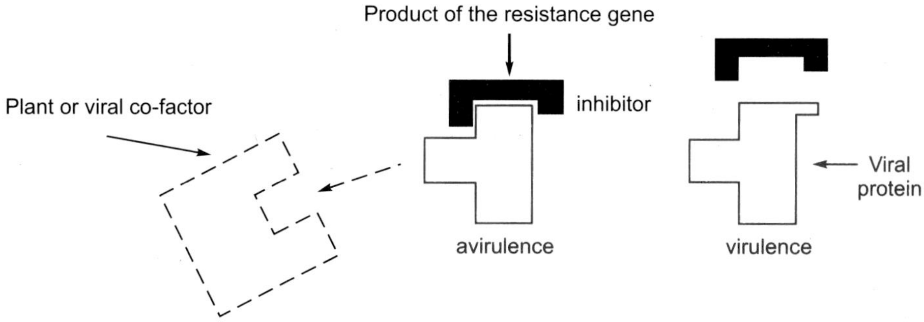

Figure 11.6 Simplified diagram of the positive model of resistance (constitutive resistance) to a plant virus. In red, the viral factor concerned.

crossed with the susceptible line, the F1 hybrid still synthesizes a dose of the inhibitor and is thus resistant. The resistance gene is thus dominant. The virulence results from a loss of recognition of the inhibitor by the mutated viral factor.

The differences in the expression of resistance governed by dominant genes make it possible to distinguish between constitutive (or direct) resistances and induced resistances.

Constitutive resistances

In constitutive resistances, the product of the resistance gene, in addition to its presumed physiological role in plant metabolism, has an inhibitor effect on a viral function. This inhibitor effect has been demonstrated in cowpea cultivar Arlington resistant to CPMV (*Cowpea mosaic virus, Comovirus*) on a protease coded by the virus (Ponz et al., 1988). The blocking of the maturation of the viral polyprotein leads to blocking of the virus infection cycle. The inhibition is highly specific: in fact, the protease of another comovirus also infecting cowpea, CPSMV (*Cowpea severe mosaic virus*), is insensitive to the effect of the inhibitor. The resistance to ToMV (*Tomato mosaic virus, Tobamovirus*) conferred in tomato by the *Tm1* gene could also be classified in this category (Table 11.2).

Induced resistances (see Chapter 6)

There are different types of induced resistance:

— Extreme resistance, or resistance induced without hypersensitivity reaction. This is the case of the gene *Rx* for resistance to *Potato virus X* (PVX, *Potexvirus*) in potato. Double inoculation of a virulent and an avirulent PVX strain triggers a resistance against the two strains. This experiment carried out on protoplasts of plants with the *Rx* gene confirms the inducible character of the resistance. In a second experiment, after double inoculation of an avirulent PVX strain and CMV, the induction and the non-specificity of the resistance were also demonstrated (Bendahmane et al., 1995).

— Induced resistance with hypersensitivity reaction (HR). One example is provided by the *N* and *N′* genes for resistance to *Tobacco mosaic virus* (TMV, *Tobamovirus*) in tobacco. This resistance with hypersensitivity is associated with systemic acquired resistance (SAR). Despite its theoretical interest, SAR is not of great interest from the agronomic point of view (Pennazio and Roggero, 1998).

Induced resistances require a specific recognition of a viral protein by the product of the resistance gene. This interaction then activates a cascade of signals that induce different defence responses, not necessarily virus-specific, that block the virus infection process in some way.

In short, virus resistance genes correspond to three major categories:

— Recessive resistance genes, which are mutated genes coding for susceptibility factors that become non-functional.

— Dominant resistance genes that give a plant a constitutive resistance specific to a virus.

— Dominant resistance genes that confer a resistance induced by the infection. This last type of resistance is more frequent and more general, applying to all types of pathogens and parasites to the extent that evolution has selected a specific recognition mechanism for triggering the resistance.

Durability of resistance genes

■ Virulence/avirulence can involve each gene of a virus

Isolation and characterization of virulent strains

Virulent virus strains can sometimes be identified from the first stages of the characterization of a resistance, when its efficiency is checked against a collection of isolates of diverse geographic origins and representative of the known virus variability. Some isolates can infect the resistant plants, and the resistance is then referred to as strain-specific. In other cases, it is only after several years of cultivation of resistant varieties that infected resistant plants are occasionally observed. From such plants, generally virulent strains can be isolated that "overcome" the resistance. These variants may be strains that existed previously but remained undetected, or mutants that appeared within the viral quasi-species and were subsequently selected by resistant plants (Chapter 13).

To analyse the molecular bases of virus pathogenicity, virulent/avirulent pairs of strains must be available. One approach is to compare the complete genome sequences of the two strains. In general, this type of analysis reveals numerous mutations on the entire genome and it is difficult to attribute the property of virulence precisely to one of them, without having quasi-isogenic strains differing only for virulence/avirulence or comparing a very large number of sequences. The second approach is to reproduce artificially the mutation or mutations leading to virulence (see box and Fig. 11.7).

The three genes for resistance to *Tomato mosaic virus* in tomato can be overcome

This experimental approach has so far been applied only to the study of a small number of models, including the ToMV/tomato pathosystem. Three genes for resistance to ToMV are used in tomato breeding, all originating from related species (*Lycopersicon hirsutum* and *L. peruvianum*) (see box below, Table 11.2).

The *Tm-1* gene is easily overcome, and strains of pathotype 1 appear very rapidly after cultivation of a variety possessing this resistance gene. The resistance conferred by this gene against pathotype 0 of ToMV (common strains) is expressed by a reduction of more than 90% of virus multiplication in comparison to a susceptible variety. In tomato protoplasts, the expression of *Tm-1* is even more pronounced and viral multiplication is completely blocked. These observations suggest that the

An approach to identify avirulence genes in plant viruses

The method generally used for the study of genetic determinism of virus pathogenicity consists of making recombinants and using reverse genetics. Most plant viruses have RNA genomes, therefore infectious cDNAs of avirulent and virulent strains are required (or at least one of them) (Fig. 11.7a). Recombinants between the two cDNAs allow a primary location of the genome regions implicated in the expression of the virus pathogenicity (if only one infectious cDNA is available, the recombinants could be obtained with cDNA fragments from the second strain) (Fig. 11.7b). When these regions are identified, sequences are carefully compared to locate a small number of mutations. The impact of each mutation found on the expression of virus pathogenicity should then be confirmed by directed mutagenesis (Fig. 11.7c). Nevertheless, this type of result must be critically analysed: mutations other than those that are thus identified, or even an entire protein domain, could also help modify the pathogenic properties of a strain.

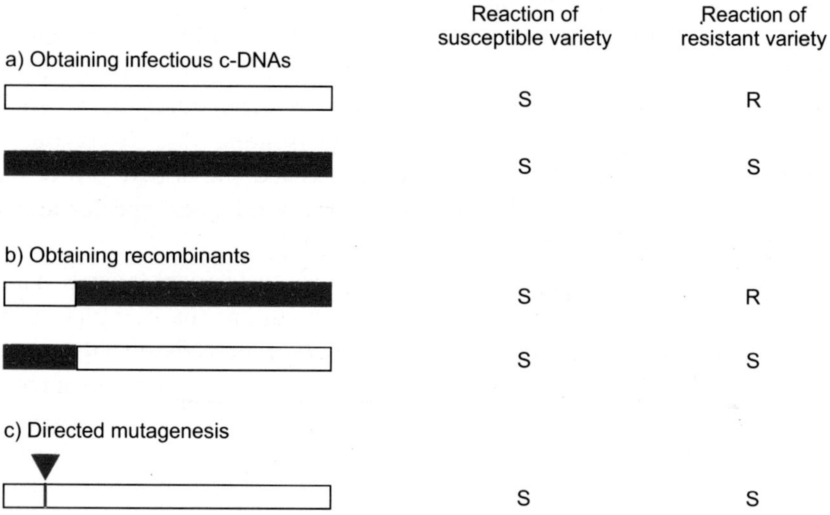

Figure 11.7 Method used for studying the genetic determinism of virus pathogenicity.

resistance inhibits viral replication. The determinant of the virulence is the gene coding for the proteins implicated in viral replication. The substitution of a single amino acid ($Glu^{979} \rightarrow Gln$) on the infectious cDNA of ToMV results in the obtaining of a virulent strain (Meschi et al., 1988) (Fig. 11.8).

The *Tm-2* and *Tm-2*^{2} genes are alleles; they are both overcome but strains of pathotype 2 are infrequent and strains of pathotype 2^2 are very rare. The resistance conferred by the *Tm-2* and *Tm-2*^{2} genes to pathotype 0 of ToMV is expressed by an inhibition of cell-to-cell movement of the virus as well as by necrotic reactions (at high temperatures). After mechanical inoculation, it can be observed by immunofluorescence that the virus is limited to a few epidermal cells. Moreover, strains of pathotype 0 multiply normally in protoplasts prepared from resistant

Specificity and management of *Tomato mosaic virus* (ToMV) resistance genes in tomato

Three genes of resistance to ToMV (previously considered a strain of TMV) are used in tomato (*Lycopersicon esculentum*), all originating from related species (Table 11.2).

These resistance genes are all overcome by virulent pathotypes and are associated with deficiencies in fertility and fruit quality in the homozygous state. Moreover, the *Tm-2* and *Tm-2*^{2} genes can lead to necrotic reactions at high temperature (which is frequent for greenhouse crops) in the heterozygous state. These constraints have led breeders to use various gene combinations to obtain resistant F1 hybrids, including the combinations *Tm-1 Tm-2*^{2}/*Tm-1*$^+$ *Tm-2*$^+$ or *Tm-1 Tm-2*/*Tm-1*$^+$ *Tm-2*^{2}.

Table 11.2 Principal characteristics of ToMV resistance genes in tomato (Lecoq et al., 1982; Fraser, 1985; Laterrot, 1989).

Locus	*Tm-1*	*Tm-2*	*Tm-2*
Chromosome	2	9	9
Source	*L. hirsutum*	*L. peruvianum*	*L. peruvianum*
Alleles	*Tm-1*	*Tm-2*	*Tm-2*^{2}
Reactions with:			
Pathotype 0	Resistance (a)	Resistance (b)	Resistance (b)
Pathotype 1	Susceptibility	Resistance (b)	Resistance (b)
Pathotype 2	Resistance (a)	Susceptibility	Resistance (b)
Pathotype 1-2	Susceptibility	Susceptibility	Resistance (b)
Pathotype 2^2	Resistance (a)	Resistance (b)	Susceptibility
Pathotype 1-2^2	Susceptibility	Resistance (b)	Susceptibility

(a) attenuated symptoms and reduced viral multiplication; (b) virus localization.

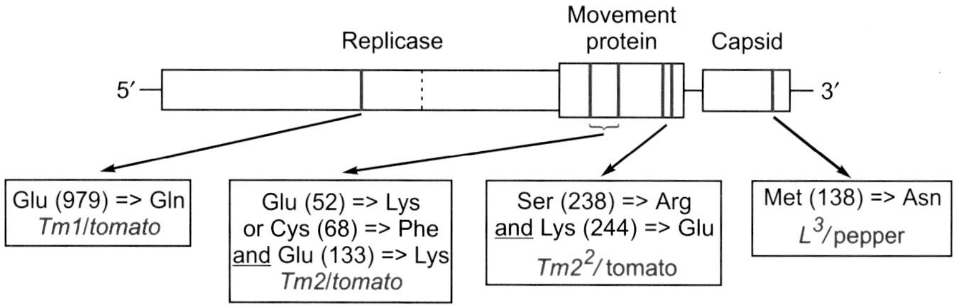

Figure 11.8 Location on *Tobamovirus* genome of mutations leading to virulence with respect to three resistance genes in tomato and one resistance gene in pepper.

tomatoes. The determinant of the virulence of pathotype 2 (overcoming the gene *Tm-2*) is the gene coding for the 30 kDa protein that is the virus movement protein. Two substitutions in the conserved regions of the N-terminal part of this protein are sufficient for the complete expression of the virulence: $Glu^{52} \rightarrow$ Lys or $Cys^{68} \rightarrow$ Phe and $Glu^{133} \rightarrow$ Lys (Fig. 11.8) (Meshi et al., 1989). The determinant of the virulence of pathotype 2^2 (overcoming the gene *Tm-2*^{2}) is also the gene coding for the movement protein of ToMV but, in this case, the two mutations $Ser^{238} \rightarrow$ Arg and $Lys^{244} \rightarrow$ Glu necessary for virulence involve the variable C-terminal part of the protein (Fig. 11.8) (Weber et al., 1993). The deletion of this C-terminal part of the movement protein also leads to virulence (Weber and Pfitzner, 1998).

For the three genes *Tm-1*, *Tm-2*, and *Tm-2*^{2}, a correspondence is observed between the resistance mechanism (reduction of viral multiplication or inhibition of cell-to-cell movement) and the functions of viral genes involved in the increase in virulence (replicase or movement protein). Nevertheless, it would be premature to conclude that there is a simple "functional" dialogue between virus and resistant plants. In fact, it seems that in the case of pathotype 2^2 of ToMV, the virulence could correspond also to an inability to induce a defence reaction in the resistant plant (Weber and Pfitzner, 1998). This latter type of molecular interaction is observed for the resistance of pepper to another tobamovirus, PMMoV (*Pepper mild mottle virus*), conferred by the L^3 gene. A mutation $Met^{138} \rightarrow$ Asn, leading to a modification of charge in the exposed C-terminal part of the capsid protein, does not allow induction of the hypersensitivity reaction (Fig. 11.8) (Berzal-Herranz et al., 1995).

Other studies are still diversifying virulence genes

In the case of CMV, a point mutation in the gene coding for one of the subunits of replicase allows the overcoming of a hypersensitivity resistance controlled by the *Cry* gene in cowpea (Karasawa et al., 1999). For several recessive resistances to various potyviruses in different hosts, a single or a few point mutations in the gene coding for the viral protein associated with the genome (VPg) confer virulence (Johansen et al., 2001, Masuta et al., 1999, Moury et al., 2004).

It seems at present that practically all viral genes could act as avirulence genes, depending on the resistance mechanism and the virus concerned, and that most often only one or two point mutations suffice to render a strain virulent. Even non-coding regions of viral genomes can act as avirulence determinants, as it has been demonstrated for MNSV (*Melon necrotic spot virus*, *Carmovirus*) and the *nsv* gene in melon (Diaz et al., 2004).

Virtually all resistance types can be overcome, including tolerance, since a single point mutation in the P3 protein gene induced an increase in ZYMV (*Zucchini yellow mosaic virus*, *Potyvirus*) aggressiveness, which caused severe symptoms in tolerant zucchini squash hybrids (Desbiez et al., 2003).

How can long-lasting resistances be obtained?

Some resistances prove efficient against all strains of a virus; they protect the crop, sometimes partially, but against all the isolates of a virus. They are called *horizontal resistances*. Even though there may be some examples of this ideal situation, more often the resistances are efficient only against some virus strains; they are referred to as *specific* or *vertical resistance* (Fraser, 1985). The major interest of this classification developed by Van der Planck (1968) is that it might be predictive, the horizontal resistances being most often polygenic and stable over time, while vertical resistances are more often monogenic and easily overcome. In fact, this interest has been called into question by the existence of many exceptions.

It has been observed, most fortunately, that many virus resistances have long-term efficiency in the field: indeed, the occurrence of virulent strains does not therefore imply the "failure" of the corresponding resistance gene. Sometimes, the virulent strains are not competitive with respect to the avirulent strains. They do not generalize and do not actually jeopardize the future of resistances (e.g., pathotype 2^2 of ToMV in tomato). In other cases, however, the virulent strains supplant the avirulent strains, the resistance is no longer effective in the field, and it will not be long-lasting (pathotype 1 of ToMV in tomato). In order to better assess the durability of a resistance gene, it therefore seems necessary to complement genetic studies on virus pathogenicity with a comparative analysis of the dynamics of virulent and avirulent strain populations (Lecoq et al., 2004). In fact, it is possible that the increase in virulence provides a "genetic burden" that renders the strain less competitive in the absence of the resistant varieties. In the case of ToMV, for example, the deployment of varieties possessing the *Tm1* gene in Great Britain was rapidly followed by a predominance of strains of pathotype 1. The producers thus chose to return to susceptible varieties, and pathotype 0 again became more common (Pelham et al., 1970). Similarly, in zucchini squash, the aggressive pathotype overcoming the ZYMV tolerance is less competitive than the "common" strains in susceptible hosts (Desbiez et al., 2003).

The durability or stability of a resistance is most often evaluated after several years of cultivation of resistant commercial cultivars. But the durability of a resistance can sometimes be predicted from laboratory studies using different methods of inoculation and successive transfers on resistant varieties. This approach, used to evaluate the stability of resistance to ZYMV in melon, revealed that the *Zym* gene could easily be overcome, even before virulent isolates were identified in nature (Lecoq and Pitrat, 1984). Predicting virus resistance durability is not an easy task. It requires sound knowledge of the nature of the resistance (inheritance and mechanisms involved), the genetic determinants of virulence (especially the number of mutations required to pass from an avirulent to a virulent pathotype), and also the fitness potential of virulent isolates (Lecoq et al., 2004).

One approach to make resistance to a virus more durable is to construct composite resistances, associating in a single genotype complete or partial resistances active at different levels of the viral infection cycle. This "pyramiding" strategy was chosen for ToMV resistance in tomato (see box earlier on specificity and management of resistance genes in tomato). In pepper, a similar approach was undertaken 30 years ago to breed for CMV-resistant varieties (Pochard, 1977). Resistance to CMV migration in the plant has already been introduced in commercial varieties; it is complemented by a tendency to escape infection and a resistance to CMV multiplication (Palloix et al., 1997). Similarly, in the melon Virgos (Fig. 11.9), the combination of resistance to CMV transmission by the aphid *Aphis gossypii* and a specific resistance to CMV "common" pathotype (which includes around two-thirds of the CMV isolates) provides excellent protection in the field. Some infected Virgos plants may be observed at the end of the crop, but these late contaminations have no significant effect on the yield (Lecoq and Pitrat, 1989).

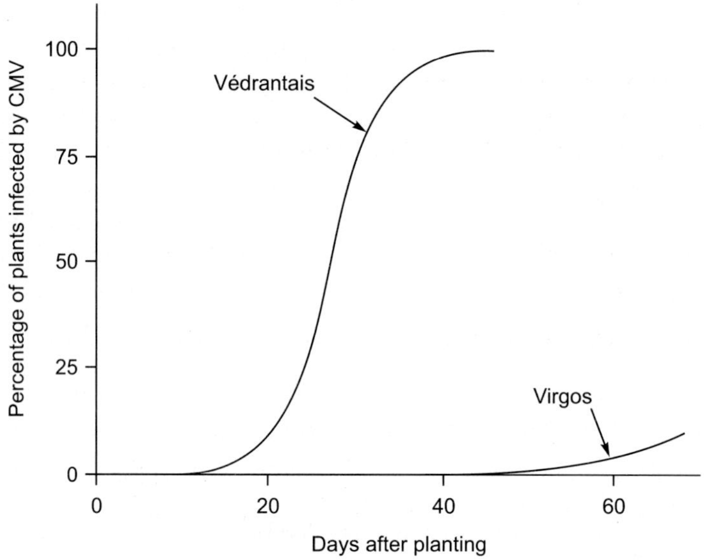

Figure 11.9 Development of CMV epidemics average curves established from 4 years of observation) in a susceptible melon variety (Védrantais) or a variety presenting a "composite" resistance to CMV (Virgos).

It is obvious that this type of approach could complicate and lengthen breeding programmes for virus resistance, which must in addition take into account many other agronomic traits or resistance to other pathogens (bacteria, fungi). This is no doubt the price to be paid for building durable resistances. Nevertheless, the discovery that numerous genes for resistance to viruses or other pathogens are often associated in gene clusters will undoubtedly facilitate the selection of varieties with multiple resistances (Kyle and Dickson, 1988; Caranta et al., 1997). Moreover, present advances in molecular marker-assisted selection will also speed up these programmes in the near future.

Resistances obtained through transgenesis, particularly resistances derived from the pathogen (Chapter 12), when they are authorized by local legislation and accepted by consumers, naturally integrate the strategy of pyramiding composite resistances. It is obvious that, in a given agro-environment, an integrated control strategy associating these composite resistances with cultural practices that can slow down or reduce viral epidemics (Chapter 10) will help keep the resistance efficient over time (Lecoq and Pitrat, 1982).

12

CHAPTER

Control of Plant Viral Diseases: Genetic Engineering for Protection

The breeding of virus-resistant plant varieties is the cheapest and most effective means of disease control, but it is a long and difficult process: the resistance genes are rare and when they are known they do not always confer a long-lasting resistance.

During processes of breeding between sexually compatible varieties or species, there is a transfer not only of genes conferring the desired resistance, but also of sometimes undesirable traits present on nearby loci; repeated back-crosses are needed to introduce a particular gene. The possibilities offered by genetic engineering to rapidly transfer a single gene or a small number of genes into a genotype while conserving its other characteristics have opened up new perspectives for control of viral diseases.

It must be emphasized that viral plant diseases are presently controlled by preventive methods that are often difficult to implement. Hence the interest in installing resistance or tolerance in plants in a constitutive manner. The sequences used for this purpose often come from the virus itself but may also have other origins.

Viruses cited

AMV (*Alfalfa mosaic virus, Alfamovirus*); BCTV (*Beet curly top virus, Curtovirus*); BMV (*Brome mosaic virus, Bromovirus*); CaMV (*Cauliflower mosaic virus, Caulimovirus*); CCMV (*Cowpea chlorotic mottle virus, Bromovirus*); CMV (*Cucumber mosaic virus, Cucumovirus*); LMV (*Lettuce mosaic virus, Potyvirus*); PVX (*Potato virus X, Potexvirus*); PVY (*Potato virus Y, Potyvirus*); RDV (*Rice dwarf virus, Phytoreovirus*); RSV (*Rice stripe virus, Tenuivirus*); RTBV (*Rice tungro bacilliform virus, Badnavirus*); RTSV (*Rice tungro spherical virus, Waikavirus*); RYMV (*Rice yellow mottle virus, Sobemovirus*); TEV (*Tobacco etch virus, Potyvirus*); TMV (*Tobacco mosaic virus, Tobamovirus*); TRSV (*Tobacco ringspot virus, Nepovirus*); TYLCV (*Tomato yellow leaf curl virus, Begomovirus*); TYMV (*Turnip yellow mosaic virus, Tymovirus*); WMV (*Watermelon mosaic virus, Potyvirus*); ZYMV (*Zucchini yellow mosaic virus, Potyvirus*).

Gene transfer

A very old history revisited with new techniques

Since life began, the transfer of genes between organisms (from bacteria to bacteria, from bacteria to Eukaryotes) has favoured the emergence and evolution of numerous species.

Since the genetic code is universal, a gene, whether viral or cellular, can be expressed outside its cell of origin, whether that cell is bacterial, vegetal, or animal. The discovery of spontaneous gene transfer and its mechanisms has inspired recent technologies of genetic engineering. The set of techniques defined as "genetic engineering" makes it possible to isolate and introduce a gene (or sequence) at will and rapidly into the genome of a cell. Subsequently, an organism possessing this new property must be regenerated from this single modified cell. Genetic transformation thus consists of limited hereditary modification of the cell genome by integration of a new gene; the transfer of a new monogenic trait cannot be considered the creation of a completely new species.

Gene transfer (or transgenesis) has already found wide fields of application in gene therapy and the obtaining of modified micro-organisms (yeasts, bacteria) for the food industry and for pharmaceutical production. Gene transfer in plants is presently a widely used tool. In basic research, obtaining transformed cells or plants makes it possible to identify genes, analyse their function, and study their role and regulation in the life of the plant or virus. For the purpose of improvement of cultivated plants, genetic engineering offers the possibility of transferring genes with a particular advantage, derived from a virus or various organisms, from one species to another in order to quickly obtain plants with a new character. The desired characters pertain to resistance to diseases, insects, or herbicides, as well as the synthesis of proteins for industrial purposes (enzymes) or therapeutic use (vaccines, plasma proteins including human albumin, hemoglobin, monoclonal antibodies for diagnosis) (De Wilde et al., 2000). A new potential is also being explored with attempts to create vaccine plants that can be ingested to provide immunization against diseases that are not presently under control (Castanon et al., 1999).

From lab to field

Plant genetic engineering began in 1983 (Herrera-Estrella et al., 1983; Zambryski, 1983) with the obtaining of transgenic tobacco varieties expressing a chimeric gene that gave them resistance to an antibiotic, kanamycin. Following these preliminary studies, research flourished in many countries and a wide variety of transgenic plants with new properties were obtained. The first commercial production of a genetically modified plant (late-maturing tomato) was carried out in the United States in 1994; it was followed in 1995 with that of a squash variety resistant to two potyviruses, WMV (*Watermelon mosaic virus, Potyvirus*) and ZYMV (*Zucchini yellow mosaic virus, Potyvirus*). Since then, millions of hectares in China, the United States, and other countries have been planted with modified crops mainly with resistance to insects or herbicides.

To reach this stage, several steps had to be cleared: in the laboratory, a foreign gene was made to penetrate the target cell, its integration in the genome and its expression were verified, and entire plants were selected and propagated from genetically modified cells. In the second phase, their agronomic interest and behaviour were verified in greenhouse assays and eventually in the field.

How is a transgenic plant obtained?

The various steps required to obtain a transgenic plant are summarized in Fig. 12.1: choice of a gene, completion of gene construct, plant transformation, regeneration, and evaluation of the modified plants.

■ Gene constructs

The target gene is inserted in a gene construct to be amplified

The genetic information (gene, gene fragment, non-coding sequence) that could be used to confer protection is defined according to the molecular data available for the target virus. The corresponding sequence is located and isolated by DNA manipulation techniques (amplification, sequencing, ligation, enzymatic restriction). In the case of RNA viruses, these manipulations are done on DNA complementary of the viral genome (cDNAs). Depending on the strategy adopted, various sequences might be used: coding sequences that will be expressed as proteins or non-coding sequences whose transcripts will be sense or antisense RNAs. Genes may also be obtained from prokaryotes or eukaryotes (plants or animals) (see page 305). In all cases, the gene product (protein or RNA) must interfere only with the viral infection and have no effect on the host physiology.

The gene is rendered functional in the plant cell by adding sequences controlling its expression. The gene construct must allow a stable expression of the information in the plant. Also, in order to be expressed in the plant cell, the sequence chosen (Fig. 12.1) must be placed between particular DNA sequences: promoter, initiation and termination codons, "leader" sequence for recognition by ribosomes, "enhancer" consensus sequence. These act as signals recognized by the plant cellular mechanism, regulating its expression (transcription into messenger RNAs, translation by ribosomes, transcription termination). The choice of sequences must also take into account the phenomenon of silencing (see Chap. 5) (De Wilde et al., 2000).

One criterion for obtaining a sufficient level of expression lies in the choice of the promoter, a sequence that triggers the expression of the gene. The promoter most widely used at present is the 35S RNA promoter of CaMV (*Cauliflower mosaic virus*,

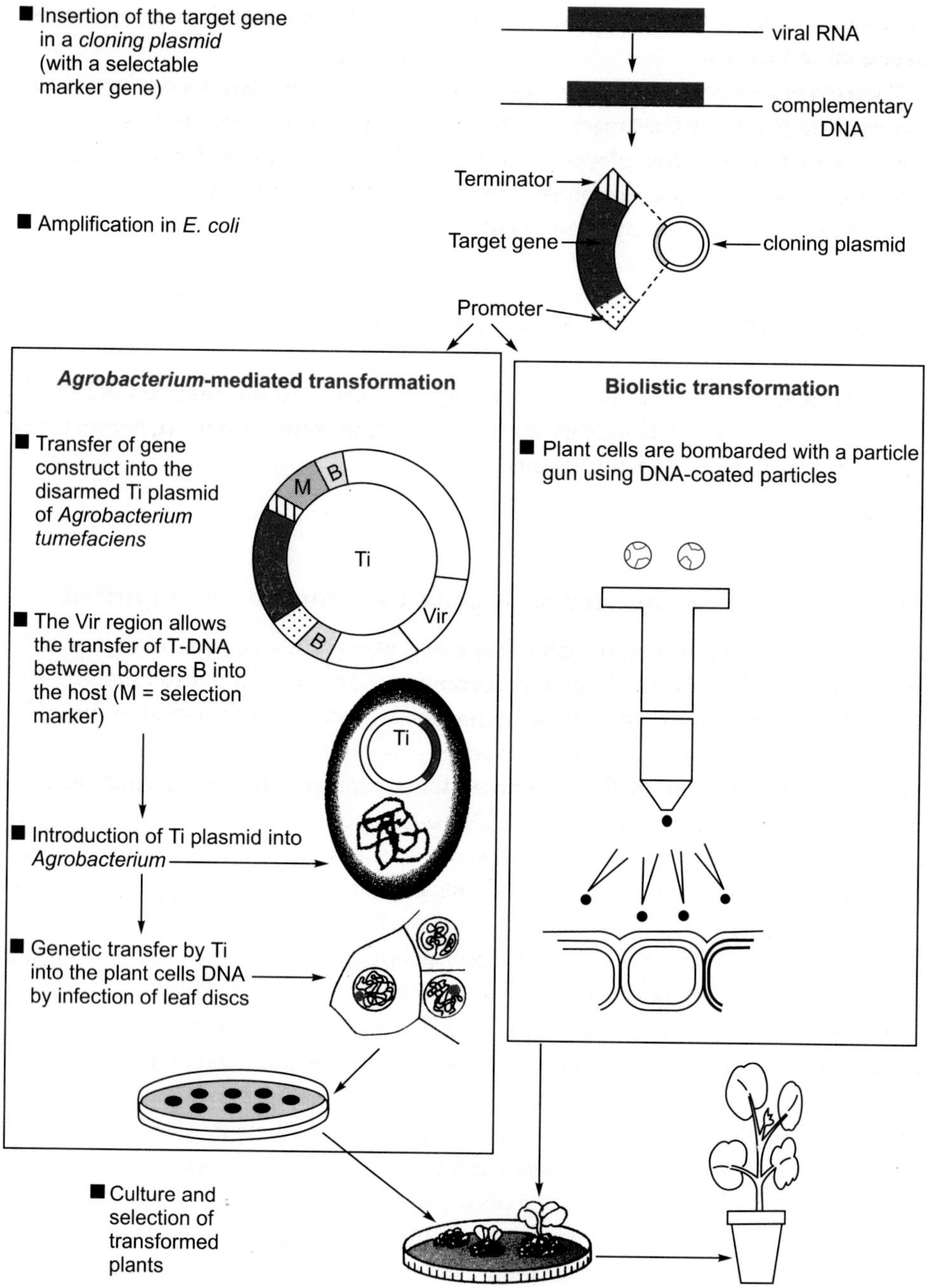

Figure 12.1 Plant transformation process: *Agrobacterium*-mediated and biolistic gene transfer.

Caulimovirus), which is expressed in various tissues. It has many derivatives; the duplication of the 5′ region of this viral promoter makes it possible to obtain an even greater efficiency (Kay et al., 1987). The promoter of the nopaline synthetase gene of *Agrobacterium tumefaciens* is also widely used. Experiments are done with other promoters that can be induced either by certain stresses or only in certain tissues or organs (flowers, roots, etc.). This possibility of orienting and controlling the expression of the transgene is important for the development of biotechnological applications.

The frequency of integration of a foreign gene is very low and therefore a large quantity of DNA must be introduced in the cells to be transformed. The gene construct is inserted in a bacterial cloning plasmid and multiplied in *E. coli.* To easily select the bacteria in which the plasmid and the target gene have been inserted, a gene for resistance to an antibiotic is used as a selection agent.

The gene construct is introduced in the nucleus of the plant cell

There are two main techniques to make the gene construct become a part of the plant cell genome: biological transfer and direct transfer.

Biological transfer through modified Agrobacterium strains

After cloning, the gene construct is generally introduced in transformation vectors carried by soil bacteria known as agrobacteria. In their natural processes of infection, agrobacteria spontaneously transfer a fragment of their genetic information directly into the nucleus of the cells they infect. *Agrobacterium tumefaciens* (causing crown gall) and *A. rhizogenes* (causing hairy root) respectively carry Ti (tumour-inducing) and Ri (root-inducing) plasmids. These plasmids are responsible for the transfer of genetic information and for tumoral transformation of the plant cell. To transfer a gene construct into the genome of a plant cell in a stable manner, modified agrobacteria that are incapable of causing tumours are used. The gene construct is inserted in a Ti plasmid between the borders of the T-DNA that will permit its integration into the host DNA during infection by the bacterium (see box). There may be errors in the copy. For that reason, several controls are done on the construct by sequencing, establishment of the restriction map, expression of the transcript *in vitro*.

Direct gene transfer

Direct gene transfer requires various chemical or physical processes to introduce DNA directly and force it to penetrate the plant cell by passing through the rigid pectocellulosic wall. These strategies, already applied in the animal kingdom, have been developed for the transformation of plant cells and particularly in the case of plants that have so far been resistant to transformation by agrobacteria (numerous monocotyledons). The processes are referred to as direct transfer because purified gene constructs penetrate without passing through the intermediary plasmids of agrobacteria. In the past, entire bacterial plasmids were often used for the transfer,

Agrobacteria are natural genetic engineers

Plant genetic engineering developed through studies and discoveries based on *Agrobacterium tumefaciens*, a common soil bacterium that is the agent of crown gall in numerous dicotyledonous plant species. Another bacterium, *A. rhizogenes*, induces a proliferation of roots (hairy root). The bacterial inoculation is carried out through contact with fresh injuries on the host plants.

In 1960, at INRA in Versailles, France, G. Morel et al. demonstrated that tumours induced by *A. tumefaciens* secrete particular substances, opines, that are specific to bacterial strains and serve as growth substrates. In 1974, in Gand, Belgium, Schell and van Montagu showed that this modification of plant cells is due to the presence of a large plasmid of around 200 kbp in the virulent strains. These are the Ti (tumour-inducing) plasmid and Ri (root-inducing) plasmid, depending on the type of bacterium, which carry numerous genes including those necessary for the replication of the plasmid. In 1977, in the United States, Chilton et al. proved that only a small part of the Ti plasmid (T-DNA or transfer DNA) is integrated in the nuclear genome of plant cells infected after inoculation of the bacterium. The genes carried by the T-DNA are not expressed in the bacterium but only in the transformed plant cell because they carry regulation signals of eukaryotes. These genes are, on the one hand, the tumoral genes (oncogenes) the expression of which, intervening in the synthesis of growth substances, leads to uncontrolled multiplication of cells and, on the other hand, genes necessary for the synthesis of opines, amino acids used as substrate only by the bacterium. Apart from the T-DNA fragment, the Ti plasmid has a region called virulence (vir). This region has genes that, with several genes from the bacterial chromosome, control the transfer of T-DNA into the genome of the plant cell. The presence of frontier sequences of 25 bp (right and left borders) on either side of the T-DNA is one of the conditions of its transfer.

This natural process of genetic transformation has been improved by suppressing the tumoral pathogenic character of the bacterium by deletion of oncogenic sequences (disarmed T-DNA) and insertion of the target gene in place of the T-DNA (Tagu, 1999). The disarmed Ti plasmids are used for various purposes, for example to study the function and regulation of cellular or viral genes or to evaluate the interest of certain genes that could carry new characters (e.g., resistance to insects or pathogens).

but it is possible and preferable to separate the target gene from plasmid sequences that are not useful for its expression and to let it to penetrate alone. This would resolve any problem, real or supposed, linked to a possible escape of the ampicillin resistance gene carried by the bacterial plasmid (Casse, 2000).

The penetration of DNA molecules carrying the gene construct into the cytoplasm of plant cells lacking a cell wall (protoplasts) is made possible by chemical means: the plasma membrane is destabilized temporarily and locally by means of polyethylene glycol (PEG), which allows DNA molecules in solution to penetrate the protoplasts. A fusion may also be carried out between the plasma membrane and liposomes containing the DNA to be transferred. A simple and effective physical means, electroporation, consists in subjecting a preparation of protoplasts and DNA to a series of electric pulses that will cause the transient formation of pores through which the DNA can be transferred.

A particle gun can be used to cross the pectocellulosic cell wall of the plant cell. Microparticles made of tungsten or gold coated with DNA are projected by means of kinetic energy generated by an explosive charge of gunpowder or by compressed air or helium. The microprojectiles can thus cross the wall and the cell membrane.

The DNA molecules thus introduced in the cytoplasm by different processes will migrate up to the nucleus and only some of them will be integrated into the plant genome.

■ Regeneration

Regenerated plants are subjected to molecular and agronomic evaluation.

Identification of primary transformants

For the detection of primary transformants (plants that have integrated the transgene), most of the experiments carried out so far have associated the target gene with a transformation marker gene that is expressed in the plant cell and is easy to select. The most frequently used is the gene coding for neomycin phosphotransferase II (NPTII), a gene that confers resistance to kanamycin, an antibiotic toxic to the cell. To avoid the disadvantages of selection by antibiotics or herbicides, other genes are developed, such as those using a positive selection based on the expression of a gene coding for a xylase that favours the regeneration of cells rather than just allowing them to survive. In the context of biosecurity problems, let us note that the transformation marker gene is a useful but not indispensable tool: advances in PCR detection methods and development of new procedures make it possible to get around the need for a transformation marker gene.

Other marker genes called reporter genes offer multiple possibilities for research: coding sequence of β-glucuronidase of *E. coli* (GUS gene), an enzyme with a measurable activity, or green fluorescent protein (GFP), implicated in bioluminescence of a jellyfish, *Aequorea victoria*. The coloured or fluorescent signals emitted allow to locate with high precision the tissue and cell site where the introduced gene is expressed. These marker genes are also widely used in experiments of transitory expression.

Regeneration of whole plants

The regeneration of whole plants from transformed cells (protoplasts or calluses) is another critical step the success of which lies in the knowledge of *in vitro* culture conditions, often specific to each plant species. The plants then undergo a certain number of molecular analyses pertaining to the integration of transgenes in the genome (integrated sequences, number of copies) as well as their expression. These analyses are done through characterization of their products: transcripts (northern blot) and proteins (ELISA, western blot). Although the insertion site of the gene in the chromosome cannot be predicted, the size of the region can be verified from PCR amplifications with primers bordering the regions to be transferred. Through analyses by Southern blot it can be ensured that, in the selected transformants, the right and left frontier borders of the T-DNA are not surpassed, and thus that no undesired sequences are issued from the plasmid vector; it can also be determined whether multiple insertions have taken place.

The various greenhouse assays conducted are meant first to ensure the conformity of the plant during the course of its development, the stability of expression of the gene, its transmission to the progeny, mendelian segregation and other factors. Tests for resistance or tolerance to the target virus are carried out on the selected lines, taking into account the appearance and type of symptoms as well as the rate of viral multiplication. Eventually, a field experiment confirms or not the expected properties in natural conditions of infection and will establish the agronomic behaviour. Crosses with traditional varieties can then be planned to obtain resistant varieties responding to precise agronomic criteria.

Transgenic protection against plant viruses

■ A broad concept: pathogen-derived resistance

Cross-protection experiments have demonstrated that the infection of a plant by a "mild" virus strain protects it from a later infection by another more severe strain of the same virus (Chapter 10). These results have suggested the idea of using the expression in a plant of a gene derived from the virus genome to limit the viral infection. This idea was put forth by Hamilton in 1980 and then formulated as a general concept for all pathogens in 1985 (Sanford and Johnston, 1985; Grumet et al., 1987): the expression of a gene derived from the virus may interfere with the replication of the virus if the product of the gene is present in a non-functional form or disturbs the cycle of the virus in the cell (Chapter 2). Transgenesis, particularly interesting in the case of viral infections, has made it possible to test these hypotheses by suggesting a specific means of inhibiting a process closely bound to cellular life (Wilson, 1993).

The first application of the concept of pathogen-derived resistance (PDR) was the constitutive expression of the capsid protein of TMV (*Tobacco mosaic virus, Tobamovirus*) in *Nicotiana tabacum* (Powell-Abel et al., 1986). It was next applied to the expression of capsid genes of different viruses, then of other viral genes (polymerase, movement protein, protease) and non-coding viral transcripts.

Overall, various resistances have been successfully generated with these approaches for all groups of RNA viruses and for a certain number of DNA viruses.

Over the course of these studies, the transformed plants expressing viral genes proved to be highly useful investigative tools in understanding the role of viral genes and the cellular functions mobilized by the virus.

The results presently allow us to differentiate three major types of PDR:

1. Protein-mediated resistance. Resistance depends on the synthesis of the transgene-encoded protein. There is direct interaction of the transgenic protein expressed previous to the infection, with a virus or with a structure necessary to the virus. We will present the resistances conferred by the expression of the capsid and other viral proteins (movement protein, replicase, Rep protein of *Geminivirus*).
2. Competitor RNA. Resistance can be due to competition between viral RNA and a competitor RNA.
3. RNA-mediated protection. The inhibition may result from a general response mechanism (gene silencing, see chap. 5) leading to the specific early degradation of the viral messenger RNA and transcripts of the transforming gene (transgene).

The expression of the viral capsid

The capsid gene is one of the most widely studied genes in numerous viruses; it was the first gene subjected to experimentation with a view to conferring antiviral resistance. Numerous viruses, belonging to more than 15 taxonomic groups, have been the subject of assays of expression of viral capsids in different plants, Dicotyledons and Monocotyledons. The level of protection is linked to the level of expression of the transgene measured by the concentration of the capsid protein. The protection is stronger if the similarity between the capsid expressed by the plant and that of the infecting virus is high. This type of protection has been shown to be effective for numerous viruses with single- or double-stranded RNA genomes (positive or negative) and double-stranded DNA, with a gradient from tolerance to quasi-immunity (Kanievsky and Lawson, 1998).

The protection conferred by the capsid gene against the virus results, in hypersensitive plants, in a reduction in the number of local lesions; in plants with systemic infection there is a delay in the appearance of symptoms, attenuation or disappearance of symptoms (recovery), reduction of the virus titre compared to the control, and sometimes quasi-immunity. This protection occurs in natural conditions of inoculation (Colour plates VIII.6 and VIII.7), especially inoculation by aphids.

The protection is often effective only against the viral strain from which the transgene comes and against similar strains (homologous protection), for example in the case of PRSV (*Papaya ringspot virus, Potyvirus*) on papaya (colour plate VIII.5). However, more extended protection has been observed. The capsid of ZYMV (*Zucchini yellow mosaic virus, Potyvirus*) strain Ct protects against numerous strains of this virus and against the similar virus WMV (*Watermelon mosaic virus, Potyvirus*) (Grumet, 1994). The capsid of LMV (*Lettuce mosaic virus, Potyvirus*) confers protection effective against other potyviruses. This is called heterologous protection (see box).

Homologous protection and heterologous protection

LMV (*Lettuce mosaic virus, Potyvirus*) is a serious problem in lettuce crops. The capsid gene, expressed in lettuce, provides tolerance to the LMV strain from which the transgene is developed as well as against other strains isolated in Europe. This is called homologous protection (Dinant et al., 1997). In tobacco, which is not susceptible to LMV, the capsid of this virus confers quasi-immunity against PVY (*Potato virus Y, Potyvirus*) (colour plate VIII.3) (Dinant et al., 1993) and protection against other potyviruses (colour plate VIII.4). This is called heterologous protection (Blaise et al., 1995). There is an entire range in the efficiency of protection observed, depending on the potyviruses and depending on the hosts.

The potential disadvantages of this strategy are related to the possibility of hetero-encapsidation or intermolecular recombination. To control these possibilities, a new transgene has been constructed. It contains neither the motif needed for transmission by aphids nor the 3′ end sequence for recognition by replication enzymes. Other deletions may still be used without altering the efficiency of the gene, in the central structural domain, which ensures the cohesion of the capsid. This concept of *minicapsid*, a minimal gene conferring protection, is sought to be developed in the context of transgene biosecurity.

In the case of TMV, the role of the transgenic capsid protein in the resistance is partly elucidated. The protection is overcome when the plant is inoculated with viral RNA. The same is true if it is inoculated with virions that, after exposure to basic pH, have lost 60 to 70 subunits of the 5′ end, or around four turns of the helix (Register and Beachy, 1988). In the first events of infection, decapsidation of the 5′ end of the RNA is necessary to permit the ribosomes to attach and begin translation. Experiments of cross-protection with capsids carrying mutations that modify the aggregation of subunits into a helix and their capacity to fix to RNA have shown that the protection is conferred by capsids that can assemble to form discs and bind to the viral RNA. These observations suggest that interference between the RNA in the process of decapsidation and the capsid already expressed in the cell will lead to a re-encapsidation, thus blocking the process of infection (Bendahmane et al., 1997; Lu et al., 1998) (Fig. 12.2).

This mechanism is probably not the only one at work in various host-virus combinations. The protection conferred by the capsid of PVX (*Potato virus X, Potexvirus*) or that of AMV (*Alfalfa mosaic virus, Alfamovirus*) is manifested after inoculation with virions as well as with viral RNA. The AMV capsid plays a complex role in the initiation of replication (Yusibov and Loesch-Fries, 1995), which suggests that infection may be blocked at later stages than decapsidation. The multiple functions that could be served by viral capsids and their role in various

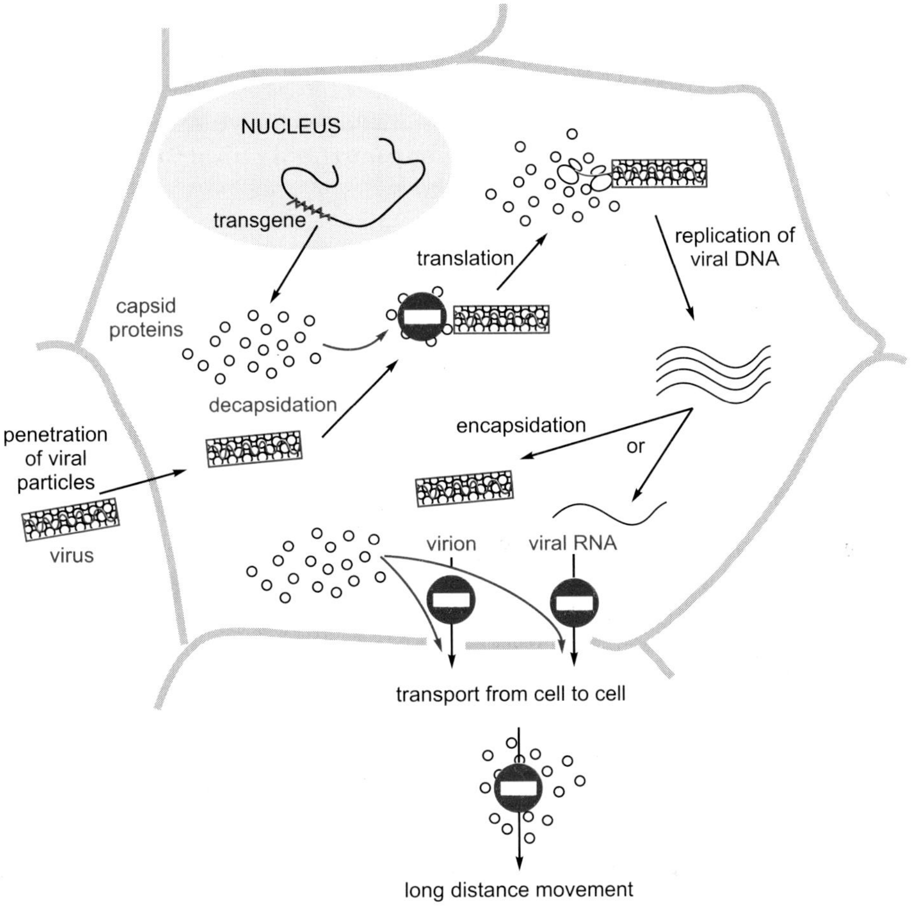

Figure 12.2 Resistance linked to the expression of a capsid protein in a transgenic plant (Berthomé et al., 2000).

stages of infection, especially movement and transmission, are now better understood (Bol, 1999; Pacot-Hiriart et al., 1997); the expression of the capsid may therefore confer resistance through various mechanisms that are still unidentified (Beachy, 1997). Let us note finally that sequences of the capsid may also confer resistance through silencing (see below) that is sometimes difficult to distinguish from resistance resulting from the expression of the capsid protein.

Expression of other viral proteins

Movement protein

Movement protein allows the virus to propagate from cell to cell in a systemic fashion, modifying the limit of exclusion of plasmodesmata (Chapter 4). Expressed in a functional form, it allows a TMV strain defective for movement to move to neighbouring cells (Holt and Beachy, 1991). On the other hand, expressed in truncated form, it confers resistance to a wide spectrum of viruses. For example, the expression of truncated movement protein of BMV (*Brome mosaic virus*, *Bromovirus*) confers resistance against TMV (Malyshenko, 1993) and the expression of truncated movement protein of TMV confers resistance against CaMV (*Cauliflower mosaic virus*, *Caulimovirus*), AMV, tobamoviruses similar to TMV, and TRSV (*Tobacco ringspot virus*, *Nepovirus*) (Cooper et al., 1995).

The levels of protection are not the same in these different examples, but competition between the viral and mutated transgenic movement proteins, either at the level of plasmodesmata or for attachment to viral RNA, is highly probable in all cases. The non-functional proteins behave like dominant negative mutants blocking the long-distance movement of the virus with high efficiency and produce broad-spectrum resistance (Beachy, 1997).

Replicase

Among the examples of transgenic plants expressing the replicase gene, some experiments describe a resistance linked to the synthesis of the protein (Palukaitis and Zaitlin, 1997). Plants expressing a non-functional AMV replication protein mutated in the polymerase motif GDD, present a resistance linked to a strong expression of the transgene (Brederode et al., 1995); the expression of the defective replicase, ahead of the viral cycle, could be the source of the resistance. However, in numerous cases, the characteristics of the resistance in plants expressing replicase gene sequences depend on the homologous resistance linked to the RNA and it is impossible to demonstrate the role of the protein. The resistance is thus limited to strains in which the sequence is very similar to that of the transgene (Carr and Zaitlin, 1991; Baulcombe, 1996; Jones et al., 1998).

Rep protein of *Geminiviridae*

The Rep protein of *Geminiviridae* is not a replicase; it is the only viral protein required for viral replication by recruiting cellular enzymes related to host DNA synthesis. It also regulates it own transcription from the C1 gene. The Rep protein of begomoviruses mutated in its active site (Sangaré et al., 1999) or truncated (Brunetti et al., 2001) is a dominant negative competitor in viral replication. The first 210 amino acids of Rep expressed in *Nicotiana benthamiana* can inhibit, but not abolish, C1 transcription and induce a strong resistance to TYLCV (*Tomato yellow leaf curl virus*, *Begomovirus*).

RNA competitors of the viral genome

The examples above describe situations in which the synthesis of a protein coded by the transgene confers resistance. However, direct inhibition of the infection can be achieved when the transgene transcripts behave like parasite molecules and compete with their assistant virus for the replicase (satellite RNA, defective interfering RNA or DNA). The viral infection results in the amplification of the transgene transcript, which subsequently limits the multiplication of the virus.

This two-fold strategy has been explored for CMV (*Cucumber mosaic virus, Cucumovirus*), an important pathogen of vegetable crops. This virus may be associated with a satellite RNA of 332 to 405 nucleotides, which depends totally on its assistant virus for replication and encapsidation (Chapter 7). In tomato, certain variants cause severe necrosis while others attenuate the symptoms (Jacquemond et al., 1998). These latter variants, expressed in tomato, may confer a high level of protection against CMV. Tomato plants expressing a beneficial CMV satellite have been extensively cultivated in China. However, as beneficial and necrogenic satellites differ only by a few mutations, this very effective strategy has been completely abandoned (Jacquemond and Tepfer, 1998; Tepfer, 2002).

The defective interfering (DI) RNA are natural forms that are derived from the viral genome through internal deletions and conserve some structural elements; they replicate with the help of their parent virus. Transgenes coding for DI RNA or DNA have proven capable of diminishing the symptoms and viral multiplication of BCTV (*Beet curly top virus, Curtovirus*) on *N. benthamiana* and TYMV (*Turnip yellow mosaic virus, Tymovirus*). These are laboratory experiments that have so far not been applied in the fields (Stanley et al., 1990; Kollar et al., 1993; Zaccomer et al., 1993).

Expression of a transgene homologous to the viral genome

Unlike situations that we have described in which the resistance is associated with high amount of mRNA of the transgene and of the corresponding protein, there are numerous cases in which the resistance is associated with a low or undetectable presence of this mRNA.

Tobacco plants strongly expressing the capsid gene of TEV (*Tobacco etch virus, Potyvirus*) manifest, after infection by this virus, a phenomenon of recovery. But a mutated form of the gene, the transcript of which is non-translatable, gives the same result, which shows that the translation product of the transgene does not intervene (Lindbo and Dougherty, 1992). After inoculation the transgenic plants present at first a phase of susceptibility followed by a state of recovery, characterized by:

- a strong resistance manifested by a low or undetectable presence of viral RNA, which is not overcome by a concentrated inoculum (virion or RNA);
- a narrow protection limited to the virus itself or to very similar strains;
- a low level of detection of the transgene mRNA in the cytoplasm.

The transgene is transcribed in the nucleus at a normal level. To explain this apparent paradox it has been hypothesized that there is activation in the cytoplasm, after transcription, of a system of specific degradation targeting the transgene as well as the homologous viral RNA, by a phenomenon of silencing (van den Boogaart et al., 1998) (Fig. 12.3) (Chapter 5).

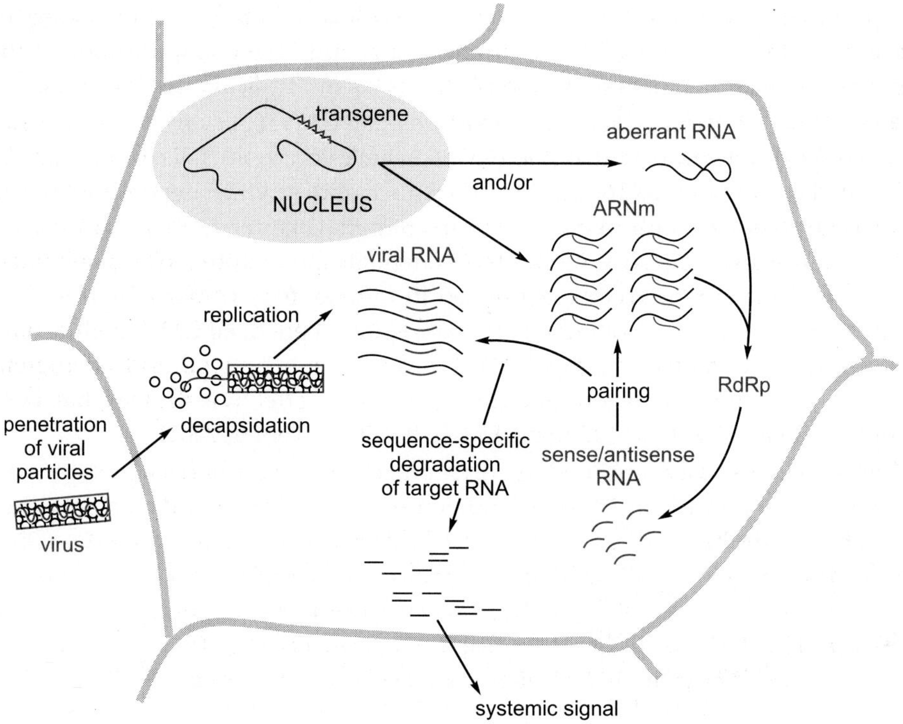

Figure 12.3 Resistance linked to the expression in a transgenic plant of RNA sequences homologous of the viral nucleic acid (silencing) (Berthomé et al., 2000).

The characters of this resistance are the following:

— The triggering of the phenomenon depends closely on the homology between the sequences of the transgene and of the infecting virus. A double-stranded RNA seems to be involved in the starting of the process.

— There is diffusion of a systemic signal from cell to cell and through the phloem pathway.

— The protection conferred may be expressed in two phenotypes: either a state of total immunity that takes place from the beginning of the infection, or a state of "recovery" in which the plant is first susceptible then becomes resistant (the leaves formed after recovery are free of the virus and resistant to a new inoculation). In fact, these two states are probably the results of a single process. The plants expressing the capsid gene of TEV in the form of a non-translatable RNA show different levels of resistance: the presence of one or two copies of the transgene produce the state of recovery, three copies or more are necessary to activate the system at first and establish a highly resistant state (Goodwin et al., 1996).

The introduction of a virus in the cell expressing a viral transgene triggers a process of silencing of the resident transgene and of the homologous infecting viral genome, leading to the co-degradation of the viral RNA and the transgene. This phenomenon of silencing induced by a virus infecting a transgenic plant expressing homologous viral sequences is very similar to silencing that manifests the natural defence reaction of the plant to viral infection. It is now considered an antiviral response mechanism that is part of the general mechanism of post-transcriptional regulation (called PTGS) of the expression of genes in the cell (Ratcliff et al., 1997, 1999; Covey, 1997; Marathe et al., 2000) (Chapter 5).

Expression of antisense viral RNA or of ribozyme structure

It has been demonstrated that the expression of cellular mRNA can be blocked by antisense sequences and this was assayed on the expression of viral genomes. The antisense sequence acts by formation of a duplex that either interferes directly with the biological activity of the "sense" nucleic acid or induces the specific degradation of the duplex (as in the case of silencing seen above). This strategy has given limited results; the principal difficulty pertains to the effective concentration of the transgene in the cytoplasm in which it must face a high concentration of viral sense RNA molecules. However, significant results have been obtained against TRSV (*Tomato ringspot virus, Nepovirus*) (Yepes et al., 1996), potyviruses (Hammond and Kamo, 1995), and RTSV (*Rice tungro spherical virus, Waikavirus*) (Huet et al., 1999).

A ribozyme is a small RNA that has a catalytic domain giving it a specific enzymatic property. It may cleave in a specific site an RNA with which it hybridizes. From natural ribozymes that exist in viroids and some satellites, antisense RNA of viral sequences have been obtained with a ribozyme structure designed to specifically cleave viral RNA. Although some encouraging results have been obtained *in vitro*, some difficulties *in vivo* (at the cell or plant levels) remain to be overcome (Tabler et al., 1998).

Some examples of non-viral genes

Resistance genes

In tobacco, the N gene confers a resistance with a hypersensitive reaction to TMV limiting the infection to a local lesion (Chapter 6). This gene has been isolated and transferred into susceptible tomatoes and gives them the expected resistance (Whitham et al., 1996). Similarly, the Rx gene has been introduced in potato and tobacco (Bendahmane et al., 1999). This approach will make it possible in the future to achieve the direct transfer of resistance genes provided they can be isolated and cloned. Nevertheless, some questions remain: how effective will the gene be after transfer, and how durable will the resistance be (Berthomé et al., 2000)?

Rice viral diseases and the contribution of genetic engineering

Rice has a particular importance in human nutrition that amply justifies efforts to improve its agronomic and nutritional qualities as well as its resistance to pests and diseases. Genetic engineering is beginning to provide a few tracks to fight rice viral diseases; fifteen viruses are known to infect rice. The expression of the capsid of RSV (*Rice stripe virus, Tenuivirus*) in two Japanese varieties protects the transformed plants and their progeny against this virus (Hayakama et al., 1992). The same is true with the outer coat protein of RDV (*Rice dwarf virus, Phytoreovirus*) (Zheng et al., 1997). In these two cases, the protein coded by the transgene is present.

On the other hand, resistance against RYMV (*Rice yellow mottle virus, Sobemovirus*) has been obtained by expressing a construct derived from the replicase gene, a gene in which the sequence is particularly conserved between different isolates. This is a resistance "by homology" between the transgene RNA and the viral RNA. The resistance seems exceptionally wide-ranging, and it is manifested against isolates of different serological groups (Pinto et al., 1999).

One of the most serious disease of rice, present in all of Southeast Asia, is the rice tungro complex: the epidemic begins with RTSV (*Rice tungro spherical virus, Waikavirus*) transmitted by a leafhopper. This virus assists the transmission by this leafhopper of RTBV (*Rice tungro bacilliform virus, Badnavirus*), which is responsible for severe symptoms of rice tungro but cannot be transmitted alone. The genes of resistance to the leafhopper are often overcome. Rice plants expressing the replicase gene of RTSV in the antisense form are protected up to 60% against infection. The plants expressing the complete or truncated replicase gene of RTSV regularly manifest a total resistance to the virus even with high doses of inoculum, by a phenomenon of silencing; moreover, these plants are no longer capable of assisting RTBV for its transmission (Huet et al., 1999). These results open up an epidemiological approach to the fight against rice tungro, in which the progression of the disease will be greatly slowed by an effective control targeted against RTSV. However, despite all these strategies, it is not possible to create in all cases a resistance transmissible to the progeny. Research is intensifying to fight viral diseases of rice, which have spread alarmingly over the past 20 years.

In mammals, there is a pathway of highly general response to viral infections: interferon. One of the components of this response is the 2-5A system activated by interferon: the synthesis of the 2′-5′ oligoadenylate synthetase (called 2-5A synthetase) is activated specifically by double-stranded RNA, then it in turn activates ribonuclease L, which, linked to the 2-5A synthetase, degrades the viral RNA. Some components of the 2-5A system are present in low amounts in plants, although a 2′-5′ synthetase function has not been found. Several attempts to transfer the 2-5A system of mammals into tobacco have given positive results (Mitra et al., 1996). The 2′-5′ synthetase and RNase L genes have been introduced separately into tobacco plants, which have then been crossed to obtain plants expressing the two genes simultaneously. These plants have proven resistant to several viruses (AMV, CMV, TMV, TEV). The inoculation causes the formation of necrotic lesions on the inoculated leaf, and the infection stops at this stage. These non-viral genes can be used to obtain protection against a wide range of viruses.

Ribosome inactivating proteins

Ribosome inactivating proteins (RIP) are basic proteins that have glycosidase activity; they inactivate the ribosomal RNA and cause cell death by arrest of protein synthesis. Their activity is exerted with a clear antiviral specificity, as established for RIP extracted from *Mirabilis jalapa, Phytolacca americana,* and *Dianthus caryophyllus.* The synthesis of such a protein put under the control of a viral promoter leads to cell death during induction of the promoter by the viral infection. This is the case with dianthine gene under the control of the promoter of a geminivirus. The synthesis of dianthine consecutive to the activation of the viral promoter causes the death of the infected cell (Hong et al., 1996). In *Phytolacca americana,* the antiviral activity of the RIP has been separated from the phytotoxic activity, which opens up avenues to a practical application.

Inductors of systemic resistance

Inductors of resistance have been extracted from *Boerhaavia diffusa* and *Clerodendron aculeatum*. These substances, if they are non-phytotoxic and have a wide range of action, are good candidates to improve natural resistance (Verma, 1997).

Potential environmental impact of virus-resistant transgenic plants

Genetically modified organisms (GMOs) first appeared in laboratories and then started to be used in some countries by food and pharmaceutical industries. The announcement of commercial production of GMOs designed for food in Europe set off public debate and controversy. In this emerging field of biological investigation, the scientific community pursues with rigour and transparency an extensive study of the agronomic interests as well as the environmental consequences of the use of transgenic plants.

It must be emphasized that any experimentation on GMOs implies a gene construct that is completely defined in which all the elements may be important (promoter, genes or gene fragments, termination sequences, other sequences). It also involves a clearly identified host in environmentally specified conditions. This is thus a case-by-case experimentation including the analysis of numerous parameters, and generalizations are ruled out in this field.

In the particular case of virus-resistant GMOs, the benefits expected are mainly the following:

- reduction in losses and improvement in quality of products;
- reduction in cost of inputs and reduction in the use of environmentally toxic pesticides;
- shorter delays in selection of resistant varieties.

The use of plants genetically modified by viral sequences will nevertheless raise new questions for which answers will have to be found. Among these questions are those relative to risk to human health. One response may be given by regulatory tests of toxicology. It must, however, be remembered that phytopathogenic viruses are very frequent, that they infect numerous cultivated species, including crop species in which the fruits, leaves, or roots are often consumed raw. In no case have indications of toxicity of virus-infected plants to humans been recorded. Another question concerns the long-term agronomic interest of resistances conferred, their spectrum of action, and their durability. The most serious questions have pertained to the impact on the environment through gene flow, on viral populations as well as on cultivated or wild plants. This impact depends, of course, on the gene and species concerned.

The following potential risks will be considered:

— transfer of genes from cultivated plants to wild species;

— complementation of a function of a foreign virus over the course of its multiplication in the transgenic plant (hetero-encapsidation);

— recombination between viral sequences of an infecting virus and a transgene expressed in the cell.

Note that these phenomena are not specific to transgenic plants; they can take place with resistant plants obtained by conventional genetics or during natural viral infections. However, experimentation is ongoing to investigate whether their impact is significantly different in the case of transgenic plants.

■ Dispersal of transgenes through pollen

Viral transgenes can be transferred from cultivated transgenic allogam species to sexually compatible wild species. The resistance thus acquired by these wild plants may eventually give them a selective advantage and thus increase their capacity for colonization.

A study was carried out in a field over five consecutive years in the United States to assess the flow of viral transgenes, alone or in combination, between transgenic squash (CMV, ZYMV, WMV) and wild related species. The transfer of CP genes into wild squash was actually observed (Fuchs and Gonsalves, 1997). In the absence of the virus, these plants did not present increased competitivity; on the other hand, in regions in which the virus is endemic, it is probable that they will have a selective advantage. This question also arises with conventional resistances. If the transgene can confer a new trait resulting in increased fitness of the wild plant, the potential impact of each virus-resistance transgene must be carefully evaluated and strategies to prevent gene flow must be proposed (Tepfer, 2000).

Functional complementation

An endogenous protein of viral origin may modify a property of a virus infecting a transgenic plant. The situation most thoroughly studied involves the phenomenon of heterologous encapsidation. When a plant is infected by several viruses at once, it may happen that the RNAs are encapsidated by a combination of their own capsids as well as capsids of another strain, of a related virus, and more rarely, of a virus belonging to another genus. The phenomenon of heterologous encapsidation known for a long time was associated with modifications of the efficiency or specificity of transmission (Chapter 8).

The capsid protein expressed by a transgenic plant may in the same way encapsidate all or part of the viral RNA (Fig. 12.4). The capsid proteins generally have domains that are involved in the specificity of vection, and a modification of the aptitude to be transmitted by a vector can thus be observed (Lecoq et al., 1993). Similarly, a movement protein expressed by a transgenic plant may allow infection by a virus that normally does not become systemic in the plant (Lomonossof, 1995).

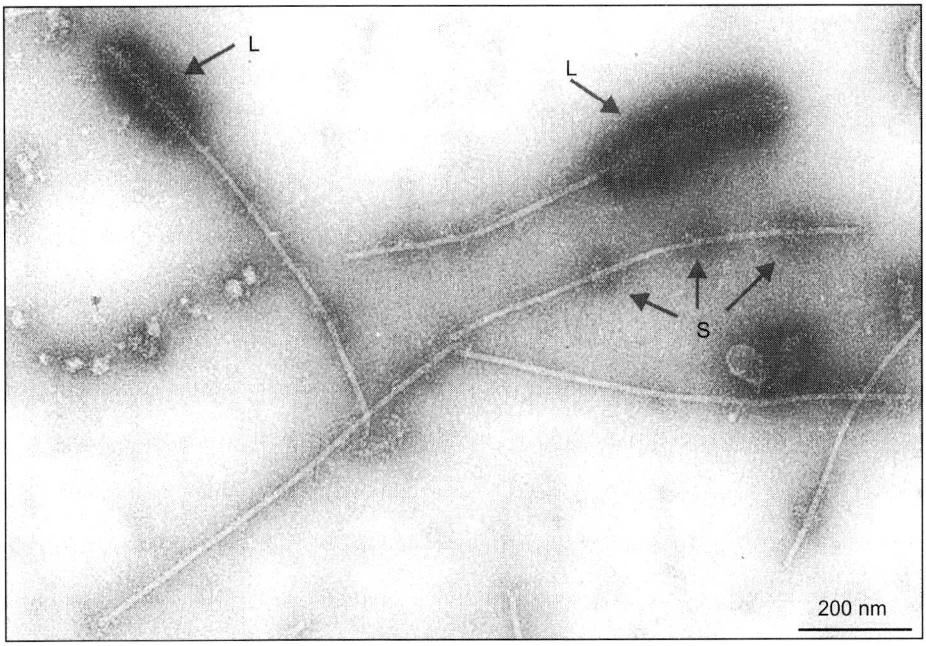

Figure 12.4 Particles of ZYMV (*Zucchini yellow mosaic virus*) from a *Nicotiana benthamiana* plant expressing the capsid protein of PPV (*Plum pox virus*) observed under electron microscope (negative staining with ammonium molybdate) after incubation with a polyclonal antiserum directed against PPV. The particles of ZYMV present a partial decoration by PPV antibodies over short (S) or long (L) segments, revealing a phenomenon of heterologous encapsidation by the capsid protein expressed by the transgenic plant (see controls in Fig. 9.18; for more details see Lecoq et al., 1993).

To avoid these situations of heterologous assistance, genes coding for mutated or truncated proteins can be used to suppress certain functions. For example, the consequences of heterologous encapsidation can be suppressed by using genes mutated on sites necessary for transmission by vectors (DAG triplet of potyviruses, for example) or on domains implicated in encapsidation (Jacquet et al., 1998). Similarly, in the case of movement protein, the use of non-functional forms does not allow the assistance of other viruses (Lomonossof, 1995). To increase the security of viral transgenes, the knowledge of molecular determinants of biological functions of viral constituents should be improved. This is particularly true for capsid protein which is involved in encapsidation, specificity of vection, movement over a short or long distance and heterologous encapsidation (Pacot-Hiriart et al., 1997).

Do phenomena of complementation have a particular impact over the long term in the context of virus resistant transgenic plants? In fact, they modify the phenotype but not the genotype of viruses infecting the transgenic plants and it has been observed in experiments that these modifications disappear as soon as the virus reaches a plant that is not transgenic because the viral genome remains unchanged (Fuchs et al., 1998).

■ Recombination of transgene with a viral genome

Risk of genetic recombination

More complex is the risk of genetic recombination between a virus infecting a transgenic plant and the information of viral origin that the plant expresses. In this case, there may be a long-term modification of the viral genome and its biological properties. In fact, in a virus infected cell expressing a transgene, are often found together the viral genome and a sequence homologous or non-homologous that may eventually exchange information. The natural process of recombination has occupied a large place in the evolution of viruses, and the precise mechanisms of various types of recombination have been widely studied to establish the sequences and signals involved (Chapter 13).

Recombination between the viral genome (most often strains deficient for a given function) and the homogolous transgene has been studied in conditions that do not permit silencing. The CCMV (*Cowpea chlorotic mottle virus*, *Bromovirus*) deleted in the non-coding 3' region of the capsid gene is not systemic in *Nicotiana benthamiana*. This function may be restored by simple recombination with a transgene expressing the capsid and the non-coding 3' region (Greene and Allison, 1994). It should be noted that none of the recombinants is found to be more competitive than the wild strain. Another study was conducted on a CaMV strain defective for the VI gene that is a transactivator of transcription, which could be restored by a transgene expressing the VI gene. A strain incapable of infecting the *Solanaceae* may thus acquire this possibility following a double recombination, in which either reverse transcriptase or a DNA-DNA recombination is involved. These results were obtained, for reasons of experimental convenience, with defective viruses, and thus with a high selection pressure in favour of recombinants. There are also examples in which recombinants have been observed in medium or low selection pressure (Wintermantel and Schoeltz, 1996; Aaziz and Tepfer, 1991).

Recombination can also take place between a viral transgene and a different viral RNA, producing a new recombinant RNA. Many experiments are done to evaluate the possibility that recombination in transgenic plants could generate new viruses with altered properties (reviewed in Rubio et al., 1999; Aaziz and Tepfer, 1999).

How can the design of transgenes be improved?

In relation to events that could occur during multiple infections, the possibility of recombination of a transgene may be a novel situation. This is the case when the genetic information of the modified plant comes from a virus that does not naturally infect this species or when this information is expressed in a tissue or cellular compartment that is not the usual site of the viral multiplication. In this case, there may be coexistence, in a single cell, of viral sequences that do not meet in nature during multiple infections.

Even though studies on the problems of recombination between transgenes and viruses are very far from having examined all the factors at work, the existence of these recombinations must lead to the search for possibilities of reducing their frequency or consequences. A first task is to take into account sequences favourable or not to recombination, as well as secondary structures involved (Simon et al., 1997; Bujarski and Nagy, 1996). We must also take into account the role of replicase and delete in the transgenes the 3' sequences that are recognized by replicase to prevent simple recombination with viral RNA. Deletion of the non-coding 3' end in a transgene of the capsid of CCMV suppresses the occurrence of recombinants (Greene and Allison, 1996). The same remark can be made with respect to sequences of recognition by reverse transcriptase. The choice of sequences used is increasingly rigorous. Their length is reduced to the minimum required and their aptitude to recombine tested experimentally (Varrelmann et al., 2000).

■ Transgenes and legislation

The acceleration of gene flows results principally from human activities, especially agricultural practices and international exchanges. Undesired effects may result, such as the introduction of new viruses or new vectors responsible for the emergence of new diseases. These effects are sought to be reduced by regulatory measures, which have only a limited efficiency because of the complexity of international exchanges.

The problem of gene flows also arises with the commercial release of transgenic varieties. But here, we have the obligation to carry out a prior evaluation on a case-by-case basis in order to reduce as far as possible the margin of uncertainty. For this purpose, any genetically modified plant is first studied in the laboratory in a confined environment; there the expression of the gene, its stability, and its effect on the plant are specified. Selected clones regenerated from genetically modified cells must subsequently be submitted to greenhouse experiments and prove to have normal growth and to be better than the non-transformed host with regard to viral resistance. Plants selected in greenhouse experiments can eventually be released for field testing under controlled conditions, to evaluate their positive features (e.g., stability of the protection against different viral strains, agricultural performance) and possible negative features concerning environmental risks. These field trials must take into account the plant biology and agricultural practices. Finally, some of the lines of virus-resistant plants can be authorized for general release and/or commercial use.

In all these stages (creation, greenhouse and field testing, cultivation, marketing), experiments and trials are subject to more or less strict regulations established in each country by governmental agencies. These regulations must address scientific, legal, and economic problems in a precise framework. No general international safety rules have been agreed upon so far and no standard expertise procedures have been established for risk assessment. Consequently, regulations for each stage of transgenic plant production vary from one country to another. The legal acceptance of the patentability of living matter modified by genetic engineering has prompted many industrial firms to create transgenic varieties of commercial interest. In this context of "patent war" and taking into account the variable consumer acceptance for genetically modified organisms (GMOs), the real social benefit for each country in cultivating existing GMOs is matter for debate, far removed from any real scientific question.

Although virus-resistant transgenic (VRT) plants have been proved to be very efficient in protecting against a number of severe virus diseases of various crops, their actual use for commercial production is marginal in comparison to the use of insect- or herbicide-tolerant transgenic crops, which were estimated to cover nearly 70 million ha in 2003 throughout the world. The commercial cultivation of VRT plants is presently mostly restricted to the United States and concern mainly two crops and four viruses.

Papaya transgenic for the PRSV-coat protein have proved to be very efficient in protecting against the severe epidemics caused by *Papaya ring spot virus* (*Potyvirus*) in Hawaii. In 2003, 46% of the papaya crop planted in Hawaii was transgenic (about 440 ha). A constraint on the generalized use of PRSV transgenic resistant papaya is that 20% of Hawaii papaya production is exported to Japan, which has not yet approved transgenic papaya for human consumption.

Transgenic squash carrying resistances to ZYMV, WMV and CMV are mainly grown commercially in Florida and Georgia, where they account for 2% and 17% respectively (about 725 ha in total) of the total acreage planted in squash. VRT squashes are mostly grown during the fall season, when the virus pressure is high.

Limitations on larger-scale cultivation are the seed price (two to four times that of regular cultivars), the diversity of squash types for which VRT cultivars are not always available, and the availability at a lower price of tolerant cultivars obtained through conventional breeding.

The future of transgenes in the control of plant viruses

Plant genetic engineering constitutes one of the effective means for varietal improvement. It has a potential particularly in cases where a highly damaging virus cannot be controlled by resistance genes existing in the host or in related species. It makes it possible to use new antiviral mechanisms that can be introduced quickly in the plant and that can be added to known resistance genes.

In the perspective of using virus-derived resistance for agronomic purposes, many parameters are to be considered. The resistance must be durable, it must have a spectrum of action as wide as possible, and it must offer the greatest possible security with respect to the environment. The resistances conferred by a protein are often quite wide-ranging, and they can be improved by structural modifications of the protein. The resistances resulting from silencing are often highly specific. They are linked to sequence homologies greater than 80%. They can be extended by linking several viral sequences in a single transcript (Pang et al., 1997). In all cases, precise knowledge of molecular structures conferring the resistance is necessary in order to optimize the resistance. There is also the possibility of creating transgenes similar to elicitors of natural resistance (Baulcombe, 1996; Marathe et al., 2000) and combining several types of resistance. More generally, the precise determination of active sequences conferring the protection will lead to minimal transgenes that do not code for a viral function and in which the undesirable traits have been eliminated.

The preliminary examination to which the transformed plants are subjected will be reinforced to the extent that the techniques of evaluation will develop (Chesson and Janes, 2000), in conformity with the principle of precaution (see box). Potential risks and benefits of using VRT plants must be evaluated in a broad context including other potential control methods as well as economical and agricultural consequences of unsatisfactory control of viral diseases (Tepfer, 2002).

However, it must be taken into account that these techniques are just beginning to be used and that future gene constructs will be significantly different (elimination of selection genes, use of specific promoters and multiple genes, minimal sequences required for the antiviral effect, etc.). New insights into the host-pathogen interactions will emerge from the study of "natural" transgenic plants, which have benefited from viral infection to keep partial viral sequences in their genome (see Chapter 13). The integrated sequences possibly provide a novel type of homology-dependent resistance.

In the longer term, homologous recombination will make it possible to modify specifically one sequence in the plant genome, and thus to modify a metabolic

The principle of precaution and genetically modified plants

The principle of precaution is a new concept related to the early accounting for risks. Contrary to the common understanding in the media, the principle does not demand abstention but attempts to define the rules of action in the presence of potential risks. It must lead to the elaboration of procedures allowing the formation of a clear judgement about the seriousness of potential damage, preventive measures, and economic impact (Godard, 2000). The procedures to frame thinking and action on this subject have been presented in the form of a coherent set of "commandments" about risks of any nature (Kourilsky and Viney, 2000):

- Any risk must be defined, evaluated, and graduated.
- Risk analysis must compare scenarios of action and inaction.
- Any risk analysis must include an economic analysis.
- The structures of risk evaluation must be independent.
- The decisions must be revisable, the solutions adopted must be reversible and proportionate.
- Research is essential to reduce uncertainties.
- The decision-making processes and the safety mechanisms must be effective, coherent, and reliable.
- The evaluations, decisions, and the way in which they are carried out must be transparent.
- The public must be informed.

The measures pertaining to GMOs constituted one of the first attempts to implement the principle of precaution. It has been possible to judge how difficult it is to take into account simultaneously *all* these aspects. The scope of research efforts already achieved must be underlined, as well as those still needed to clarify the different stakes linked to the modification of plant genes and its impact on the future of agriculture.

function without input of foreign DNA (Gura, 2000). These different approaches will result in genetically modified plants of the second generation, which could be part of a new context of evaluation in which possible sources of risk will be accurately identified.

Evolution and Classification of Viruses

13

CHAPTER

Evolution of Viruses

"The preservation of favourable variations and the rejection of injurious variations, I call Natural Selection."

C. Darwin, 1859

"Nothing makes sense in biology except in the light of evolution."

T. Dobzhansky, 1964

Viruses show two faces to the observer: they have numerous and important variations, and on the other hand the diseases they cause prove to be stable over long periods. The analysis of phenotypic viral variability has long been a major theme in the study of viruses. Sequencing of viral genomes has revealed the very wide diversity not only in the nature of the support of information, but also in the organization and modes of expression of the genes. How is this variation generated? How can viral diseases persist over time? These are the questions that are now raised to address the evolution of viruses, thus placing the problem in the vast and exhaustively documented context of the evolution of living entities.

First observation: Variation is produced in each generation, and the molecular mechanisms at work in viruses are not different from those that act in the evolution of organisms. They are essentially mutation and recombination, reassortment, duplication, and rearrangement of genes.

Second observation: Selection operates on this ever-changing genetic material and causes a relatively stable organism to emerge, particularly capable of reproducing in its immediate environment. These observations and their interpretation are presently placed in the conceptual perspectives of the synthetic theory of evolution (or neo-Darwinism), in which the contributions of Darwin and Mendel are combined, enriched with elements from the molecular study of genomes that have clarified the source of variation (Morse, 1994a). The generation time of viruses being much shorter than that of infected organisms, their evolution is a perceptible process in our own time scale.

Mutation

Mutation, a primary force in evolution

During the replication of viral genomes, polymerases make occasional errors: limited substitutions, insertions, or deletions. The *error rate* or *substitution frequency* quantifies biochemical events: it is the number of misincorporation events per nucleotide site and per replicative cycle. It is difficult to measure; for DNA polymerases, which possess a reading control function and a function for repairing synthesis errors, it can be estimated at around 10^{-9}, and for RNA polymerases, which do not possess such proof reading system, it can be estimated at 10^{-4} (Drake, 1993; Ward et al., 1988). Thus, each copy of an RNA genome of 10 kb will statistically include one error.

Another variable, the *mutation frequency* or *mutant frequency*, measures the proportion of mutants that can be detected in a population. This variable integrates all the aspects of the relationship between the virus and its host at a given time and in a given environment, and the mutation can confer a selective advantage, be unfavourable, or simply be neutral. The mutation frequency of an RNA virus has been evaluated in some examples as 10^5 to 10^6 times that of the eukaryote cell that harbours it (Domingo and Holland, 1994).

The emergence of variants: hazard and/or necessity

It is now clear that RNA genomes have a high error rate, much higher than the mutation frequency. The divergence of sequences cannot increase indefinitely. What are the processes that will cause certain variations to emerge while eliminating others? Here, opinions differ. Those who believe the variations are not equivalent attribute the principal role to selection. Others observe that many substitutions of nucleotides in coding regions are silent (they do not lead to modification of the amino acid sequence) and believe that this fact confirms the neutralist theory of molecular evolution formulated by Kimura (1989). These authors postulate that the variations occur at random, without inherent advantage or disadvantage; evolution will then effect the elimination of unfavourable mutations and the more or less random survival of neutral variations or rare favourable mutations. This neutralist-selectionist debate is attenuated by comparative analysis of sequence polymorphism showing that selectively important sites are clearly influenced by natural selection, and less important sites follow a more neutral pattern of evolution (Ohta, 1996).

In viruses, it is difficult to ensure that a variation is actually neutral in all circumstances. A mutation that does not change the sequence of amino acids but only the nucleotide sequence may modify the secondary and tertiary structures of the nucleic acid, which are the interface between the genome and cellular factors. There will be elimination of mutations (silent or amino acid changing) which prove unfavourable in a given environment at a given time for replication, movement, and transmission. Two principal factors combine to shape viral populations: the selection of the best-adapted variants and the chance that favours multiplication and dispersal of some of these variants (Domingo and Holland, 1994).

Mutation and selection: a constant adjustment of the host-virus relationship

Eventhough the error rate is quite constant and high for all RNA viruses, it is observed that the speed of evolution may differ according to the virus, the viral strain, and the gene considered because of the different selection pressures that are exerted. TMV (*Tobacco mosaic virus, Tobamovirus*) has an error rate of around 10^{-3} to 10^{-5}, like most RNA viruses. However, the genus *Tobamovirus* contains several viruses that are highly stable on their entire genome; each virus is adapted to a host or group of hosts in which it varies little. For example, the adaptation of TMGMV (*Tobacco mild green mosaic virus, Tobamovirus*) with its natural host *Nicotiana glauca* has probably played an important role in selecting variants of very similar sequence even though they are distant in terms of time and space (Fraile et al., 1995). In the genus *Potyvirus*, it is observed that certain regions of the genome are particularly variable (non-coding 5′ region, P1), while the polymerase gene is only slightly variable (Marie-Jeanne et al., 1995).

By means of the PCR technique, it is easy to establish sequences of genomic segments to make comparisons: between species or within a species, between strains or within a strain, between isolates as a function of biological, geographical, and other characters. The possibilities of investigation are wide-ranging and a better knowledge of molecular variation may considerably clarify questions in the field of epidemiology.

Only some variations in the sequences will result in variations easy to observe in the phenotype: e.g., host range, symptoms, transmissibility, serological properties. The determinant of a phenotypic variation may turn out to be a single point mutation (Weiland and Edwards, 1996; Granier et al., 1993).

In order to survive in animals, viruses must escape specific defences put in place by the immune system since the appearance of a new antigen. Very often, the sequence of antigenic motifs of surface proteins is particularly variable. Animal viruses, which have a high mutation frequency of their surface antigens (for example, the human immunodeficiency virus or HIV and the influenza virus) pose major problems of public health. Vaccines are hardly effective against viruses that vary constantly in an evasion strategy. In contrast, they effectively protect against viruses in which the antigenic determinants vary little, such as measles. Mass vaccination led to eradication of smallpox and poliomyelitis in the Northern hemisphere.

Recombination

Recombination is observed in DNA and RNA viruses

Recombination is the joining of genetic material from two different DNA or RNA chains, producing a new biologically active structure. The importance of recombination phenomena in viruses (and especially in RNA viruses) emerged late. However, genome sequencing revealed numerous examples of recombination and now this phenomenon is seen to play a major role in the evolution of all viruses (Worobey and Holmes, 1999).

The first example of recombination in RNA viruses was described in poliovirus, in the family *Picornaviridae*. When cells are infected with two strains carrying a phenotypic marker that differentiates them (resistance to guanidine, sensitivity to temperature), recombinants carrying the two markers can be isolated in the progeny at a frequency (a few per hundred) higher than that of spontaneous mutations. Recombination is encountered commonly in the natural evolution of poliovirus strains and numerous wild-type and vaccine recombinants are identified (Dahourou et al., 2002). This phenomenon is called *homologous recombination*, in which changes in a chain are produced at homologous sites between nearly identical chains. Most frequently, the recombination is exact to the nearest nucleotide. In some cases it may be less precise and generate duplications, rearrangements, or deletions (Lai, 1992). Recombination may also occur between RNAs that are unrelated, for example, distant viral RNA or even a viral RNA and a cellular RNA. This is called *non-homologous recombination*.

Recombination is linked to replication

In studying recombination in poliovirus, it is seen that recombinants appear during the synthesis of minus-sense chains (Kirkegaard and Baltimore, 1986). Some experimental results suggest that replicase shifts from the template chain, the process involving several structures: the replicase itself, the first template chain (donor strand), and its nascent strand, the second template chain (acceptor strand). The replicase will be brought to cause a pause favouring the switch from the donor strand towards the acceptor strand, either in a misincorporation site that liberates the enzyme or in fixed sites corresponding to secondary structures (Nagy and Simon, 1997). However, this model cannot explain other results showing that RNA fragments not linked to replicase can be incorporated by recombination in new chains, in such a condition that they hybridize the negative-sense chain and that they will have a 3'-OH end: there is thus extension of a fragment serving as primer (Pierangeli et al., 1999).

Other mechanisms can be invoked to explain the recombinations: premature termination, breakage and ligation; this last possibility will serve to incorporate non-viral sequences at the 5' end of subgenomic RNA of TSWV (*Tomato spotted wilt virus*, *Tospovirus*).

An experimental recombination system: BMV

BMV (*Brome mosaic virus*, *Bromovirus*) has a genome divided into three segments that, at their 3′ ends, have identical sequences of around 200 nucleotides. Bujarski and Kaesberg introduced a deletion of 20 nucleotides in this region of RNA 3, then infected plants with normal RNA 1 and 2 and deleted RNA 3. In the progeny, it was observed that deletion was repaired by homologous and precise recombination between the 3′ segments of RNA 1, 2, and 3, and that the recombination points were not distributed at random. In donor RNA, a sequence rich in A and U precedes a sequence rich in G and C (Nagy and Bujarski, 1997). Regions in which recombination events are frequently observed are called hot-spots. The subgenomic promoter for RNA 4, situated internally on the minus-strand of BMV RNA 3, is a recombination hot-spot: the polymerase complex pauses and reinitiates on another RNA 3 minus-strand (Wierzchoslawski et al., 2003). In other experiments, recombination is produced at sites non-homologous but presenting a possibility of hybridization between the two strands, which could lead to significant rearrangements in the sequences, and notably duplications. A short sequence capable of forming a heteroduplex (imperfect double-stranded structure) of 15 to 60 nucleotides facilitates the recombination by creating an obstacle for replicase (Nagy and Bujarski, 1995). The template-switching of the replicase seems to be the mechanism at work in these examples of recombination: mutations in the polymerase or in the helicase can affect the frequency and nature of recombinants (Inoue-Nagata et al., 1997; Dzianott et al., 2001; Kim and Kao, 2001).

The homology of sequence between segments of parental chains can be insufficient to ensure the recombination and other factors may favour it. In TCV (*Turnip crinkle virus*, *Carmovirus*), the presence of a stem-loop structure is the major determinant of recombination events (Simon and Nagy, 1996; Carpenter et al., 1995).

These examples show that recombination may facilitate restoration of defective sequences and introduce rearrangements in sequences, notably duplications, a source of new genes.

Recombination may be observed in nature

Recombination occurs within species as well as between species, genera, and families. The study of sequences of natural isolates frequently reveals or suggests examples of recombination. It is possible that plants infected simultaneously by several viruses have been the site of these exchanges.

Within species

Numerous examples have been described in species belonging to genera *Potyvirus, Nepovirus, Alfamovirus, Tombusvirus, Tobravirus, Hordeivirus*, and *Cucumovirus* (MacFarlane, 1997). The frequency of recombination events in natural conditions within a species can be evaluated by examining a large number of sequences with the help of computer programs (Candresse et al., 1997). Various situations are observed: PVY (*Potato virus Y, Potyvirus*) presents in the 5' part numerous recombinations between strains in nearly a third of isolates studied. On the other hand, the potyviruses PPV (*Plum pox virus*) and PRSV (*Papaya ring spot virus*) show no example of recombination in this region. In yam, multiple recombinations have been observed in natural populations of YMV (*Yam mosaic virus*), resulting from exchanges between variants co-infecting the same host (Bousalem et al., 2000).

Between species and genera

Examples of recombination are numerous. In the family *Luteoviridae*, genomes of *Luteovirus* and *Polerovirus* resemble each other in the 3' part and are very different in the 5' part where the polymerase gene is located. A phenomenon of recombination introduced a polymerase of the carmovirus type into an ancestor of poleroviruses (Gibbs and Cooper, 1995). The recombination site is in the intergenic region, where the promoter (recognized by the replicase on the minus strand) of the subgenomic RNA synthesis is found. SbDV (*Soybean dwarf virus, Luteoviridae*) has been considered a recombinant between a luteovirus and a polerovirus (Rathjen et al., 1994). Frequent recombination events within the subgenomic promoter led to rearrangements and modular exchanges contributing to diversification in the family *Closteroviridae* (Bar-Joseph et al., 1997).

In the family *Geminiviridae*, viruses with circular single-stranded DNA, transmission by extremely polyphagous vectors results in numerous mixed infections. This situation is highly favourable to recombination events that have been observed in nature with a high frequency within species as well as between species (Padidam et al., 1999). The origin of a viral epidemic on cotton in Pakistan was attributed to recombination between endemic begomoviruses infecting okra, giving rise to CLCuV (*Cotton leaf curl virus, Begomovirus*) capable of infecting cotton (Zhou et al., 1998). Interspecific recombination could contribute significantly to the emergence of new species producing new diseases (Zhou et al, 1997).

Between families

Known recombination events between viruses belonging to different families remain rare. In association with begomovirus AYVV (*Ageratum yellow vein virus*), a chimeric circular DNA was found, originating from recombination between this begomovirus and sequences borrowed from a nanovirus. This component could contribute to the vein-clearing phenotype observed in infected plants (Saunders and Stanley, 1999).

More surprising, sequences very similar to a gene of calicivirus (animal viruses with RNA genome) are part of the circular DNA genome of circovirus. This acquisition is perhaps linked to the emergence of circovirus, which infects vertebrates, from nanovirus, which infects plants. Indeed, these two groups have highly significant sequence homologies involving the protein Rep associated with replication and the origin of DNA replication, which suggests the hypothesis of a common origin followed by a change in host (Gibbs and Weiller, 1999).

■ Recombination between viral and cellular nucleic acids

In viral genomes there are sequences very similar to non-viral sequences; they were probably acquired a short time ago by the virus in the course of its evolution. For example, an isolate of the PLRV (*Potato leafroll virus, Polerovirus*) is differentiated from the wild type by the addition of 119 nucleotides homologous to a chloroplast DNA segment (Mayo and Jolly, 1991); short complementary sequences that are present in the wild type and chloroplast DNA made recombination possible. A Canadian isolate of TuMV (*Turnip mosaic virus, Potyvirus*) contains 467 nucleotides strongly homologous to the sequence coding for a chloroplast ribosomal protein (Sano et al., 1992).

However, the sequence similarity is often lower and limited to functional domains. The most probable hypothesis here is that the similarity is reduced as much as the recombination event is old in the history of the viruses concerned, the sequences having then undergone more or less significant adaptations. These observations concern notably helicases, proteases, polymerases, heat shock proteins HSP 70 and 90, and the tRNA-like 3' ends.

■ Recombination between viral genomes and transgenes

In CaMV (*Cauliflower mosaic virus, Caulimovirus*), recombination between the viral DNA and a transgene expressed by the host can take place during reverse transcription (Wintermantel and Schoelz, 1996). Defective viruses can be repaired by functional sequences from a transgene (Greene and Allison, 1994; Rubio et al., 1999). This possibility is widely exploited to study the functional characteristics of a gene by operating mutations on the viral genome and inoculating them into transgenic plants expressing the non-mutant gene. Recombination poses problem carefully studied in the case of transgenes conferring resistance to viruses (Chapter 12).

■ Recombination, homologous or non-homologous, is a major agent in the evolution of viruses

In conclusion, it is presently observed that recombination of viral chains, in its different modalities, is a continuous source of new variants and plays a major role in the evolution of viruses:

— It makes it possible to create new molecules from existing fragments and especially motifs that are the sites of recognition between nucleic acids and proteins. It could be the source of the construction of viral nucleic acids.
— Between distant genomes it makes possible the transfer of modules, which are sets of genes involved in a function: replication, movement, transmission by insects. This is called modular evolution, more rapid than gradual evolution ensured by mutations.
— It allows the replacement of defective segments, restoration of the overall fitness of the virus, duplication of sequences and genes, and thus creation of new genes. It can modify virulence.
— Sometimes it renders difficult or impossible a classification that reflects the phylogeny of all the genes of a virus; an innovative approach to this problem is required.
— It has implications in the consequences of the expression of transgenes.

"Surpluses" of recombination: defective interfering RNA

The mechanisms that operate in recombination are similar to those that produce defective interfering (or DI) RNA, generated by deletions and rearrangements of the genomic viral RNA. These are truncated forms that have lost genes essential for replication, movement, and encapsidation. They can therefore replicate only with the help of an assistant virus that provides the necessary elements in *trans*. DIs have a replicative advantage that allows them to develop at the expense of the assistant virus, which produces generally an attenuation of symptoms. Although they are quite rare in natural conditions, they become frequent by repeated passages of abundant inoculum in experimental conditions. It has been observed that slight modifications in the polymerase or helicase can modify the amplitude of this phenomenon (Inoue-Nagata et al., 1997).

Reassortment, an additional possibility of exchanges for divided genomes

In plant viruses with a divided genome, some components may sometimes be autonomous enough to be functional in the presence of components of related viruses. This is called reassortment or pseudo-recombination. For example, the RNA 3 of CMV (*Cucumber mosaic virus, Cucumovirus*) and RNA 1 and 2 of PSV (*Peanut stunt virus, Cucumovirus*) can give in natural conditions a functional pseudo-recombinant (White et al., 1995). In this example, the RNA 3 has acquired by recombination the 3' and 5' ends of PSV. The amplitude of the phenomenon within the species is not known, except in a few well-studied cases (Fraile et al., 1997), which indicate a counter-selection of pseudo-recombinants.

Viral sequences are unfrequently integrated in plant DNA

Integrated viral sequences

Until recently, integration of plant viral sequences into host genome was considered a very rare event. Several examples of such integration events are now described in families *Geminiviridae* and *Caulimoviridae*:

Begomovirus-related DNA sequences

Begomovirus-related sequences are integrated in a unique locus of *Nicotiana tabacum* genome, in the form of multiple direct repeats. Multiple copies of these elements are present in three *Nicotiana* species (section *Tomentosae*) but not in other *Nicotiana* species or other Solanaceae. DNA sequence analysis shows that all elements contain sequences similar to the rolling-circle replication origin and the adjacent Rep gene encoding the replication protein of bipartite begomovirus. Results suggest that a unique integration event by illegitimate recombination followed by rearrangements has occurred in meristematic tissues and in the germline during *Nicotiana* evolution (Bejerano et al., 1996; Ashby et al., 1997).

Caulimoviridae sequences

PVCV (*Petunia vein clearing virus, Petuvirus*) is transmitted by seed and grafting, and not by mechanical inoculation or by insect vectors; vein clearing and stunting symptoms are particularly visible under stress conditions. Viral DNA hybridize with petunia genomes; the entire virus genome is present in a cultivar of *Petunia hybrida* and most likely is the source of episomal infections; PVCV sequences are detected in many petunia species (Harper et al., 2002). PVCV polyprotein contains two short sequences similar to integrase motifs of retroviruses and retrotransposons (Richert-Poggeler and Shepherd, 1997).

Another member of the family *Caulimoviridae*, TVCV (*Tobacco vein clearing virus*), is a previously undescribed virus. Viral particles are only found, under certain conditions, in *Nicotiana edwardsonii* (*N. glutinosa* × *N. clevelandii*) where the virus is seed-transmitted. Genomic DNA of TVCV hybridize with genomic DNA of *N. edwardsonii*, but no entire TVCV genome has been identified. This suggests recombination events to construct putative full-length infectious genomes (Lockhart et al., 2000). TVCV genome has 78% identity with dispersed repetitive sequences present in very high copy number in *N. tabacum, N. glutinosa*, datura, and tomato genomes. These sequences contain defective ORFs and are nearly not transcribed; no episomal virus has yet been found (Jakowitsch et al., 1999).

BSV (*Banana streak virus, Badnavirus, Caulimoviridae*) naturally infects banana and plantain. *Musa balbisiana* genome contains integrated BSV sequences that can give rise to episomal infectious genomes via homologous recombination (Ndowora et al.,

1999). The proliferation stage of the micropropagation procedure of virus-free hybrids (*Musa balbisiana* × *Musa acuminata*) is determinant for episomal infection. BSV is widely distributed in *Musa* hybrids; control of disease requires to study the genetic composition of *Musa* genomes concerning the presence and distribution of viral integrants (Geering et al., 2001).

These three examples of integrated viral sequences have several features in common: the viruses can give rise to episomal infections in certain hybrid hosts and under stress conditions, but more frequently they are present in the genome as "inactive" sequences, without virus infection; the viruses are members of the family *Caulimoviridae* (i.e., they replicate via a circular intermediate in the nucleus and need the action of a reverse transcriptase). The family *Caulimoviridae* belongs to the "pararetrovirus" cluster in which genomes code for a reverse transcriptase, but not for an integrase gene; they multiply without integration in the host genome, in contrast to *Retroviridae*.

The demonstration that pararetroviral DNA sequences might be inserted into the host genome by illegitimate recombination or other integrative process gives new insight into the natural interaction between viruses and their hosts. The sequence conservation and maintenance over time of non-pathogenic elements in plant genomes suggest a beneficial function; integration sequences possibly induce a host-pathogen equilibrium, protecting the host against infection by a homologous virus (Jakowitsch et al., 1999; Mette et al., 2002). In this context, post-transcriptional gene silencing (Chapter 5) could explain some aspects of the behaviour and evolution of integrated viral sequences in these natural transgenic plants; advances in genome sequencing reveal that pararetroviral sequences are present in rice, tomato, and other plants (review in Harper et al., 2002).

Retrotransposons

Retrotransposons are mobile genetic elements replicating through an RNA intermediate and reverse transcription. Some of them form virus-like particles and are ubiquitous in plant genomes, frequently in high copy number (Voytas et al., 1992). They are classified into two families: the family *Metaviridae* contains the SIRE-1 element present in soybean; the family *Pseudoviridae* contains the Athila element present in *Arabidopsis*. SIRE-1 and Athila encode sequences similar to reverse transcriptase and integrase motifs, but also an "env" sequence reminiscent of the "env" gene of *Retroviridae*, involved in fusion with the cell membrane and budding allowing retroviruses to be infectious; the significance of this gene in plant retrotransposons is not understood (Harper et al., 2002). The boundaries between pararetroviruses, retrotransposons, and retroviruses are increasingly blurred (Kumar, 1998; Peterson-Burch et al., 2000). Integration of viral sequences in host DNA occurs frequently in animals; until recently, only very few integration events were reported in plants. However, an increasing number of observations now support the possibility of more common events.

Viral quasi-species

A probabilist and evolutionist concept

A virus usually presents good stability and identity in time and space, even if it evolves and proves capable of new properties. These are the two seemingly contradictory aspects of the "dynamic identity" of a viral population that are included in the term "quasi-species". Before being a highly useful biological concept in understanding the evolution of viruses, the quasi-species had a mathematical and physico-chemical definition.

To give a mathematical and physical representation of quasi-species, Eigen (1993, 1996) proposed that we imagine a space in which all possible sequences would be represented. Theoretically, a given sequence is associated with one point in this space. Another sequence is placed at one unit of distance if it differs by a single nucleotide, at two units for two nucleotides, etc. In this space, the quasi-species forms a point cloud centred on the sequence of origin. The cloud does not have a regular form, and points appear and disappear constantly.

If the frequency of mutation per nucleotide is around 10^{-4}, it can be estimated that an RNA genome of 10 kb undergoes on average one error per copy. The progeny of a single genome is thus made up of a set of closely related but not identical sequences, differing by one or several mutations. Here the notion of quasi-species associates the idea of a very close genetic proximity resulting from a common ancestor and a micro-heterogeneity of sequences. There is a pool of variants present in a cell or an organism, even in the absence of selection pressure. This collection of sequences behaves "as if" it were homogeneous, reproducing with a reliability sufficient to conserve its collective identity. The virus thus cannot be defined by a single genomic sequence but rather as a population of genomes centred on an average (Domingo et al., 1995). Experimentally, sequencing after cloning gives a result similar to the sequence most frequently represented in the population (master sequence), with loss of viral heterogeneity during the *in vitro* process. Direct sequencing after PCR approaches the consensus sequence, an average of all the sequences of the sample at each nucleotide site. Hybridization of PCR-amplified cDNA with microchips containing a set of oligonucleotides covering the sequence under study is a rapid and efficient method able to distinguish single nucleotide polymorphisms in a heterogeneous sample representative of a viral population.

The concept of quasi-species was elaborated to take into account the evolution of RNA viruses in which the polymerases are subjected to a high error rate. It appears that this notion can be extended to *Geminiviridae*, whose single-stranded DNA genome shows a wide variability that could be described in terms of quasi-species; these viruses, which use DNA polymerase of the host, seem to have a high potentiality of variation because they escape the repair system proper to the double-stranded DNA. With regard to plant viruses with a double-stranded DNA genome

(*Caulimoviridae*), their replication has a reverse transcription stage with a high error rate.

■ The quasi-species is stabilized by selection

The biological characteristics of a virus and its potentialities of evolution are those of a swarm of mutants that constitute the quasi-species. Since the best-adapted mutants are the most frequent, they will generate a numerous progeny; there is thus a bias that accelerates the emergence of mutants that are increasingly better adapted. The adaptation of the set of individuals determines the adaptation of the species, which is thus found to be the framework of natural selection. While being adapted to its ecological niche, the species can evolve very quickly, thanks to its permanent pool of variants, its dynamic equilibrium if the environment changes. For example, in plants, the environment changes during transmission to another part of the host-plant (Sacristan et al., 2003) or to another host plant of the same species or to a different species; this emphasizes the importance of the sampling when isolates are taken to be studied. In animals, it may be for example the immune response that modifies the equilibrium. The evolution of the quasi-species depends greatly on its pool of variability, and thus on the size of the population; repeated passages of inoculum containing little virus (bottlenecks) may lead to a loss of efficiency by elimination of high-performing individuals.

The genomic variations that are constantly produced in the quasi-species ought to lead to a chaotic divergence of sequences. However, a certain stability of species is observed, which is explained by constraints imposed by the functionality of proteins and viral nucleic acids; the mutants deficient for essential functions are eliminated from the pool. Variants continuously emerge and disappear. These phenomena are also present in cellular organisms; but a virus, characterized by enormous population size and short generation time, may produce a very high number of mutants over a single generation of the cellular organism that harbours it. This difference validates the quasi-species concept (Domingo and Holland, 1997). In plants, these processes are inscribed in a particular context because of the "fluidity" of the genome, which results notably from the activity of meristematic zones (Graham et al., 2000). This continuous post-embryonic shaping of the plant is manifested in a great plasticity and intra-plant heterogeneity (Hallé, 1999), which favours the emergence of viral variants.

Vectors, a field that is constantly explored by viruses

To survive, viruses must be transmitted from one individual plant to another, sometimes from one species to another (Chapter 8). Most often they use a vector—insect, mite, fungus, or nematode—and the virus thus evolves within the host-virus-vector triangle.

Each acquisition by the vector in the viral population represents a highly restricted sampling of genomes present in the quasi-species, which could lead to an irreversible loss of high-performing variants. In viruses that have a non-circulating transmission, the virus adapts to its vector through the intermediary of the capsid protein and, depending on the case, of the helper component protein. The helper component, which can allow the transmission of several similar viruses by a single vector, could diminish the bottleneck effect that accompanies each transmission event.

Vectors have played a very important role in the evolution of viruses. Polyphagous vectors could cause the encounter of different viral genomes, similar or very distant, and allow a new round of diversification. Viruses multiplying in their insect vector (propagative mode) can be considered phytophagous insect viruses having acquired secondarily the possibility of multiplying in the plant; the "intermediary" state in which the virus multiplies in the insect and is transmitted horizontally between insects that feed on the same plant without multiplying in the plant has been found in rice (Nakashima and Noda, 1995). Most of the members of the family *Bunyaviridae* are transmitted by insects; this family comprises several genera of animal viruses and two genera infecting plants (*Tospovirus* infecting Dicotyledons and transmitted by thrips, *Tenuivirus* infecting Monocotyledons and transmitted by planthoppers); it is probable that these viruses have a common ancestor infecting insects and animals.

New viral diseases and emerging viruses

In the field of animal viruses, researchers are often confronted with viral diseases the agent of which is not yet described. It may be new (a new genome that has evolved from viruses already known) or emerging (a genome already known but causing a new disease following a change in host, for example). History relates the emergence of new epidemic diseases; but it seems that the frequency of emergences has increased recently. In the last 15 years, more than 40 viruses affecting humans have been described for the first time. These are generally RNA viruses. Present for a long time in limited ecological niches (humans, domestic animals, wild animals), they have seized the opportunity to extend their ecological niche and have become capable of causing an epidemic from which they are identified (e.g., *Arenavirus, Bunyavirus, Hantavirus, Lyssavirus*, hepatitis, *Herpesvirus*) (Morse, 1994b). Very recently, SARSCoV (*Severe acute respiratory syndrome coronavirus*) was identified during an outbreak in Southern China, spreading through Hong Kong, Vietnam, Canada, and Singapore (Ng et al., 2003). It can be supposed that new epidemics resulting from viruses exploring humans as potential hosts will be produced with an increasing frequency, in parallel with the growth of the human population and changes in living conditions that bring virus reservoirs close to potential targets (Griscelli, 1997; Holland and Domingo, 1998).

In plants, there are numerous examples of viruses that emerge following favourable circumstances. The increasing incidence throughout the world of diseases caused by members of genera *Begomovirus* (*Geminiviridae*) and *Crinivirus* (*Closteroviridae*) is directly linked to the recent dissemination of a particularly polyphagous biotype of the whitefly vector *Bemisia tabaci*. The tospoviruses, confined to limited geographic zones for about 50 years, extend their attack to a wide range of crops (ornamentals, vegetables, field crops) in relation with the spread of their particularly effective thrips vector *Frankliniella occidentalis* (in which the virus multiplies) (Moury et al., 1998). The interspecific recombination between the DNA of the two cassava viruses ACMV (*African cassava mosaic virus, Begomovirus*), and EACMV (*East African cassava mosaic virus, Begomovirus*) led to the emergence of a recombinant causing a very serious epidemic in Uganda; 12 cases of interspecies recombination were detected in cassava, potential sources of epidemics (Zhou et al., 1997). BaYMV (*Barley yellow mosaic virus, Bymovirus*) became an agronomic problem since the autumn varieties were preferred, which allowed the virus to develop during the winter. PPV (*Plum pox virus, Potyvirus*), which was first reported in Macedonia, extended within a few decades throughout central and southern Europe and led to measures of eradication.

The evolution of a virus already present in a population can also cause epidemics. Influenza A virus, which is observed each winter, evolves by occasional mutations in the genes of glycoproteins of the envelope (neuraminidase and haemaglutinin) against which the hosts produce protective antibodies. To this slow evolution (antigenic drift), which allows an equilibrium between the virus and partly immunized populations, are opposed the sudden evolutions (antigenic shift) that follow the passage from a bird host to a mammal host. Indeed, the history of the influenza virus shows that over about a century an avian strain became capable of infecting pig. The pig became a sort of "mixer" of human and avian strains (Webster et al., 1992). New strains appeared by reassortment among the eight segments of the divided genome derived from strains coming from different hosts. These mixed infections caused variants to emerge over time carrying new antigenic motifs of avian origin, which found non-immunized populations and caused worldwide pandemics as in 1957 and 1968. In 1997, the H5N1 strain in chicken, responsible for an epidemic in Hong Kong, proved directly transmissible to humans, which led to systematic destruction of poultry in producing regions. New outbreaks were recorded in 2004 and in subsequent years.

Although diseases caused by viruses are most often the cause of their discovery, we must not therefore conclude that all viruses induce a pathological state in their hosts. It seems, on the contrary, that the virus, in order to multiply and be transmitted to other hosts, "has a benefit", in this sense, in maintaining the health of its host, because the survival of the virus in nature requires interactions with living hosts. Aggressiveness is not a character subject to selection, except when it is correlated with higher reproductive performances. Selection favours maximum multiplication and transmission, rather than the diseased state and death of the host. Antiviral control must try to displace the equilibrium in favour of the host by exerting selection pressures that reduce the impact of the disease without necessarily preventing the virus from multiplying (Chapters 10, 11, and 12).

Molecular phylogenies

The evolution of viruses has resulted in the nearly 2000 viral species described to date, plus all those to be discovered. Most are classified into genera and families. Species include numerous subspecies, strains, and variants. How are these taxa related? This is the subject of phylogeny. To establish phylogenies, there must be *comparable characters*, then a diagram must be proposed that best translates the relationships of proximity established for the selected characters. This diagram is the phylogenetic tree, which illustrates hereditary relationships. In principle, such a tree comprises the ancestor (the root) and its descendants. But in the absence of knowledge of the direction of an evolutionary process by which the anteriority of sequences with respect to one another can be established, trees without root are often traced for viruses in which the outermost element of the group holds that place.

The *comparable characters* on which molecular phylogenies are based are differences in the nucleotide or protein sequence, established most often by sequencing or restriction profile. The proximity can be measured by the comparison of sequences. They are considered to have accumulated as much differences as they diverged long ago. An entire genome may be compared, but most often the comparison covers a region, a gene, a segment, or a functional motif. The proximity is expressed as a percentage of identity between nucleotide sequences. If the protein sequences are compared in which the amino acids are equivalent in their biochemical group, it is called percentage of similarity. The comparison is based on alignment of homologous sequences, taking into account deletions and insertions. Each alignment position, nucleotide or amino acid, is one site.

The differences indicated by the alignment of sequences are quantified by two types of approach, each of which implies a certain number of hypotheses. *Phenetic* methods establish *overall similarity* by exploiting a distance matrix by two-by-two comparison of all the sites. *Cladistic* methods (from the Greek *klados* for branch; the clade is the group formed by an ancestor and all its descendents) retain only the *informative sites*, i.e., those that present differences, and interpret these differences according to the principle of maximum parsimony. If it is considered that evolution takes the shortest route, the best tree is that which minimizes the number of substitutions at informative sites. Cladistic methods display monophyletic groups of phylogenetic relationship (the clades). The members of a clade are identified by characters that unite them and those that distinguish them from others.

It must be emphasized that the sequence alignments must be examined for possible recombinations before the phylogenetic trees are constructed (Worobey and Holmes, 1999). The most robust trees often involve a segment or a gene. It is always risky to extend them to entire genomes that could have undergone recombination and consist of segments having different phylogenies. This explains the difficulty, or impossibility, of describing the evolution of viruses in the form of a phylogenetic tree (or a small number of trees). In contrast, the estimation of genetic variation in a viral population, established for a gene or for a genome segment, is very useful for characterizing it and comparing it with other populations differing in host, location, biological properties, or other characteristics. The phylogenetic trees selected as the most probable by statistical methods (for example, bootstrapping) remain theoretical constructions, extremely useful as study tools, and elements of conjecture that can constantly be improved (Leigh Brown, 1994).

Origin of viruses and viral genes: modular evolution

Despite the advances made in comprehension of mechanisms that operate in the evolution of viruses, their origin remains a subject of speculation. Three principal hypotheses have been put forward:

- Viruses are derived from more complex micro-organisms by a regressive evolution.
- Viruses are vestiges of a pre-cellular world.
- Viruses are genetic elements derived from cellular material, and their origin is endogenous.

The first hypothesis supposes a continuity between bacteria and viruses; this continuity was invalidated by the molecular characterization of viruses and the hypothesis is presently discarded. The two other possibilities remain, which are not mutually exclusive, for which arguments are advanced.

Research on the first forms of life, before the appearance of prokaryotic and eukaryotic cells, led to the hypothesis of an "RNA world" that preceded the "DNA world" that we know today. These first molecules of RNA would have played a double role, as genetic material and as enzyme. Such molecules still exist, the ribozymes, discovered by Cech and Altman in 1980. These could be the ancestors of present RNAs that are capable of replication: RNA viruses, viroids.

The theory of endogenous origin was advanced following studies by Temin (1980), which showed that retroviruses could originate in cellular elements. Retrotransposons are mobile genetic elements capable of integrating themselves in the genome and are structurally very close to retroviruses, which have acquired the additional potential to multiply autonomously and infect other cells by the acquisition of a capsid (Becker, 1996). Numerous sequence data have confirmed the idea that DNA viruses and retroviruses originate from cellular DNA (Xiong and

Eickbush, 1990). For RNA viruses, which replicate by the intermediary of viral polymerases, sequence data have made it possible to demonstrate similarities between the functional motifs of viral and cellular polymerases, whether they are enzymes copying DNA or RNA (Argos, 1988). Cryptic viruses are another type of potential precursor. These small double-stranded RNA lack movement from cell to cell. They are present in pollen, ovule, and seed and are diffused in the plant by cell division (Roossink, 1997). The movement protein may also have an ancestor in cellular proteins, such as *Cucurbita* protein (CmPP16), which is presently phylogenetically classified in the 30 K superfamily of viral movement proteins (Xonocostle-Cazares et al., 1999) (Chapter 4). We have seen above that the acquisition and horizontal diffusion of cellular genes could operate by recombination between DNA or between RNA, which gives viruses constant access to a vast pool of functional modules, genes and gene fragments, which then undergo adaptations.

These adaptations of genes for new functions lead sometimes to a new creation. For example, the tymoviruses have at the 5' end an ORF that has methyl-transferase and polymerase motifs homologous throughout the supergroup of alpha-like viruses (Chapter 3). However, only the tymoviruses have an overprinted 5' ORF, read in a different phase. An analogous observation pertains to the capsid protein of PLRV (Gibbs and Keese, 1995). The viruses offer numerous examples of production of several proteins using a polynucleotide segment in various ways.

A phylum: positive-sense RNA viruses

Positive-sense RNA viruses constitute the largest and most diversified group. It comprises various families of viruses that infect bacteria, plants, and animals and present a wide variety in genome structure and replication strategies. The unifying element is the RNA-dependent RNA polymerase, which realizes the elongation of RNA chains. A monophyletic origin has been proposed for this gene for all positive-sense RNA viruses (Gorbalenya, 1995). An analysis of motifs conserved shows three major groups of families and genera (Chapter 3) (Koonin, 1991a; Candresse et al., 1990). Each branch of the phylogenetic tree contains viruses infecting bacterial, plant, and animal hosts. This raises questions about the acquisition of this gene, which could have been anterior to the division of kingdoms or more probably have been realized by horizontal transfer, especially through insects.

Some authors have extended the comparisons to all RNA viruses. But the methods and interpretation of results of sequence comparisons over vast groups remain difficult, and reservations have been expressed about the solidity of certain similarities pertaining to the polymerase genes of all RNA viruses, for example between RNA polymerases and reverse transcriptases (Zanotto et al., 1996). Positive-sense RNA, negative-sense RNA, and double-stranded RNA are probably separate groups whose phylogenic links are no longer clear.

A provisional conclusion: How is a virus produced?

Cells have existed simultaneously with or before viruses, which are by definition intracellular parasites. There is therefore little doubt (although that remains difficult to prove) that the genetic material of viruses derives from cellular elements or common elements of a precellular world. F. Jacob (1981) put forth the idea of tinkering ("do-it-yourself or bricolage") of evolution working as an engineer with the available materials: genes, gene fragments, motifs that fit together to form a regulated whole. Disparate elements are combined and sometimes evolve to create a new molecule with specific properties. The following sequence of events can thus be imagined: A DNA or RNA genetic element becomes capable of replication in a host cell. It thus resembles a plasmid and can acquire, because of its reduced size, a protective capsid that allows it to remain in an inactive but protected virion state. It can then face the external environment and become transmissible. This extremely rare series of events could have occurred on different independent occasions, which explains why viruses do not have a monophyletic origin. The mechanisms of diversification, mutation, exchange of modules or fragments, acquisitions, duplications, and rearrangements occur simultaneously. These molecular changes are, however, limited by several factors. The sometimes extreme compactness of viral genomes imposes serious constraints, particularly evident for overprinted genes, and induces strong selective pressure on genes to become multifunctional. The relationships with cellular elements and regulations between viral proteins must continuously ensure a functional cohesion of the genome (Mayr, 1994); in fact, there is co-evolution of the host and the viral genetic message.

The result of this recycling can be observed, but it is very difficult or impossible to describe the stages of it with certainty and to associate them with a time scale. Viruses do not constitute a group that shares a common phylogeny. According to A. Gibbs, there are units of genetic information that have in common a lifestyle of parasite of cellular life. Viruses are probably as old as life itself, with a subtle balance between order, i.e., identical replication, and disorder, i.e., variation, which allows survival and creates diversity.

14

CHAPTER

Classification of Plant Viruses

Virology is a relatively recent science that is rapidly widening and deepening its scope of investigation. The classification of viruses has greatly evolved over the past twenty years to respond to the need to name viral entities in a pertinent and universal way, and to group them as rationally as possible.

Once the characters that differentiate viruses were known, these new entities had to be named. Viruses are inaccessible to direct observation and were described at first simply as agents of a disease. Their names refer in general to the principal symptoms they cause (e.g., mosaic, yellowing, leafroll, stunting) in the host on which they were first found, and sometimes to their geographic origin. It was soon observed, however, that different viruses may induce similar or identical symptoms on a single host, and that strains of a single virus may induce different symptoms. It was thus necessary to find other, more pertinent characters.

From 1930 onward, information on the structure and composition of viral particles as well as their antigenic properties accumulated and, around the year 1950, Bawden proposed that viruses be regrouped according to the serological properties of particles. In 1960, the development of the negative staining technique made it possible to visualize easily virus particles with an electron microscope. Numerous new viruses were discovered and described (form, size, symmetry, surface structure, possible presence of an envelope), and researchers became aware of the need to classify viruses already discovered and to include new viruses in the classification as they were discovered. The first classification system for all viruses was initiated in 1966. It was based on data about the characteristics of virions and the biological properties of viruses, appraised pragmatically by the virologists of the International Committee on Taxonomy of Viruses (ICTV). Plant viruses were placed in non-hierarchical "groups" and without regard to the notion of species.

More recently, with new information contributed by the molecular studies of viral genomes and the possibility of making comparisons on their molecular features (sequence, genome structure, replication strategies), it became evident that viral taxonomy would have to integrate an evolutionary (phylogenetic) dimension relating the members of a group and relating the groups to one another. The classification of plant viruses was thus revised in 1993. The old groups, generally validated by sequencing data, were considered families or genera depending on the case, which allowed a rational integration with classification categories already adopted for animal viruses. Four taxa are presently used to name and classify viruses: species, genus, family, and order.

Species

The first object of study is "the" virus, observed and described on a plant. For a long time, scientists hesitated to call it a species, because it was difficult to find a definition of species satisfying all virologists. The chief difficulty was an excessively narrow definition, adapted essentially to cellular organisms, while in the virus the concept of "biological tinkering" expressed by F. Jacob is strong. In 1993, the ICTV adopted the definition proposed by Van Regenmortel (1989): "A virus species is defined as a polythetic class of viruses that constitutes a replicating lineage and occupies a particular ecological niche."

This definition includes several complementary aspects:

- The term *polythetic* is particularly important here. It implies that membership of a species is not defined by a single character but by a set of characters none of which is necessary taken separately. The members of a species have numerous properties in common, but isolates or strains may differ by one or several properties. It is not necessary that all the members of a species possess any one property.
- This definition takes into account the fact that viruses possess genes, replicate, are subject to selection pressures in their ecological niche, and thus vary and evolve. It is this continuous possibility of variation during replication that leads to the emergence of new species (Chapter 13).
- The viral species is also defined by its ecological niche: relationships with the host, infected tissues, relationships with the vector, host range, and geographical location.

It must be noted that this taxonomic definition considers the viral species a biological and genomic entity. This point differentiates it from the concept of quasi-species (Chapter 13), which emphasizes the heterogeneity of sequences of a line in replication.

To mark the frontiers of each virus species, a set of characters is chosen from a range of properties: genomic, structural, and physicochemical properties of the virion, serological properties of the capsid, host range, characteristics of vection, and relationships with the host. The sequence of amino acids and the serological properties often have considerable significance within the species. Even if the species is polythetic by definition, it is observed that the principal characters (for example, the modalities of vection) are constant or modified following mutations or loss of part of the genome. The classification must make it possible to identify and rapidly name an isolate by a process of comparison with already defined species on the basis of a set of properties that are useful as diagnostic, determined case by case by consensus between virologists who are specialists in a virus group (Van Regenmortel et al., 1997, 2000).

Within a species, virologists observe a variability that they describe in various terms. An isolate is the sample taken from an infected plant. It may contain several viruses. Each virus must be characterized in order to be identified as a known species or to be declared a species that is not yet described. A strain is a set of natural isolates that have in common one or several characteristic properties. It becomes useful to designate them by a common term. A strain generally has some stability over time, i.e., numerous isolates have the characteristics of the strain. A pathotype is a collection of isolates of a single virus that have a similar behaviour with respect to host resistance; this is revealed by the inoculation of a collection of genotypes of the host. A serotype is a set of isolates characterized by antigenic properties of the capsid that differentiate them from other isolates. The study of intraspecific variations is a major topic of interest for the plant virologist.

Genera

Genera (with the suffix *–virus*) are groups of species that share certain characteristics, with differences depending on the case. The notion of phylogeny is here very important. Two viruses with different phylogenies cannot belong to the same genus. Within a genus, the number of genes, the presence of specific coding regions, and the types of vectors are common; usually the evolution of the different genes has been linked. Unlike the species, the genus and family are defined by a limited number of necessary and sufficient characters. In each genus, the ICTV designates a type species, which is generally the first species that necessitated the original creation and naming of the genus, but is not always the best characterized species of the genus. It shares most of its properties with the other species of the genus, but it is not necessarily representative of all the species of the genus. In large families, there is a tendency to create new genera.

Families

Families (with the suffix *–viridae*) are groups of genera, characterized by the structure of the genome and with some similarities in replication strategy. This taxon has some reality in phylogeny, even though it represents quite a distant relationship based on a small number of characters that reflect the properties of the entire genome. With the accumulation of molecular data, it has been possible to trace phylogenetic trees and to observe that the trees elaborated from the single polymerase gene are quite relevant for the purposes of classification. However, because of the number of recombinations in viral genomes, there is no general rule and the taxonomist may be faced with difficult choices. The example of the family *Luteoviridae* is particularly illuminating. The former genus *Luteovirus* contained *Barley yellow dwarf virus* and *Potato leafroll virus*, very similar in vection and physicochemical characters but distant in the characteristics of the polymerase (Gibbs and Cooper, 1995). At present, two genera are distinguished, *Luteovirus* and *Polerovirus*, grouped in the family *Luteoviridae* (D'Arcy and Mayo, 1997). The modular acquisition of polymerase genes of different origins, following processes of recombination, does not challenge the taxonomic unity of this family, which is based especially on characteristics of vection. The phenomena of recombination, the extent of which is revealed by sequence comparison, pose complex problems in the context of classification.

Orders

Orders have the suffix *–virales*. At present there are only three orders (*Nidovirales, Mononegavirales, Caudovirales*). Only the order *Mononegavirales* contains plant viruses (*Rhabdoviridae*). These orders were created to group families sharing a common phylogeny. Some authors (Ward, 1993) consider superfamilies to be a prototype of future orders.

An example of classification: potato virus Y

The various characters are used in a hierarchical order. At each stage, new identifying characters are added to the preceding ones.

1: non-reverse-transcribing RNA virus (without DNA phase)
2: positive-sense single strand RNA
3: RNA with VPg and polyA, codes for a polyprotein

→ ***Comoviridae and Potyviridae***

4: the replication proteins form a group in the order helicase-VPg-protease-polymerase
5: the particles are flexuous

→ **Family: *Potyviridae***

6: the capsid gene is located at the 3′ end
7: the cytoplasm contains cylindrical inclusion (CI) bodies
8: the virus is transmitted by aphids
9: the genome is monopartite and codes for a polyprotein (P1-HC-P3-6K1-CI-6K2-VPg-NIaPro-NIbPol-Cp)

→ **Genus: *Potyvirus***

10: numerous strains that infect potato, tomato, pepper, and other Solanaceae: three principal strains Y^O, Y^N, Y^C
11: cytoplasmic inclusion bodies of type IV
12: genome of around 9700 nucleotides

→ **Species: PVY**

13: percentage of identity in the sequences, serology, host range, symptoms

→ **various strains**

Plant virus classification

With these criteria, the ICTV established in 2005 a eight version of the classification (Fauquet et al., 2005). The ICTV functions as a forum of constant discussion that takes the evolution of knowledge into account and makes the changes official every four years. New families and genera are continuously defined, and unassigned genera are inserted in existing families.

The ICTV establishes a list, continuously updated, of about 2000 virus species (with approved names) and develops a universally available virus database describing all known viruses at all taxonomic levels (ICTVdB on the Web). More than 1000 virus species infect plants. They are grouped in 18 families and 81 genera (Fauquet et al., 2005) (Tables 14.1 and 14.2).

Table 14.1 Classification of viruses.

Positive-sense single-stranded RNA viruses	*Astroviridae, Barnaviridae, Bromoviridae* (***Alfamovirus, Bromovirus, Cucumovirus, Ilarvirus, Oleavirus***), *Caliciviridae, Closteroviridae* (***Ampelovirus, Closterovirus, Crinivirus***), *Comoviridae* (***Comovirus, Fabavirus, Nepovirus***), *Dicistroviridae, Flaviviridae, Flexiviridae* (***Allexivirus, Capillovirus, Carlavirus, Foveavirus, Mandarivirus, Potexvirus, Trichovirus, Vitivirus***), *Leviviridae, Luteoviridae* (***Enamovirus, Luteovirus, Polerovirus***), *Marnaviridae*, order *Nidovirales* [*Arteriviridae, Coronaviridae, Roniviridae*], *Narnaviridae, Nodaviridae, Picornaviridae, Potyviridae* (***Bymovirus, Ipomovirus, Macluravirus, Potyvirus, Rymovirus, Tritimovirus***), *Sequiviridae* (***Sequivirus, Waikavirus***), *Tetraviridae, Togaviridae, Tombusviridae* (***Aureusvirus, Avenavirus, Carmovirus, Dianthovirus, Machlomovirus, Necrovirus, Panicovirus, Tombusvirus***), *Tymoviridae* (***Tymovirus, Marafivirus, Maculavirus***) + 15 genera not related to a family including ***Benyvirus, Cheravirus, Furovirus, Hordeivirus, Idaeovirus, Pecluvirus, Pomovirus, Ourmiavirus, Sadwavirus, Sobemovirus, Umbravirus, Tobamovirus, Tobravirus.***
Negative-sense (or ambisense) single-stranded RNA viruses	*Arenaviridae, Bunyaviridae* (*Orthobunyavirus, Hantavirus, Nairovirus, Phlebovirus*, ***Tospovirus***), order *Mononegavirales* [*Bornaviridae, Filoviridae, Paramyxoviridae, Rhabdoviridae* (***Cytorhabdovirus***, *Ephemerovirus, Lyssavirus, Novirhabdovirus*, ***Nucleorhabdovirus***, *Vesiculovirus*)], *Orthomyxoviridae* + 3 genera not related to a family, including ***Ophiovirus, Tenuivirus, Varicosavirus.***
Double-stranded RNA viruses	*Birnaviridae, Chrysoviridae, Cystoviridae, Hypoviridae, Partitiviridae* (***Alphacryptovirus, Betacryptovirus***, *Partitivirus*), *Reoviridae* (*Aquareovirus, Coltivirus, Cypovirus*, ***Fijivirus***, *Idnoreovirus, Mycoreovirus, Orbivirus*, ***Oryzavirus***, *Orthoreovirus*, ***Phytoreovirus***, *Rotavirus, Seadornavirus*), *Totiviridae.* + 1 genus not related to a family, ***Endornavirus.***
Single-stranded DNA viruses	*Circoviridae, Geminiviridae* (***Begomovirus, Curtovirus, Mastrevirus Topocuvirus***), *Inoviridae, Microviridae, Nanoviridae* (***Nanovirus, Babuvirus***), *Parvoviridae* + 1 genus not related to a family.
DNA or RNA reverse-transcribing viruses	*Caulimoviridae* (***Badnavirus, Caulimovirus, Cavemovirus, Petuvirus, Soymovirus, Tungrovirus***), Hepadnaviridae, *Metaviridae* (*Errantivirus*, ***Metavirus***, *Semotivirus*), *Pseudoviridae* (*Hemivirus*, ***Pseudovirus, Sirevirus***), *Retroviridae.*

Non-reverse-transcribing double-stranded DNA viruses

Adenoviridae, Ascoviridae, Asfarviridae, Baculoviridae, Corticoviridae, order *Caudovirales* [*Myoviridae, Podoviridae, Siphoviridae*], *Fuselloviridae, Guttaviridae, Herpesviridae, Iridoviridae, Lipothrixviridae, Nimaviridae, Papillomaviridae, Phycodnaviridae, Plasmaviridae, Polydnaviridae, Polyomaviridae, Poxviridae, Rudiviridae, Tectiviridae.*

+ 2 genera not related to a family.

The virus families comprising plant viruses are in red, the plant virus genera in bold.

Table 14.2 Classification of plant viruses

Family	Genus	Type species	No.*
Positive-sense single-stranded RNA viruses			
* Mainly monopartite genome, isometric particles			
Luteoviridae	*Luteovirus*	*Barley yellow dwarf virus-PAV* (BYDV-PAV)	1
	Polerovirus	*Potato leafroll virus* (PLRV)	2
	Enamovirus	*Pea enation mosaic virus-RNA1* (PEMV-RNA1)	3
Sequiviridae	*Sequivirus*	*Parsnip yellow fleck virus* (PYFV)	4
	Waikavirus	*Rice tungro spherical virus* (RTSV)	5
Tombusviridae	*Aureusvirus*	*Pothos latent virus* (PoLV)	6
	Avenavirus	*Oat chlorotic stunt virus* (OCSV)	7
	Carmovirus	*Carnation mottle virus* (CarMV)	8
	Dianthovirus	*Carnation ringspot virus* (CRSV)	9
	Machlomovirus	*Maize chlorotic mottle virus* (MCMV)	10
	Necrovirus	*Tobacco necrosis virus* (TNV)	11
	Panicovirus	*Panicum mosaic virus* (PMV)	12
	Tombusvirus	*Tomato bushy stunt virus* (TBSV)	13
Tymoviridae	*Tymovirus*	*Turnip yellow mosaic virus* (TYMV)	14
	Marafivirus	*Maize rayado fino virus* (MRFV)	15
	Maculavirus	*Grapevine fleck virus* (GFkV)	16
Unassigned genus	*Sobemovirus*	*Southern bean mosaic virus* (SBMV)	17
* Bipartite genome, isometric particles			
Comoviridae	*Comovirus*	*Cowpea mosaic virus* (CPMV)	18
	Fabavirus	*Broad bean wilt virus 1* (BBWV-1)	19
	Nepovirus	*Tobacco ringspot virus* (TRSV)	20
Unassigned genera	*Cheravirus*	*Cherry raspleaf virus* (CRLV)	20a
	Idaeovirus	*Raspberry bushy dwarf virus* (RBDV)	21
	Sadwavirus	*Satsuma dwarf virus* (SDV)	21a
* Tripartite genome, isometric particles			
Bromoviridae	*Alfamovirus*	*Alfalfa mosaic virus* (AMV)	22
	Bromovirus	*Brome mosaic virus* (BMV)	23
	Cucumovirus	*Cucumber mosaic virus* (CMV)	24
	Ilarvirus	*Tobacco streak virus* (TSV)	25
	Oleavirus	*Olive latent virus 2* (OLV-2)	26
Unassigned genus	*Ourmiavirus*	*Ourmia melon virus* (OuMV)	27

Family	Genus	Type species	No.*
* Helical rod-shaped particles			
Unassigned genera	*Benyvirus*	*Beet necrotic yellow vein virus* (BNYVV)	28
	Furovirus	*Soil-borne wheat mosaic virus* (SBWMV)	29
	Hordeivirus	*Barley stripe mosaic virus* (BSMV)	30
	Pecluvirus	*Peanut clump virus* (PCV)	31
	Pomovirus	*Potato mop-top virus* (PMTV)	32
	Tobamovirus	*Tobacco mosaic virus* (TMV)	33
	Tobravirus	*Tobacco rattle virus* (TBRV)	34
* Helical filamentous particles			
Closteroviridae	*Closterovirus*	*Beet yellows virus* (BYV)	35
	Crinivirus	*Lettuce infectious yellows virus* (LIYV)	36
	Ampelovirus	*Grapevine leafroll-associated virus 3* (GLRaV-3)	37
Potyviridae	*Bymovirus*	*Barley yellow mosaic virus* (BaYMV)	38
	Ipomovirus	*Sweet potato mild mottle virus* (SPMMV)	39
	Macluravirus	*Maclura mosaic virus* (MMV)	40
	Potyvirus	*Potato virus Y* (PVY)	41
	Rymovirus	*Ryegrass mosaic virus* (RGMV)	42
	Tritimovirus	*Wheat streak mosaic virus* (WSMV)	43
Flexiviridae	*Allexivirus*	*Shallot virus X* (ShVX)	44
	Capillovirus	*Apple stem grooving virus* (ASPV)	45
	Carlavirus	*Carnation latent virus* (SLV)	46
	Foveavirus	*Apple stem pitting virus* (ASPV)	47
	Potexvirus	*Potato virus X* (PVX)	48
	Trichovirus	*Apple chlorotic leaf spot virus* (ACLSV)	49
	Vitivirus	*Grapevine virus A* (GVA)	50
	Mandarivirus	*Indian citrus ringspot virus* (ICRSV)	50a
No definite particle	*Umbravirus*	*Carrot mottle virus* (CMoV)	51
Negative-sense single-stranded RNA viruses			
Bunyaviridae	*Tospovirus*	*Tomato spotted wilt virus* (TSWV)	52
Rhabdoviridae	*Cytorhabdovirus*	*Lettuce necrotic yellows virus* (LNYV)	53
	Nucleorhabdovirus	*Potato yellow dwarf virus* (PYDV)	54
Unassigned genera	*Ophiovirus*	*Citrus psorosis virus* (CPsV)	55
	Tenuivirus	*Rice stripe virus* (RSV)	56
	Varicosavirus	*Lettuce big-vein virus* (LBVV)	57
Double-stranded RNA viruses			
Partitiviridae	*Alphacryptovirus*	*White clover cryptic virus 1* (WCCV1)	58
	Betacryptovirus	*White clover cryptic virus 2* (WCCV2)	59
Reoviridae	*Fijivirus*	*Fiji disease virus* (FDV)	60
	Oryzavirus	*Rice ragged stunt virus* (RRSV)	61
	Phytoreovirus	*Wound tumor virus* (WTV)	62
Unassigned genus	*Endornavirus*	*Vicia faba endornavirus* (VFV)	62a

Family	Genus	Type species	No.*
Single-stranded DNA viruses			
Geminiviridae	*Begomovirus*	*Bean golden mosaic virus* (BGMV)	63
	Curtovirus	*Beet curly top virus* (BCTV)	64
	Mastrevirus	*Maize streak virus* (MSV)	65
	Topocuvirus	*Tomato pseudo-curly top virus* (TPCTV)	66
Nanoviridae	*Nanovirus*	*Subterranean clover stunt virus* (SCSV)	67
	Babuvirus	*Banana bunchy top virus* (BBTV)	68
DNA or RNA reverse-transcribing viruses			
Caulimoviridae	*Badnavirus*	*Commelina yellow mottle virus* (ComYMV)	69
	Tungrovirus	*Rice tungro bacilliform virus* (RTBV)	70
	Caulimovirus	*Cauliflower mosaic virus* (CaMV)	71
	Cavemovirus	*Cassava vein mosaic virus* (CsVMV)	72
	Petuvirus	*Petunia vein clearing virus* (PVCV)	73
	Soymovirus	*Soybean chlorotic mottle virus* (SbCMV)	74
Pseudoviridae	*Pseudovirus*	*Saccharomyces cerevisiae Ty 1 virus* (SCeTY1V)	75
Metaviridae	*Metavirus*	*Saccharomyces cerevisiae Ty 3 virus* (SceTY3V)	76
	Sirevirus	*Glycine max SIRE1 virus* (GmaSIRV)	76a

*The numbers refer to descriptions of genera in Chapter 15.

Legend of genome diagrams

	Nucleic acid (DNA or RNA)
	Open reading frame
	Readthrough
	Frameshift
	Transcription
Cap	Cap
VPg	Genome-linked viral protein
AAA	PolyA 3' end
	tRNA-like 3' end
	Synthesized proteins
	Translation
Cp	Capsid subunit
Pol	Polymerase
Hel	Helicase
Me	Methyltransferase
Pro	Protease

15

CHAPTER

Description of Viral Genera

The families and genera of viruses are listed according to the following criteria:

— nature of genome, RNA or DNA;
— single-stranded or double-stranded genome;
— positive-sense or negative-sense nucleic acid.

The groups constituted according to these criteria are divided into subgroups corresponding to particle morphology and the division of the genome; these criteria have no taxonomic value. A number relates the descriptions to the list of genera of plant viruses (Chapter 14, Table 14.2).

Positive-sense single-stranded RNA viruses

mainly monopartite genome, isometric particles

Family *Luteoviridae* (from the Latin *luteus*, yellow)

Depending on the genomic organization, phylogenetic origin of the polymerase gene, and biological properties, the family is divided into three genera: *Enamovirus*, *Luteovirus*, and *Polerovirus*.

Virion Particle of 25 nm with isometric symmetry.

Genome One positive-sense single-stranded RNA molecule (5.6 to 6 kb). The ORFs are grouped in two blocks separated by a non-coding region of 100 to 200 nucleotides. The expression of genes uses various mechanisms: e.g., readthrough, frameshift, subgenomic RNA, transactivator of translation located at the 3' end.

Phylogenetic relationships

The polymerase gene, located in the 5' part of the genome, presents homologies with the corresponding gene of *Tombusviridae* in the case of *Luteovirus*, and with that of *Sobemovirus* in the case of *Polerovirus* and *Enamovirus*. The family has 11 species not classified in genera.

Luteoviridae

Genus *LUTEOVIRUS* (from the Latin *luteus*, yellow) —— 1

Type species *Barley yellow dwarf virus-PAV* (BYDV-PAV)

Genome The RNA has 6 ORFs. ORFs 1 and 2 are expressed on the genomic RNA; P1 could be helicase, even though known motifs are not found in it; P2 is expressed by a change in the reading frame of ORF 1, it has polymerase motifs of the carmovirus type of *Tombusviridae*. ORFs 3, 4, and 5 are expressed through the intermediary of subgenomic RNAs: gene 3 corresponds to the capsid; ORF 5, which is directly contiguous with it, is expressed by readthrough. In plants and virions the fusion protein 3 + 5 is found in small quantities. Protein 5 is found on the surface of viral particles and plays a role in the recognition of the particle during transmission by aphids. ORF 4, located within ORF 3 but in a different reading frame, will be expressed by internal initiation and could correspond to the movement protein. ORF 6 is expressed by the intermediary of subgenomic RNA 2. A subgenomic RNA 3 is also detected.

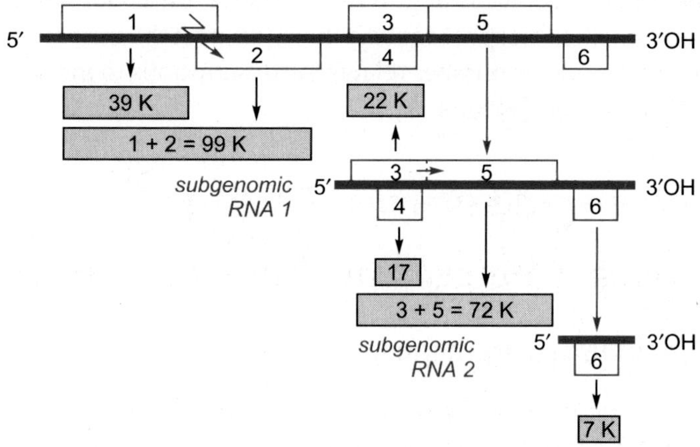

Biology See *Polerovirus*.

Luteoviridae

Genus *POLEROVIRUS* (from POtato LEaf ROll virus) —— 2

Type species *Potato leafroll virus* (PLRV)

Genome ORFs 0, 1 and 2 are expressed on the genomic RNA; P0 is implicated in the expression of symptoms. P1 has the sequence coding for the protein VPg, and next motifs characteristic of serine proteases. P2 has polymerase motifs similar to those found in picornaviruses (VPg-protease-polymerase domain) of the sobemovirus type.

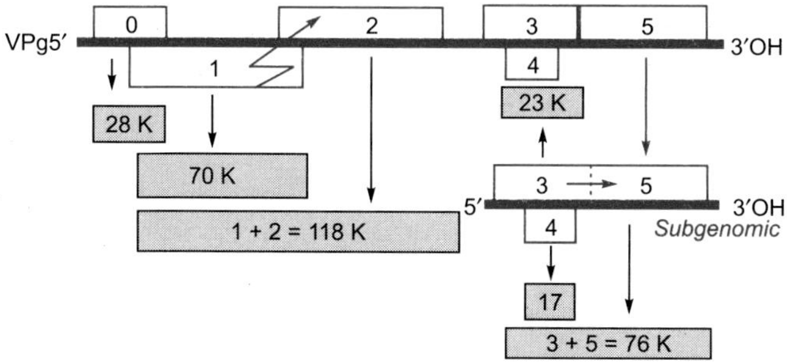

Biology of *Luteovirus* and *Polerovirus*

Respectively 5 and 9 described species. These viruses are specific to phloem tissues, disturb nutrition, and cause yellowing, leafroll, and stunting of the plant.

Mechanical transmission is not possible, agro-infection is possible. Transmitted by aphids (more than 12 genera) in the persistent circulative mode. Each virus is transmitted preferentially by one species (or a small number of species). BYDV causes serious diseases on cereals and maize throughout the world. These species are named with initials indicating their principal vector species in North America: *Macrosiphum avenae* (MAV), *R. padi* and *M. avenae* (PAV).

Host range BYDV on Gramineae. *Beet western yellows virus* (BWYV) on 23 families of dicotyledons; symptoms of yellowing on sugar beet and lettuce.

BRUYERE, A., BRAULT, V., ZIEGLER-GRAFF, V., SIMONIS, M.T., VAN DEN HEUVEL, J.F., RICHARD, K., GUILLEY, H., JONARD, G., HERRBACH, E., 1997. Effects of mutations in the beet western yellows virus readthrough protein on its expression and packaging and on virus accumulation, symptoms and aphid transmission. *Virology*, 230, 323-334.

CHALHOUB, B.A., KELLY, L., ROBAGLIA, C., LAPIERRE, H.D. 1994. Sequence variability in the genome-3'-terminal region for 10 geographically distinct PAV-isolates of barley yellow dwarf virus. Analysis of the ORF 6 variation. *Arch. Virol.*, 139, 403-416.

KOEV, G., MOHAN, B.R., MILLER, W.A. 1999. Primary and secondary structural elements required for synthesis of barley yellow dwarf virus subgenomic RNA1. *J. Gen. Virol.*, 78, 2876-2885.

WANG, J.Y., CHAY, C., GILDOW, F.E., GRAY, S.M. 1995. Readthrough protein associated with virions of barley yellow dwarf luteovirus and its potential role in regulating the efficiency of aphid transmission. *Virology*, 206, 954-962.

VAN DER WILK, F., VERBEEK, M., DULLEMANS, A.M., VAN DEN HEUVEL, J.F. 1997. The genome linked protein of potato leaf roll virus is located downstream of the putative protease domain of the ORF 1 product. *Virology*, 234, 300-303.

Luteoviridae

Genus *ENAMOVIRUS* (from ENAtion MOsaic virus) —— 3

Type species *Pea enation mosaic virus* (PEMV-RNA1)

Virion 2 partly independent particles, forming a symbiotic association.

Genome 2 molecules of single-stranded RNA of 5.7 and 4.2 kb. Each codes for a polymerase, of a polerovirus type in the case of RNA 1 and luteovirus type in the case of RNA 2. ORF 2 is translated by frame shift giving a product of 132 K (RNA 1) and 95 K (RNA 2). In RNA1, ORF 3 is translated via a subgenomic RNA, and ORF 3 and 5 by read through (not shown on diagram). Each RNA can be replicated autonomously, but they form a stable association.

RNA 1 greatly resembles a polerovirus complemented by RNA 2 for systemic movement and mechanical transmission. RNA 2 resembles an umbravirus, and it is complemented by RNA 1 for the capsid (and thus transmission by vectors).

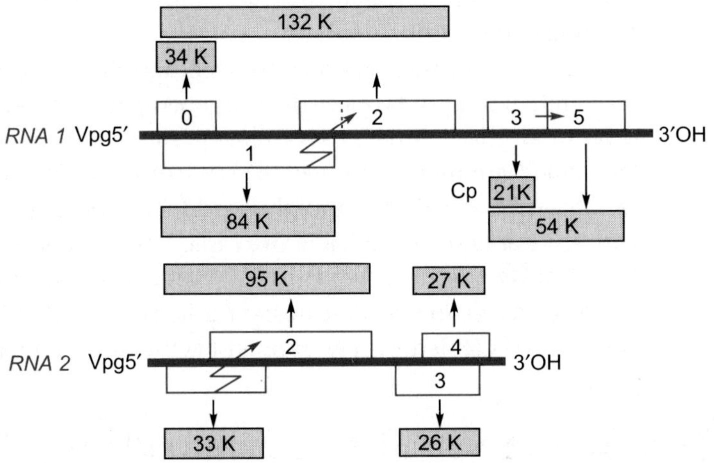

Biology One described species. The association of two viral entities as a stable complex (which can be considered defective viruses) makes possible the systemic invasion of the host, mechanical transmission and transmission by aphids in the circulative mode for the two particles. The virus multiplies in all tissues of the host.

Host range Leguminosae. Symptoms of leaf curl mosaic, stunting, hyaline patches, and enation. PEMV is widespread in legume and pea crops in North America, Asia, and Europe.

DEMLER, S.A., RUCKER, D.G., DE ZOETEN, G.A. 1993. The chimeric nature of the genome of pea enation mosaic virus: the independent replication of RNA 2. *J. Gen. Virol.*, 74, 1-14.

DEMLER, S.A., BORKHSENIOUS, O.N., RUCKER, D.G., DE ZOETEN, G.A. 1994. Assessment of the autonomy of replicative and structural functions encoded by the luteo-phase of pea enation mosaic virus. *J. Gen. Virol.*, 75, 997-1007.

DEMLER, S.A., RUCKER-FEENEY, D.G., SKAF, J.S., DE ZOETEN, G.A. 1997. Expression and suppression of circulative aphid transmission in pea enation mosaic virus. *J. Gen. Virol.*, 78, 511-523.

Family *Sequiviridae*

The family *Sequiviridae* comprises two genera: *Sequivirus* and *Waikavirus*.

Phylogenetic relationships

The non-structural proteins of *Sequiviridae* have sequence homologies with those of *Comoviridae* and *Picornaviridae*. The presence of three structural proteins in the N-terminal region recalls the arrangement observed in the genomic map of *Picornaviridae*.

Sequiviridae

Genus *SEQUIVIRUS* (from the Latin *sequi*, dependent) — 4

Type species *Parsnip yellow fleck virus* (PYFV).

Virion Isometric; the capsid is assembled from three subunits (32, 26, 23 kDa).

Genome A positive-sense single-stranded RNA (10 kb), translated into a polyprotein that carries in its N-terminal part the three capsid proteins, and in the C-terminal part domains containing the protease, polymerase, and helicase motifs.

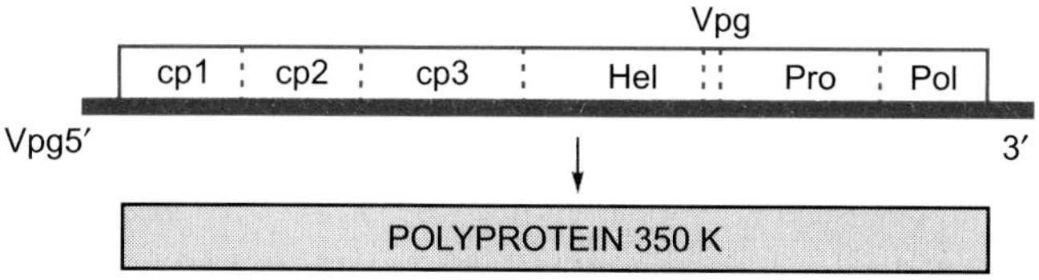

Biology Two species described. The transmission of PYFV by aphids in a semi-persistent mode is dependent on the presence of a helper component produced by a very similar virus, *Anthriscus yellows virus*, *Waikavirus*. Species transmissible mechanically. Symptoms of mosaics, chlorotic or necrotic spots. *Dandelion yellow mosaic virus* causes a lettuce disease in Western Europe.

Reavy, B., Mayo, M.A., Turnbull-Ross, A.D., Murant, A.F. 1993. Parsnip yellow fleck and rice tungro spherical viruses resemble picornaviruses and represent two genera in a proposed new plant picornavirus family (*Sequiviridae*). *Arch. Virol.*, 131, 441-446.
Turnbull-Ross, A.A., Mayo, M.A., Reavy, B., Murant, A.F. 1993. Sequence analysis of parsnip yellow fleck virus polyprotein evidence of affinities with picornaviruses. *J. Gen. Virol.*, 74, 555-561.

Sequiviridae

Genus *WAIKAVIRUS* (from the Japanese) — 5

Type species *Rice tungro spherical virus* (RTSV).

Genome The genetic map of the 11 kb RNA is very similar to that of *Sequivirus*; it has in the 3' region a supplementary ORF that is probably translated through a subgenomic RNA. RTSV is transmitted by leafhoppers (*Cicadellidae*) in a semi-persistent mode and depends on a helper component coded by the virus. This component allows the transmission of a much more severe disease caused by a ds DNA virus, RTBV (*Rice tungro bacilliform virus, Badnavirus*). Very mild symptoms of mosaic. Three species described.

DRUKA, A., BURNS, T., ZHANG, S., HULL, R. 1996. Immunological characterization of rice tungro spherical virus coat protein and differentiation of isolates from the Philippines and India. *J. Gen. Virol.*, 77, 1975-1983.

SHEN, P., KANIEWSKA, M.B., SMITH, C., BEACHY, R.N. 1993. Nucleotide sequence and genetic organisation of rice tungro spherical virus. *Virology*, 193, 621-623.

Family *Tombusviridae*

The family *Tombusviridae* has eight genera: *Aureusvirus, Avenavirus, Carmovirus, Dianthovirus, Machlomovirus, Necrovirus, Panicovirus, Tombusvirus.*

Virion Particle of 30 nm with isometric symmetry. The capsid is made up of 180 subunits. Each subunit has an outer domain that, by adhering to a homologous domain of the neighbouring subunit, forms one of the 90 protrusions of the particle (Fig. 1.12).

Phylogenetic relationships

Apart from very clear sequence homologies of the polymerase gene throughout the family *Tombusviridae*, more marked homologies are noted in the polymerase gene of *Carmovirus, Necrovirus,* and *Luteovirus* and in the capsid gene of *Dianthovirus* and *Tombusvirus* and, to a lesser degree, *Carmovirus*.

Tombusviridae

Genus *AUREUSVIRUS* (from the name of pothos, *Scindapsus aureus*) —— 6

Type species *Pothos latent virus* (PoLV).

Tombusviridae

Genus *AVENAVIRUS* (from the Latin *avena*, oat) —— 7

Type species *Oat chlorotic stunt virus* (OCSV).

Tombusviridae

Genus *CARMOVIRUS* (from CARnation MOttle virus) —— 8

Type species *Carnation mottle virus* (CarMV).

Virion and genome They are very similar to those of *Tombusvirus* (4 kb), as are the translation strategies. One difference is that the ORF coding for the capsid is found on the 3' end on the genomic RNA. Size of proteins synthesized in relation with the five ORFs: 28, 88 (ORF 1 + ORF 2), 8, 9 and 40 kDa.

Biology 14 species are described, 8 are tentative. They are transmitted by contact, through soil, by beetles, or by the soil fungus *Olpidium*, as well as through vegetative means and by mechanical inoculation. CarMV is widespread in carnation crops throughout the world. *Melon necrotic spot virus* (MNSV) is a serious problem on melons and cucumbers grown in greenhouses in Europe and in Japan. Symptoms: chlorotic or necrotic spots, necrosis on stems and petioles.

CAMPBELL, R.N., WIPF-SCHEIBEL, C., LECOQ, H. 1996. Vector-assisted seed transmission of melon necrotic spot virus in melon. *Phytopathology*, 86, 1294-1298.
CARRINGTON, J.C., HEATON, L.A., UIDEMA, D., HILLMAN, B.I., MORRIS, T.J. 1989. The genome structure of turnip crinkle virus. *Virology*, 170, 219-226.
HACKER, D.L., PETTY, I.R., WEI, N., MORRIS, T.J. 1992. Turnip crinkle genes required for RNA replication and virus movement. *Virology*, 186, 1-8.

Tombusviridae

Genus *DIANTHOVIRUS* (from the Latin *dianthus*, carnation) —— 9

Type species *Carnation ringspot virus* (CRSV).

Genome 2 positive-sense single-stranded RNA molecules. It is probable that the viral particles contain either one RNA 1 molecule or 3 RNA 2 molecules. RNA 1 (3.9 kb) can replicate in protoplasts alone and produces virions. It has three ORFs. The polymerase is synthesized by frameshift. The capsid (ORF 2) is expressed through a subgenomic RNA. It is necessary for long-distance movement. RNA 2 (0.5 kb) codes for the movement protein.

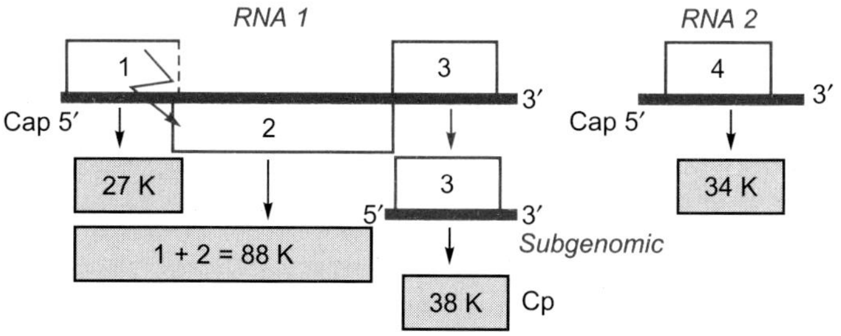

Biology 3 species are described, 3 are tentative.

Host range Dicotyledons.

Transmission Transmitted by mechanical means, through soil and possibly through irrigation water. No vector known. Symptoms of mosaic, ring spot, dwarfing, and flower curl. CRSV, which was found on carnation worldwide because of vegetative propagation, has practically disappeared in production regions in Europe and the United States with the advance of rigorous sanitary certification systems.

XIONG, Z., KIM, K.H., KENDALL, T.L., LOMMEL, S.A. 1993. Synthesis of the putative red clover necrosis virus RNA polymerase by ribosomal frameshifting. *Virology*, 193, 213-221.
XIONG, Z., KIM, K.H., GIESMAN-COOKERMEYER, D., LOMMEL, S.A. 1993. The role of the RCNMV capsid and cell-to-cell movement proteins in systemic infection. *Virology*, 192, 27-32.

Tombusviridae

Genus *MACHLOMOVIRUS* (from MAize CHLOrotic MOttle virus) —— 10

Type species *Maize chlorotic mottle virus* (MCMV).

Genome One positive-sense single-stranded RNA molecule (4.4 kb) carrying 4 ORFs. The capsid is expressed through the intermediary of a subgenomic RNA.

Biology A single species described in America. Its natural host is maize, which reacts with a mild mosaic. In co-infection with a potyvirus, serious necrosis is observed. The virus is transmitted through the seed, by mechanical inoculation, and by insects (beetles or thrips).

LOMMEL, S.A., KENDALL, T.L., SIN, N.F., NUTTER, R.C. 1991. Characterization of maize chlorotic mottle virus. *Phytopathology*, 81, 819-823.

Tombusviridae

Genus *NECROVIRUS* (from the Greek *necros*, death) —— 11

Type species *Tobacco necrosis virus* (TNV).

Genome One positive-sense single-stranded RNA molecule (3.7 kb) carrying 4 ORFs. The genomic RNA allows the expression of ORF 1. The product obtained by readthrough is replicase. Subgenomic RNA 1 allows the expression of ORF 2 and perhaps of ORF 3. Subgenomic RNA 2 allows the expression of ORF 4, the capsid. The products of ORF 2 and 3 are implicated in cell-to-cell movement. Strain A has a sixth ORF.

Biology 6 described species, 2 tentative species. The host range is large (Mono- and Dicotyledons). In natural conditions, the virus is transmitted by contact and through the soil fungus *Olpidium brassicae*: the virus is often limited to roots. Mechanical inoculation causes necrotic lesions on the infected leaves, and sometimes

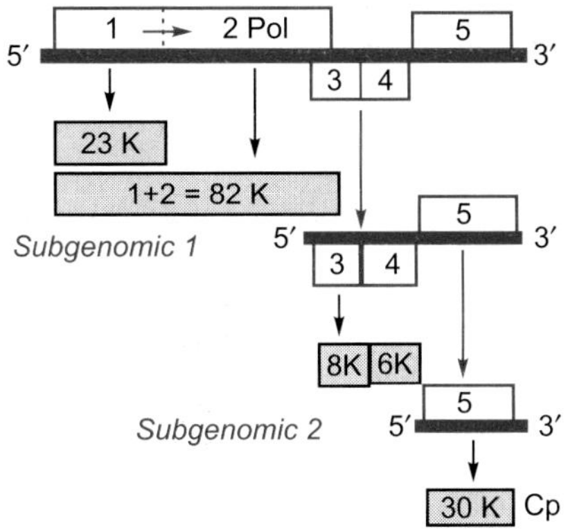

generalized necrosis. TNV has a worldwide distribution and causes locally serious diseases on cucumber, tulip, and bean. TNV may help a satellite virus, the genome of which is an RNA coding for the satellite capsid protein. The 17 nm particle contains 60 subunits.

MEULEWAETER, F., SEURINCK, J., VAN EMMELO, J. 1990. Genome structure of tobacco necrosis virus strain A. *Virology*, 177, 669-679.
OFFEI, S.K., COFFIN, R.S., COUTTS, R.H. 1995. The tobacco necrosis virus p7 protein is a nucleic-acid binding protein. *J. Gen. Virol.*, 76, 1493-1496.

Tombusviridae

Genus *PANICOVIRUS* (from PANICum mOsaic virus) — 12

Type species *Panicum mosaic virus* (PMV).

Tombusviridae

Genus *TOMBUSVIRUS* (from TOMato BUshy Stunt virus) — 13

Type species *Tomato bushy stunt virus* (TBSV)

Genome One positive-sense single-stranded RNA molecule (4.7 kb) carrying 4 ORFs. The genomic RNA directs the synthesis of protein 1 and of the readthrough protein that carries the polymerase motif and two helicase motifs. The other proteins, including the capsid subunit, are expressed through a subgenomic RNA. Proteins 3 and 4 are involved in movement.

Biology 15 described species. The virus accumulates in paracrystalline structures in the cytoplasm and in the vacuoles. It is transmitted by mechanical means, by contact, through the seed, through vegetative propagation, sometimes by the soil fungus *Olpidium*. Some tombusviruses have been isolated from river water, suggesting the possibility of transmission through irrigation. Symptoms: necrotic

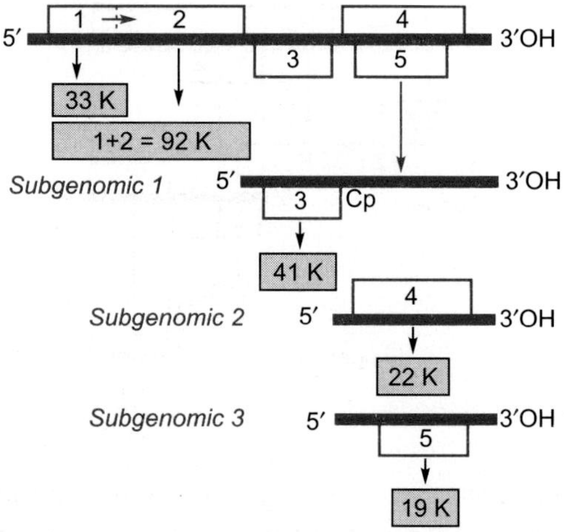

mosaic, chlorotic or necrotic spots, leaf curl, and dwarfing. TBSV particularly infects tomato in greenhouses and fruit trees in various regions of the world.

McLean, M.A., Campbell, R.N., Hamilton, R.I., Rochon, D.M. 1994. Involvement of the cucumber necrosis virus coat protein in the specificity of fungus transmission by *Olpidium bornovanus*. *Virology*, 204, 840-842.

Russo, M., Burgyan, J., Martelli, G.P. 1994. Molecular biology of *Tombusviridae*. *Adv. Virus Res.*, 44, 381-428.

Wu, B., White, K.A. 1998. Formation and amplification of a novel tombusvirus defective RNA which lacks the 5' nontranslated region of the viral genome. *J. Virol.*, 72, 9897-9905.

Family *Tymoviridae*

The family comprises three genera: *Tymovirus, Marafivirus, Maculavirus*.

Tymoviridae

Genus *TYMOVIRUS* (from Turnip Yellow MOsaic virus) — 14

Type species *Turnip yellow mosaic virus* (TYMV).

Virion Particle of 30 nm with isometric symmetry. The capsid is formed of 180 identical subunits of 20 kDa.

Genome One positive-sense single-stranded RNA molecule (6.3 kb) with 3 ORFs: ORF 1 codes for a protein of 206 kDa that contains polymerase motifs and protease activity. ORF 2 overlaps with ORF 1; it codes for cell-to-cell movement protein. ORF 3 codes for the capsidial subunit. It is expressed through a subgenomic RNA. Replication occurs in the vesicles at the periphery of the chloroplasts. It leads to destruction of chlorophyll and yellowing of leaves.

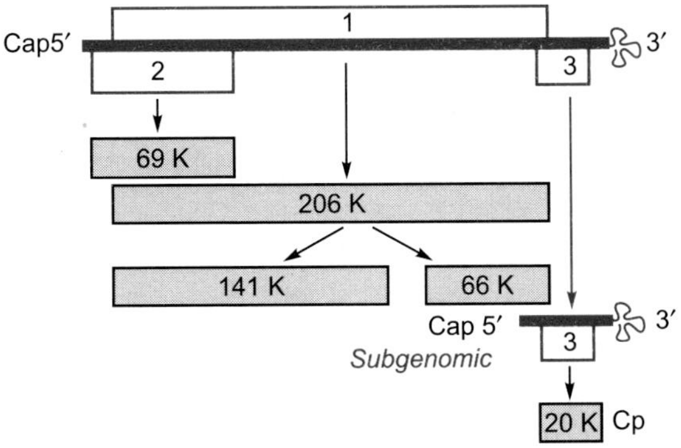

Biology 23 described species. Transmission by mechanical means, by contact, by beetles in the semi-persistent mode. Symptoms of sometimes bright yellow mosaic, dwarfing, and leaf curl.

Phylogenetic relationships

The polymerase sequences of *Potexvirus, Marafivirus*, and *Carlavirus* are similar to those of *Tymovirus*.

DEIMAN, B.A., KORTLEVER, R.M., PLEIJ, C.W.A. 1997. The role of the pseudoknot at the 3' end of turnip yellow mosaic virus RNA in minus-strand synthesis by the viral RNA-dependent RNA polymerase. *J. Gen. Virol.*, 71, 5990-5996.

MORCH, M.D., JOSHI, R.L., DENIAL, T.M., HAENNI, A.L. 1988. Overlapping reading frame revealed by complete nucleotide sequencing of turnip yellow mosaic virus genomic RNA. *Nucl. Acids Res.*, 16, 6157-6173.

WEILAND, J.J., DREHER, T.W. 1993. Cis-preferential replication of the turnip yellow mosaic virus RNA genome. *Proc. Natl. Acad. Sci. USA*, 90, 6095-6099.

Tymoviridae

Genus ***MARAFIVIRUS*** (from MAize RAyado FIno virus) —— 15

Type species *Maize rayado fino virus* (MRMV).

Virion Particle of around 30 nm with isometric symmetry.

Genome One positive-sense single-stranded RNA molecule.

Biology 3 species described. Multiplication in the phloem. Transmission by leafhoppers on the persistent mode, with multiplication in the insect. No mechanical transmission. Symptoms of streaks, stunting, and enation.

Host range Essentially the family *Poaceae*, maize, and teosinte. MRFV causes a sometimes serious disease in the southern United States, Central America, and South America.

ESPINOZA, A.M., RAMIREZ, P., LEON, P. 1988. Cell-free translation of maize rayado fino virus genomic RNA. *J. Gen. Virol.*, 69, 757-762.

Genus *MACULAVIRUS* — 16

Type species *Grapevine fleck virus* (GFkV).

■ Unassigned genus

Genus *SOBEMOVIRUS* (from SOuthern BEan MOsaic virus) — 17

Type species *Southern bean mosaic virus* (SBMV).

Virion Particle of 30 nm diameter with isometric symmetry, present in the cytoplasm and nucleus.

Genome Positive-sense single-stranded RNA (4.2), which carries a VPg protein at the 5' end. 4 ORFs: ORF 2 is expressed as a polyprotein (VPg-protease-polymerase); the capsid protein (ORF 4) is expressed through a subgenomic RNA.

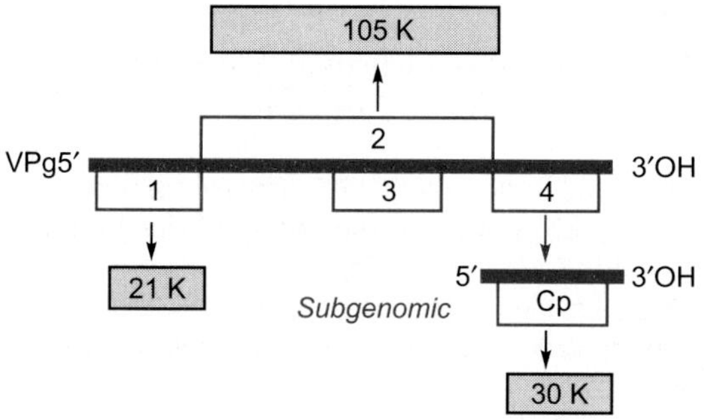

Biology 13 described species, 4 tentative species.

Mechanical transmission is possible, transmission through the seed is frequent (SBMV). It is disseminated by beetles (SBMV), myrids (*Velvet tobacco mottle virus*, VTMoV), and aphids according to the semi-persistent mode. Symptoms of mosaic, leaf curl, yellowing, stunting. *Rice yellow mottle virus* (RYMV) causes one of the most serious rice diseases in Africa.

Phylogenetic relationships

Related to the family *Tombusviridae*.

HACKER, D.L., SIVAKUMARAN, K., 1997. Mapping and expression of southern bean mosaic virus genomic RNA and subgenomic RNA. *Virology*, 234, 317-327.

SIVAKUMARAN, K., FOWLER, B.C., HACKER, D.L. 1998. Identification of viral genes required for cell-to-cell movement of southern bean mosaic virus. *Virology*, 252, 376-386.

Positive-sense single-stranded RNA viruses

bipartite genome, isometric particles

■ Family *Comoviridae*

The family *Comoviridae* comprises three genera: *Comovirus, Nepovirus, Fabavirus*. The genera *Comovirus* and *Nepovirus* are related to *Picornavirus* in capsid morphology. In genomic organization, strategy of expression (RNA with VPg and polyA, proteolysis), and sequence homologies of replication proteins, *Comoviridae* are related to *Potyviridae* and *Picornaviridae*.

Comoviridae

Genus *COMOVIRUS* (from COwpea MOsaic virus) **18**

Type species *Cowpea mosaic virus* (CPMV).

Virion 2 particles (B and M) of 28 to 30 nm diameter with isometric symmetry. The particles comprise an RNA molecule and 60 copies of each of two capsid proteins. There are also particles without RNA.

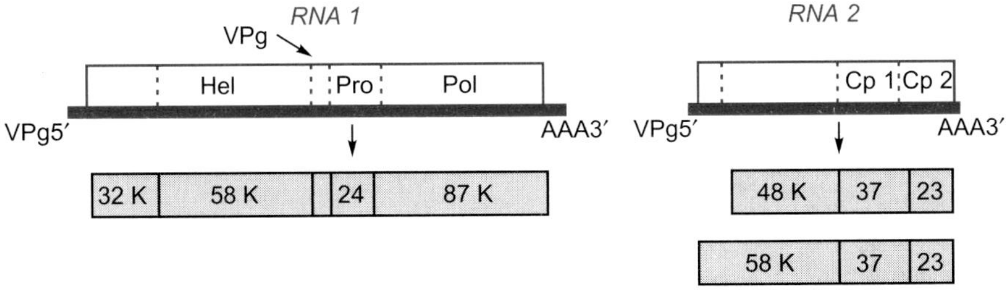

Genome 2 positive-sense single-stranded RNAs. RNA 1 (5.9 kb) is found within B particles. When it is inoculated alone, it may cause the infection of protoplasts, without producing viral particles. It thus carries information for replication. It codes for a polyprotein that *in vitro* is cleaved into two polypeptides of 32 and 170 kDa; the 170 kDa is hydrolysed and the final products are the following: one 58 kDa protein that has a domain for fixation of nucleotides, the protein VPg, a proteinase (of cysteine type) of 24 kDa, and a polymerase of 87 kDa. The 32 kDa protein remains associated with the 170 kDa protein and slows its self-hydrolysis. RNA 2 (3.8 kb) is expressed in two polypeptides of 95 and 105 kDa initiated at two different AUG, but having the same termination. They are cleaved at the same internal site. The N-terminal region (of 48 and 58 kDa)

produces proteins associated with movement. The C-terminal region is the precursor of 37 and 23 kDa proteins, which constitute the capsid. The hydrolysis is realized by the 24 kDa protease and accelerated by the 32 kDa protein coded by RNA 1.

Biology 15 described species.

Host range Mostly legumes (*Vigna unguiculata, Phaseolus vulgaris*), where it reaches a high concentration. Transmitted mechanically, through beetles (*Chrysomelidae*) and also through contact; several species are transmitted through the seed. Symptoms of mosaic and stunting. *Squash mosaic virus* (SqMV) causes a serious disease of cucurbits in warm regions in which its vectors develop.

HAUDENSHIELD, J., PALUKAITIS, P. 1998. Diversity among isolates of squash mosaic virus. *Virology*, 79, 2331-2341.

VAN LENT, J.W.M., WELLINK, J., GOLDBACH, R. 1990. Evidence for the involvement of the 58K and 48K proteins in the intercellular movement of cowpea mosaic virus. *J. Gen. Virol.*, 71, 219-223.

WELLINK, J., VAN LENT, J., VERVER, J., SIJEN, T., GOLDBACH, R., VAN KAMMEN, A. 1993. The cowpea mosaic virus M RNA-encoded 48 K protein is responsible for induction of tubular structures in protoplasts. *J. Virol.*, 67, 3660- 3664.

Comoviridae

Genus *FABAVIRUS* (from the Latin *faba*, broad bean) — 19

Type species *Broad bean wilt virus 1* (BBWV-1).

Biology 4 described species. These viruses are transmitted by aphids in the non-persistent mode. Their other characteristics are similar to those of *Comovirus*.

Comoviridae

Genus *NEPOVIRUS* (from NEmatode POlyhedral virus) — 20

Type species *Tobacco ringspot virus* (TRSV).

Virion 2 particles of 28-30 nm with isometric symmetry. The capsidial subunit is particularly large, 57 kDa.

Genome Each particle contains a positive-sense single-stranded RNA molecule. The organization of the genome is very similar to that of *Comovirus*. RNA 1 (7.4 kb) codes for a polyprotein of 254 kDa cleaved by the internal protease (23 kDa). The 72 kDa protein carries a motif for nucleotide fixation, the 87 kDa protein carries the polymerase motifs. RNA 2 (4.7) codes for a polyprotein cleaved into three proteins. The first has no known function, the second has sequence homologies with movement proteins of other viruses, and the third corresponds to the capsid.

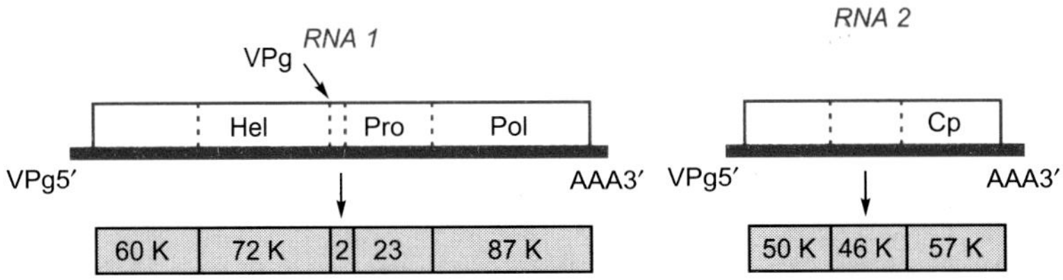

Biology 32 described species. Viruses distributed through the temperate zone in vegetable, fruit tree and small fruit crops, giving symptoms of sometimes necrotic mosaic or ring spots followed by a recovery phase. 11 species are transmitted by nematodes (*Xiphinema, Longidorus*). They can be transmitted through pollen and seed (in the case of three of them) and all are transmitted by mechanical infection. *Nepovirus* can cause very serious diseases, such as *Tomato black ring virus* in artichoke, tomato, lettuce, potato, raspberry, and strawberry. *Grapevine fanleaf virus* infects grapevine throughout the world.

Brault, V., Hilbrand, L., Candresse, T., LeGall, O., Dunez, J. 1989. Nucleotide sequence and genetic organization of Hungarian grapevine chrome mosaic nepovirus RNA 2. *Nucl. Acids Res.*, 17, 7809-7819.

Ritzenthaler, C., Viry, M., Pinck, M., Margis, R., Fuchs, M., Pinck, L. 1991. Complete nucleotide sequence and genetic organization of grapevine fanleaf nepovirus. *J. Gen. Virol.*, 72, 2357-2365.

Zalloua, P.A., Buzayan, J.M., Bruening, G. 1996. Chemical cleavage of 5'-linked protein from tobacco ringspot genomic RNAs and characterization of the protein-RNA linkage. *Virology*, 219, 1-8.

■ Unassigned genera

Genus *CHERAVIRUS* (from CHErry RAsp leaf virus) — 20a

Type species *Cherry rasp leaf virus* (CRLV)

Biology 2 described species, 2 tentative species. Symptoms mild or absent on infected trees. CRLV is transmitted by the nematode *Xiphinema index*.

Genus *IDAEOVIRUS* (from *Rubus idaeus*, raspberry) — 21

Type species *Raspberry bushy dwarf virus* (RBDV), the only species in the genus.

Genome It is bipartite (5.5 and 2.2 kb). Sequence homologies relate this genus to the tripartite *Bromoviridae*.

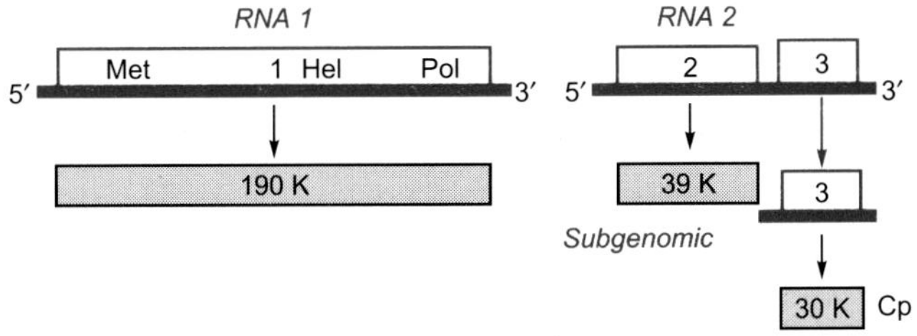

Biology Transmission through pollen, seed, and vegetative propagation. Very mild symptoms. RBDV is probably present on raspberry throughout the world.

Genus *SADWAVIRUS* (from SAtsuma DWArf virus) —— **21a**

Type species *Satsuma dwarf virus* (SDV)

Positive-sense single-stranded RNA viruses

tripartite genome, isometric particles

■ Family *Bromoviridae*

Five genera: *Alfamovirus, Bromovirus, Cucumovirus, Ilarvirus, Oleavirus.*

Brome mosaic virus (BMV) and *cucumber mosaic virus* (CMV) are among the most widely studied RNA viruses: genome structure, translation, replication.

Bromoviridae

Genus *ALFAMOVIRUS* (from ALFalfa MOsaic virus) —— **22**

Type species *Alfalfa mosaic virus* (AMV).

Virion 4 bacilliform particles of 18 nm diameter and 30-57 nm length. Each particle contains one RNA molecule (for genomic RNA 1 and 2) or two (genomic RNA 3 + genomic RNA 4).

Genome 3 positive-sense single-stranded RNA: 3.6 kb, 2.6 kb, 2 kb. RNA 4 (0.9 kb) codes for the capsidial subunit, which binds to the 3' end of genomic RNA, allowing recognition by replicase and activation of replication.

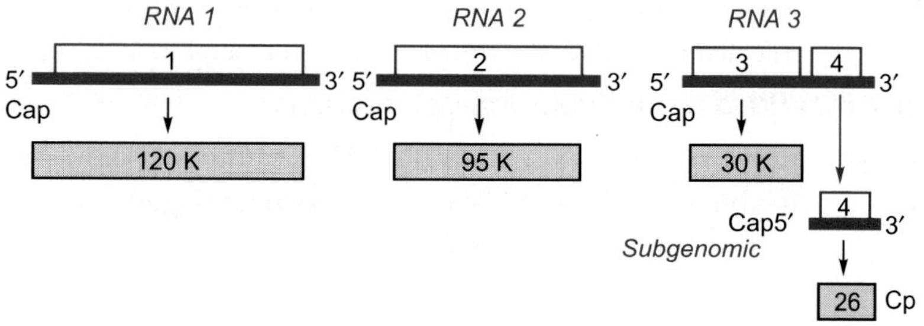

Biology A single species is described. Very wide host range (more than 300 species). Easy mechanical transmission, transmission by aphids in the non-persistent mode and through seed in some hosts, including lucerne. Symptoms of sometimes white mosaic, necrosis, and stunting. AMV has a worldwide distribution and causes serious diseases on lucerne, clover, carrot, eggplant, pepper, tomato, and celery.

BOL, J. 1999. Alfalfa mosaic virus and ilarvirus: involvement of coat protein in multiple steps of the replication cycle. *J. Gen. Virol.*, 80, 1089-1102.
VAN DER VOSSEN, E.A., NEELMAN, L., BOL, J.F. 1994. Early and late functions of alfalfa mosaic virus coat protein can be mutated separately. *Virology*, 202, 891-903.

Bromoviridae

Genus *BROMOVIRUS* (from the Latin *bromus*, bromegrass) — 23

Type species *Brome mosaic virus* (BMV).

Virion 3 particles of 27 nm with isometric symmetry.

Genome 3 positive-sense single-stranded RNA (3.2 kb, 2.9 kb, 2.1 kb). RNA 1 codes for a single protein that carries helicase motifs. RNA 2 codes for a single protein that carries the polymerase motif. RNA 3 is dicistronic: cell-to-cell movement protein (ORF 3) and capsid (ORF 4). The capsid is expressed through the intermediary of a subgenomic RNA (RNA 4, 0.9 kb) encapsidated with RNA 3. The infection of protoplasts can occur with RNA 1 and 2. Systemic infection requires RNA 3 as well.

Biology 6 described species. Easy mechanical transmission. It can also be transmitted by beetles and aphids. Symptoms of mosaic and chlorotic spots. In natural conditions, the host range of BMV is limited to grasses; infections have no economic impact.

FIGLEROWICZ, M., NAGY, P.D., TANG, N., KAO, C.C., BUJARSKI, J.J. 1998. Mutations in the N terminus of the brome mosaic virus polymerase affect genetic RNA-RNA recombination. *J. Virol.*, 72, 9192-9200.
KAO, C.C., AHLQUIST, P. 1992. Identification of the domains required for direct interaction of the helicase-like and polymerase-like RNA replication proteins of brome mosaic virus. *J. Virol.*, 66, 7293-7302.
O'REILLY, E., WANG, Z., FRENCH, R., CHENG KAO, C. 1998. Interactions between the structural domains of the RNA replication proteins of plant-infecting RNA viruses. *J. Virol.*, 72, 7160-7169.
QUADT, R., KAO, C.C., BROWNING, K.S., HERSHBERGER, R.P., AHLQUIST, P. 1993. Characterization of a host protein associated with brome mosaic virus RNA-dependant RNA polymerase. *Proc. Natl. Acad. Sci. USA*, 90, 1498-1502.

Bromoviridae

Genus *CUCUMOVIRUS* (from CUCUmber MOsaic virus) — 24

Type species *Cucumber mosaic virus* (CMV).

Virion Particles of 25 nm with isometric symmetry. The capsid is made up of 180 subunits.

Genome 3 positive-sense single-stranded RNAs (3.4 kb, 3 kb, 2.2 kb). RNA 2 codes for two overlapping genes, 2a and 2b. Gene 2b is expressed by the intermediary of a subgenomic RNA 4A. This gene, which has no equivalent in Bromoviruses, is important for long-distance movement, and it acts through interference with the defence system, which could explain the exceptionally wide host range of Cucumoviruses.

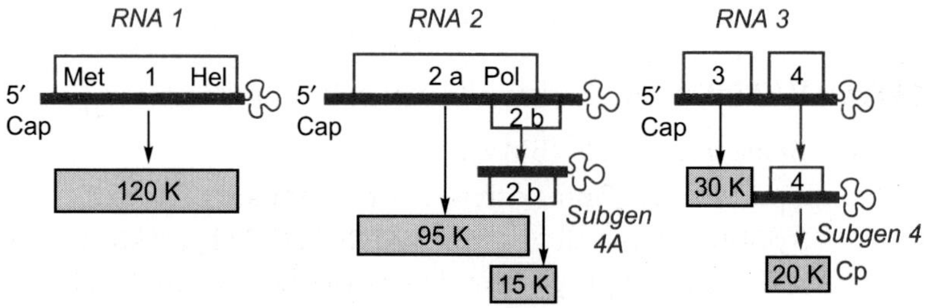

Biology 3 species are described. Mechanical transmission is possible. It can be transmitted by aphids (more than 75 species) in the non-persistent mode, as well as through the seed in some species. The host range of CMV is extremely wide, undoubtedly the most vast of plant viruses (more than 1000 species in 85 botanical families); the depressive effect and the symptoms are often marked (mosaic, leaf curl, filiformism, necrosis), which makes CMV a virus of great economic importance. CMV is often associated with a satellite RNA of around 350 nt, which depends on its helper virus for replication and encapsidation. The satellite sequence (with no coding capacity) determines the effect on the symptoms: accentuation, lethal necrosis, sometimes attenuation.

Ding, S.W., Anderson, B.J., Haase, H.R., Symons, R.H. 1994. New overlapping gene encoded by the cucumber mosaic virus genome. *Virology*, 198, 593-601.

Hu, C.C., Aboul-Ata, A.E., Naidu, R.A., Ghabrial, S.A. 1997. Evidence for the occurrence of two distinct subgroups of peanut stunt cucumovirus strains: molecular characterization of RNA 3. *J. Gen. Virol.*, 78, 929-939.

Hu, C.C., Ghabrial, S.A. 1998. Molecular evidence that strain BV-15 of peanut stunt cucumovirus is a reassortant between subgroup I and II strains. *Phytopathology*, 88, 92-97.

Mayers, C.N., Palukaitis, P., Carr, J.P. 2000. Subcellular distribution analysis of the cucumber mosaic virus 2b protein. *J. Gen. Virol.*, 81, 219-226.

Militao, V., Moreno, I., Rodrigez-Cerezo, E., Garcia-Arenal, F. 1998. Differential interactions among isolates of peanut stunt cucumovirus and its satellite RNA. *J. Gen. Virol.*, 79, 177-184.

Moreno, E., Bernal, J.J., de Blas, B.G., Rodriguez-Cerezo, E., Garcia-Arenal, F. 1997. The expression level of the 3a movement protein determines differences in severity of symptoms between two strains of tomato aspermy cucumovirus. *Mol. Plant-Microbe Inter.*, 10, 171-179.

NAIDU, R.A., HU, C.C., PENNINGTON, R.E., GHABRIAL, S.A. 1995. Differentiation of eastern and western strains of peanut stunt cucumovirus based on satellite RNA support and nucleotide sequence homology. *Phytopathology*, 85, 502-507.
PALUKAITIS, P., ROOSSINCK, M.J., DIETZGEN, R.G., FRANCKI, R.I.B. 1992. Cucumber mosaic virus. *Adv. Virus Res.*, 41, 281-348.
PERRY, K.L., ZHANG, L., PALUKAITIS, P. 1998. Amino acid changes in the coat protein of cucumber mosaic virus differentially affect the transmission by the aphids *Myzus persicae* and *Aphis gossypii*. *Virology*, 242, 204-210.
ROOSSINCK, M.J., KAPLAN, I., PALUKAITIS, P. 1997. Support of a cucumber mosaic virus satellite RNA maps to a single amino acid proximal to the helicase domain of the helper virus. *J. Virol.*, 71, 608-612.
ZHANG, L., HANADA, K., PALUKAITIS, P. 1994. Mapping local and systemic symptoms determinants of cucumber mosaic cucumovirus in tobacco. *J. Gen. Virol.*, 75, 3185-3191.

Bromoviridae

Genus ***ILARVIRUS*** (from Isometric LAbile Ringspot virus) —— 25

Type species *Tobacco streak virus* (TSV).

Virion Particles of 30 nm with isometric symmetry.

Genome 3 genomic RNAs (3.4 kb, 2.9 kb, 2.2 kb). One subgenomic RNA of 0.8 kb. The corresponding proteins present similarities with those of other genera of *Bromoviridae* and particularly with *Alfamovirus*.

Biology 16 species are described, infecting principally the woody plants. Mechanical transmission, through pollen, through vegetative propagation, and sometimes through the seed. Symptoms of mosaic, necrosis, and leaf curl. Responsible for often serious diseases in perennial crops, such as *Prunus necrotic ringspot virus* (PNRSV) on almond, cherry, plum, and rose.

XIN, H.W., JI, L.H., SCOTT, S.W., SYMONS, R.H., DING, S.W. 1998. Ilarviruses encode a cucumovirus-like 2b gene that is absent in other genera within the Bromoviridae. *J. Virol.*, 72, 6956-6959.

Bromoviridae

Genus ***OLEAVIRUS*** (from the Latin *olea*, olive tree) —— 26

Type species *Olive latent virus-2* (OLV-2).

Biology No known vector, mechanically transmitted. Does not cause symptoms on its natural host, the olive tree.

MARTELLI, G.P., GRIECO, F. 1997. *Oleavirus*, a new genus in the family *Bromoviridae*. *Arch. Virol.*, 142, 1933-1936.

■ Unassigned genus

Genus ***OURMIAVIRUS*** (from Ourmia, a place in Iran) —— 27

Type species *Ourmia melon virus* (OuMV).

Virion Bacilliform particles of variable size. 3 genomic RNAs. 3 described species.

Positive-sense single-stranded RNA viruses

helical rod-shaped particles

The genera *Benyvirus, Furovirus, Pomovirus, Pecluvirus* are derived from the earlier Furovirus group. The genome is monopartite (*Tobamovirus*), bipartite (*Furo-, Peclu-, Tobravirus*), tri- or multipartite (*Hordei-, Pomo-, Benyvirus*). The RNA carries a triple gene block (TGB) implicated in movement in *Beny-, Pomo-, Peclu-, Hordeivirus*.

■ Unassigned genera

Genus *BENYVIRUS* (from BEet NEcrotic Yellow vein virus) — 28

Type species *Beet necrotic yellow vein virus* (BNYVV).

Virion 4 rod-shaped particles of circa 85, 100, 265 and 390 nm in length and 20 nm in diameter.

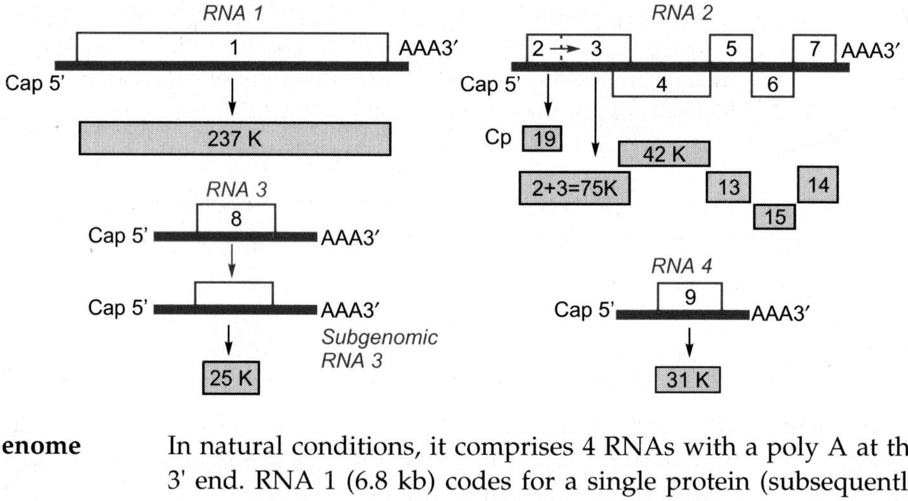

Genome In natural conditions, it comprises 4 RNAs with a poly A at the 3' end. RNA 1 (6.8 kb) codes for a single protein (subsequently cleaved) that carries the helicase and polymerase motifs. RNA 2 (4.6 kb) carries 6 ORFs. The first codes for the capsid; it is contiguous with ORF 3 read by readthrough, and the 75 kDa protein thus synthesized intervenes in virus-vector interactions. Next, RNA 2 has the triple gene block 4-5-6 implicated in cell-to-cell movement. The 42 kDa protein has helicase motifs. These proteins are probably synthesized by the intermediary of subgenomic RNAs. The products of RNA 3 (1.8 kb) intervene in movement and the expression of symptoms. The product of RNA 4 (1.5 kb) intervenes in the virus transmission by its vector, with the readthrough capsid protein. A fifth RNA is also involved in symptom expression.

Biology 2 described species. Mechanical transmission and transmission by zoospores of *Polymyxa betae*. Conservation in soil for several years in fungal resting spores. Symptoms of yellowing, stunting, and development of hairy root in the case of BNYVV. The virus is found in the roots. RNA 1 and 2 are sufficient to infect the leaves of chenopodium. RNA 1, 2, 3, 4, and sometimes 5 are associated with a serious disease of sugarbeet, rhizomania, present in the major production areas in the world.

BLEYKESTEN-GROSSHANS, C., GUILLEY, H., BOUZOUBA, A.S., RICHARDS, K.E., JONARD, G. 1997. Independent expression of the first two triple gene block proteins of beet necrotic yellow vein virus complements virus defective in the corresponding gene but expression of the third protein inhibits viral cell-to-cell movement. *Mol. Plant-Microbe Interaction*, 10, 240-246.

GILMER, D., BOUZOUBAA, S., HEHN, A., GUILLEY, H., RICHARDS, K., JONARD, G. 1992. Efficient cell-to-cell movement protein of beet necrotic yellow vein virus requires 3' proximal genes located on RNA 2. *Virology*, 189, 40-47.

JUPIN, I., RICHARDS, K., GUILLEY, H., PLEIJ, C.W.A. 1990. Mapping sequences required for productive replication of beet necrotic yellow vein virus RNA 3. *Virology*, 78, 273-280.

TAMADA, T., ABE, H., 1989. Evidence that beet necrotic yellow vein virus RNA-4 is essential for efficient transmission by the fungus *Polymyxa betae*. *J. Gen. Virol.*, 70, 3391-3396.

TAMADA, T., SCHMITT, T., SAITO, C., GUILLEY, M., GUILLEY, H., RICHARDS, K., JONARD, G. 1996. High resolution analysis of the readthrough domain of beet necrotic yellow vein virus readthrough protein: a KTER motif is important for efficient transmission of the virus by the fungus *Polymyxa betae*. *J. Gen. Virol.*, 77, 1359-1367.

Genus *FUROVIRUS* (from FUngus-borne ROd-shaped virus) —— 29

Type species *Soil-borne wheat mosaic virus* (SBWMV).

Virion 2 rod-shaped particles of 20 × 92-390 nm, with helical symmetry.

Genome 2 positive-sense single-strand RNA molecules. RNA 1 (6.5 kb) codes for three proteins: P1 has helicase motifs, P2 (read by readthrough) carries the polymerase motif; P3 intervenes in cell-to-cell movement. RNA 2 codes for the capsid and a readthrough protein probably implicated in transmission by the fungal vector, as well as for a small cysteine-rich protein.

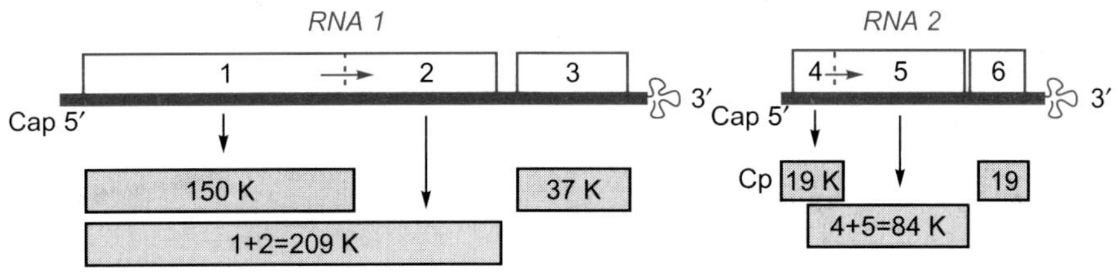

Biology 5 described species. Mechanical transmission possible. In natural conditions, transmission by mobile spores of the soil fungus *Polymyxa graminis* (*Plasmodiophoromycetes*). The resting spores remain infectious for several years. Symptoms of mosaic, rosetting, and reduction of tillering are expressed more intensively in spring.

Phylogenetic relationships

The replicase of SBWMV has sequence homologies with those of *Tobamovirus, Tobravirus*, and *Hordeivirus*, homologies that are greater than with that of *Benyvirus*.

DIAO, A., CHEN, J., GITTON, F., ANTHONIW, J.F., MULLINS, J., HALL, A.M., ADAMS, M.J. 1999. Sequences of European wheat mosaic virus and oat golden stripe virus and genome analysis of the genus Furovirus. *Virology*, 261, 331-339.

SHIRAKO, Y., WILSON, T.M.A. 1993. Complete nucleotide sequence and organization of the bipartite RNA genome of soil-borne wheat mosaic virus. *Virology*, 195, 16-32.

TAMADA, T., SCHMITT, C., SAITO, M., GUILLEY, H., RICHARDS, K., JONARD, G. 1999. High resolution analysis of the readthrough domain of beet necrotic yellow vein virus readthrough protein: a KTER motif is important for efficient transmission of the virus by *Polymyxa betae*. *J. Gen. Virol.*, 77, 1359-1367.

TORANCE, L., MAYO, M.A. 1997. Proposed re-classification of furoviruses. *Arch. Virol.*, 142, 435-439.

Genus *HORDEIVIRUS* (from the Latin *hordeum*, barley) —— 30

Type species *Barley stripe mosaic virus* (BSMV).

Virion 3 rigid, rod-shaped particles of 20 × 110-150 nm with helical symmetry. Virions are found in the nucleus and in the cytoplasm, where they are associated with vesicles of the outer wall of chloroplasts.

Genome 3 positive-sense single-stranded RNA molecules. RNA α and γ are sufficient for multiplication in protoplasts, and RNA γ is necessary for systemic infection.

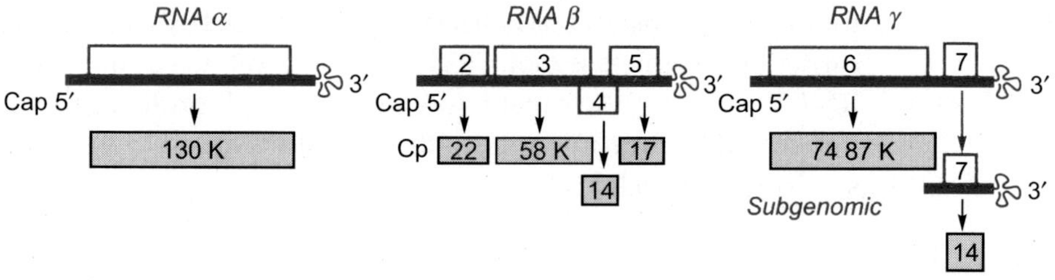

RNA α (3.8 kb) codes for a protein carrying a methyltransferase motif in the C-terminal, and a helicase motif for nucleotide fixation in the N-terminal. RNA β (3.3 kb) carries 4 ORFs; the first gene in the

5' end is the capsid. The subsequent ORFs are expressed by the intermediary of 2 subgenomic RNAs. After an intergenic region are found ORF 3 (helicase), then a triple gene block intervening in movement. RNA γ (2.8 or 3.2 kb) carries 2 ORFs: the polymerase with the GDD motif is translated on the genomic RNA; a cysteine-rich protein (which may have a regulatory effect on the expression of RNA β) is translated via a subgenomic RNA.

Biology 4 described species. No known vectors, easy mechanical transmission. BSMV is transmitted by the seed and the pollen. Symptoms of stripe mosaic and necrosis. BSMV has probably a worldwide distribution because of the high seed transmission rate. It infects grasses and barley crops in which the damage can be considerable.

DONALD, R.G.K., JACKSON, A.O. 1996. RNA-binding activities of barley stripe mosaic virus γb fusion proteins. *J. Gen. Virol.*, 77, 879-888.
PETTY, I.T.D., EDWARDS, M.C., JACKSON, A.O. 1990. Systemic movement of an RNA plant virus determined by a point substitution in a 5' leader sequence. *Proc. Natl. Acad. Sci. USA*, 87, 8894-8897.
PETTY, I.T.D., FRENCH, R., JONES, R.W., JACKSON, A.O. 1990. Identification of barley stripe mosaic virus proteins involved in viral replication and movement. *EMBO J.* 9, 3453-3457.

Genus *PECLUVIRUS* (from PEanut CLUmp virus) — 31

Type species *Peanut clump virus* (PCV).

Virion Particles of around 190 and 250 nm contain 2 RNA of 5.9 and 4.3 kb, comprising respectively 3 and 5 ORF. RNA 1 can replicate alone in protoplasts.

Biology 2 species described. Transmitted mechanically, by zoospores of *Polymyxa graminis*, and through seed. Conservation in soil in resting spores for several years. Symptoms of mosaic and stunting. PCV causes a serious peanut disease in Western Africa and India.

HERZOG, E., GUILLEY, H., MANOHAR, S.K., DOLLET, M., RICHARDS, K., FRITSCH, C., JONARD, G. 1994. Complete nucleotide sequence of peanut clump virus RNA1 and relationships with other fungus-transmitted rod-shaped viruses. *J. Gen. Virol.*, 75, 3147-3155.
HERZOG, E., GUILLEY, H., FRITSCH, C. 1995. Translation of the second gene of peanut clump virus RNA 2 occurs by leaky scanning *in vitro*. *Virology*, 208, 215-225.
HERZOG, E., HEMMER, O., MEYER, G., BOUZOUBAA, S., FRITSCH, C. 1998. Identification of genes involved in replication and movement of peanut clump virus. *Virology*, 248, 312-322.

Genus *POMOVIRUS* (from POtato MOp top virus) — 32

Type species *Potato mop-top virus* (PMTV).

Virion Particles of around 150 and 300 nm contain 3 RNA molecules of 6.4 kb, 3 kb, and 2.5 kb.

Biology 4 species are described. Mechanical transmission, by zoospores of *Spongospora subterranea* (PMTV), or *Polymyxa betae* (*Beet soil-borne virus*, BSBV) as well as by vegetative propagation. Conservation in soil in resting spores for several years. PMTV causes leaf discoloration, stunting, and necrosis in tubers.

COWAN, G.H., TORRANCE, L., REAVY, B. 1997. Detection of potato mop-top virus capsid readthrough protein in virus particles. *J. Gen. Virol.*, 78, 1779-1783.

GERMUNDSSON, A., SANDGREEN, M., BARKER, H., VALKONEN, J. 2002. Initial infection of roots and leaves reveals different resistance phenotypes associated with coat protein gene-mediated resistance to *Potato mop-top virus*. *J. Gen. Virol.*, 83, 1201-1209.

KASHIWASAKI, S., SCOTT, S., REAVY, B., HARRISON, B.D. 1995. Sequence analysis and gene organization of potato mop-top virus RNA-3. *Virology*, 206, 701-706.

REAVY, B., ARIF, M., COWAN, G.H., TORRANCE, L. 1998. Association of sequences in the coat protein readthrough domain of potato mop-top virus with transmission by *Spongospora subterranea*. *J. Gen. Virol.*, 79, 2343-2347.

Genus *TOBAMOVIRUS* (from TOBAcco MOsaic virus) — 33

Type species *Tobacco mosaic virus* (TMV).

Virion One rod-shaped particle of around 300 nm × 18 nm with helical symmetry. Capsidial subunit of 18 kDa.

Genome One positive-sense single-stranded RNA molecule (6.4 kb) carries 4 ORFs. The genomic RNA allows the synthesis of 2 replication proteins: one 126 kDa protein that carries methyltransferase and helicase motifs and one 183 kDa protein expressed by readthrough that carries the polymerase motif. Subgenomic RNA 1 is bicistronic. It allows the synthesis of the 30 kDa movement protein. Subgenomic RNA 2 allows the synthesis of the capsid. The virus multiplies very rapidly in infected cells (up to 60×10^6 virions per cell), forming paracrystalline inclusion bodies.

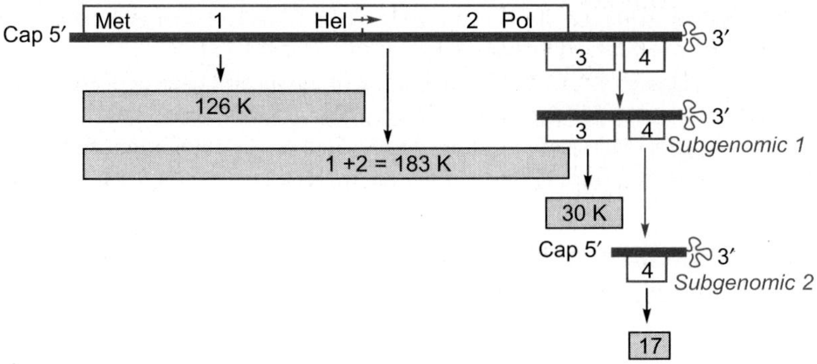

Biology 22 described species, 1 tentative species. Transmitted by mechanical means, by contact between plants, sometimes by seeds because of the external contamination of teguments. The virus remains infectious for a long time in plant debris and in the soil. It can be disseminated by irrigation water. Symptoms of sometimes necrotic mosaic. TMV causes a serious disease of tobacco, tomato, and pepper, particularly in greenhouse cultivation, where plants are frequently handled.

Goelet, P., Lomonossoff, G.P., Butler, P.J., Akam, M.E., Gait, M.J., Karn, J. 1982. Nucleotide sequence of tobacco mosaic virus RNA. *Proc. Natl. Acad. Sci. USA*, 79, 5818-5822.
McLean, G.G., Zupan, J., Zambryski, P.C. 1995. Tobacco mosaic virus movement protein associates with the cytoskeleton in tobacco cells. *Plant Cell*, 7, 2101-2114.
Padgett, H.S., Watanabe, Y., Beachy, R.N. 1997. Identification of the TMV replicase sequence that *activates* the N-gene mediated hypersensitive response. *Mol. Plant-Microbe Interaction*, 10, 709-715.
Rodriguez-Cerezo, E., Elena, S.F., Moya, A., Garcia-Arenal, F. 1991. High genetic stability in natural populations of the plant RNA virus tobacco mild green mosaic virus. *J. Mol. Evol.*, 32, 328-332.

Genus *TOBRAVIRUS* (from TOBacco RAttle virus) — 34

Type species *Tobacco rattle virus* (TRV).

Virion 2 rod-shaped particles of 22 × 180-215 and 46-115, with helical symmetry.

Genome 2 positive-sense single-stranded RNA molecules. RNA 1 (6.8 kb) can produce by itself systemic infection. It codes for 4 non-structural proteins: P1 and P2 are replication proteins translated from the genomic RNA; P2 derives from P1 by readthrough. P3 is implicated in movement and P4 has no known function. P3 and P4 are translated from subgenomic RNAs. RNA 2 (1.8 to 4.5 kb) has no messenger activity. It gives rise to a messenger subgenomic RNA

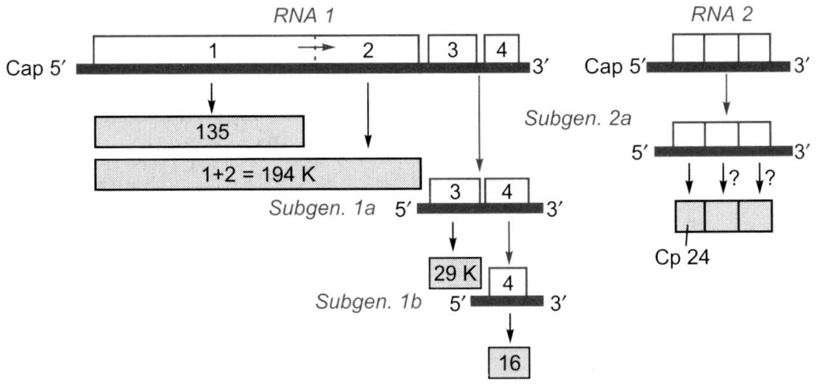

coding for the capsid and sometimes for other proteins of unknown function, but possibly implicated in transmission by nematodes.

Biology 3 described species. Mechanical transmission and by nematodes of genera *Trichodorus* and *Paratrichodorus*. Symptoms of mosaic, necrosis, and stunting. Wide host range: more than 400 mono- or dicotyledonous botanical species. TRV causes serious diseases in tobacco, potato, pepper, tomato, sugar beet, and various ornamental plants. Some isolates (NM) are not transmitted by nematodes; they contain only RNA 1 and do not produce particles.

Phylogenetic relationships

Closely related to *Hordeivirus* and *Tobamovirus*.

MacFarlane, S.A., Brown, D.J.F., Bot, J.F. 1995. The transmission by nematodes of tobraviruses is not determined exclusively by the virus coat protein. *Eur. J. Plant Pathol.*, 101, 535-539.

MacFarlane, S.A. 1999. Molecular biology of tobraviruses. *J. Gen. Birol.*, 80, 2799-2807.

MacFarlane, S.A., Vassilakos, N., Brown, D.J. 1999. Similarities in the genome organization of tobacco rattle virus and pea early browning virus isolates that are transmitted by the same vector nematode. *J. Gen. Virol.*, 80, 273-276.

Vellios, E., Duncan, G., Brown, D., MacFarlane, S. 2002. Immunogold localization of tobravirus 2b nematode transmission helper protein associated with virus particles. *Virology*, 300, 118-124.

Positive-sense single-stranded RNA viruses

helical filamentous particles

■ Family *Closteroviridae*

The family *Closteroviridae* comprises *Ampelovirus*, *Closterovirus* and *Crinivirus*. The virions are filamentous and extremely flexuous. The genome of *Closterovirus* is monopartite, that of *Crinivirus* is bipartite.

Closteroviridae

Genus *CLOSTEROVIRUS* (from the Greek *closter*, filament) — 35

Type species *Beet yellows virus* (BYV).

Genome These viruses have genomes between 15 and 20 kb, the longest RNA viruses, and extremely flexuous particles.

Virions Very long flexuous filaments of 12 × 1200-2000 nm with helical symmetry. The capsidial subunits of 22 kDa surround the RNA, except at one end, at which subunits of 24 kDa form a distinct zone.

Genome A positive-sense single-stranded RNA molecule of 15.5 kDa contains 9 ORFs. ORF 1 codes for a protein of around 300 kDa that carries cysteine protease, methyltransferase, protease, and helicase functions. ORF 2 codes for polymerase. It overlaps ORF 1 and is expressed by a frameshift of ribosomes (+1) that is unique in viruses. ORF 2 is (weakly) expressed in the form of a fusion protein.

The subsequent ORFs are expressed through a set of subgenomic RNA molecules (at least 7 coterminal RNA molecules at the 3' end). ORFs 3, 4, and 5 form a partly overlapping triple gene block, which play a role in transport. ORF 4 codes for a protein of 65 kDa that is strongly homologous with the family of stress proteins (chaperons) HSP70. ORF 5 codes for a protein that presents a domain homologous to stress proteins HSP90. ORFs 6 and 8 code for the capsid, one for a protein of 24 kDa (minor component forming a short region at one end of the virion), the other for a protein of 22 kDa. These are two slightly divergent redundant genes. ORF 9 codes for a protein of 21 kDa that increases the amplification of RNA.

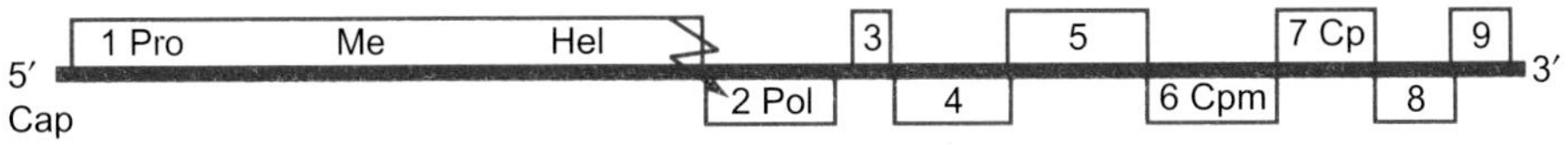

The difficulties in maintaining the long genomes of *Closterovirus* over the course of replication are overcome by the particularly high rate of recombination that allows the elimination of some copy errors during RNA replication, the presence of potentially protective proteins (chaperons), and the presence of insertions between Met and Hel in the replication module.

Biology The genus comprises 8 described species and 4 tentative species. Mechanical transmission is often difficult. Transmission through vegetative propagation and by vectors: aphids in the semi-persistent mode. The infection is generally limited to the phloem. Closteroviruses are responsible for serious diseases in perennial species (*Citrus tristeza virus*, CTV) and in annual crops (*Beet yellows virus*, BYV).

Phylogenetic relationships

Coronaviruses in animals have a similar organization complexity. It is believed that closteroviruses evolved from an ancestor resembling a tobamovirus and increased their genome by acquisition and duplication. Several species earlier included in *Closterovirus* are now classified in different genera.

AGRANOVSKI, A.A. 1996. Principles of molecular organization, expression and evolution of closterovirus: over the barriers. *Adv. Virus Res.*, 47, 119-157.

BOYKO, V.P., KARASEV, A.V., AGRANOVSKY, A.A., KOONIN, E.V., DOLJA, V.V. 1992. Coat protein gene duplication in a filamentous RNA virus of plants. *Proc. Natl. Acad. Sci. USA*, 89, 9156-9160.

DOLJA, V.V., KARASEV, A.V., KOONIN, E.V. 1994. Molecular biology and evolution of Closteroviruses: sophisticated build-up of large RNA genome. *Annu. Rev. Phytopathol.*, 32, 261-285.

EROKHINA, T.N., ZINOVKIN, A., VITUSHKINA, M.V., JELKMANN, W., AGRANOVSKY, A.A. 2000. Detection of beet yellows closterovirus methyltransferase-like and helicase-like proteins in vivo using monoclonal antibodies. *J. Gen. Virol.*, 81, 597-603.

GOWDA, S., AYLLON, M.A., SATYANARAYANA, T., BAR-JOSEPH, M., DAWSON, W.O. 2003. Transcription strategy in a closterovirus: a novel 5'-proximal controller element of *Citrus tristeza virus* produces 5' and 3'-terminal subgenomic RNAs. *J. Gen. Virol.*, 77, 340-352.

KARASEV, A.V., NIKOLAEVA, O.V., MUSHEGIAN, A.R., LEE, R.F., DAWSON, W.O. 1996. Organisation of the 3' terminal half of beet yellow stunt virus genome and implication for the evolution of closteroviruses. *Virology*, 221, 199-207.

Karasev, A. 2000. Genetic diversity and evolution of *Closteroviruses. Annu. Rev. Phytopathol.*, 293-324.

PEREMYSLOV, V.Y., HAGIWARA, Y., DOLJA, V.V. 1998. Genes required for replication of the 15.5 kilobase RNA genome of a plant closterovirus. *J. Virol.*, 72, 5870-5876.

Closteroviridae

Genus *CRINIVIRUS* (from the Latin *crinis*, hair) — 36

Type species *Lettuce infectious yellows virus* (LIYV).

Virion 2 filamentous particles of 12 x 800-850 nm.

Genome 2 positive-sense RNA of 8.1 and 7.2 kb. RNA 1 can replicate alone in protoplasts. It carries a replication module analogous to that of *Closterovirus* and codes for a protein of 30 kDa coded by ORF 3. RNA 2 resembles the 3' half of the RNA of *Citrus tristeza virus*, with differences in the order of genes.

Biology 8 described species, 2 tentative species. The virus is found in cells of the phloem (phloem parenchyma and companion cells). Transmission by whiteflies (*Bemisia tabaci*) according to the semi-persistent mode. The virus cannot be transmitted mechanically. Symptoms of yellowing of older leaves, sometimes necrosis. Criniviruses cause presently severe epidemics in vegetable crops, in relation with the considerable development of whitefly populations in subtropical and Mediterranean regions in field crops (LIYV) or in greenhouses (*Cucurbit yellow stunting disorder virus*, CYSDV).

AGUILAR, J.M., FRANCO, M., MARCO, C.F., BERDIALES, B., RODRIGUES-CEREZO, E., TRUNIGER, V., ARANDA, M.A. 2003. Further variability within the genus *Crinivirus*, as revealed by determination of the complete RNA genome sequence of *Cucurbit yellow stunting disorder virus. J. Gen. Virol.*, 84, 2555-2564.

KLASSEN, V.A., BOESHORE, M.L., KOONIN, E.V., TIAN, T., FALK, B.W. 1995. Genome structure and phylogenetic analysis of lettuce infectious yellows virus, a whitefly-transmitted, bipartite closterovirus. *Virology*, 208, 99-110.
MEDINA, V., TIAN, T., WIERZCHOS, J., FALK, B.W. 1998. Specific inclusion bodies are associated with replication of lettuce infectious yellows virus RNAs in *Nicotiana benthamiana* protoplasts. *J. Gen. Virol.*, 79, 2325-2329.

Closteroviridae

Genus *AMPELOVIRUS* (from the Greek *ampelos*, grapevine) — 37

Type species *Grapevine leafroll-associated virus 3* (GLRaV-3).

Virion Very long flexuous filaments (1400-2200 nm).

Biology 6 described species, 5 tentative species. Transmitted semi-persistently by mealybugs, but not by mechanical inoculation. Several species infect grapevine.

■ Family *Potyviridae*

The family *Potyviridae* comprises the genera *Bymovirus, Ipomovirus, Macluravirus, Potyvirus, Rymovirus, Tritimovirus.*

Virion Filamentous particles of 11 to 15 nm diameter with helical symmetry. The length is 650 to 900 nm for monopartite viruses (*Potyvirus, Rymovirus*) and between 250 and 600 nm for bipartite viruses (*Bymovirus*).

Genome The RNA carries a VPg protein of around 24 kDa linked covalently to the 5' end and a polyA of variable length at the 3' end. Each RNA comprises a single ORF, translated into a polyprotein that undergoes proteolysis by viral proteases (Fig. 2.7). The viral protein of around 70 kDa, which has ATPase and helicase activity, forms in the cytoplasm cylindrical inclusion bodies that are highly characteristic of the family (pinwheels, Fig. 4.11).

Phylogenetic relationships

In families *Potyviridae, Comoviridae, Picornaviridae*, the replication module genes have sequence homologies and are placed in the same order.

Potyviridae

Genus *BYMOVIRUS* (from Barley Yellow MOsaic virus) — 38

Type species *Barley yellow mosaic virus* (BaYMV).

Virion 2 flexuous particles of 13 nm × 250-300 nm and 500-600 nm with helical symmetry.

Genome 2 positive-sense single-stranded RNA. RNA 1 (7.6 kb) is translated in the form of a polyprotein that carries domains presenting sequence homologies with known proteins in *Potyvirus*: P3, 6K1, cytoplasmic inclusion (CI helicase), 6K2, nuclear inclusion *a* (VPg-Pro), nuclear inclusion *b* (polymerase), capsid. RNA 2 (3.6 kb) is translated in the form of a polyprotein that has domains having sequence homologies with the helper component of *Potyvirus* (P1) and the capsid protein of rod-shaped viruses (P2).

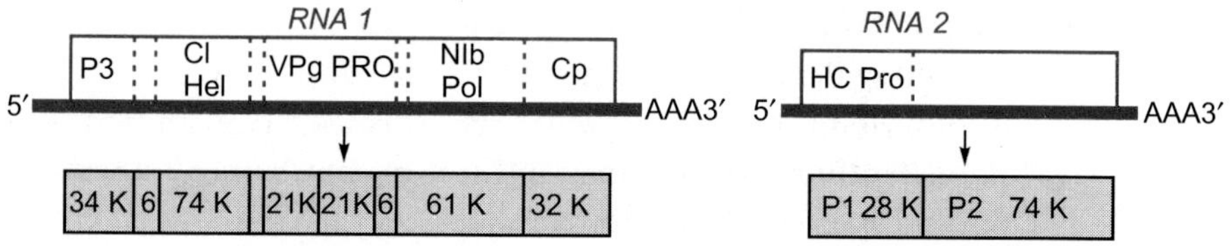

Biology 6 described species. The cytoplasm of infected cells contains inclusion bodies (pinwheels) characteristic of *Potyviridae*. The host range is limited to the family *Gramineae*. The virus is transmitted by a soil fungus, *Polymyxa graminis*.

DESSENS, J.T., NGUYEN, N., MEYER, M. 1995. Primary structure and sequence analysis of RNA 2 of a mechanically transmitted barley mild mosaic virus isolate: an evolutionary relationship between bymo- and furoviruses. *Archiv. Virol.*, 140, 325-333.

MEYER, M., DESSENS, J.T. 1996. The complete nucleotide sequence of barley mild mosaic virus RNA 1 and its relationship with other members of the Potyviridae. *Virology*, 219, 268-273.

SCHENK, P.M., STEINBIB, H.H., MULLER, B., SCHMITZ, K. 1993. Association of two barley yellow mosaic virus (RNA 2) encoded proteins with cytoplasmic inclusion bodies revealed by immunogold localisation. *Protoplasma*, 173, 113-122.

Potyviridae

Genus *IPOMOVIRUS* (from *Ipomea*, sweet potato) — 39

Type species *Sweet potato mild mottle virus* (SPMMV).

Biology 3 described species, 1 tentative species. Transmission by the whitefly *Bemisia tabaci* in the semi-persistent mode and by vegetative propagation. SPMMV is one of numerous viruses infecting sweet potato in Africa. Its host range is wide.

Potyviridae

Genus *MACLURAVIRUS* (from MACLURA mosaic virus) — 40

Type species *Maclura mosaic virus* (MacMV).

Biology 3 described species, 1 tentative species. Transmission by aphids in the non-persistent mode.

Potyviridae

Genus *POTYVIRUS* (from POTato virus Y) —— 41

Type species *Potato virus Y* (PVY).

Virion Flexuous filaments of 12 nm x 680-900 nm.

Genome One RNA molecule (9.7 kb) comprising a single ORF translated into a polyprotein of 3063 amino acids, which undergoes a complex proteolysis (Fig. 2.7) achieved by three viral proteins: P1-pro, HC-pro, and NIa. The final products are P1-pro, HC-pro, P3, CI-Hel, 6 kDa, NIa, NIb, and CP.

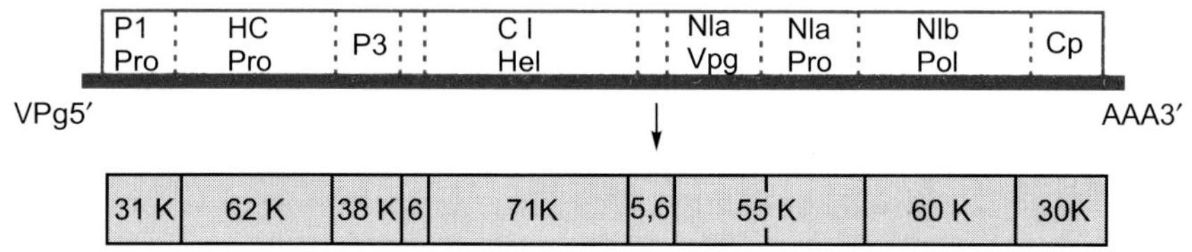

P1-pro carries a protease function that effects cleavage in *cis* of the polyprotein between P1 and HC-pro; it carries RNA fixation motifs and exerts in *trans* a function in the amplification of the genome. HC-pro is the helper component for transmission by aphids. It carries a protease function (with cysteine) that cleaves between HC-pro and P3. P3 has no known function. It carries two highly hydrophobic regions characteristic of transmembrane molecules. Among the proteins translated from the RNA of TVMV (*Tobacco vein mottling virus*), it is possible to detect P3 and a small product of 6 kDa that is a late product of hydrolysis of a precursor. CI-Hel (cytoplasmic inclusion helicase) forms in infected cells, cytoplasmic inclusion bodies (Fig. 4.11) characteristic of all *Potyviridae*. It carries 7 conserved motifs characteristic of a helicase function. 6 kDa is a highly hydrophobic protein that could connect the replication complex to the membranes. NIa (nuclear inclusion *a*) is a protein found first in the cytoplasm, where it achieves proteolysis of the entire N-terminal region of the polyprotein. It effects an auto-catalytic cleavage that liberates the VPg protein. It accumulates in the nucleus. NIb (nuclear inclusion *b*) carries motifs characteristic of the polymerase function and notably GDD. Like NIb, it carries a sequence of nuclear addressing and forms inclusion bodies in the nucleus (Fig. 9.1), sometimes in association with NIa. CP is a capsid that, apart from its role as structural protein, intervenes notably in transmission by aphids (presence of a DAG triplet in the N-terminal region) and in movement.

Biology 111 described species, 86 tentative species. Most have a host range limited to a few species, while others, such as PVY, can infect a large number of species in 30 different families. These viruses are transmitted by aphids in the non-persistent mode and by mechanical inoculation. Transmission by aphids requires a helper component (HC-Pro). Some species are transmitted through the seed (*Pea seed-borne mosaic virus*, PSbMV; *lettuce mosaic virus*, LMV; *Bean common mosaic virus*, BCMV). The symptoms are highly varied because of the existence of numerous strains for many species.

Potyviruses cause serious diseases in numerous crops and in highly diverse environments: temperate or tropical regions, traditional or intensive agriculture, commercial or ornamental horticulture. All in all, they are responsible for most of the losses associated with viral disease in plants. In fruit trees, PPV is responsible for sharka in Southeastern Europe. The virus is spreading in Western Europe and in the United States. PVY is present in all the potato production zones. Some recently emerging strains cause necrosis on the tubers, others infect tomato or pepper. *Zucchini yellow mosaic virus* (ZYMV) is undoubtedly one of the major emerging viruses in cucurbits. In about 25 years, it has invaded the principal production zones in the world and caused considerable damage.

ARBATOVA, J., LEHTO, K., PEHU, E., PEHU, T. 1998. Localization of the P1 protein of potato Y potyvirus in association with cytoplasmic inclusion bodies and in the cytoplasm of infected cells. *J. Gen. Virol.*, 79, 2319-2323.

ATREYA, C.D., ATREYA, P.L., THORNBURY, D.W., PIRONE, T.P. 1992. Site-directed mutations in the Potyvirus HC-Pro gene affect helper component activity, virus accumulation and symptom expression in infected plants. *Virology*, 191, 106-111.

CARRINGTON, J.C., FREED, D.D., CHAN-SEOK, O.H. 1990. Expression of potyviral polyproteins in transgenic plants reveal three proteolytic activities required for complete processing. *The EMBO J.*, 9, 1347-1353.

DOLJA, V.V., MCBRIDE, H.J., CARRINGTON, J.C. 1992. Tagging of plant potyvirus replication and movement by insertion of β-glucuronidase into the viral polyprotein. *Proc. Natl. Acad. Sci. USA*, 89, 10208-10212.

DOMIER, L.L., SHAW, J.G., RHOADS, R.E. 1987. Potyviral proteins share amino acid sequence homology with picorna-, como-, and caulimovirus proteins. *Virology*, 158, 20-27.

DOUGHERTY, W.G., PARKS, T.D. 1991. Post-translational processing of the tobacco etch virus 49 k small nuclear inclusion polyprotein. Identification of an internal cleavage site and delimitation of VPg and proteinase domains. *Virology*, 183, 449-456.

RIECHMAN, J.L., LAIN, S., GARCIA, J.A. 1992. Highlights and prospects of potyviral molecular biology. *J. Gen. Virol.*, 73, 1-16.

REVERS, F., LEGALL, O., CANDRESSE, T., MAULE, A.J. 1999. New advances in understanding the molecular biology of Plant/Potyvirus interactions. *Mol. Plant Microbe Interact.* 12, 367-376.

VERCHOT, J., CARRINGTON, J.C. 1995. Evidence that the potyvirus P1 proteinase functions in *trans* as an accessory factor for genome amplification. *J. Virol.*, 69, 3668-3674.

VERCHOT, J., HERNDON, K.L., CARRINGTON, J.C. 1992. Mutational analysis of the tobacco etch potyviral 35 kDa proteinase: identification of essential residues and requirements for autoproteolysis. *Virology*¸190, 298-306.

Potyviridae

Genus ***RYMOVIRUS*** (from RYegrass MOsaic virus) — 42

Type species *Ryegrass mosaic virus* (RGMV).

Virion Virion and genome are very similar to those of Potyvirus.

Biology 3 described species, one tentative.
The hosts belong to the family *Gramineae* (*Poaceae*); these viruses are transmitted by eriophyd mites and by mechanical inoculation.

Potyviridae

Genus ***TRITIMOVIRUS*** (from the Latin *triticum*, wheat) — 43

Type species *Wheat streak mosaic virus* (WSMV).

Biology 3 described species. Mechanical transmission, transmission by mites according to the semi-persistent mode and sometimes through the seed. The host range is restricted to *Gramineae*. WSMV causes a serious stripe mosaic, stunting, and yield losses in North America.

RABENSTEIN, F., SEIFERS, D., SCHUBERT, J., FRENCH, R., STENGER, D.C. 2002. Phylogenetic relationships, strain diversity and biogeography of tritimoviruses. *J. Gen. Virol.*, 83, 895-906.

■ Family *Flexiviridae*

The recently established family *Flexiviridae* comprises 8 genera: *Allexivirus*, *Capillovirus*, *Carlavirus*, *Foveavirus*, *Mandarivirus*, *Potexvirus*, *Trichovirus* and *Vitivirus*. Virions are flexuous filaments usually circa 13 nm in diameter and 470-1000 nm in length.

Flexiviridae

Genus ***ALLEXIVIRUS*** (from the Latin *allium*, shallot and X) — 44

Type species *Shallot virus X* (ShVX).

Virion Flexuous particle, one RNA molecule about 9 kb.

Biology 8 described species, 3 tentative. Transmission by eriophyd mites and by mechanical inoculation.
Host range restricted; naturally infect *Allium sp.*

SUMI, S.I., TSUNEYOSHI, T., FURUTANI, H. 1993. Novel rod-shaped viruses isolated from garlic, *Allium sativum*, possessing an unique genome organization. *J. Gen. Virol.*, 74, 1879-1885.

Flexiviridae

Genus *CAPILLOVIRUS* (from the Latin *capillus*, hair) — 45

Type species *Apple stem grooving virus* (ASGV).

Virion Flexuous particle of 12 × 640 nm, with helical symmetry.

Genome One positive-sense single-stranded RNA molecule (6.5 kb) carries 5 ORFs. ORF 1 codes for a protein of 240 kDa with a polymerase motif. ORFs 2, 3 and 4 constitute a triple gene block. In its C-terminal part, ORF 5 codes for the capsid protein.

Biology 3 described species, one tentative. No known vector. Mechanical transmission through farm tools and sometimes through the seed. During grafting, ASGV induces incompatibility between stock and scion.

Phylogenetic relationships

Very closely related to *Trichovirus*.

OHIRA, K., NAMBA, S., ROZANOV, M., KUSUMI, T., TSUCHIZAKI, T. 1995. Complete sequence of an infectious full-length clone of citrus tatter leaf capillovirus: comparative sequence analysis of capillovirus genomes. *J. Gen. Virol.*, 76, 2305-2309.

YOSHIKAWA, N., SASAKI, E., KOTO, M., TAKAHASHI, T. 1992. The nucleotide sequence of apple stem grooving capillovirus genome. *Virology*, 191, 98-105.

Flexiviridae

Genus *CARLAVIRUS* (CARnation LAtent virus) — 46

Type species *Carnation latent virus* (CLV).

Virion Particles are flexuous filaments of 12 x 610-700 nm, with helical symmetry. Capsidial subunit: 34 kDa.

Genome One positive-sense single-stranded RNA molecule (7.6 kb) carries 6 ORFs. ORF 1 represents around 70% of the genome. The protein contains several domains: at the N-terminal a methyltransferase domain, a domain with helicase motifs, and at the C-terminal the polymerase motifs. This protein undergoes autoproteolysis. The overlapping ORFs 2, 3, and 4 form a triple gene block (TGB). P2 contains a helicase motif, it is necessary for cell-to-cell movement. P3 and P4 contain highly hydrophobic, probably transmembrane regions. These proteins are expressed by a subgenomic RNA. The partly overlapping ORFs 5 and 6 code for the capsidial subunit and for a protein rich in cysteine residues capable of binding to the RNA. P5 and P6 are expressed by the intermediary of a subgenomic RNA. P5 could result from an internal initiation or a frameshift.

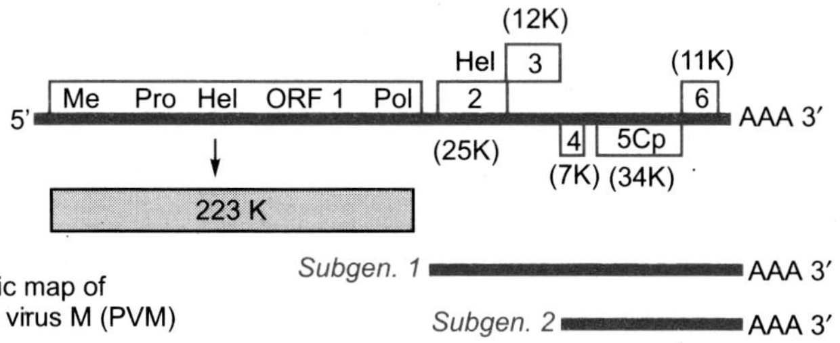

Genetic map of potato virus M (PVM)

Biology 35 described species, 29 tentative species. Transmission by mechanical inoculation and by aphids in a non-persistent mode. Some species are transmitted by *Bemisia tabaci* according to the semi-persistent mode (*Cowpea mild mottle virus*, CPMMV) or through the seed. These viruses often cause latent infections (CLV) as well as mosaic (*Poplar mosaic virus*, PopMV), necrosis (*Pea streak virus*, PeSV) and occasionally significant yield losses (*Potato virus M*, PVM).

Phylogenetic relationships

The polymerase has homologies with that of *Potexvirus*, *Closterovirus*, and *Tymovirus*. The triple gene block has homologies with the corresponding genes of *Beny-*, *Pomo-*, *Pecluvirus*, and *Hordeivirus* (P2) and of *Potexvirus* (P3 and P4).

Foster, G.D. 1992. The structure and expression of the genome of Carlaviruses. *Res. Virol.*, 1992, 103-112.

Zavriev, S.K., Kanyuka, K.V., Levay, K.E. 1991. The genome organization of potato virus M RNA. *J. Gen. Virol.*, 72, 9-14.

Flexiviridae

Genus *FOVEAVIRUS* (from the Latin *fovea*, hole) —— 47

Type species *Apple stem pitting virus* (ASPV).

Virion Extremely flexuous particle circa 900 nm; one RNA molecule (9.3 kb).

Biology 3 described species. Transmission through vegetative propagation. No known vector and mechanical inoculation difficult. Cause diseases of trees. ASPV is present on apple and pear trees throughout the world.

Jelkmann, W. 1994. Nucleotide sequence of apple stem pitting virus and the coat protein gene of similar virus from pear associated with vein yellows disease and their relationships with potex and carla viruses. *J. Gen. Virol.*, 75, 1535-1542.

Flexiviridae

Genus *POTEXVIRUS* (from POTato virus X) —— 48

Type species *Potato virus X* (PVX).

Virion Flexuous particle of 13×470-580 nm, with helical symmetry.

Genome One RNA molecule (6.4 kb) containing 5 ORFs. The genomic RNA allows the expression of ORF 1. The protein has methyltransferase, helicase, and polymerase motifs. ORFs 2 to 5 are expressed through several subgenomic RNAs. ORF 2 codes for a protein that presents a second helicase motif. The proteins 3 and 4 are implicated in cell-to-cell movement. The triple gene block (TGB) 2 + 3 + 4 is found also in *Carlavirus, Benyvirus, Pomovirus, Pecluvirus, Hordeivirus, Allexivirus, Foveavirus*. ORF 5 codes for the capsid.

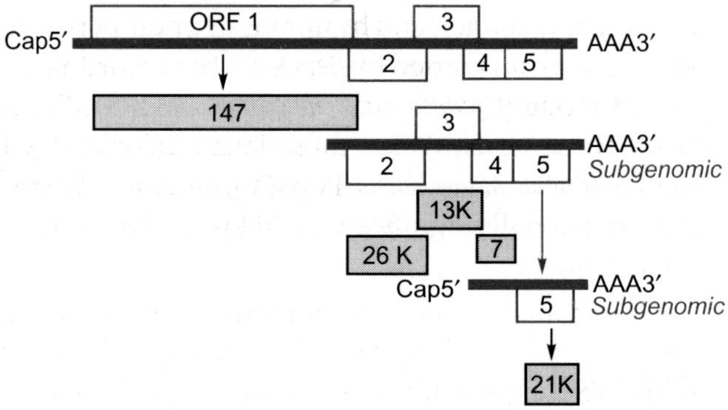

Biology 28 described species, 18 tentative species. Transmission by mechanical means, by leaf-to-leaf contact, sometimes through the seed (*White clover mosaic virus*, WClMV). Symptoms sometimes have no impact on yield (most strains of PVX) and sometimes have a severe impact (*Cactus virus X*, CVX).

BECK, D.L., GUILFORD, P.J., VOOT, D.M., ANDERSEN, M.T., FORSTER, R.L. 1991. Triple gene block proteins of white clover potexvirus are required for transport. *Virology*, 183, 695-702.

CHAPMAN, S.N., KAVANAGH, T.A., BAULCOMBE, D.C. 1992. Potato virus X as a vector for gene expression in plants. *Plant J.*, 2, 549-557.

MOROZOV, S.Y., FEDORKIN, O.N., JUTINER, G., SCHIEMANN, J., BAULCOMBE, D.C., ATABEKOV, J.G. 1997. Complementation of a potato X mutant mediated by bombardment of plant tissues with cloned viral movement protein genes. *J. Gen. Virol.*, 78, 2077-2083.

MOROZOV, S.Y., SOLOVIEV, A.G., 2003. Triple gene block: modular design of a multifunctional machine for plant virus movement. *J. Gen. Virol.*, 84, 1351-1366.

SKRYABIN, K.G., MOROZOV, S., YU, S., KRAEV, A.S., ROZANOV, M.N., CHERNOV, B.K., LUKASHEVA, L.I., ATABEKOV, J.G. 1988. Conserved and variable elements in RNA genomes of potexviruses. *FEBS Lett.*, 240, 33-40.

VERCHOT, J.M., ANGELL, S.M., BAUCOMBE, D.C. 1998. In vivo translation of the triple block of potato virus X requires two subgenomic mRNAs. *J. Virol.*, 72, 8316-8320.

Flexiviridae

Genus *TRICHOVIRUS* (from the Greek *thrix*, hair) —— 49

Type species *Apple chlorotic leaf spot virus* (ACLSV).

Virion Flexuous filaments of 12 nm diameter, 600-800 nm length, with helical symmetry.

Genome One positive-sense single-stranded RNA molecule (7.5 kb) that presents 3 ORFs that partly overlap. The first ORF is expressed by translation of genomic RNA. It codes for a protein of 216 kDa carrying methyltransferase, helicase, and polymerase motifs. The second and third ORFs are probably expressed through subgenomic RNAs, giving a protein of 50 kDa that could be the movement protein, and the capsid protein.

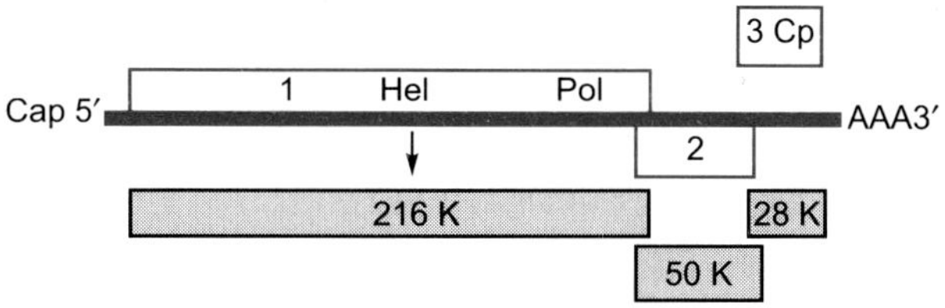

Biology 4 described species. The host range is narrow and symptoms are hardly visible. Dissemination by grafting and vegetative propagation. Transmission by mechanical inoculation and for some species, by mites. ACLSV infects numerous fruit species (*Malus, Prunus, Pirus*) throughout the world.

Phylogenetic relationships

ORF 1 is homologous with ORF 5' of the Alphavirus supergroup (more particularly genera *Tymovirus, Potexvirus, Carlavirus, Capillovirus*).

ABOU-GHANEM, N., SALDARELLI, P., MINAFRA, A., BUZKAN, N., CASTELLANO, M.A., MARTELLI, G.P. 1997. Properties of grapevine virus D, a novel putative trichovirus. *J. Plant Pathol.*, 78, 15-25.

GERMAN, S., CANDRESSE, T., LANNEAU, M., HUET, J.C., PERNOLLET, J.C., DUNEZ, J. 1990. Nucleotide sequence and genomic organization of apple chlorotic leaf spot closterovirus. *Virology*, 179, 104-112.

JAMES, D., JELKMANN, W., UPTON, C. 2000. Nucleotide sequence and genome organisation of cherry mottle leaf and its relationship to members of the *Trichovirus* genus. *Arch. Virol.*, 145, 995-1007.

MARTELLI, G.P., CANDRESSE, T., NAMBA, S. 1994. Trichovirus, a new genus of plant viruses. *Arch Virol.*, 134, 451-455.

YOSHIKAWA, N., IIDA, H., GOTO, S., MAGOME, H., TAKAHASHI, T., TERAI, Y. 1997. Grapevine berry inner necrosis, a new trichovirus: comparative studies with several known trichoviruses. *Arch. Virol.*, 131, 1351-1363.

Flexiviridae

Genus *VITIVIRUS* (from the Latin *vitis*, grapevine) —— 50

Type species *Grapevine virus A* (GVA).

Virion Flexuous particle. Positive-sense single-stranded RNA (7.6 kb). Transmitted by mechanical inoculation, grafting and vegetative propagation. GVA can be transmitted by mealybugs in a semi-persistent manner, *Heracleum latent virus* (HLV) is transmitted by aphids assisted by a helper virus. These viruses can be transmitted mechanically.

La Notte, P., Burzan, N., Choueiri, E., Minafra, A., Martelli, G.P. 1997. Acquisition and transmission of grapevine virus A by the mealybug *Pseudococcus longispinus*. *J. Plant Pathol.*, 78, 79-85.

Flexiviridae

Genus *MANDARIVIRUS* (from mandarin, host of the type species) —— **50a**

Type species *Indian citrus ringspot virus* (ICRSV)

■ Unassigned genus

Genus *UMBRAVIRUS* (from the Latin *umbra*, shadow) —— **51**

Type species *Carrot mottle virus* (CMoV).

Virion No viral particles, but rather infectious membranous structures of circa 50 nm, present in vacuoles and cytoplasm.

Genome One single-stranded RNA molecule of 4.2 kb. There is also a large quantity of double-stranded RNA in infected cells. The polymerase gene sequence has 63% identity with that of RNA 2 of *Enamovirus*. The RNA of *Groundnut rosette virus* (GRV) has 4 ORFs: on the 5' side, ORFs 1 and 2 code for replication proteins; on the 3' side, ORFs 3 and 4 overlap, and ORF 4 intervenes in cell-to-cell movement.

Biology 7 described species, 3 tentative species.

Viruses confined to one or a few host plants. Mechanical transmission is possible, transmission by aphids occurs through a specific helper polerovirus, which provides the capsid protein. The virus multiplies in all the cells of the plant. There are highly variable symptoms: mosaic, yellowing, stunting, rosetting. GRV is responsible for a serious disease of peanut in Africa. *Lettuce speckles mottle virus* (LSMV) in synergy with *Beet western yellows virus* (BWYV) causes a serious lettuce disease in California.

Umbraviruses may assist mechanical transmission of poleroviruses.

Ryabov, E.V., Oparka, K.J., Santa-Cruz, S., Robinson, D.J., Taliansky, M.E. 1998. Intracellular location of two groundnut rosette proteins delivered by PVX and TMV vectors. *Virology*, 242, 303-313.

Taliansky, M.E., Robinson, D.J., Murant, A.F. 1996. Complete nucleotide sequence and organization of the RNA genome of groundnut rosette umbravirus. *J. Gen. Virol.*, 77, 2335-2345.

Negative-sense single-stranded RNA viruses

This group of viruses contains major pathogens: e.g., influenza, rabies, haemorrhagic fevers (Ebola, Marburg). Their genome with negative or ambisense RNA is always protected in a ribonucleoprotein form.

Family *Bunyaviridae* (from Bunya, a place in Uganda)

The family *Bunyaviridae* comprises the genus *Tospovirus* infecting plants and insects as well as the genera *Hantavirus, Bunyavirus, Nairovirus, Phlebovirus*, which cause haemorrhagic fevers in mammals and are transmitted by arthropods (except *Hantavirus*, transmitted by rodents), in which the virus multiplies without apparently being pathogenic.

Bunyaviridae

Genus *TOSPOVIRUS* (from TOmato SPOtted wilt virus) — 52

Type species *Tomato spotted wilt virus* (TSWV).

Virion Globular pleomorphic particle of 80-100 nm. The virus contains 3 ribonucleoproteic particles surrounded by a lipid membrane (see Chapter 3, Fig. 3.17). The lipid membrane coming from the host (Golgi apparatus) carries viral glycoproteins G1 and G2 exposed outwards. The ribonucleoproteic particles are formed of viral RNA encapsidated by numerous copies of protein N (nucleocapsid). The viral RNAs present complementary sequences at the ends: the ribonucleoproteins form pseudo-circular structures. They are associated with an RNA-dependent RNA polymerase L (15-20 molecules per virion).

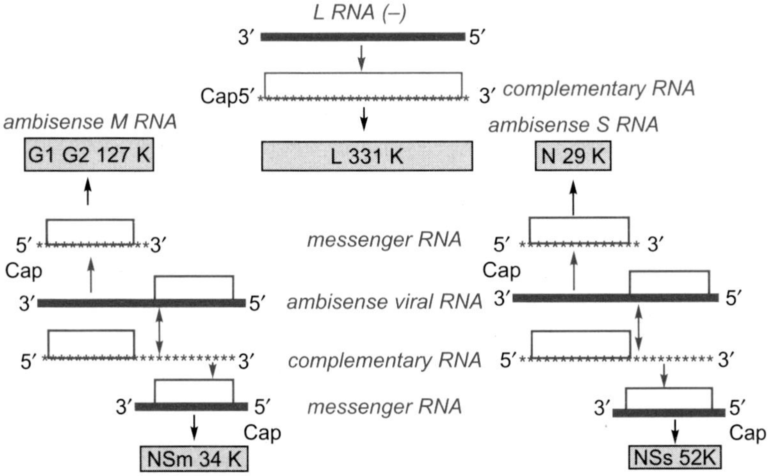

Genome 3 negative-sense or ambisense single-stranded RNA molecules (called L for large, M for medium, S for small), the terminal sequences of which are complementary. Each RNA can form a pseudo-circular structure by non-covalent bonds. These terminal sequences are conserved for the three RNA molecules. Negative-sense L RNA (8.9 kb) codes for protein L, which is the viral transcriptase associated with nucleocapsids in the virion. M RNA (4.8 kb) is ambisense. It codes for a non-structural protein NSm (which probably intervenes in movement) in the genomic sense and for the precursor of G1 and G2 in the complementary sense, expressed by a subgenomic RNA. S RNA (2.9 kb) is ambisense. It codes for a non-structural protein NSs in the genomic sense and for protein N in the complementary sense, expressed by a subgenomic RNA. The ribonucleoproteins serve as a model for transcription into messenger RNA as well as for replication. The subgenomic RNAs are delimited by a double-stranded hairpin structure (A-U) located in the intergenic region of S and M RNA. The 5' ends have a cap and are constituted by a sequence of 12 to 20 nucleotides coming from cellular messenger RNA, used as a primer during synthesis of viral messenger RNA by transcription of genomic RNA.

Biology 8 described species, 6 tentative species. Dense paracrystalline viral masses are observed in the cavities of the endoplasmic reticulum. The glycosylation of G1 and G2 and acquisition of the lipid membrane occur in the Golgi apparatus. Mechanical transmission is possible. Transmission in natural conditions in the persistent mode by numerous species of thrips (*Thysanoptera, Thripidae*), notably *Frankliniella occidentalis* and *Thrips tabaci*. The virus multiplies in the insect, which acquires the virus at the larval state, and remains infectious throughout its cycle. It is estimated that TSWV can infect more than 925 species in 70 families, in temperate and tropical regions. The virus seems to be able to infect most species colonized by thrips vectors. It is an emerging virus that has rapidly expanded over the past 15 years and caused damage throughout the world. It has a considerable economic impact on vegetable crops (tomato, pepper, aubergine, lettuce), ornamental crops (dahlia, chrysanthemum), and field crops (peanut, tobacco). The symptoms are often severe, mostly mosaics with partial or total necrosis of the plant.

De Haan, P., Kormelink, R., Resende, R., Van Poelwij, F., Peters, D., Goldbach, R. 1991. Tomato spotted wilt virus L RNA encodes a putative RNA polymerase. *J. Gen. Virol.*, 72, 2207-2216.

Kikkert, M., Van Lent, J., Storms, M., Bodegom, P., Kormelink, R., Goldbach, R. 1999. Tomato spotted wilt virus particle morphogenesis in plant cells. *J. Virol.*, 73, 2288-2297.

KORMELINK, R., DE HAAN, P., PETERS, D., GOLDBACH, R. 1992. The nucleotide sequence of the M segment of tomato spotted wilt virus: a plant-infecting bunyavirus with two ambisense RNA segments. *J. Gen. Virol.*, 73, 2795-2804.

MOURY, B., PALLOIX, A., GEBRE-SELASSIE, K., MARCHOUX, G. 1998. L'émergence des tospovirus. *Virologie*, 2, 357-367.

RICHMOND, K.E., CHENAULT, K., SHERWOOD, J.L., GERMAN, T.L. 1998. Characterization of the nucleic acid binding properties of tomato spotted wilt virus nucleocapsid protein. *Virology*, 248, 6-11.

WIJKAMP, I., VAN LENT, J., KORMELINK, R., GOLDACH, R.W., PETERS, D. 1993. Multiplication of tomato spotted wilt virus in its insect vector *Frankliniella occidentalis*. *J. Gen. Virol.*, 74, 341-349.

Order Mononegavirales

Family *Rhabdoviridae*

This family comprises three genera of viruses infecting vertebrates (*Vesicular stomatitis virus, Rabies virus, Bovine fever virus*) and two genera of viruses infecting plants: *Cytorhabdovirus* and *Nucleorhabdovirus*.

Virion Bacilliform particle of 100-430 nm length and 40-94 nm diameter, composed of two structures (Chapter 1, Fig. 1.9). One central part, the nucleocapsid, is constituted of an RNA associated with a protein N, both twisted in a spiral, and proteins NS (a phosphoprotein) and L (a viral transcriptase). The virion contains only 1% of RNA. The second structure is a lipidic envelope in two strata coming from vesicles associated with the endoplasmic reticulum during the viral infection. This envelope has glycoprotein projections (protein G) on the outside and a hexagonal structure (protein M) on the inside. It can be destroyed by non-ionic detergents without affecting the infectivity. The virion contains 15-25% lipids, the composition of which is similar to that of cell membranes.

Genome One negative-sense RNA molecule (around 13 kb) with reverse complementary extremities. It is transcribed into 6 messenger RNAs by a viral transcriptase associated with the virion. These messengers are generally monocistronic, and their abundance decreases from the 3' end to the 5' end. The 5' end is transcribed (leader I) but not translated. The synthesized proteins are all associated with the virion. Protein N (50 kDa) associated with the genomic full length RNA, positive or negative, modulates the transcription. Protein NS, component of polymerase, intervenes also in transcription. Protein M (20-30 kDa), a basic protein that regulates transcription, is linked to the nucleocapsid and protein G. Protein G (70 kDa) forms external projections that are linked to the host cell. It is associated in a covalent manner with a lipid chain.

Protein L (200 kDa) effects the transcription with associated methyltransferase and polyA polymerase functions. On the template of the parental chain, the viral proteins effect the synthesis of complete positive chains, then of complete negative chains.

Biology Major feature: the same virus can multiply in animal and plant cells. Transmission by aphids or leafhoppers with multiplication in the insect vectors and transmission to the progeny. The symptoms are yellowing, mosaic, leaf curl, and vein clearing.

Note Because of the morphology of their virions, 58 species of plant viruses are included in the family *Rhabdoviridae* without being associated with a genus; 3 viruses without envelope are also placed in this category. 2 genera are described, *Cytorhabdovirus* and *Nucleorhabdovirus*.

Rhabdoviridae

Genus *CYTORHABDOVIRUS* (from the Greek *rhabdos*, rod) —— 53

Type species *Lettuce necrotic yellows virus* (LNYV).

Virion The replication occurs in cytoplasmic viroplasms, with an early nuclear phase for LNYV. The maturation of particles occurs in vesicles of the endoplasmic reticulum.

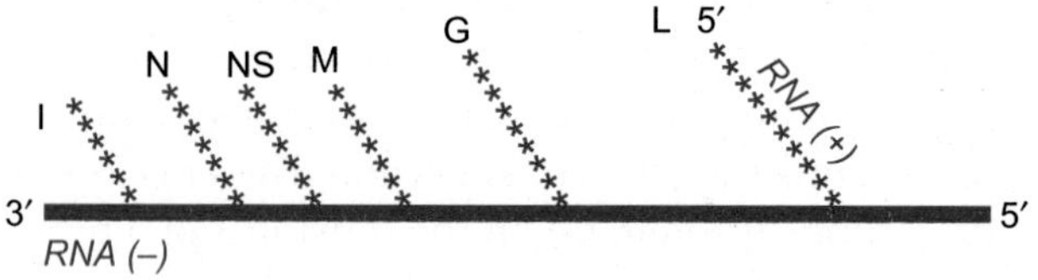

Biology 8 species are described. LNYV causes important epidemics on lettuce in Australia. *Barley yellow striate mosaic virus* (BYSMV) infects cereals in Europe, Africa, the Middle East, and Australia.

Rhabdoviridae

Genus *NUCLEORHABDOVIRUS* —— 54

Type species *Potato yellow dwarf virus* (PYDV).

Virion The viral proteins, synthesized in the cytoplasm, transit through the nucleus (nuclear addressing sequences), where the nucleocapsid is assembled. The mature virions accumulate in the space between the two nuclear membranes by budding through the internal nuclear membrane.

Biology 7 species are described. PYDV infects potato in North America. *Eggplant mottled dwarf virus* (EMDV) infects several vegetable species (eggplant, tomato, cucumber) in the Mediterranean basin.

HEATON, L.A., HILLMAN, B.I., HUNTER, B.G., ZUIDEMA, D., JACKSON, A.O. 1989. Physical map of the genome of sonchus yellow net virus, a plant rhabdovirus with six genes and conserved junction sequences. *Proc. Natl. Acad. Sci. USA*, 86, 8665-8668.
WAGNER, J.D., JACKSON, A.O. 1997. Characterization of components and activity of Sonchus yellow net rhabdovirus polymerase. *J. Gen. Virol.*, 71, 2371-2382.
WETZEL, T., DIETZGEN, R.G., GEERING, A.D.W., DALE, J.L. 1994. Analysis of the nucleocapsid gene of lettuce necrotic yellows rhabdovirus. *Virology*, 202, 1054-1057.

Unassigned genera

Genus *OPHIOVIRUS* (from the Greek *ophis*, serpent) — 55

Type species *Citrus psorosis virus* (CPsV).

Virion Extremely thin particles (3 nm) forming coiled circles of different sizes.

Genome Negative-sense RNA molecule in 4 segments.

Biology 5 described species, 1 tentative species. Transmitted by mechanical inoculation, vegetative propagation and for some species by zoospores of the chytrid fungus *Olpidium brassicae*.

VAN DER WILK, F., DULLEMANS, A.M., VERBEEK, M., VAN DEN HEUVEL, J. 2002. Nucleotide sequence and genomic organization of an ophiovirus associated with lettuce big-vein disease. *J. Gen. Virol.*, 83, 2869-2877.

Genus *TENUIVIRUS* (from the Latin *tenuis*, small) — 56

Type species *Rice stripe virus* (RSV).

Virion Filamentous particles of 3 to 5 nm width and variable length (up to 2000 nm), sometimes massed in pellets, branched or circular, without an envelope. These nucleocapsids are formed of RNA, capsidial subunits of 34 kDa, and a 230 kDa protein that is a polymerase.

Genome 4 or more single-stranded RNA molecules (10 kb, 3.4 kb, 2.3 kb, and 2 kb). The terminal sequences are complementary over a length of 20 bases. RNA 1 is negative-sense and codes for polymerase. The other

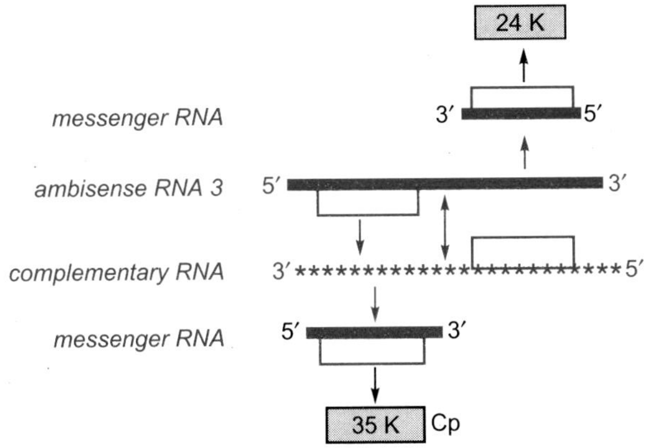

RNAs are ambisense: one ORF is open at the 5' end of the negative viral strand and the other at the 5' end of the positive strand, and the RNA presents an intergenic region with a strong secondary hairpin structure of 126 bases.

Biology 6 described species, 5 tentative species. Transmission by planthoppers (*Delphacidae*) in the persistent mode. The virus multiplies in the insect and is transmitted to the progeny. The genus *Tenuivirus* shares with the genus *Tospovirus* the capacity to replicate in the cells of insects and plants.

Hosts *Gramineae*. The symptoms are marked, with confluent chlorotic stripes. Mechanical transmission is difficult or impossible. RSV (Far East) and *Rice hoja blanca virus* (RHBV) (Central and South America) cause serious diseases in rice.

Phylogenetic relationships

The sequence analyses of protein L show a quite marked relationship between *Tenuivirus* and *Tospovirus*. These two genera of plant viruses (which share genome organization, strategy of expression (ambisense RNA), and multiplication in the insect vector) could be the widely diverging descendants of an ancestor of the family *Bunyaviridae* infecting animals and having acquired the capacity to infect plants.

KAKUTANI, T., HAYANO, Y., HAYASHI, T., MINOBE, Y. 1991. Ambisense segment 3 of rice stripe virus: the first instance of a virus containing two ambisense segments. *J. Gen. Virol.*, 72, 465-468.

RAMIREZ, B.C., MACAYA, G., CALVERT, L.A., HAENNI, A.L. 1992. Rice hoja blanca virus genome characterization and expression in vitro. *J. Gen. Virol.*, 73, 1457-1464.

RAMIREZ, B.C., HAENNI, A.L. 1994. Molecular biology of Tenuiviruses, a remarkable group of plant viruses. *J. Gen. Virol.*, 75, 467-475.

TORIYAMA, S., KIMISHIMA, T., TAKAHASHI, M., SHIMIZU, T., MINAKA, N., AKUTSU, K. 1998. The complete nucleotide sequence of the rice grassy stunt virus genome and genomic comparisons with viruses of the genus *Tenuivirus*. *J. Gen. Virol.*, 79, 2051-2058.

TORIYAMA, S., TAKAHASHI, M., SANO, Y., SHIMIZU, T., ISHIHAMA, A. 1994. Nucleotide sequence of RNA 1, the largest genomic RNA segment of the rice grassy stunt virus the prototype of the tenuiviruses. *J. Gen. Virol.*, 75, 3569-3579.

VAN POELWIJK, F., PRINS, M., GOLDBACH, R. 1997. Completion of the impatiens necrotic spot virus genome sequence and genetic comparison of the L protein within the family Bunyaviridae. *J. Gen. Virol.*, 78, 543-546.

Genus *VARICOSAVIRUS* (from the Latin *varicosus*, dilated vein) — 57

Type species *Lettuce big-vein associated virus* (LBVaV).

Virion Fragile, non-envelopped rod-shaped virion circa 18×320-360 nm.

Biology The virus is transmitted by spores of the soil fungus *Olpidium brassicae*. One described species, 1 tentative species.

LOT, H., CAMPBELL, R.N., SOUCHE, S., MILNE, R.G., AND ROGGERO, P. 2002. Transmission by *Olpidium brassicae* of Mirafiori lettuce virus and *Lettuce big-vein virus*, and their roles in lettuce big-vein etiology. *Phytopathology*, 92, 288-293.

Double-stranded RNA viruses

Family *Partitiviridae*

The family comprises three genera: two infecting plants (*Alphacryptovirus, Betacryptovirus*), and one infecting fungi (*Partitivirus*).

Virion Isometric particle of 30-40 nm.

Genome 2 double-stranded RNA segments of 1.4 to 3 kbp. An RNA-polymerase activity is associated with the virion. It transcribes the viral RNA into messenger RNA.

Biology These viruses have no known vectors, the transmission is vertical. They are transmitted by division of the mycelium in fungi and by transmission to the ovule and pollen (and thus to the seed) in plants. They are transmitted from cell to cell by cell division and cause latent infections without known pathogenic effect and symptoms.

Partitiviridae

Genus *ALPHACRYPTOVIRUS* (from the Greek *alpha* and *cryptos*, hidden) – 58

Type species *White clover cryptic virus 1* (WCCV-1).

Biology 16 described species, 10 tentative species. No vector known and no mechanical transmission. Infect Dicotyledons.

Partitiviridae

Genus *BETACRYPTOVIRUS* (from the Greek *beta* and *cryptos*, hidden) — 59

Type species *White clover cryptic virus 2* (WCCV-2).

Biology 4 described species, 1 tentative species. As alphacrytoviruses.

ACCOTTO, G.P., MARZACHI, C., LUISONI, E., MILNE, R.G. 1990. Molecular characterization of alfalfa cryptic virus 1. *J. Gen. Virol.*, 71, 433-437.

BOCCARDO, G., MILNE, R.G., LUISONI, E., LISA, V., ACCOTTO, G.P. 1985. Three seed-borne cryptic viruses containing double-stranded RNA isolated from white clover. *Virology*, 147, 29-40.

Family *Reoviridae*

The family *Reoviridae* comprises 12 genera including 3 plant viruses (*Fijivirus, Oryzavirus, Phytoreovirus*) confined to the phloem and inducing its proliferation (galls, tumours).

Virion Particle of isometric symmetry of 60-70 nm formed of a central part (which contains 10-12 segments of double-stranded RNA and 4 viral proteins) surrounded by an "external capsid" (several protective layers of two viral proteins P2 and P8).

Reoviridae

Genus *FIJIVIRUS* (from the Fiji Islands) — 60

Type species *Fiji disease virus* (FDV).

Virion Paraspherical particle with 2 layers of proteins and 12 spicules.

Genome 10 double-stranded RNA segments.

Biology 8 described species. Transmission by planthoppers (*Delphacidae*) in a persistent mode. No transovarial transmission.

Host range *Gramineae*. The virus multiplies in the phloem cells, producing enations and galls on the underside of leaves. The virus causes nanism in sugarcane in Thailand, the Philippines, and Australia. *Maize rough dwarf virus* (MRDV) causes a serious maize disease in Europe and Argentina.

KUDO, H., UYEDA, I., SHIKATA, E. 1991. Viruses in the Reoviridae family have the same conserved terminal sequences. *J. Virol.*, 72, 2857-2866.

LU, G., HONG ZHOU, Z., BAKER, M.L., JAKANA, J., CAI, D., WEI, X., CHEN, S., GU, X., CHIU, W. 1998. Structure of double-shelled rice dwarf virus. *J. Virol.*, 92, 8541-8549.

NUSS, D.L., DALL, D.J. 1990. Structural and functional properties of plant reovirus genome. *Adv. Virus Res.*, 38, 249-306.

SOO, H.M., HANDLEY, J.A., MAUGERI, M.M., BURNS, P., SMITH, G.R., DALE, J.L., HARDING, R.M. 1998. Molecular characterization of Fiji disease fijivirus genome segment 9. *J. Gen. Virol.*, 79, 3155-3161.

SUZUKI, N., SUGAWARA, M., NUSS, D.L., MATSUURA, Y. 1996. Polycistronic (tri- or bicistronic) phytoreoviral segments translatable in both plant and insect cells. *J. Virol.*, 70, 8155-8159.

SUZUKI, N., WATANABE, Y., KUSANO, T., KITAGAWA, Y., 1990. Sequence analysis of rice dwarf phytoreovirus genome segments S4, S5, and S6: comparison with the equivalent wound tumor virus segments. *Virology*, 179, 446-454.

UEDA, S., MASUTA, C., UYEDA, I. 1997. Hypothesis on structure and assembly of rice dwarf phytoreovirus: interactions among multiple structural proteins. *J. Virol.*, 78, 3135-3140.

UPADHYAYA, N.M., RAMM, K., GELLATLY, J.A., LI, Z., KOSITRATANA, W., WATERHOUSE, P.M. 1997. Rice ragged stunt oryzavirus genome segments S7 and S10 encodes non-structural proteins of Mr 68025 (Pns7) and 32363 (Pns 10). *Arch. Virol.*, 142, 1719-1726.

Reoviridae

Genus *ORYZAVIRUS* (from the Latin *oryza*, rice) — 61

Type species *Rice ragged stunt virus* (RRSV).

Virion Paraspherical particle of 75-80 nm carrying 12 truncated protrusions covering the central core of 50 nm in diameter; 10 double-stranded RNA segments.

Biology 2 described species. Transmission by planthoppers (*Delphacidae*). The virus multiplies in the insect and in the phloem cells.

Host range Gramineae. RRSV is responsible for a rice disease in the Far East.

Reoviridae

Genus *PHYTOREOVIRUS* —— **62**

(from the Greek *phyton*, plant, and Respiratory Enteric Orphan virus)

Type species *Wound tumor virus* (WTV).

Virion Isometric particle of 70 nm, which comprises a central core of 50 nm, and a double layer of capsomeres.

Genome 12 double-stranded RNA segments each having a single ORF. Proteins synthesized: 7 structural proteins (36 to 155 kDa) and 5 non-structural proteins. The different RNAs have identical sequences, characteristic of the genus *Phytoreovirus*, at their 3' end. The 12 RNA segments of *Rice dwarf virus* (RDV) are sequenced and functions can be proposed for some of the corresponding proteins. P1 carries the polymerase consensus motif and is part of the central part of the virion, like P3 and P7. P5 plays a role in the transmission. P2 and P8 are the components of the external capsid.

Biology 3 described species, 1 tentative species. Transmission by leafhoppers (*Cicadellidae*) in the persistent mode. The insect remains infectious throughout its life and transmits the virus to its eggs (2%). The virus is confined to the phloem and cannot be transmitted mechanically. Symptoms are dwarfing and enations, darkened green leaves.

■ Unassigned genus

Genus *ENDORNAVIRUS* (from the Greek *endo*, within and RNA) —— **62a**

Type species *Vicia faba endornavirus* (VFV)

Virion None reported. Endornaviruses do not produce virions.

Genome One molecule (14-18 kbp) of double-stranded RNA.

Biology 4 described species. Seed transmission, but no mechanical transmission, no vector known and no evidence for horizontal transmission. No symptoms associated with infection except VFV associated with cytoplasmic male sterility

Single-stranded DNA viruses

■ Family *Geminiviridae* (from the Latin *gemini*, twin)

Four genera in this family: *Begomovirus, Curtovirus, Mastrevirus, Topocuvirus*. The earlier term *Geminivirus* (in place of *Geminiviridae*) is still often used.

Virion The particle is made up of two geminate icosahedrons, of 30 × 18 nm, containing a single DNA molecule. The genome is constituted of one single-stranded circular DNA molecule (monopartite *Geminiviridae*) or two (bipartite *Geminiviridae*), with an intergenic region of around 300 bases. In the intergenic region the conserved sequence TAATATTAC is the starting point of the bi-directional transcription, which occurs either in the viral sense (v) or in the complementary sense (c). The virus is replicated in the nucleus by cell enzymes by a rolling circle mechanism on a double-stranded replicative form (Chapter 3, Fig. 3.18).

Phylogenetic relationships

Comparisons of sequence and organization of genomes have made it possible to propose an evolutionary diagram within the family *Geminiviridae*. The ancestral type is represented by a virus having a monopartite DNA, infecting *Poaceae* and transmitted by leafhoppers (thus similar to *Mastrevirus*). It diversified by acquiring the possibility of infecting dicotyledons and of being transmitted by whiteflies; with the doubling of DNA and then the specialization of each molecule (one molecule devoted to multiplication and the other to transport), the viruses reached the present state, in which highly diverse situations are observed.

Biology Some of these viruses are limited to the phloem. The symptoms may be dwarfism, yellow mosaic, leaf curl, enations. According to the host range, vector, and genome structure, four genera are distinguished: *Begomovirus, Curtovirus, Mastrevirus, Topocuvirus.*

BENDAHMANE, M., SCHALK, H.J., GRONENBORN, B. 1995. Identification and characterisation of wheat dwarf virus from France using a rapid method for geminivirus DNA preparation. *Mol. Plant Pathol.*, 85, 1449-1455.

BRIDDON, R. 2003. Cotton leaf curl disease, a multicomponent begomovirus complex. *Mol. Plant Pathol.*, 4, 427-434.

CZOSNEK, H., LATERROT, H. 1997. A worldwide survey of tomato yellow leaf curl viruses. *Arch. Virol.*, 142, 1391-1406.

GHANIM, M., MORIN, S., ZEIDAN, M., CZOSNEK, H. 1998. Evidence for transovarial transmission of tomato yellow leaf curl virus by its vector, the whitefly *Bemisia tabaci*. *Virology*, 240, 295-303.

GUTTIEREZ, C. 2000. DNA replication and cell cycle in plants: learning from geminiviruses. *EMBO J.*, 19, 792-799.

HEFFERON, K.L., DUGDALE, B. 2003. Independent expression of Rep and RepA and their roles in regulating *Bean yellow dwarf virus* replication. *J. Gen. Virol.*, 84, 3465-3472.

LIU, L., DAVIES, J.W., STANLEY, J. 1998. Mutational analysis of bean yellow dwarf virus, a geminivirus that is adapted to dicotyledonous plants. *J. Gen. Virol.*, 79, 2265-2274.

NORIS, E., VAIRA, A.M., CACIAGLI, P., MASENGA, V., GRONENBORN, B., ACCOTO, G.P. 1998. Amino-acids in the capsid protein of tomato yellow leaf curl virus that are crucial for systemic infection, particle formation, and insect transmission. *J. Virol.*, 72, 10050-10057.

PADIDAM, M., BEACHY, R.N., FAUQUET, C.M. 1995. Classification and identification of geminiviruses using sequence comparison. *J. Gen. Virol.*, 76, 249-263.

RYBICKI, E.P. 1998. A proposal for naming geminiviruses: a reply by the *Geminiviridae*. Study group chair. *Arch. Virol.*, 143, 441-444.

WARD, B., MEDVILLE, R., LAZAROWITZ, S.G., TURGEON, R. 1997. The geminivirus BL1 movement protein is associated with endoplasmic reticulum-derived tubules in developing phloem cells. *J. Virol.*, 71, 3726-3733.

WARTIG, L., KHEYR-POUR, E., NORIS, E., DE KOUTCHOVSKY, F., JOUANNEAU, B., GRONENBORN, B., JUPIN, I. 1997. Genetic analysis of the monopartite tomato yellow leaf curl geminivirus: roles of V1, V2, and C2 ORFs in viral pathogenesis. *Virology*, 228, 132-140.

Geminiviridae

Genus *BEGOMOVIRUS* (from BEan GOlden MOsaic virus) — 63

Type species *Bean golden mosaic virus* (BGMV).

Genome 2 DNA molecules (A and B) of around 2.6 kb. DNA A has 4 genes (AV1, capsid; AC1, replication; AC2, transactivation; AC3, accumulation of DNA). DNA B has 2 genes (BV1 implicated in movement and BC1 implicated in nucleus-cytoplasm transfers). DNA A is capable of autonomous replication in protoplasts.

Apart from bipartite *Begomovirus* that constitute the majority of members of the genus, this genus contains viruses in which the genome is made up of a single DNA molecule, such as *Tomato yellow leaf curl* (TYLCV), the single Thai isolate of which has a bipartite genome. *Cotton leaf curl virus* is a single-component *Begomovirus*, but symptoms in cotton require a satellite component (DNA β).

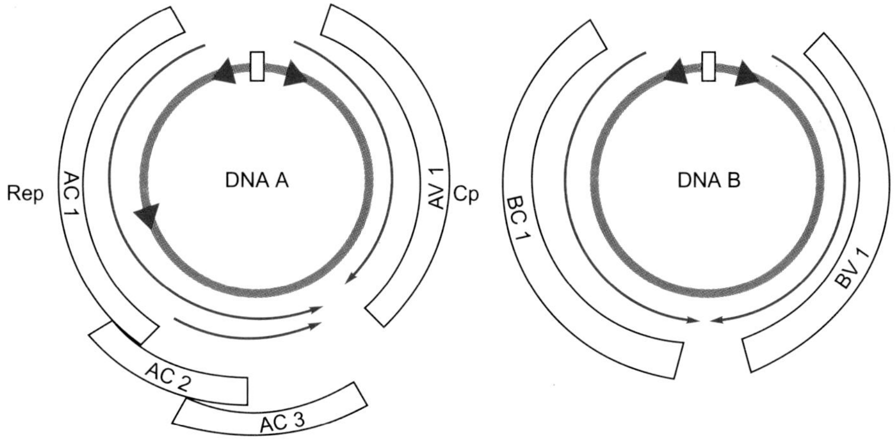

Biology 117 described species, 54 tentative species. Many of these species are similar and the species/strain distinction is difficult. They are transmitted by the whitefly *Bemisia tabaci* in a persistent mode. Mechanical transmission is often impossible; agro-infection is possible. Multiplication in the vector and transmission to eggs have been demonstrated for TYLCV-Is.

Host range Few families (dicotyledons) are sensitive. TYLCV attacks tomato and more rarely bean. *Begomovirus* causes very serious diseases in tropical, subtropical, and Mediterranean regions. The emergence of the virus generally follows the extension of the vector. *African cassava mosaic virus* (ACMV) causes a serious disease on cassava and devastating epidemics have been observed following recombination between strains. On cucurbits, *Squash leaf curl virus* (SLCV) is present in North America and *Watermelon chlorotic stunt virus* (WmCSV) in Africa.

Geminiviridae

Genus *CURTOVIRUS* (from beet CURly TOp virus) —— 64

Type species *Beet curly top virus* (BCTV).

Genome One DNA molecule of 2.9 kb. The DNA codes for 6 or 7 genes. The movement protein V3 and the capsid V1 are translated on the "viral sense" transcript. The protein C1 associated with replication is translated from the transcript in the complementary sense. The virus is restricted to phloem cells.

Biology 4 described species, one tentative species. Transmission by leafhoppers in the circulative persistent mode. Mechanical transmission is very difficult, agro-infection possible.

Host range Very broad for BCTV (more than 300 species of dicotyledons). It causes economically serious diseases on sugarbeet, tomato, and melon in America, the Mediterranean basin, and India.

Geminiviridae

Genus *MASTREVIRUS* (from MAize STREak virus) —— 65

Type species *Maize streak virus* (MSV).

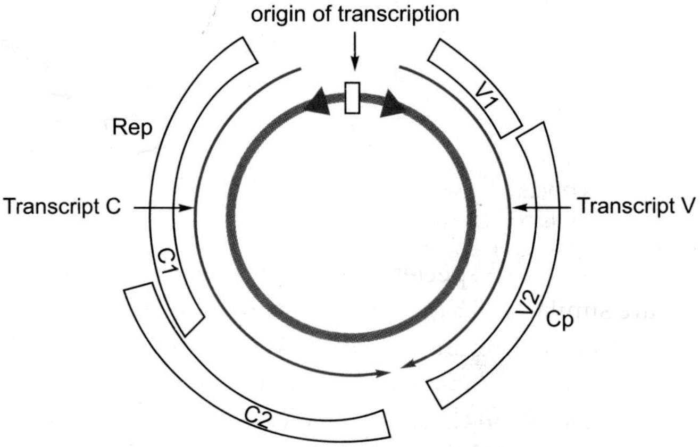

Genome One DNA molecule of 2.6 kb. The DNA carries two non-coding intergenic regions. The transcription diverges from the longest, which contains the conserved motif and the origin of transcription, and ends in the short intergenic region. The DNA codes for 4 genes; 2 are read on the transcript in the viral sense (ORF V1 and V2) and 2 on the transcript in the complementary sense (ORF C1-C2 with an overlapping intron C1 and C2).

The protein V1 is involved in cell-to-cell movement. V2 is the capsid. Protein C1-C2 (Rep protein) is the only viral protein indispensable for replication. Rep derepresses the enzymes that replicate the host DNA.

Biology 11 described species, 6 tentative species. Transmission by leafhoppers (*Cicadellidae*) in the circulative persistent mode. Mechanical transmission is impossible, agro-infection is possible.

Host range These viruses are found mostly on monocotyledons, but at least two viruses of dicotyledons are known (*Tobacco yellow dwarf virus*, TYDV; *Bean yellow dwarf virus*, BeYDV). MSV causes serious epidemics on maize in Africa; *Wheat dwarf virus* (WDV) can have a severe impact on wheat in Europe.

Geminiviridae

Genus *TOPOCUVIRUS* (from TOmato PseudO-CUrly top virus) — 66

Type species *Tomato pseudo-curly top virus* (TPCTV). The only species of the genus.

Family *Nanoviridae* (from the Greek *nanos*, dwarf)

Two genera make up this family: *Nanovirus* and *Babuvirus*.

Virion 6 to 12 isometric particles 17-20 nm in diameter, each containing one single-stranded circular DNA molecule.

These viruses have in common their genomic structure, composed of several (6 to 12) single-stranded circular DNA molecules of around 1 kb each. In several of these DNA molecules, there is evidence of an intergenic region that presents numerous analogies with the corresponding region of *Geminiviridae*. The protein associated with replication presents homologies with those of *Geminiviridae*, which suggests that replication occurs by the rolling circle mechanism. *Nanoviridae* are related to *Circovirus*, which infect animals (pig, chicken).

Biology Transmission by aphids in the persistent mode, no mechanical transmission possible. *Nanoviridae* cause dwarfing, yellowing, or necrosis. *Faba bean necrotic yellows virus* (FBNYV) seriously affects legumes in Africa. *Banana bunchy top virus* (BBTV) seriously affects banana. It is disseminated throughout Africa and in the Pacific through the import of infected plant material.

ARONSON, M.N., MEYER, A.D., GYORGYEY, J., KATUL, L., VETTEN, H.J., GRONENBORN, B., TIMCHENKO, T. 2000. Clink, a nanovirus encoded protein. *J. Virol.*, 74, 2967-2972.

DUGDALE, B., BEETHAM, P.R., BECKER, D., HARDING, R.M., DALE, J.L., 1998. Promoter activity associated with the intergenic regions of banana bunchy top virus DNA-1 to -6 in transgenic tobacco and banana cells. *J. Gen. Virol.*, 79, 2301-2311.

KATUL, L., MAISS, E., MOROZOV, S., VETTEN, H.J. 1997. Analysis of six DNA components of the faba bean necrotic yellows virus genome and their structural affinity to related plant virus genomes. *Virology*, 233, 247-259.

KATUL, L., TIMCHENKO, T., GRONENBORN, B., VETTEN, H.J. 1998. Ten circular ss-DNA components, foru of which encode putative replication-associated proteins, are associated with the faba bean necrotic yellows virus genome. *J. Gen. Virol.*, 79, 3101-3109.

SANO, Y., WADA, M., HASHIMOTO, Y., MATSUMO, T., KOJIMA, M. 1998. Sequence of ten circular ssDNA components associated with the milk vetch dwarf virus genome. *J. Gen. Virol.*, 79, 3111-3118.

SAUNDERS, K., BEDFORD, I.D., STANLEY, J. 2002. Adaptation from whitefly to leafhopper transmission of an autonomously replicating nanovirus-like DNA component associated with ageratum yellow vein disease. *J. Gen. Virol.*, 83, 907-916.

TIMCHENKO, T., DE KOUCHKOVSKY, F., KATUL, L., DAVID, C., VETTEN, H.J., GRONENBORN, G. 1999. Single rep protein initiates replication of multiple genome components of faba bean necrotic yellows virus, a single-stranded DNA virus of plants. *J. Virol.*, 73, 10173-10182.

Nanoviridae

Genus *NANOVIRUS* (from the Greek *nanos*, dwarf) —————— **67**

Type species *Subterranean clover stunt virus* (SCSV).
Biology 3 described species.

Nanoviridae

Genus *BABUVIRUS* (from BAnana BUnchy top virus) —————— **68**

Type species *Banana bunchy top virus* (BBTV).
Biology 1 described species.

DNA or RNA reverse-transcribing viruses

Family *Caulimoviridae*

Phylogenetic relationships

Caulimoviridae use reverse transcriptase for their replication and thus form, with *Hepadnaviridae* of animals, the supergroup of pararetroviruses (called so to distinguish them from retroviruses in which the virion contains RNA and the DNA transcript is integrated in the host genome).

Caulimoviridae

Genus ***BADNAVIRUS*** (from BAcilliform DNA virus) — 69

Type species *Commelina yellow mottle virus* (ComYMV).

Virion One bacilliform particle of 30 × 130 nm (different lengths are sometimes observed).

Genome One double-stranded circular DNA molecule (7.5 kbp), with one interruption on each strand. The complete transcript of DNA carries 3 ORFs coding for I, the transmission protein, and for II and III, the polyprotein carrying the protease, ribonuclease H, and reverse transcriptase functions. This transcript is also the pregenomic RNA that serves as a template for reverse transcriptase for the synthesis of viral DNA. ORF 4 exists in certain badnaviruses (absent in ComYMV).

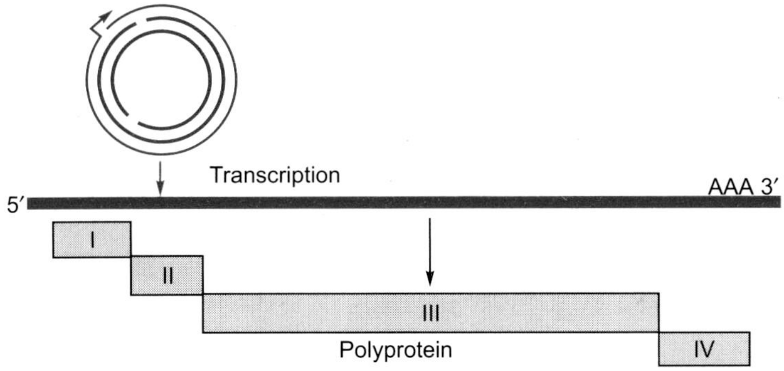

Biology 8 described species, 5 tentative species. Transmission in natural conditions by various vectors (mealybug, leafhopper) according to the semi-persistent mode, and during vegetative propagation.

Hosts Monocotyledons, dicotyledons. *Cacao swollen shoot virus* (CSSV) infects cacao in West Africa. It causes serious deterioration of trees accompanied by characteristic symptoms of trunk and stem swelling. *Banana streak virus* (BSV), the DNA of which can be integrated in the DNA of its host, infects banana in numerous regions in which it has been introduced by contaminated plants.

JACQUOT, E., HAGEN, L.S., JACQUEMOND, M., YOT, P. 1996. The open reading frame product of the cacao swollen shoot badnavirus is a nucleic acid-binding protein. *Virology*, 225, 191-195.

MEDBERRY, S.L., LOCKART, B.E., OLSZEWSKI, N.E., 1990. Properties of Commelina yellow mottle virus's complete sequence, genomic discontinuities and transcript suggest that it is a pararetrovirus. *Nucl. Acids Res.*, 18, 5505-5513.

NDOWORA, T., LAFLEUR, D., HARPER, G., HULL, R., OLSZEWSKI, N.E., LOCKART, B. 1999. Evidence that badnavirus infection in *Musa* can originate from integrated pararetroviral sequences. *Virology*, 255, 214-220.

Caulimoviridae

Genus *TUNGROVIRUS* (from rice TUNGRO bacilliform virus) ———— **70**

Type species *Rice tungro bacilliform virus* (RTBV).

Biology Transmitted by a leafhopper (*Nephottetix virescens, Cicadellidae*), in association with *Rice tungro spherical virus* (RTSV), which is necessary for the transmission; infection limited to the phloem.

HERZOG, E., GUERRA-PERAZA, O., HOHN, T. 2000. The rice tungro bacilliform virus gene II product interacts with the coat protein domain of the viral gene III polyprotein. *J. Virol.*, 74, 2073-2083.

HULL, R. 1996. Molecular biology of rice tungro viruses. *Annu. Rev. Phytopathol.*, 34, 275-297.

JONES, M.C., GOUGH, K., DASGUPTA, I., SUBBA RAO, B.L., CLIFFE, J., SHEN, P., KANIEVSKA, M., BLACKEBROUGH, M., DAVIES, J.W., BEACHY, R.N., HULL, R. 1991. Rice tungro disease is caused by an RNA and a DNA virus. *J. Gen. Virol.*, 72, 757-761.

Caulimoviridae

Genus *CAULIMOVIRUS* (from CAULIflower MOsaic virus) ———— **71**

Type species *Cauliflower mosaic virus* (CaMV).

Virion One isometric particle of 50 nm. Capsid of 420 subunits of 57 kDa.

Genome (See Fig. 3.20.) One double-stranded circular DNA molecule (8 kb), having a discontinuity on the α strand (transcribed strand) and two on the β strand (complementary). These discontinuities result from the mode of replication by reverse transcription. Nuclear transcription of the α strand by the cellular RNA polymerase II gives rise to two messenger RNA translated in the cytoplasm. RNA 35S is a complete transcript of the α strand, increased by 180 nucleotides corresponding to the repeated end. It has two intergenic regions of regulation and 6 major ORFs. I, cell-to-cell movement; II, transmission helper component; III, needed for transmission by aphids; IV, capsidial subunit; V, reverse transcriptase, protease, ribonuclease H. RNA 35S is also the pregenomic RNA that serves as template for the reverse transcriptase for the synthesis of viral DNA. RNA 19S is a partial transcript. It corresponds to ORF VI and codes for the protein of viroplasms, cytoplasmic inclusion bodies (characteristic of the genus) in which viral DNA is synthesized and viral particles accumulate. This protein serves as transactivator for the synthesis of proteins coded by RNA 35S. It contains determinants of the host range and symptomology.

Biology 8 described species, 4 tentative species. Transmission by aphids in the semi-persistent mode (viral factor of transmission by aphids of 18 kDa and protein P3), by vegetative propagation (*Dahlia mosaic virus*, DMV), and by mechanical inoculation. Causes mosaic sometimes accompanied by necrosis.

Host range Often restricted.

BONNEVILLE, J.M., ŚANFACON, H., FUTTERER, J., HOHN, T. 1989. Post-transcriptional trans-activation in cauliflower mosaic virus. *Cell*, 59, 1135-1153.

JACQUOT, E., DAUTEL, S., LEH, V., GELDREICH, A., YOT, P., KELLER, M. 1997. Les pararetrovirus de plantes. *Virologie*, 1, 111-120.

JACQUOT, E., GELDREICH, A., KELLER, M., YOT, P. 1998. Mapping regions of the cauliflower mosaic virus ORF III product required for infectivity. *Virology*, 242, 395-402.

SCHMIDT, I., BLANC, S., ESPERANDIEU, P., KUHL, G., DEVAUCHELLE, G., LOUIS, C., CERUTTI, M. 1994. Interaction between the aphid transmission factor and virus particles is a part of the molecular mechanism of cauliflower mosaic aphid transmission. *Proc. Natl. Acad. Sci. USA*, 91, 8885-8889.

Caulimoviridae

Genus *CAVEMOVIRUS* (from CAssava VEin MOsaic virus) — 72

Type species *Cassava vein mosaic virus* (CsVMV).

Caulimoviridae

Genus *PETUVIRUS* (from PETUnia vein clearing virus) — 73

Type species *Petunia vein clearing virus* (PVCV).
PVCV makes a transition with retrotransposons; the latter code for a transposase and are integrated in the host genome.

RICHERT-POGGELER, K.R., SHEPHERD, R.J. 1997. Petunia vein-clearing virus: a pararetrovirus with the core sequence for an integrase function. *Virology*, 236, 137-146.

Caulimoviridae

Genus *SOYMOVIRUS* (from SOYbean chlorotic MOttle virus) — 74

Type species *Soybean chlorotic mottle virus* (SbCMV).

Family *Pseudoviridae*

This family contains three genera: *Hemivirus* (no species in plants) *Sirevirus* and *Pseudovirus*. *Pseudoviridae* is a family of retrotransposable elements, primarly identified by genome sequencing. They replicate via a virus-like intermediate, virus-like particles (VLPs) round to ovoid of circa 40-60 nm.

Pseudoviridae

Genus *PSEUDOVIRUS* (from the Greek *pseudo*, false) —————— **75**

Type species *Saccharomyces cerevisiae Ty1 virus* (SceTY1V).

Genome Virus-like particles contain positive-sense single-stranded RNA (5.6 kb). The RNA is reverse-transcribed, and integration into the host DNA is mediated by the integrase and long terminal repeat sequences. Encoded functions: 5'-aspartate protease-integrase-reverse transcriptase-ribonucleaseH-3'. 20 described species.

Pseudoviridae

Genus *SIREVIRUS* (from glycine max SIRE 1 virus) —————— **75a**

Type species *Glycine max SIRE1 virus* (GmaSIRV)

Biology 5 described species, 1 tentative species. SIRE1 is present in several hundreds copies in soybean and Endovir 1 in *Arabidopsis thaliana.*

■ Family *Metaviridae*

Metaviridae is a family of retrotransposable elements that have been found in all types of eukaryotes. This family contains three genera: *Errantivirus* and *Semotivirus* (no species in plants) and *Metavirus*.

Metaviridae

Genus *METAVIRUS* (from the Greek *metathesis*, transposition) —————— **76**

Type species *Saccharomyces cerevisia Ty 3 virus* (Sce TY3V).

Encoded functions: 5'-aspartate protease-reverse transcriptase-ribonucleaseH-integrase-3'.

Athila is present in *Arabidopsis thaliana* (0.3% genome), Cyclops in peas and legumes, Calypso in soybean.

Harper, G., Hull, R., Lockhart, B., Olszewski, N. 2002. Viral sequences integrated into plant genomes. *Annu. Rev. Phytopathol.*, 40, 119-136.
Hull, R. 2002. *Matthew's Plant Virology*, 4th ed. Academic Press, pp. 40-43.

Glossary

Acquired resistance: A resistance response developed by a normally susceptible host following a predisposing treatment, such as inoculation by a virus, bacterium, or fungus or treatment with certain chemicals. Acquired resistance is not inherited.

Aggressiveness: Component of the viral pathogenicity that is expressed as the symptom intensity in a susceptible host.

Ambisense: Single-stranded nucleic acid containing genetic information in each sense.

Amino acid: Building block of proteins, holding an amino group, a carboxylic group and a variable side chain. The linear succession of the 20 different amino acids is the protein sequence. A protein has an **NH2 end** with a free basic amino-group, and a **COOH end** with a free carboxylic acid group.

Antibody: A specific protein of the immunoglobulin (Ig) class formed in the blood of vertebrate animals in response to the injection of a foreign protein or polysaccharide.

Anticodon: Three bases in a tRNA complementary to a codon in the messenger RNA. The codon-anticodon pairing allows the translation of messenger RNA into a protein chain.

Antigen: Substance (generally a protein or a carbohydrate) capable of stimulating the immune response and the production of antibodies.

Antiserum: The blood serum containing antibodies.

Bacteriophage: Viruses that multiply in bacteria. Their genome is either DNA or RNA.

Capsid: Protein coat composed of capsid sub-units (or capsomeres) surrounding and protecting the viral nucleic acid.

cDNA, cRNA: Abbreviations for complementary DNA and complementary RNA.

Cluster: A group of virus isolates or virus species that have sequence homologies such that they aggregate in phylogenetic studies. Not an official taxonomic concept.

Codon: Three adjacent nucleotides in an mRNA, coding for an amino acid by codon-anticodon pairing. **Start codon**: the trinucleotide AUG coding for methionine usually starts the translation. **Stop codon** or termination codon: UGA, UAA, UAG have no corresponding tRNA and are chain-terminating codons.

Complementary chain: Two nucleic acid chains are complementary when their sequences follow the rule of base-pairing: adenine with thymidine or uracil, cytosine with guanine.

Complementation: Occurs when a virus is helped by another virus or strain to move within a plant, to replicate or to be transmitted by a vector.

Cubic symmetry: A solid with cubic symmetry has three series of symmetry axes passing through faces, edges and apexes.

Defective interfering RNA (DI RNA): Replicating RNA deriving from genomic RNA by deletions.

Deletion: Loss of a piece of the nucleic acid chain.

DNA (deoxyribonucleic acid): Chain of deoxyribonucleotides, each containing a five-carbon deoxyribose sugar + phosphoric acid + one puric base (adenine A, guanine G) or one pyrimidic base (thymidine T, cytosine C). The linear chain is made up of sugar residues covalently linked by phophor ions; each sugar is covalently linked to a base (A, T, C or G). The sequence is the succession of bases.

Double-stranded nucleic acid: It is formed with two complementary single strands held together by hydrogen bonds: A-T and C-G in DNA, A-U and C-G in RNA. This bonding is the most important feature in the replication process.

ELISA: Enzyme-linked immunosorbent assay. A serological test in which the sensitivity of the antibody-antigen reaction is increased by fixing the antibody on an inert support and by revealing the reaction through an enzyme attached to the antibody.

Ends of nucleic acid chains: Nucleic acid chains are oriented. In a RNA molecule, one end carries an -OH group on the ribose 3′ carbon (**3′ end**) while at the other end there is a phosphate group on the ribose 5′ carbon (**5′ end**). The **C-ter** or **C-terminus** of a protein chain is the end of the chain bearing a free carboxylic group. The **N-ter** or **N-terminus** of a protein chain is the end of the chain bearing a free amino group.

Envelope: Bilayer lipoprotein membrane structure surrounding the capsid or the nucleocapsid of some viruses.

Eukaryotes: Uni- or multicellular organisms whose cells contain a nucleus delimited by a membrane, as opposed to **Prokaryotes**, in which the nucleus is not delimited.

Hybridization test: Formation of an artificial hybrid double-stranded structure by mixing single-stranded DNA or RNA with a probe (DNA or RNA) in appropriate conditions of temperature and ionic strength. Formation of hydrogen bonds demonstrates the complementarity between the sample to be tested and the probe.

Immune response: The ability of an animal to produce antibodies as a result of antigens, such as proteins, being introduced in its body.

Immunity: A form of resistance in which there is no virus replication. If all varieties of a plant species are immune to a virus, it is a non-host resistance.

Integration: Covalent insertion of viral DNA in the host DNA.

Isometric: Used to describe virus particles that are approximately spherical in shape.

Kilobase (kb): Unit of measurement of the length of a single-stranded nucleic acid, corresponding to 1000 bases.

Kilobase pair (kbp): Unit of measurement of a double-stranded nucleic acid, corresponding to 1000 base pairs.

Kilodalton kDa: The dalton is equal to the mass of the hydrogen atom ($1/6.23 \times 10^{23}$ g). The kilodalton equals 1000 daltons.

Mechanical inoculation: A technique of artificial virus transmission in which an infectious preparation is rubbed onto test plants. A few plant viruses may also be transmitted in the field through leaf rubbing or contact between plants.

Messenger RNA, mRNA: RNA that can be translated by the host ribosomes into proteins

Mutant: A virus that shows one or more nucleotide differences from a standard type.

Nanometre (nm): One thousand millionth part of a metre.

Negative-sense strand, minus strand, (–) strand: RNA strand with a nucleotide sequence complementary to the positive strand coding for proteins.

Nucleoside, nucleotide: A nucleoside consists of a sugar (ribose or deoxyribose) and a puric base (A or G) or pyrimidic base (C or T/U). The phosphate derivatives of nucleosides are called nucleotides and form the nucleic acid chains.

Oncogene: A gene coding for a protein modifying cell divisions.

ORF (open reading frame): Region in an mRNA with a start codon at the 5′ end and a termination codon at the 3′ end, coding for a polypeptide or a protein.

Pathotype: Group of strains or isolates having the same pathogenicity characteristics.

Phylum: Line of living beings sharing the same origin.

Plasmid: Mobile extrachromosomal replicating genetic element.

Plasmodesmata: Cytoplasmic connections between neighbouring plant cells.

Polycistronic: Nucleic acid coding for several adjacent genes.

Positive RNA, plus-sense RNA, (+) RNA: An RNA strand coding for proteins.

Resistance: An inherited character of a plant. A plant is considered resistant if it has the ability to suppress (or interfere with) the virus life cycle. Resistance is the opposite of susceptibility and may be identified as high (extreme) or low (partial) depending on the effectiveness of the protective mechanism.

Retrotransposon: A genetic information that can become integrated in eukaryotic DNA through its repeat terminal sequences and an integrase function. Retrotransposons are similar to retroviruses, although they lack their *env* gene coding for a capsid protein.

RNA (ribonucleic acid): Chain of ribonucleotides containing a ribose sugar + phosphoric acid + a nitrogenous puric base (adenine A, guanine G) or a nitrogenous pyrimidic base (uracil U, cytosine C). The backbone of RNA is made up of sugars linked by phosphates; each sugar is linked to a base A, U, C or G. The sequence is the succession of bases.

Southern blot: DNA fragments immobilized on a membrane (nitrocellulose or nylon) to be detected by hybridization with a labelled probe.

Strain: A virus isolate with specific biological, serological, or molecular properties and, by extension, a distinct group of virus isolates sharing some biological, serological, or molecular properties.

Subgenomic RNA: A partial copy of a genomic RNA, synthesized during replication to allow complete translation of viral RNA.

Synergy: Association of two or more viruses leading to more severe symptoms or to more efficient replication of one of these viruses.

Tolerance: A host response to virus infection that results in mild or no symptoms and no significant effect on yields, although virus movement and replication are similar to those in susceptible plants.

Transcription: Enzymatic copy of a nucleic acid sequence into messenger RNA. For instance, viral DNA is transcribed into mRNAs by cellular transcriptases; negative viral RNA is transcribed into mRNA by a virion-associated viral transcriptase.

Transfer RNA, tRNA: Small RNA molecules to which amino acids are attached; they adapt the different amino acids to the mRNA sequence by specific base pairing of the mRNA codons and the tRNA anticodon.

Translation: Synthesis of protein directed by ribosomes moving along the mRNA and reading the successive triplets (see **anticodon**).

Virion: A complete and infectious viral particle.

Virulence: A component of virus pathogenicity expressed as the ability to infect a plant possessing a gene for resistance to some strains of this virus.

VPg (viral protein genome-linked): Viral protein covalently linked to the 5′ end of some viral RNAs.

References

Aarts, N., Metz, M., Holub, E., Staskawicz, B.J., Daniels, M.J., 1998. Different requirements for ESD1 and NDR1 by disease resistance genes define at least two R gene-mediated signaling pathways in *Arabidopsis*. *Proc. Natl. Acad. Sci. USA*, 95, 10306-10311.

Aaziz, R., Dinant, S., Epel, B.L., 2001. Plasmodesmata and plant cytoskeleton. *Trends Plant Sci.*, 6(7), 326-330.

Aaziz, R., Tepfer, M., 1999a. Recombination between genomic RNA of two cucumoviruses under conditions of minimal selection pressure. *Virol.*, 263, 282-289.

Aaziz, R., Tepfer, M., 1999b. Recombination in RNA viruses and in virus-resistant transgenic plants. *J. Gen. Virol.*, 80, 1338-1346.

Abbink, T.E.M., Tjernberg, P.A., Bol, J.F., Linthorst, H.J.M., 1998. Tobacco mosaic virus helicase domain induces necrosis in N-gene-carrying tobacco in the absence of virus replication. *Mol. Plant-Microbe Interact.*, 11, 1242-1246.

Adlerz, W.C., Elmstrom, G.W., Purcifull, D.E., 1985. Response of "Multipick" squash to mosaic virus infection. *Hortsci.*, 20, 892-893.

Ahl, P., Gianinazzi, S., 1982. "β-protein" as a constitutive component in highly TMV resistant interspecific hybrids of *Nicotiana glutinosa* × *Nicotiana debneyi*. *Plant Sci. Lett.*, 26, 173-181.

Ahlquist, P., Noueiry, A., Lee, W., Kushner, D., Dye, B., 2003. Host factors in positive-strand RNA virus genome replication. *J. Virol.*, 77, 8181-8186

Ahlquist, P., 2002. RNA-dependent RNA polymerases, viruses, and RNA silencing. *Science*, 296, 1270-1273.

Akad, F., Teverovsky, E., David, A., Czosnek, H., Gidoni, D., Gera, A., Loebenstein, G., 1999. A cDNA from tobacco codes for an inhibitor of virus replication (IVR)-like protein. *Plant Mol. Biol.*, 40, 969-976.

Al Moudallal, Z., Briand, J.P., Van Regenmortel, M.H.V., 1985. The major part of the polypeptide chain of tobacco mosaic virus protein is antigenic. *EMBO J.*, 4, 1231-1235.

Alam, S.L., Atkins, J.F., Gesteland, R.F. 1999. Programmed ribosomal frameshifting: much ado about knotting! *Proc. Natl. Acad. Sci. USA*, 96, 14177-14179.

Albouy, J., Kusiak, C., Louanchy, M., Wang, W., 1994. Evaluation of rapid ELISA procedures for detecting viruses in propagation stock of pelargonium, dahlia, and orchids. *Acta Hort.*, 377, 189-195.

Albouy, J., Flouzat, C., Kusiak, C., Tronchet, M., 1988., Eradication of orchid viruses by chemotherapy from *in vitro* cultures of *Cymbidium*. *Acta Hort.*, 234, 413-420.

ALBOUY, J., MORAND, J.C., POUTIER, J.C., 1980. Les méthodes de mise en évidence des virus de pélargonium (plantes indicatrices et ELISA) dans le cadre d'une production de plantes saines. *Meded. Fac. Landbouwwet. Rijksuniv. Gent*, 45, 359-367.

ALCONERO, R., PROVVIDENTI, R., GONSALVES, D., 1986. Three pea seed-borne mosaic virus pathotypes from pea and lentil germplasm. *Plant Dis.*, 70, 783-786.

AL-KAFF, N.S., COVEY, S.N., KREIKE, M.M., PAGE, A.M., PINDER, R., DALE, P.J., 1998. Transcriptional and posttranscriptional plant gene silencing in response to a pathogen. *Science*, 279, 2113-2115.

ALLEN, E., WANG, S., MILLER, W.A., 1999. Barley yellow dwarf virus RNA requires a cap-independent translation sequence because it lacks a 5' cap. *Virol.*, 253, 139-144.

ALTSCHUH, D., AL MOUDALLAL, Z., BRIAND, J.P., VAN REGENMORTEL, M.H.V., 1985. Immunochemical studies of tobacco mosaic virus. VI. Attempts to localize viral epitopes with monoclonal antibodies. *Mol. Immunol.*, 22, 329-337.

AMBROS, S., HERNANDEZ, C., FLORÈS, R., 1999. Rapid generation of genetic heterogeneity in progenies from individual cDNA clones of peach latent mosaic viroid in its natural host. *J. Gen. Virol.*, 80, 2239-2252.

ANANDALAKSHMI, R., MARATHE, R., GE, X., HERR, J.M. JR., MAU, C., MALLORY, A., PRUSS, G., BOWMAN, L., VANCE, V.B., 2000. A calmodulin-related protein that suppresses posttranscriptional gene silencing in plants. *Science*, 290, 142-144.

ANDRET-LINK, P., SCMITT-KEICHINGER, C., DEMANGEAT, G., KOMAR, V., FUCHS, M., 2004. The specific transmission of Grapevine fanleaf virus by its nematode vector Xiphinema index is solely determined by the viral coat protein. *Virology*, 320: 12-22.

ANTIGNUS, Y., MOR, N., BEN-JOSEPH, R., LAPIDOT, M., COHEN, S., 1996. UV-absorbing plastic sheets protect crops from insect pests and from virus diseases vectored by insects. *Env. Entomol.*, 25, 919-924.

ANTONIW, J.F., WHITE, R.F., 1986. Changes with time in the distribution of virus and PR protein around single local lesions of TMV infected tobacco. *Plant Mol. Biol.*, 6, 145-149.

AOKI, S., TAKEBE, I., 1969. Infection of tobacco mesophyll protoplasts by tobacco mosaic virus ribonucleic acid. *Virology*, 39, 439-448.

ARANDA, M.A., ESCALER, M., WANG, D., MAULE, A., 1996. Induction of HSP70 and polyubiquitine expression associated with plant virus replication. *Proc. Natl. Acad. Sci. USA*, 93, 15289-15293.

ARGOS, P., 1988. A sequence motif in many polymerases. *Nucl. Acids Res.*, 16, 9909-9919.

ARONSON, M.N., MEYER, A.D., GYORGYEY, J., KATUL, L., VETTEN, H.J., GRONENBORN, B., TIMCHENKO, T. 2000. Clink, a nanovirus-encoded protein, binds both pRB and SKP1. *J. Virol.*, 74, 2967-2972.

ASHBY, M., WARRY, A., BEJERANO, E., KHASHOGGI, A., BURREL, M., LICHSTENSTEIN, C., 1997. Analysis of multiple copies of geminiviral DNA in the genome of four closely related *Nicotiana* species suggest a unique integration event. *Plant Mol. Biol.*, 35, 313-321.

ASJES, C.J., 1974. Control of the spread of the brown ring formation virus disease in the lily Mid-Century hybrid "Enchantment" by mineral-oil sprays. *Acta Hort.*, 36, 85-91.

ASJES, C.J., 1985. Control of field spread of non-persistent viruses in flower-bulb crops by synthetic pyrethroid and pirimicarb insecticides and mineral oils. *Crop Prot.*, 4, 485-493.

ASTIER-MANIFACIER, S., CORNUET, P., 1971. RNA-dependent RNA polymerase in Chinese cabbage. *Biochim. Biophys. Acta*, 232, 484-493.

Astier-Manifacier, S., Cornuet, P., 1978. Purification et poids moléculaire d'une ARN polymérase ARN-dépendante de *Brassica oleracea* var. *Botrytis*. *C.R. Acad. Sci. Paris*, 287, 1904-1906.

Atabekov, J.G., 1975. Host specificity of plant viruses. *Annu. Rev. Phytopathol.*, 13, 127-145.

Atabekov, J.G., Taliansky, M.E., 1990. Expression of plant virus specific transport function by various viral genomes. *Adv. Virus Res.*, 38, 201-248.

Atkins, D., Hull, R., Wells, B., Roberts, K., Moore, P., Beachy, R.N., 1991. The Tobacco mosaic virus 30K movement protein in transgenic tobacco plants is localized to plasmodesmata. *J. Gen. Virol.*, 72, 209-211.

Atkinson, P.H., Matthews, R.E.F., 1970. On the origin of dark green tissue in tobacco leaves infected by tobacco mosaic virus. *Virology*, 40, 344-356.

Atreya, C.D., Pirone, T.P., 1993. Mutational analysis of the helper component-proteinase gene of a potyvirus: effects of amino acid substitutions, deletions and gene replacement on virulence and aphid transmissibility. *Proc. Natl. Acad. Sci. USA*, 90, 11919-11923.

Atreya, P.L., Lopez-Moya, J.J., Chu, M., Atreya, C.D., Pirone, T.P., 1995. Mutational analysis of the coat protein N-terminal amino acids involved in potyvirus transmission by aphids. *J. Gen. Virol.*, 76, 265-270.

Avrameas, S., 1969. Coupling of enzymes to proteins with glutaraldehyde. Use of the conjugates for the detection of antigens and antibodies. *Immunochemistry*, 6, 43-52.

Ban, N., Larson, S.B., McPherson, A., 1995. Structural comparison of plant satellite viruses. *Virology*, 214, 571-583.

Bar-Joseph, M., Yang, G., Mawass, M., 1997. Subgenomic RNAs: the possible building blocks for modular recombination of *Closteroviridae* genomes. *Sem. Virol.*, 8, 113-119.

Barker, H., 1987. Invasion of non-phloem tissue in *Nicotiana clevelandii* by Potato leafroll luteovirus is enhanced in plants also infected with Potato Y potyvirus. *J. Gen. Virol.*, 68, 1223-1227.

Barker, H., 1989. Specificity of the effect of sap-transmissible viruses in increasing the accumulation of luteoviruses in co-infected plants. *Ann. Appl. Biol.*, 115, 71-78.

Bartels, R.,1954. Serologische Untersuchungen über das Verhalten des Kartoffel-A-virus in Tabakpflanzen. *Phytopathol. Z.*, 21, 395-406.

Basso, J., Dallaire, P., Charest, P.J., Devantier, Y., Laliberte, J.F. 1994. Evidence for an internal ribosome entry site within the 5′ non-translated region of turnip mosaic potyvirus RNA. *J. Gen. Virol.*, 75, 3157-3165.

Baulcombe, D. C., 2004. RNA silencing in plants. *Nature*, 431, 356-363.

Baulcombe, D.C., 1996. Mechanism of pathogen-derived resistance to viruses in transgenic plants. *Plant Cell*, 8, 1833-1844.

Baulcombe, D.C., Chapman, S.N., Santa Cruz, S., 1995. Jellyfish green fluorescent protein as a reporter for virus infections. *Plant J.*, 7, 1045-1053.

Bawden, F.C., Pirie, N.W., Bernal, J.D., Fankuchen, I., 1936. Liquid crystalline substances from virus-infected plants. *Nature*, 138, 1051-1055.

Beachy, R.N., 1997. Mechanisms and applications of pathogen-derived resistance in transgenic plants. *Curr. Opin. Biotechnol.*, 8, 215-220.

Becker, Y., 1996. A short introduction to the origin and molecular evolution of viruses. *Virus genes*, 11, 73-77.

BEFFA, R.S., HOFER, R.M., THOMAS, M., MEINS, F., JR., 1996. Decreased susceptibility to viral disease of β-1,3-glucanase-deficient plants generated by antisense transformation. *Plant Cell*, 8, 1001-1011.

BEIJERINCK, M.W., 1898. Üeber ein contagium vivum fluidum als Ursache der Fleckenkrankheit der Tabaksblätter. *Verhand. Kon.Akad. Wetensch. Amsterdam*, 65, 3-21. (English translation: Concerning a contagium vivum fluidum as cause of the spot disease of tobacco leaves, in Johnson J., ed., Phytopathological Classics N°7, *American Phytopathological Society*, Saint Paul, Minnesota, 1942, 33-52).

BEJERANO, E.R., KHASHOGGI, A., WITTY, M., LICHTEINSTEIN, C., 1996. Integration of multiple repeats of geminiviral DNA into the nuclear genome of tobacco during evolution. *Proc. Natl. Acad. Sci. USA*, 93, 769-774.

BENDAHMANE, A., KANYUKA, K., BAULCOMBE, D.C., 1999. The Rx gene from potato controls separate virus resistance and cell death responses. *Plant Cell*, 11, 781-791.

BENDAHMANE, A., KOHM, B.A., DEDI, C., BAULCOMBE, D.C., 1995. The coat protein of potato virus X is a strain specific elicitor of Rx1-mediated virus resistance in potato. *Plant J.*, 8, 933-941.

BENDAHMANE, A., QUERCI, M., KANYUKA, K, BAULCOMBE, D.C., 2000. *Agrobacterium* transient expression system as a tool for the isolation of disease resistance genes: application to the Rx2 locus. *Plant J.*, 21, 73-81.

BENDAHMANE, M., FITCHEN, J.H., ZHANG, G., BEACHY, R.N., 1997. Studies on coat protein-mediated resistance to tobacco mosaic tobamovirus. Correlation between assembly of mutant coat protein and resistance. *J. Virol.*, 71, 7942-7950.

BENT, A.F., 1996. Plant disease resistance genes: function meets structure. *Plant Cell*, 8, 1757-1771.

BERTHOMÉ, R., TEYCHENEY, P.Y., TEPFER, M., 2000. Mécanismes de résistance aux virus dans les plantes transgéniques. *Virologie*, 4, 49-60.

BERZAL-HERRANZ, A., DE LA CRUZ, A., TENILLADO, F., DIAZ-RUIZ, J.R., LOPEZ, L., SANZ, A.I., VAQUERO, C., SERRA, M.T., GARCIA-LUQUE, I., 1995. The *Capsicum* L^3 gene-mediated resistance against the tobamoviruses is elicited by the coat protein. *Virol*ogy, 209, 498-505.

BISARO, D.M., 1996. Geminivirus DNA replication. In: *DNA Replication in Eucaryotic Cells*. Cold Spring Harbour Laboratory Press, USA, 833-854.

BLANC, S., LOPEZ-MOYA, J.J., WANG, R., GARCIA-LAMPASONA, S., THORNBURY, D.W., PIRONE, T.P., 1997. A specific interaction between coat protein and helper component correlates with aphid transmission of a potyvirus. *Virology*, 231, 141-147.

BLANCARD, D., LECOQ, H., PITRAT, M., 1994. *A Colour Atlas of Cucurbit Diseases; Observation, Identification and Control*. Mansion Publishing, London, 375 p.

BLANCARD, D., LOT, H., MAISONNEUVE, B., 2004. *A Colour Atlas of Diseases of Lettuce and Related Salad Crops Observation, Biology and Control*. Mansion Publishing, London, 375 p.

BLEYKESTEN-GROSSHANS, C., GUILLEY, H., BOUZOUBA, A.S., RICHARDS, K.E., JONARD, G., 1997. Independent expression of the first two triple gene block proteins of beet necrotic yellow vein virus complements virus defective in the corresponding gene but expression of the third protein inhibits viral cell-to-cell movement. *Mol. Plant-Microbe Interact.*, 10, 240-246.

BLOM-BARNHOORN, G.J., VAN AARTRIJK, J., LINDE, P.C., 1986. Effect of virazole on the production of hyacinth plants free from hyacinth mosaic virus (HMV) by meristem culture. *Acta Hort.*, 177, 571-574.

BLUMENTHAL, T., CARMICHAEL, G.G., 1979. RNA replication: function and structure of Qβ replicase. *Annu. Rev. Biochem.*, 48, 525-548.

Boccard, F., Baulcombe, D.C., 1993. Mutational analysis of cis-acting sequences and gene function in RNA3 of cucumber mosaic virus. *Virology*, 193, 563-578.

Bokhoven, H.V., Le Gall, O., Kasteel, D., Verver, J., Wellink, J., Van Kamen, A.B., 1993. *Cis-* and *trans*-acting elements in cowpea mosaic virus RNA replication. *Virology*, 195, 377-386.

Bol, J.F., 1999. Alfalfa mosaic virus and ilarviruses: involvement of coat protein in multiple steps of the replication cycle. *J. Gen. Virol.*, 80, 1089-1102.

Bol, J.F., Linthorst, H.J.M., Cornelissen, B.J.C., 1990. Plant-pathogenesis-related proteins induced by virus infection. *Annu. Rev. Phytopathol.*, 28, 113-138.

Bonneville, J.M., Sanfaçon, H., Fütterer, J., Hohn, T., 1989. Post-transcriptional trans-activation in cauliflower mosaic virus. *Cell*, 59, 1135-1153.

Boonham, N., Barker, I., 1998. Strain specific recombinant antibodies to potato virus Y. *J. Virol. Meth.*, 193-199.

Bos, L., 1983. *Introduction to Plant Virology*. Pudoc, Wageningen, The Netherlands 160 pp.

Boudazin, G., Vergnet, C., Gélie, B., Meyer, M., Grosclaude, J., Maury, Y., 1994. Reactivity of two monoclonal antibodies to the cylindrical inclusion protein of *Potato virus Y*. *J. Phytopathol.*, 141, 186-194.

Boulton, M.I., Pallaghy, C.K., Chatany, M., MacFarlane, S., Davies, J.W., 1993. Replication of maize streak virus mutants in maize protoplasts: evidence for a movement protein. *Virology*, 192, 85-93.

Bourdin, D., Lecoq, H., 1991. Evidence that heteroencapsidation between two potyviruses is involved in aphid transmission of a non aphid-transmissible isolate from mixed infection. *Phytopathology*, 81, 1459-1464.

Bousalem, M., Douzery, E.J.P., Fargette, D., 2000. High genetic diversity, distant phylogenetic relationship and intraspecies recombination events among natural populations of Yam mosaic virus: a contribution to understanding potyvirus evolution. *J. Gen. Virol.*, 81, 243-255.

Bowers, G.R., Goodman, R.M., 1979. Soybean mosaic virus: infection of soybean seed parts and seed transmission. *Phytopathology*, 69, 569-572.

Bowling, S.A., Guo, A., Cao, H., Gordon, A.S., Klessig, D.F., Dong, X., 1994. A mutation in *Arabidopsis* that leads to constitutive expression of systemic acquired resistance. *Plant Cell*, 6, 1845-1857

Boyko, V.P., Karasev, A.V., Agranovsky, A.A., Koonin, E.V., Dolja, V.V., 1992. Coat protein gene duplication in a filamentous RNA virus of plants. *Proc. Natl. Acad. Sci. USA*, 89, 9156-9160.

Brandes, J., Wetter, C., 1959. Classification of elongated plant viruses on the basis of particle morphology. *Virology*, 8, 99-115..

Brault, V., Miller, W.A. 1992. Translational frameshifting mediated by a viral sequence in plant cells. *Proc. Natl. Acad. Sci. USA*, 89, 2262-2266.

Brault, V., Périgon, S., Reinbold, C., Erdinger, M., Scheidecker, D., Herrbach, E., Richards, K., Ziegler-Graff, V., 2005. The polerovirus minor capsid protein determines vector specificity and internal tropism in the aphid. *J. Virol.*, 79, 9685-9693.

Brault, V., van den Heuvel, J.F.J.M., Verbeek, M., Ziegler-Graff, V., Reutenauer, A., Herrbach, E., Garaud, J.C., Guilley, H., Richards, K., Jonard, G., 1995. Aphid transmission of beet western yellows luteovirus requires the minor capsid readthrough protein P74. *EMBO J.*, 14, 650-659.

Bressanelli, S., Tomei, L., Roussel, A., Incitti, I., Vitale, R., Mathieu, M., De Francesco, R., Rey, F., 1999. Crystal structure of the RNA-dependent RNA polymerase of hepatitis C virus. *Proc. Natl. Acad. Sci. USA*, 86, 13034-13039.

Briddon, R.W., Pinner, M.S., Stanley, J., Markham, P.G., 1990. Geminivirus coat protein gene replacement alters insect specificity. *Virology*, 177, 85-94.

Brisco, M.J., Hull, R., Wilson, T.M.A., 1985. Southern bean mosaic virus specific proteins are synthesized in an *in vitro* system supplemented with intact treated virions. *Virology*, 143, 392-398.

Brisco, M.J., Hull, R., Wilson, T.M.A., 1986. Swelling of isometric and bacilliform plant virus nucleocapsids is required for virus specific protein synthesis *in vitro*. *Virology*, 143, 210-217.

Broadbent, L., 1964. Control of plant virus diseases. In: *Plant Virology*. Corbett, M.K., Sisler, H.D., eds., University of Florida Press, Gainesville, 330-360.

Broadbent, L.H. 1963. The epidemiology of tomato mosaic. III. Cleaning virus from hands and tools. *Ann. Appl. Biol.*, 52, 225-232.

Broadbent, L.H., 1976. Epidemiology and control of tomato mosaic virus. *Annu. Rev. Phytopathol.*, 14, 75-96.

Brown, C.M., Dinesh-Kumar, S.P., Allen Miller, W. 1996. Local and distant sequences are required for efficient readthrough of barley yellow dwarf virus PAV coat protein gene stop codon. *J. Virol.*, 70, 5884-5892.

Brown, D., Gold, L. 1996. RNA replication by the Qβ replicase: a working model. *Proc. Natl. Acad. Sci. USA*, 93, 11558-11562.

Browning, K.S. 1996. The plant translational apparatus. *Plant Mol. Biol.*, 32, 107-114.

Brunetti, A., Tavazza, M., Noris, E., Lucioli, A., Acotto, G., Tavazza, M., 2001. Transgenically expressed T-Rep of tomato yellow leaf curl Sardinia virus acts as a *trans*-dominant-negative mutant, inhibiting viral transcription and replication. *J. Virol.*, 75, 10573-10581.

Brunetti, A., Tavazza, M., Noris, E., Tavazza, R., Ancora, G., Crespi, S., Accotto, G.P., 1997. High expression of truncated viral Rep protein confers resistance to tomato yellow leaf curl virus in transgenic tomato plants. *Mol. Plant-Microbe Interact.*, 10, 571-579.

Bruyere, A., Brault, V., Ziegler-Graff, V., Simonis, M.T., Van den Heuvel, J.F., Richard, K., Guilley, H., Jonard, G., Herrbach, E., 1997. Effects of mutations in the beet western yellows virus readthrough protein on its expression and packaging and on virus accumulation, symptoms and aphid transmission. *Virology*, 230, 323-334.

Bujarski, J.J., Nagy, P.D., 1996. Different mechanisms of homologous and nonhomologous recombination in brome mosaic virus: role of RNA sequences and replicase proteins. *Semin. Virol.*, 7, 363-372.

Campbell, R.N., 1996. Fungal transmission of plant viruses. *Annu. Rev. Phytopathol.*, 34, 87-108.

Campbell, R.N., Wipf-Scheibel, C., Lecoq, H., 1996. Vector-assisted seed transmission of melon necrotic spot virus in melon. *Phytopathology*, 86, 1294-1298.

Candresse, T., Hammond, R.W., Hadidi, A., 1998. Detection and identification of plant viruses and viroids using polymerase chain reaction (PCR). In: *Plant Virus Disease Control*, Hadidi, A., Khetarpal, R.K., Koganezawa, H., eds., APS Press, St. Paul, Minnesota, 399-416.

Candresse, T., Morch, M.D., Dunez, J., 1990. Multiple alignment and hierarchical clustering of conserved amino acid sequences in the replication-associated proteins of plant RNA viruses. *Res. Virol.*, 141, 315-329.

CANDRESSE, T., MOUCHÈS, C., BOVÉ, J.M., 1986. Characterization of the virus encoded subunit of turnip yellow mosaic virus RNA replicase. *Virology*, 152, 322-330.

CANDRESSE, T., REVERS, F., LEGALL, O., KOFALVI, S.A., MARCOS, J., PALLAS, V., 1997. Systematic search for recombination events in plant viruses and viroids. In: *Virus Resistant Transgenic Plants: Potential Ecological Impact*. Tepfer, M., Balazs, E., eds., Springer-INRA, Berlin, 20-24.

CANTO, T., PALUKAITIS, P., 1999. Are tubules generated by the 3a protein necessary for cucumber mosaic virus movement? *Mol. Plant-Microbe Interact.*, 12, 985-993.

CANTO, T., PRIOR, D.A.M., HELLWALD, K.H., OPARKA, K.J., PALUKAITIS, P., 1997. Characterization of cucumber mosaic virus. IV. Movement protein and coat protein are both essential for cell to cell movement of CMV. *Virology*, 237, 237-248.

CAO, H., BOWLING, S.A., GORDON, A.S., DONG, X., 1994. Characterisation of an *Arabidopsis* mutant that is non responsive to inducers of systemic acquired resistance. *Plant Cell*, 9, 1573-1592.

CARANTA, C., LEFEBVRE, V., PALLOIX, A., 1997. Polygenic resistance of pepper to potyviruses consists of a combination of isolate-specific and broad-spectrum quantitative trait loci. *Mol. Plant-Microbe Interact.*, 10, 872-878.

CARANTA, C., PALLOIX, A., LEFEBVRE, V., DAUBÈZE, A.M., 1997. QTLs for a component of partial resistance to cucumber mosaic virus in pepper: restriction of virus installation in host cells. *Theor. Appl. Genet.*, 94, 431-438.

CARANTA, C., RUFFEL, S., DUSSAULT, M.H., 2003. Gènes naturels de résistance aux virus chez les plantes: relation entre structure et function. *Virologie*, 7, 165-175.

CARPENTER, C.D., OH, J.W., ZHANG, C., SIMON, A.W., 1995. Involvement of a stem-loop structure in the location of junction sites in viral RNA recombination. *J. Mol. Biol.*, 245, 608-622.

CARR, J.P., ZAITLIN, M., 1991. Resistance in transgenic tobacco plants expressing a non-structural gene sequence of tobacco mosaic virus is a consequence of markedly reduced virus replication. *Mol. Plant-Microbe Interact.*, 4, 579-585.

CARRINGTON, J.C., FREED, D.D., CHAN-SEOK, O.H., 1990. Expression of potyviral polyproteins in transgenic plants reveal three proteolytic activities required for complete processing. *The EMBO J.*, 9, 1347-1353.

CARRINGTON, J.C., HEATON, L.A., UIDEMA, D., HILLMAN, B.I., MORRIS, T.J., 1989. The genome structure of turnip crinkle virus. *Virology*, 170, 219-226.

CARRINGTON, J.C., JENSEN, P.E., SCHAAD, M.C., 1998. Genetic evidence for an essential role for the potyvirus CI protein in cell to cell movement. *Plant J.*, 14, 393-400.

CARRINGTON, J.C., KASSCHAU, K.D., MAHAJAN, S.K., SCHAAD, M.C., 1996. Cell to cell and long distance transport of viruses in plants. *Plant Cell*, 8, 1669-1681.

CARRINGTON, J.C., MORRIS, T.J., 1985. Characterization of the cell-free translation products of carnation mottle genomic and subgenomic RNAs. *Virology*, 144, 1-10.

CARROLL, T.W., MAYHEW, D.E., 1976a. Anther and pollen infection in relation to the pollen and seed transmissibility of two strains of barley stripe mosaic virus in barley. *Can. J. Bot.*, 54, 1604-1621.

CARROLL, T.W., MAYHEW, D.E., 1976b. Occurrence of virions in developing ovules and embryo sacs in relation to the seed transmissibility of barley stripe mosaic virus in barley. *Can. J. Bot.*, 54, 2497-2525.

CASPAR, D.L., KLUG, A., 1962. Physical principles in the construction of regular viruses. *Cold Spring Harbor Symp. Quant. Biol.*, 27, 1-27.

CASSE, F., 2000. Le maïs et la résistance aux antibiotiques. *La Recherche*, 327, 35-39.

CASTANON, S., MARIN, M.S., MARTIN-ALONSO, J.M., BOGA, J.A., CASAIS, R., HUMARA, J.M., ORDAS, R.J., PARRA, F. 1999. Immunization with potato plants expressing VP60 protein protects against rabbit hemorrhagic disease virus. *J. Virol.*, 73, 4452-4455.

CHANDRA-SHEKARA, A.C., NAVARRE, D., KACHROO, A., KANG, H.G., KLESSIG, D., KACHROO, P., 2004. Signaling requirements and role of salicylic acid in HRT- and rrt-mediated resistance to turnip crinkle virus in *Arabidopsis*. *Plant J.*, 40, 647-659.

CHAPMAN, E.J., PROKHNEVSKY, A.I., GOPINATH, K., DOLJA, V.V., CARRINGTON, J.C., 2004. RNA silencing suppressors inhibit the microRNA pathway at an intermediate step. *Genes Dev.*, 18, 1179-1186.

CHAPMAN, S.N., HILLS, G., WATTS, J., BAULCOMBE, D.C., 1992. Mutational analysis of the coat protein gene of potato virus X: effects on virion morphology and viral pathogenicity. *Virology*, 191, 223-230.

CHAPMAN, S.N., KAVANAGH, T.A., BAULCOMBE, D.C., 1992. Potato virus X as a vector for gene expression in plants. *Plant J.*, 2, 549-557.

CHAY, C.A., GUNASINGE, U.B., DINESH-KUMAR, S.P., MILLER, W.A., GRAY, S.M., 1996. Aphid transmission and systemic plant infection determinants of barley yellow dwarf luteovirus-PAV are contained in the coat protein readthrough domain and 17 kDa protein respectively. *Virology*, 219, 57-65.

CHEN, B., FRANCKI, R.I.B., 1990. Cucumovirus transmission by the aphid *Myzus persicae* is determined solely by the viral coat protein. *J. Gen. Virol.*, 71, 939-944.

CHEN, Z., MALAMY, J., HENNING, J., CONRATH, U., SANCHEZ-CASAS, P., SILVA, H., RICIGLIANO, J., KLESSIG, D.F., 1995. Induction, modification and transduction of the salicylic acid signal in plant defence responses. *Proc. Natl. Acad. Sci. USA*, 92, 4134-4137.

CHENG, R.H., OLSON, N.H., BAKER, T.S., 1992. Cauliflower mosaic virus: a 420 subunit (T = 7) multiplayer structure. *Virology*, 186, 665-668.

CHERIF, C., RUSSO, M., 1983. Cytological evidence of the association of the a geminivirus with the tomato yellow leaf curl disease in Tunisia. *Phytopathol. Z.*, 108, 221-225.

CHESSON, A., JAMES, P., 2000. Les aliments avec OGM sont-ils sans danger? *La Recherche*, 327, 27-35.

CHISHOLM, S.T., MAHAJAN, S.K., WHITHAM, S.A., YAMAMOTO, M.L., CARRINGTON, J.C., 2000. Cloning of the *Arabidopsis* RTM1 gene, which controls restriction of long-distance movement of tobacco etch virus. *Proc. Natl. Acad. Sci. USA*, 97, 489-494.

CHISHOLM, S.T., PARRA, M.A., ANDERBERG, R.J., CARRINGTON, J.C., 2001. *Arabidopsis* RTM1 and RTM2 genes function in phloem to restrict long-distance movement of tobacco etch virus. *Plant Physiol.*, 127, 1667-1675.

CHOI, Y., RAO, A., 2003. Packaging of brome mosaic virus RNA3 is mediated through a bipartite signal. *J. Virol.*, 77, 9750-9757.

CHRISTIE, R.G., EDWARDSON, J.R., 1986. Light microscopy techniques for detection of plant virus inclusions. *Plant Dis.*, 70, 273-279.

CITOVSKY, V., KNORR, D., SCHUSTER, G., ZAMBRYSKI, P., 1990. The P30 movement protein of tobacco mosaic virus is a single strand nucleic acid binding protein. *Cell*, 60, 637-647.

CITOVSKY, V., MCLEAN, B.G., ZUPAN, J.R., ZAMBRYSKI, P., 1993. Phosphorylation of tobacco mosaic virus cell to cell movement protein by a developmentally regulated plant cell wall associated protein kinase. *Genes & Dev.*, 7, 904-910.

CITOVSKY, V., WONG, M.L., SHAW, A.L., VENKATARAM PRASAD, B.V., ZAMBRYSKI, P., 1992. Visualization and characterization of tobacco mosaic virus movement protein binding to single stranded nucleic acids. *Plant Cell*, 4, 397-411.

CITOVSKY, V., ZAMBRYSKI, P., 2000. Systemic transport of RNA in plants. *Trends Plant Sci.*, 5, 52-54.

CLARK, M.F., ADAMS, A.N., 1977. Characteristics of the microplate method of enzyme linked immunosorbent assay for the detection of plant viruses. *J. Gen. Virol.*, 34, 475-483.

CLAUZEL, J.M., KUSIAK, C., DEOGRATIAS, J.M., LEFEBVRE, A., ALBOUY, J., 1994. DIBA: une technique de diagnostic simple et rapide. *Phytoma*, 462, 14-16.

COGONI, G., MACINO, G., 1999. Gene silencing in *Neurospora crassa* requires a protein homologous to RNA-dependent RNA polymerase. *Nature*, 399, 166-169.

COHEN, S., DUFFUS, J.E., LARSEN, R.C., LIU, H.Y., FLOCK, R.A., 1983. Purification, serology and vector relationships of squash leaf curl virus, a whitefly-transmitted geminivirus. *Phytopathology*, 73, 1669-1673.

COLLINS, R.F., GELLATLY, D.L., SEHGAL, O.P., ABOUHAIDAR, M.G., 1998. Self-cleaving circular RNA associated with rice yellow mottle virus is the smallest viroid-like RNA. *Virology*, 241, 269-275.

COOPER, B., LAPIDOT, M., HEICK, J.A., DODDS, J.A., BEACHY, R.N., 1995. A defective movement protein of TMV in transgenic plants confers resistance to multiple virus whereas the functional analog increases susceptibility. *Virology*, 206, 307-317.

CORNUET, P., 1987. *Eléments de virologie végétale*. INRA, Paris, 206 pp.

COTTEN, J., 1979. The effectiveness of soil sampling for virus vector nematodes in MAFF certification schemes for fruit and hops. *Plant Pathol.*, 28, 40-44.

COVEY, S.N., AL-KAFF, N.S., LANARA, A., TURNER, D.S., 1997. Plants combat infection by gene silencing. *Nature*, 385, 781-782.

CRICK, F.C., WATSON, J.D., 1956. Structure of small viruses. *Nature*, 177, 473-475.

CRONIN, S., VERCHOT, J., HALDEMAN-CAHILL, R., SCHAAD, M.C., CARRINGTON, J.C., 1995. Long distance movement factor: a transport function of the potyvirus helper component proteinase. *Plant Cell*, 7, 549-559.

CROWTHER, R.A., GEELEN, J.L., MELLEMA, J.E., 1974. A three dimensional image reconstruction of cowpea mosaic virus. *Virology*, 57, 20-27.

CSILLERY, G., TOBIAS, I., RUSKO, J., 1983. A new pepper strain of tomato mosaic virus. *Acta Phytopathol. Acad. Sci. Hung.*, 18, 195-200.

CULVER, J.N., 1996. Tobamovirus cross protection using a potexvirus vector. *Virology*, 226, 228-235.

CULVER, J.N., DAWSON, W.O., 1991. Tobacco mosaic virus elicitor coat protein genes produce an hypersensitive phenotype in transgenic *Nicotiana sylvestris* plants. *Mol. Plant-Microbe Interact.*, 4, 458-463.

CZOSNEK, H., GHANIM, M., 1999. Sex-mediated transmission and propagation of tomato yellow leaf curl geminivirus by whiteflies. Summary of VII International Symposium of Plant Virus Epidemiology, 11-16 Apr. 1999, Almeria, Spain, 37-38.

D'Arcy, C.J., Mayo, M.A., 1997. Proposals for changes in luteovirus taxonomy and nomenclature. *Arch. Virol.*, 142, 1285-1287.

Dahourou, G., Guillot, S., Le Gall, O., Crainic, R., 2002. Genetic recombination in wild-type poliovirus. *J. Gen. Virol.*, 83, 3103-3110.

Dalmay, T., Hamilton, A., Rudd, S., Angell, S., Baulcombe, D.C., 2000. An RNA-dependent RNA polymerase gene in *Arabidopsis* is required for posttranscriptional gene silencing mediated by a transgene but not by a virus. *Cell*, 101, 543-553.

Danks, C., Barker, I., 2000. On-site detection of plant pathogens using lateral flow devices. *EPPO Bull.*, 30, 421-426.

Daros, J.A., Marcos, J.F., Hernandez, C., Florès, R., 1994. Replication of avocado sunblotch viroid: evidence for a symmetric pathway with two rolling circles and hammerhead ribozyme processing. *Proc. Natl. Acad. Sci. USA*, 91, 12813-12817.

Darwin, C., 1859. *On the Origin of Species*. John Murray, London.

Dasgupta, R., Garcia, B.H. 2nd, Goodman, R.M., 2001. Systemic spread of an RNA insect virus in plants expressing plant viral movement protein genes. *Proc. Natl. Acad. Sci. USA*, 98, 4910-4915.

Dawson, W.O., 1976. Synthesis of TMV RNA at restrictive high temperatures. *Virology*, 73, 319-326.

Dawson, W.O., Painter, P.R., 1978. Estimation of the cell-to-cell spread of Tobacco Mosaic Virus infection in mechanically inoculated leaves. *InterVirology*, 9, 310-315.

De Graaf, M., Coscoy, L., Jaspars, E.M.J., 1993. Localization and biochemical characterization of alfalfa mosaic virus replication complexes. *Virology*, 194, 878-881.

De Haan, P., Kormelink, R., Resende, R., Van Poelwij, F., Peters, D., Goldbach, R., 1991. Tomato spotted wilt virus L RNA encodes a putative RNA polymerase. *J. Gen. Virol.*, 72, 2207-2216.

De Jong, W., Ahlquist, P., 1992. A hybrid plant RNA virus made by transferring the noncapsid movement protein from a rod-shaped to an icosahedral virus is competent for systemic infection. *Proc. Natl. Acad. Sci. USA*, 89, 6808-6812.

De Wilde, C., Van Houdt, H., De Buck, S., Angenon, G., De Jaegger, G., Depicker, A., 2000. Plants as bioreactors for protein production: avoiding the problem of transgene silencing. *Plant Mol. Biol.*, 43, 347-359.

Deiman, B.A., Koenen, A.K., Verlaan, P.W., Pleij, C.W. 1998. Minimal template requirements for initiation of minus-strand synthesis *in vitro* by the RNA-dependent RNA polymerase of turnip yellow mosaic virus. *J. Virol.*, 72, 3965-3972.

Deiman, B.A., Kortlever, R.M., Pleij, C.W.A., 1997. The role of the pseudoknot at the 3' end of turnip yellow mosaic virus RNA in minus-strand synthesis by the viral RNA-dependent RNA polymerase. *J. Gen. Virol.*, 71, 5990-5996.

Demler, S.A., Rucker-Feeney, D.G., Skaf, J.S., de Zoeten, G.A., 1996. Pea enation mosaic virus: properties and aphid transmission. In: *The Plant Viruses*, Vol. 5: *Polyhedral Virions and Bipartite RNA Genomes*. Harrison, B.D., Murant, A.F., eds. Plenum Press, New York, 303-344.

Deogratias, J.M., Dosba, F., Lutz, A., 1989b. Eradication of prune dwarf virus, prunus necrotic ringspot virus and apple chlorotic leaf spot virus in sweet cherries by a combination of chemotherapy, thermotherapy and *in vitro* culture. *Can. J. Plant Pathol.*, 11, 337-342.

DEOGRATIAS, J.M., LUTZ, A., DOSBA, F., 1989a. *In vitro*, micrografting of shoot tips from juvenile and adult *Prunus avium* (L.) and *Prunus persica* (L.) Batsch. to produce virus-free plants. *Acta Hort.*, 193, 139-145.

DEOM, C.M., SHAW, M.J., BEACHY, R.N., 1987. The 30 kilodalton gene product of tobacco mosaic virus potentiates virus movement. *Science*, 327, 389-394.

DERRICK, K.S., 1973. Quantitative assay for plant viruses using serologically specific electron microscopy. *Virology*, 56, 652-653.

DERRICK, P.M., CARTER, S.A., NELSON, R.S., 1997. Mutation of the tobacco mosaic tobamovirus 126- and 183-kDa proteins: effects on phloem dependent virus accumulation and synthesis of viral proteins. *Mol. Plant-Microbe Intearct.*, 10, 589-596.

DESBIEZ, C., GAL-ON, A., GIRARD, M., WIPF-SCHEIBEL, C., LECOQ, H., 2003. Increase in Zucchini yellow mosaic virus symptom severity in tolerant zucchini squash cultivar is related to a point mutation in P3 protein and is associated with a loss of relative fitness on susceptible plants. *Phytopathology*, 93, 1478-1484.

DESBIEZ, C., WIPF-SCHEIBEL, C., LECOQ, H., 1999. Reciprocal assistance for aphid transmission between non-transmissible strains of zucchini yellow mosaic potyvirus in mixed infections. *Arch. Virol.*, 144, 2213-2218.

DESSENS, J.T., NGUYEN, N., MEYER, M., 1995. Primary structure and sequence analysis of RNA 2 of a mechanically transmitted barley mosaic virus isolate: an evolutionary relationship between bymo- and furoviruses. *Arch. Virol.*, 140, 325-333.

DEVERGNE, J.C., CARDIN, L., BURCKARD, J., VAN REGENMORTEL, M.H.V., 1981. Comparison of direct and indirect ELISA for detecting antigenically related cucumoviruses. *J. Virol. Methods*, 3, 193-200.

DEWAR, A.M., SMITH, H.G., 1999. Forty years of forecasting virus yellows incidence in sugar beet. In: *The Luteoviridae*. Smith, H.G., Barker, H., eds. CABI Publishing, Wallingford, UK, 229-243.

DIAZ, J.A., NETO, C., MORIONES, E., TRUNIGER, V., ARANDA, M.A., 2004. Molecular characterization of a Melon necrotic spot virus strain that overcome the resistance in melon and non-host plants. *Mol. Plant Microbe Interact.*, 17, 668-675.

DIAZ-PENDON, J.A., FERNANDEZ-MUNOZ, R., GOMEZ-GUILLAMON, M.L., MORIONES, E., 2005. Inheritance of resistance to Watermelon mosaic virus in Cucumis melo that impairs virus accumulation, symptom expression and aphid transmission. *Phytopathology*, 95, 840-846.

DIENER, T.O., 1971. Potato spindle tuber virus. IV. A replicating, low molecular weight RNA. *Virology*, 45, 411-428.

DINANT, S., BLAISE, F., KUSIAK, C., ASTIER-MANIFACIER, S., ALBOUY, J., 1993. Heterologous resistance to potato virus Y in transgenic tobacco plants expressing the coat protein gene of lettuce mosaic potyvirus. *Phytopathology*, 83, 818-824.

DINANT, S., LOT, H., 1992. Lettuce mosaic virus (Review). *Plant Pathol.*, 41, 528-542.

DINANT, S., MAISONNEUVE, J., ALBOUY, J., CHUPEAU, Y., CHUPEAU, M.C., BELLEC, Y., GAUDEFROY, F., KUSIAK, C., SOUCHE, S., ROBAGLIA, C., LOT, H., 1997. Coat protein-mediated protection in *Lactuca sativa* against lettuce mosaic virus strains. *Mol. Breeding*, 3, 75-86.

DINESH-KUMAR, S.P., BRAULT, V., MILLER, W.A., 1992. Precise mapping and *in vitro* translation of a tri-functional subgenomic RNA of barley yellow dwarf virus. *Virology*, 187, 711-722.

DING, B., 1998. Intercellular protein trafficking through plasmodesmata. *Plant Mol. Biol.*, 38, 279-310.

DING, B., HAUDENSHIELD, J.S., HULL, R.J., WOLF, S., BEACHY, R.N., LUCAS, W.J., 1992a. Secondary plasmodesmata are specific sites of localisation of the tobacco mosaic virus movement protein in transgenic tobacco plants. *Plant Cell*, 4, 915-928.

DING, B., KWON, M.O., HAMMOND, R., OWENS, R., 1997. Cell-to-cell movement of potato spindle tuber viroid. *Plant J.*, 12, 931-936.

DING, B., KWON, M.O., WARNBERG, L., 1996a. Evidence that actin filaments are involved in controlling the permeability of plasmodesmata in tobacco mesophyll. *Plant J.*, 10, 157-164.

DING, B., LI, Q., NGUYEN, L., PALUKAITIS, P., LUCAS, W.J., 1995a. Cucumber mosaic virus 3a protein potentiates cell to cell trafficking of CMV RNA in tobacco plants. *Virology*, 207, 345-353.

DING, B., TURGEON, R., PARTHASARATHY, M.V., 1992b. Substructure of freeze-substituted plasmodesmata. *Protoplasma*, 169, 28-41.

DING, S.W., LI, W.X., SYMONS, H., 1995b. A novel naturally occurring hybrid gene encoded by a plant RNA virus facilitates long distance virus movement. *EMBO J.*, 14, 5762-5772.

DING, X.S., CARTER, S.A., DEOM, C.M., NELSON, R.S., 1998. Tobamovirus and potyvirus accumulation in minor veins of inoculated leaves from representatives of the *Solanaceae* and *Fabaceae*. *Plant Physiol.*, 116, 125-136.

DING, X.S., SHINTAKU, M.H., ARNOLD, S.A., NELSON, R.S., 1995c. Accumulation of mild and severe strains of tobacco mosaic virus in minor veins of tobacco. *Mol. Plant-Microbe Interact.*, 8, 32-40.

DING, X.S., SHINTAKU, M.H., CARTER, S.A., NELSON, R.S., 1996b. Invasion of minor veins of tobacco leaves inoculated with tobacco mosaic virus mutants defective in phloem movement. *Proc. Natl. Acad. Sci. USA*, 93, 11155-11160.

DINMAN, J.D., RUIZ-ECHEVARRIA, M.J., PELTZ, S.W., 1998. Translating old drugs into new treatments: ribosomal frameshifting as target for antiviral agents. *Trends Biotechnol.*, 16, 190-196.

DODDS, J.A., 1998. Satellite tobacco mosaic virus. *Annu. Rev. Phytopathol.*, 36, 295-310.

DOHI, K., MISE, K., FURUSAWA, I., OKUNO, T., 2002. RNA-dependent RNA polymerase complex of *Brome mosaic virus*: analysis of the molecular structure with monoclonal antibodies. *J. Gen. Virol.*, 83, 2879-2890.

DOKE, N., OHASHI, Y., 1998. Involvement of an O_2-generating system in the induction of necrotic lesions on tobacco leaves infected with tobacco mosaic virus. *Physiol. Mol. Plant Pathol.*, 32, 163-175.

DOLJA, V.V., HALDEMAN, R., ROBERTSON, N.L., DOUGHERTY, W.G., CARRINGTON, J.C., 1994. Distinct functions of capsid protein in assembly and movement of tobacco etch potyvirus in plants. *EMBO J.*, 13, 1482-1491.

DOLJA, V.V., HALDEMAN-CAHILL, R., MONTGOMERY, A.E., VANDENBOSCH, K.A., CARRINGTON, J.C., 1995. Capsid protein determinants involved in cell to cell and long distance movement of tobacco etch potyvirus. *Virology*, 206, 1007-1016.

DOLJA, V.V., MCBRIDE, H.J., CARRINGTON, J.C., 1992. Tagging of plant potyvirus replication and movement by insertion of β-glucuronidase into the viral prolyprotein. *Proc. Natl. Acad. Sci. USA*, 89, 10208-10212.

DOMIER, L.L., SHAW, J.G., RHOADS, R.E., 1987. Potyviral proteins share amino acid sequence homology with picorna-, como-, and caulimovirus proteins. *Virology*, 158, 20-27.

DOMINGO, E., HOLLAND, J.J., 1994. Mutation rates and rapid evolution of RNA viruses: In: *The Evolutionary Biology of Viruses*. Morse, S., ed. Raven Press, New York, 161-184.

DOMINGO, E., HOLLAND, J.J., 1997. RNA virus mutations and fitness for survival. *Annu. Rev. Microbiol.*, 51, 151-178.

DOMINGO, E., HOLLAND, J.J., BIEBRICHER, C., EIGEN, M., 1995. Quasispecies: the concept and the word. In: *Molecular Evolution of the Viruses*. Gibbs, C., Calisher, C., Garcia-Arenal, F., eds. Cambridge University Press, Cambridge, UK, 181-191.

DOOLITTLE, S.P., 1916. A new infectious mosaic disease of cucumber. *Phytopathology*, 6, 145-147.

DOOLITTLE, S.P., WALKER, M.N., 1925. Further studies on the overwintering and dissemination of cucurbit mosaic. *J. Agric. Res.*, 31, 1-57.

DOROKHOV, Y.L., ALEXANDROVA, N.M., MIROSCHNICHENKO, N.A., ATABEKOV, J.G., 1984. The informosome-like virus-specific ribonucleoprotein (vRNP) may be involved in the transport of tobacco mosaic virus infection. *Virology*, 137, 127-134.

DRAKE, J.W., 1993. Rates of spontaneous mutation among RNA viruses. *Proc. Natl. Acad. Sci. USA*, 90, 4171-4175.

DREHER, T., 1999. Functions of the untranslated regions of positive strand RNA viral genomes. *Annu. Rev. Phytopathol.*, 37, 151-174.

DRIJFHOUT, E., 1978. Genetic interaction between *Phaseolus vulgaris* and bean common mosaic virus with implications for strain identification and breeding for resistance. *Agric. Res. Rep.*, Wageningen, 872, 1-98.

DRUCKER, M., FROISSART, R., HÉBRARD, E., UZEST, M., RAVALLEC, M., ESPÉRANDIEU, P., MANI, J.C., PUGNIÈRE, M., ROQUET, F., FERERES, A., BLANC, S., 2001. Intracellular distribution of viral gene products regulates a complex mechanism of Cauliflower mosaic virus acquisition by its aphid vector. *Proc. Natl. Acad. Sci. USA*, 99, 2422-2427.

DRUMOND, M., GORDON, M., NESTER, E., CHILTON, M.D., 1977. Foreign DNA of bacterial plasmid is transcribed in crown gall tumours. *Nature*, 269, 535-536.

DUBS, M.C., ALTSCHUH, D., VAN REGENMORTEL, M.H.V., 1992. Mapping of viral epitopes with conformationally specific monoclonal antibodies using biosensor technology. *J. Chromatogr.*, 597, 391-396.

DUFFUS, J.E., 1971. Role of weeds in the incidence of virus diseases. *Annu. Rev. Phytopathol.*, 9, 319-340.

DUNOYER, P., THOMAS, C., HARRISON, S., REVERS, F., MAULE, A., 2004. A cysteine-rich plant protein potentiates Potyvirus movement through an interaction with the virus genome-linked protein VPg. *J. Virol.*, 78, 2301-2309.

DUPRAT, A., CARANTA, C., REVERS, F., MENAND, B., BROWNING, K.S., ROBAGLIA, C., 2002. The *Arabidopsis* eukaryotic initiation factor (iso)4E is dispensable for plant growth but required for susceptibility to potyviruses. *Plant J.*, 32, 927-934.

DZIANNOT, A., RAUFFER-BRUYERE, N., BUJARSKI, J., 2001. Studies on functional interaction between Brome mosaic virus replicase proteins during RNA recombination, using combined mutants *in vivo* and *in vitro*. *Virology*, 289, 137-149.

EDWARDS, M.C., 1995. Mapping of the seed transmission determinants of barley stripe mosaic virus. *Mol. Plant-Microbe Interact.*, 8, 906-915.

EDWARDSON, J.R., 1984. Cytoplasmic cylindrical inclusions of Potyviruses. *Fla. Agric. Exp. Stn. Monogr.*, Gainesville (Florida), 190 pp.

EHLERS, K., KOLLMANN, R., 2001. Primary and secondary plasmodesmata: structure, origin, and functioning. *Protoplasma*, 216, 1-30.

EIGEN, M., 1993. Les quasi-espèces virales. *Pour la Science*, 191, 36-45.

EIGEN, M., 1996. On the nature of viral quasispecies. *Trends in Microbiol.*, 4, 212-214.

ENGLISH, J., MUELLER, E., BAULCOMBE, D., 1996. Suppression of virus accumulation in transgenic plants exhibiting silencing of nuclear genes. *Plant Cell*, 8, 179-188.

ENGLISH, J.J., DAVENPORT, G.F., ELMAYAN, T., VAUCHERET, H., BAULCOMBE, D.C., 1997. Requirement of sense transcription for homology-dependent virus resistance and trans-inactivation. *Plant J.*, 12, 597-603.

ERHARDT, M., STUSSI-GARAUD, C., GUILLEY, H., RICHARDS, K.E., JONARD, G., BOUZOUBAA, S., 1999. The first triple gene block protein of peanut clump virus localizes to the plasmodesmata during virus infection. *Virology*, 264, 220-229.

ESAU, K., THORSCH, J., 1985. Sieve plate pores and plasmodesmata, the communication channels of the symplast: ultrastructural aspects and developmental relations. *Am. J. Bot.*, 72, 1641-1653.

ETSCHEID, M., TOUSIGNANT, M.E., KAPER, J.M., 1995. Small satellite of arabis mosaic virus: autolytic processing of *in vitro* transcripts of (+) and (–) polarity and infectivity of (+) strand transcripts. *J. Gen. Virol.*, 76, 271-282.

EVANS, R.K., HALEY, B.E., ROTH, D.A., 1985. Photoaffinity labeling of a viral induced protein from tobacco. Characterization of nucleotide-binding properties. *J. Biol. Chem.*, 260, 7800-7804.

FABRE, F., DEDRYVER, C.A., LETERRIER, J.L., PLANTAGENEST, M., 2004. Aphid abundance on cereals in autumn predicts yield losses caused by Barley yellow dwarf virus. *Phytopathology*, 93, 1217-1222.

FACCIOLI, G., MARANI, F., 1998. Virus elimination by meristem tip culture and tip micrografting. In: *Plant Virus Disease Control*. Hadidi, A., Khetarpal, R.K., Koganezawa, H., eds. APS Press, St. Paul, Minnesota, 346-380.

FALK, B.W., 1997. Lettuce big-vein. In: *Compendium of Lettuce Diseases*. Am. Phytopathol. Soc., ed. APS Press, St. Paul, Minnesota.

FALK, B.W., DUFFUS, J.E., 1981. Epidemiology of helper-dependent persistent aphid transmitted virus complexes. In: *Plant Diseases and Vectors: Ecology and Epidemiology*. Maramorosch, K., Harris, K.F., eds. Academic Press, New York, 161-179.

FARABAUGH, P., 1996. Programmed translational frameshifting. *Microbiol. Rev.*, 60, 103-134.

FAUQUET, C.M., MAYO, M.A., MANILOFF, J., DESSELBERGER, U., BALL, L.A., 2005. Virus taxonomy. VIIIth Report of the International Committee on taxonomy of viruses. Elsevier Academic Press, San Diego, 1259 pp.

FELLERS, J., WAN, J., HONG, H., COLLINS, G.B., HUNT, A.G., 1998. *In vitro* interaction between a potyvirus-encoded, genome-linked protein and RNA-dependent RNA polymerase. *J. Gen. Virol.*, 79, 2043-2049.

FERNANDEZ, A.S., LAIN, S., GARCIA, J.A., 1995. RNA helicase activity of the plum pox potyvirus CI protein expressed in *Escherichia coli*. Mapping of an RNA binding domain. *Nucl. Acids Res.*, 23, 1327-1332.

FERRISS, R.S., BERGER, P.H., 1993. A stochastic simulation model of epidemics of arthropod-vectored plant viruses. *Phytopathology*, 83, 1269-1278.

FISHER, D.B., WU, Y., KU, M.S.B., 1992. Turnover of soluble proteins in the wheat sieve tube. *Plant Physiol.*, 100, 1433-1441.

FLOR, H.H., 1942. Inheritance of pathogenicity in *Melampsora lini*. *Phytopathology*, 32, 653-669.

FLORÈS, R., DI SERIO, F., HERNANDEZ, C., 1997. Viroids: the noncoding genomes. *Semin. Virol.*, 8, 65-73.

FLORÈS, R., RANDLES, J.W., BAR-JOSEPH, M., DIENER, T.O., 1998. A proposed scheme for virioid classification and nomenclature. *Arch. Virol.*, 143, 622-629.

FOSTER, T.M., LOUGH, T.J., EMERSON, S.J., LEE, R.H., BOWMAN, J.L., FORSTER, R.L., LUCAS, W.J., 2002. A surveillance system regulates selective entry of RNA into the shoot apex. *Plant Cell*, 14, 1497-1508.

FRAENKEL-CONRAT, H., 1986. RNA-directed RNA polymerases of plants. *Crit. Rev. Plant Sci.*, 4, 213-226.

FRAENKEL-CONRAT, H., SINGER, B., 1957. Virus reconstitution. II. Combination of protein and nucleic acid from different strains. *Biochim. Biophys. Acta*, 24, 540-548.

FRAENKEL-CONRAT, H., WILLIAMS, R.C., 1955. Reconstitution of active mosaic virus from its inactive protein and nucleic acid components. *Proc. Natl. Acad. Sci. USA*, 41, 690-698.

FRAILE, A., ALONSO-PRADOL., J.L., ARANDA, M.A., BERNAL, J.J., MALPICA, J.M., GARCIA-ARENAL, F., 1997. Genetic exchange by recombination or reassortment is infrequent in natural population of a tripartite RNA plant virus. *J. Virol.*, 71, 934-940.

FRAILE, A., ARANDA, M.A., GARCIA-ARENAL, F., 1995. Evolution of Tobamovirus. In: *The Molecular Basis of Virus Evolution*. Gibbs, A.J., Calisher, C.H., Garcia-Arenal, F., eds. Cambridge University Press, Cambridge, UK, 338-350.

FRANCKI, R.I.B., MILNE, R.G., HATTA, T., 1987. *Atlas of Plant Viruses*, Vol. 1 (222 pp.), Vol. 2 (284 pp.). CRC Press, Boca Raton, Florida.

FRANKLIN, R., 1955. Structure of tobacco mosaic virus. *Nature*, 177, 379-381.

FRANZ, A.W., VAN DER WILK, F., VERBEEK, M., DULLEMANS, A.M., VAN DEN HEUVEL, J.F., 1999. Faba bean necrotic yellows virus (genus *Nanovirus*) requires a helper factor for its aphid transmission. *Virology*, 262, 210-219.

FRASER, R.S., 1985. Genes for resistance to plant viruses. *Crit. Rev. Plant Sci.*, 3, 257-294.

FRASER, R.S., 1990. The genetics of resistance to plant viruses. *Annu. Rev. Phytopathol.*, 28, 179-200.

FRENCH, R., AHLQUIST, P., 1987. Intercistronic as well as terminal sequences are required for efficient amplification of brome mosaic virus RNA3. *J. Virol.*, 61, 1457-1465.

FUCHS, M., FERREIRA, S., GONSALVES, D., 1997. Management of virus diseases by classical and engineered cross protection. *Mol. Plant Pathol. On-Line*, [http://www.bspp.org.uk/mppol/] 1997/0116fuchs.

FUCHS, M., GONSALVES, D., 1997. Risk assessment of gene flow associated with the release of virus resistant transgenic crop plants. In: *Virus Resistant Transgenic Plants: Potential Ecological Impact*. Tepfer, M., Balazs, E., eds. Springer-INRA, Berlin, 114-119.

FUCHS, M., KLAS, F.E., MCFERSON, J.R., GONSLAVES, D., 1998. Transgenic melon and squash expressing coat protein genes of aphid-borne virus do not assist the spread of an aphid non-transmissible strain of *Cucumber mosaic virus* in the field. *Transgen. Res.*, 7, 449-462.

FULTON, R.W., 1986. Practices and precautions in the use of cross protection for plant virus disease control. *Annu. Rev. Phytopathol.*, 24, 67-81.

FURUKI, I., 1981. Epidemiological studies on melon necrotic spot. Tech. Bull. 14, Shizuoka Agric. Exp. Stn., Shizuoka-ken, Japan, 94 pp.

FURUSAWA, I., OKUNO, T., 1978. Infection with BMV of mesophyll protoplasts isolated from five plant species. *J. Gen. Virol.*, 40, 489-491.

GAEDIGK, K., ADAM, G., MUNDRY, K.W., 1986. The spike protein of potato yellow dwarf virus and its functional role in the infection of insect vector cells. *J. Gen. Virol.*, 67, 2763-2773.

GAFFNEY, T., FRIEDRICH, L., VERNOOIJ, B., NEGROTTO, D., NYE, G., UKNESS, S., WARD, E., KESSMANN, H., RYALS, J., 1993. Requirement of salicylic acid for the induction of systemic acquired resistance. *Science*, 261, 754-756.

GALLIE, D., 2001. Cap-independent translation conferred by the 5′ leader of tobacco etch virus is eucaryotic initiation factor 4G dependent. *J. Virol.*, 75, 12141-12152.

GALLIE, D.R., 1996. Translational control of cellular and viral mRNA. *Plant Mol. Biol.*, 32, 145-148.

GALLIE, D.R., 1998. A tale of two termini: a functional interaction between the termini of an mRNA is a prerequisite for efficient translation initiation. *Gene*, 216, 1-11.

GALLIE, D.R., SLEAT, D.E., WATTS, J.W., TURNER, P.C., WILSON, T.M.A., 1987. The 5′ leader sequence of tobacco mosaic virus RNA enhances the expression of foreign genes transcripts *in vitro* and *in vivo*. *Nucl. Acids Res.*, 15, 3257-3273.

GALLIE, D.R., WALBOT, V., 1992. Identification of the motifs within the tobacco mosaic virus 5′ leader responsible for enhancing translation. *Nucl. Acids Res.*, 20, 4631-4638.

GAL-ON, A., KAPLAN, I., ROOSSINCK, M.J., PALUKAITIS, P., 1994. The kinetics of infection of Zucchini squash by cucumber mosaic virus indicates a function for RNA1 in virus movement. *Virology*, 205, 280-289.

GAL-ON, A., RACCAH, B., 2000. A point mutation in the FRNK motif of the Potyvirus helper component-protease gene alters symptom expression in cucurbits and elicits protection against the severe homologous virus. *Phytopathology*, 90, 467-473.

GAMEZ, R., LEON, P., 1986. Maize rayado fino and related viruses. In: *The Plant Viruses*, Vol. 3, *Polyhedral Virions with Monopartite RNA Genomes*. Koenig, R., ed. Plenum Press, New York, 213-233.

GAO, Z., JOHANSEN, E., EYERS, S., THOMAS, C.L., NOEL ELLIS, T.H., MAULE, A.J., 2004. The potyvirus recessive resistance gene, sbm1, identifies a novel role for translation initiation factor eIF4E in cell-to-cell trafficking. *Plant J.*, 40, 376-385.

GARGOURI, R., JOSHI, R.L., ASTIER-MANIFACIER, S., HAENNI, A.L., 1989. Mechanism of synthesis of turnip yellow mosaic virus coat protein subgenomic RNA *in vitro*. *Virology*, 171, 386-393.

GARNIER, M., CANDRESSE, T., BOVE, J.M., 1986. Immunocytochemical localization of TYMV-coded structural and non structural proteins by the protein A-gold technique. *Virology*, 151, 100-109.

GARRET, A., KERLAN, C., THOMAS, D., 1993. The intestine is a site of passage for potato leaf roll virus from the gut lumen into the haemocoel in the aphid vector *Myzus persicae* Sulz. *Arch. Virol.*, 131, 377-392.

GAZO, B.M., MURPHY, P., GATCHEL, J.R., BROWNING, K.S., 2004. A novel interaction of Cap-binding protein complexes eukaryotic initiation factor (eIF)4F and eIF(iso)4F with a region in the 3'-untranslated region of satellite tobacco necrosis virus. *J. Biol. Chem.*, 279, 13584-13592.

GEERING, A., OLSZEVSKI, N., DAHAL, G., THOMAS, J., LOCKART, B., 2001. Analysis of the distribution and structure of integrated *Banana streak virus* DNA in a range of *Musa* cultivars. *Mol. Plant. Pathol.* 2, 207-213.

GERA, A., LOEBENSTEIN, G., RACCAH, B. 1978. Protein coats of two strains of cucumber mosaic virus affect transmission by *Aphis gossypii*. *Phytopathology*, 69, 396-399.

GIANINAZZI, S., AHL, P. 1983. The genetic and molecular basis of "β-proteins" in the genus *Nicotiana*. *Neth. J. Plant Pathol.*, 89, 275-281.

GIBBS, A., 1976. Viruses and plasmodesmata. In: *Intercellular Communications in Plants: Studies on Plasmodesmata*. Gunning, B.E.S., Robards, A.W., eds. Springer-Verlag, Berlin, 149-164.

GIBBS, A., HARRISON, B., 1976. *Plant Virology, The Principles*. Edward Arnold Ltd., London, 292 pp.

GIBBS, A., KEESE, P.K., 1995. In search of the origin of viral genes. In: *The Molecular Basis of Virus Evolution*. Gibbs, A.J., Calisher, C.H., Garcia-Arenal, F., eds. Cambridge University Press, Cambridge, UK, 76-90.

GIBBS, A., MACKENZIE, A., 1997. A primer pair for amplifying part of the genome of all potyvirids by RT-PCR. *J. Virol. Meth.* 63, 9-16.

GIBBS, M.J., COOPER, J.I., 1995. A recombinational event in the history of luteoviruses probably induced by base-pairing between the genomes of two distinct viruses. *Virology*, 206, 1129-1132.

GIBBS, M.J., WEILLER, G.F., 1999. Evidence that a plant virus switched hosts to infect a vertebrate and then recombine with a vertebrate-infecting virus. *Proc. Natl. Acad. Sci. USA*, 96, 8022-8027.

GIEGÉ, R., 1996. Interplay of tRNA-like structures from plant viral RNAs with partners of translation and replication machineries. *Proc. Natl. Acad. Sci. USA*, 93, 12078-12081.

GIERER, A., SCHRAMM, G., 1956. Infectivity of ribonucleic acid from tobacco mosaic virus. *Nature*, 177, 702-703.

GILBERTSON, R.L., LUCAS, W.J., 1996. How do viruses traffic on the "vascular highway"? *Trends Plant Sci.*, 1, 260-268.

GILDOW, F.E., 1982. Coated vesicle transport of luteovirus through the salivary gland of *Myzus persicae*. *Phytopathology*, 72, 1289-1296.

GILDOW, F.E., 1985. Transcellular transport of barley yellow dwarf virus into the haemocoel of the aphid vector *Rhopalosiphum padi*. *Phytopathology*, 75, 292-297.

GILDOW, F.E., 1999. Luteovirus and mechanisms regulating vector specificity. In: *The Luteoviridae*. Smith, H.G., Barker, H., eds. CABI Publishing, Wallingford, UK, 88-112.

GILDOW, F.E., GRAY, S.M., 1993. The aphid salivary gland basal lamina as a selective barrier associated with a vector-specific transmission of barley yellow dwarf luteovirus. *Phytopathology*, 83, 1293-1302.

GILMER, D., BOUZOUBAA, S., HEHN, A., GUILLEY, H., RICHARDS, K., JONARD, G., 1992. Efficient cell-to-cell movement protein of beet necrotic yellow vein virus requires 3' proximal genes located on RNA 2. *Virology*, 189, 40-47.

GODARD, O. (DIR.), 1997. *Le principe de précaution dans la conduite des affaires humaines*. INRA-MSH Co-publication, Paris, 351 pp.

GODARD, O., 2000. Le principe de précaution est un principe d'action. *La Recherche*, 330, 111.

GOELET, P., LOMONOSSOFF, G.P., BUTLER, P.J., AKAM, M.E., GAIT, M.J., KARN, J., 1982. Nucleotide sequence of tobacco mosaic virus RNA. *Proc. Natl. Acad. Sci. USA*, 79, 5818-5822.

GOLDBACH, R., DE HAAN, P., 1995. RNA viral supergroups and the evolution of RNA viruses. In: *The Evolutionary Biology of Viruses*. Morse, S., ed. Raven Press, New York, 105-119.

GOLDBACH, R., LE GALL, O., WELLINK, J., 1991. Alphavirus in plants. *Sem. Virol.*, 2, 19-25.

GOLDBACH, R., WELLINK, J. 1988. Evolution of plus-strand RNA viruses. *InterVirology*, 29, 260-267.

GOODWIN, J., CHAPMAN, K., SWANEY, S., PARKS, D., WERNSMAN, E.A., DOUGHERTY, W.G., 1996. Genetical and biochemical dissection of transgenic RNA-mediated virus resistance. *Plant Cell*, 8, 95-105.

GORA-SOCHACKA, A., KIERZEK, A., CANDRESSE, T., ZAGORSKI, W., 1997. The genetic stability of potato spindle tuber viroid (PSTVd) molecular variants. *RNA*, 3, 68-74.

GORBALENYA, A.E., 1995. Origin of RNA viral genomes; approaching the problem by comparative sequence analysis. In: *Molecular Basis of Virus Evolution*. Gibbs, A.J., Calisher, C.H., Garcia-Arenal, F., eds. Cambridge University Press, Cambridge, UK, 49-66.

GORBALENYA, A.E., KOONIN, E.V., 1989. Viral proteins containing the purine NTP-binding sequence pattern. *Nucl. Acids Res.*, 17, 8413-8440.

GORBALENYA, A.E., KOONIN, E.V., WOLF, Y.I., 1990. A new superfamily of putative NTP-binding domains encoded by genome of small DNA and RNA viruses. *FEBS Lett.*, 262, 145-148.

GOREGAOKER, S., CULVER, J., 2003. Oligomerization and activity of the helicase domain of the Tobacco mosaic virus 126- and 183-kilodalton replicase protein. *J. Virol.* 77, 3549-3556

GOVIER, D.A., KASSANIS, B., 1974. A virus component of plant sap needed when aphids acquire potato virus Y from purified preparations. *Virology*, 61, 420-426.

GOVIER, D.A., KASSANIS, B., PIRONE, T.P., 1977. Partial purification and characterization of the potato virus Y helper component. *Virology*, 78, 306-314.

GRAHAM, L.E., COOK, M.E., BUSSE, J.S., 2000. The origin of plants: body plan changes contributing to a major evolutionary radiation. *Proc. Natl. Acad. Sci. USA*, 97, 4535-4540.

GRANDBASTIEN, M.-A., 1998. Activation of plant retrotransposons under stress conditions. *Trends Plant Sci.*, 3, 181-187.

GRANIER, F., DURAND-TARDIF, M., CASSE-DELBARD, F., LECOQ, H., ROBAGLIA, C., 1993. Mutations in zucchini yellow mosaic virus helper component protein associated with loss of aphid transmissibility. *J. Gen. Virol.*, 74, 2737-2742.

GRAY, S.M., BANERJEE, N., 1999. Mechanisms of arthropod transmission of plant and animal viruses. *Microbiol. Mol. Biol. Rev.*, 63, 128-148.

GRAY, S.M., MOYER, J.W., KENNEDY, G.G., LEE CAMPBELL, C., 1986. Virus suppression and aphid resistance effects on spatial and temporal spread of Watermelon mosaic virus 2. *Phytopathology*, 76, 1254-1259.

GRAY, S.T., 1996. Plant virus proteins involved in natural vector transmission. *Trends Microbiol.*, 4, 259-264.

GREENE, A.E., ALLISON, R.F., 1994. Recombination between viral RNA and transgenic plant transcripts. *Science*, 263, 1423-1425.

GREENE, A.E., ALLISON, R.F., 1996. Deletions in the 3′ untranslated region of cowpea chlorotic mottle virus transgene reduce recovery of recombinant viruses in transgenic plants. *Virology*, 225, 231-234.

GRIEP, R., VAN TWISK, C., SCHOTS, A., 1999. Selection of beet necrotic yellow vein virus specific single-chain Fv antibodies from a semi synthetic combinatorial antibodies library. *Eur. J. Pl. Pathol.*, 105, 147-156.

GRIMSLEY, N., HOHN, T., DAVIES, J.W., HOHN, B., 1987. Agrobacterium mediated delivery of infectious maize streak virus into maize plants. *Nature (London)*, 325, 177-179.

GRISCELLI, C., 1997. Les virus, acteurs d'une complexité encore insoupçonnée. *Virologie*, 1, 451-452.

GROGAN, R.G., 1980. Control of lettuce mosaic with virus-free seed. *Plant Dis.*, 64, 446-449.

GRUMET, R., 1994. Development of virus resistant plants via genetic engineering. *Plant Breed. Rev.*, 12, 47-79.

GRUMET, R., SANFORD, J.C., JOHNSTON, S.A., 1987. Pathogen-derived resistance to viral infection using a negative regulatory molecule. *Virology*, 161,561-569.

GUESDON, J.L., TERNYNCK, T., AVRAMEAS, S., 1979. Use of avidin-biotin interaction in immunoenzymatic techniques. *J. Histochem. Cytochem.*, 27, 1131-1139.

GUO, H.S., GARCIA, J.A., 1997. Delayed resistance to Plum Pox potyvirus mediated by a mutated RNA replicase gene: involvement of a gene silencing mechanism. *Mol. Plant-Microbe Interact.*, 10, 160-170.

GUO, L., ALLEN, E., MILLER, W.A., 2001. Base-pairing between untranslated regions facilitates translation of uncapped, nonpolyadenylated viral RNA. *Mol. Cell*, 7, 1103-1109.

GURA, T., 1999. Repairing the genome's spelling mistakes. *Science*, 285, 316-318.

GUTIEREZ, C., 2000. DNA replication and cell cycle in plants: learning from geminiviruses. *EMBO J.*, 19(5), 792-799.

HALDEMAN-CAHILL, R., DAROS, J.A., CARRINGTON, J.C., 1998. Secondary structures in the capsid protein coding sequence and 3′ nontranslated region involved in amplification of the tobacco etch virus genome. *J. Virol.*, 72, 4072-4079.

HALK, E.L., DE BOER, S.H., 1985. Monoclonal antibodies in plant disease research. *Annu. Rev. Phytopathol.*, 23, 321-350.

HALLÉ, F., 1999. *Eloge de la plante.* Seuil, Paris, 347 pp.

HAMILTON, A.J., BAULCOMBE, D.C., 1999. A species of small antisense RNA in post-transcriptional gene silencing in plants. *Science*, 286, 950-952.

HAMILTON, R.I., 1980. Defenses triggered by previous invaders: viruses. In: *Plant Diseases: an Advanced Treatise*, Vol. 5. Horsfall, J.G., Cowling, E.B., eds. Academic Press, New York. 279-299.

HAMILTON, R.I., NICHOLS, C., 1977. The influence of Bromegrass Mosaic Virus on the replication of Tobacco Mosaic Virus in Hordeum vulgare. *Phytopathology*, 67, 484-489.

HAMMOND KOSACK, K.E., JONES, J.D.G., 1996. Resistance gene-dependent plant defense responses. *Plant Cell*, 8, 1773-1791.

HAMMOND KOSACK, K.E., JONES, J.D.G., 1997. Plant disease resistance genes. *Annu. Rev. Plant Physiol. Plant Mol. Biol.*, 48, 575-607.

HAMMOND, J., KAMO, K.K., 1995. Effective resistance to potyviruses infection conferred by expression of antisense RNA in transgenic plants. *Mol. Plant-Microbe Interact.*, 8, 674-682.

HAMMOND, J., LECOQ, H., RACCAH, B., 1999. Epidemiological risks from mixed virus infections and transgenic plants expressing viral genes. *Adv. Virus Res.*, 54, 189-314.

HAMMOND, R.W., ZHAO, Y., 2000. Characterization of a tomato protein kinase induced by infection by Potato spindle tuber viroid. *Mol. Plant-Microbe Interact.*, 13, 903-910.

HAMPTON, R.O., FRANCKI, R.I., 1992. RNA-1 dependent seed transmissibility of cucumber mosaic virus in *Phaseolus vulgaris*. *Phytopathology*, 82, 127-130.

HAN, S., SANFACON, H., 2003. *Tomato ringspot virus* proteins containing the nucleoside triphosphate binding domain are transmembrane proteins that associate with the endoplasmic reticulum and cofractionate with replication complexes. *J. Virol.*, 77, 523-534

HANADA, K., HARRISON, B.D., 1977. Effects of virus genotype and temperature on seed transmission of nepoviruses. *Ann. Appl. Biol.*, 85, 79-92.

HARPER, G., HULL, R., LOCKHART, B., OLSZEWSKI, N., 2002. Viral sequences integrated into plant genomes. *Annu. Rev. Phytopathol.*, 40, 119-136.

HARPER, G., OSUJI, J.O., (PAT) HESLOP-HARRISON, J.S., HULL, R., 1999. Integration of Banana streak badnavirus into the *Musa* genome: molecular and cytogenetic evidence. *Virology*, 255, 207-213.

HARRIS, K.F., 1977. An ingestion-egestion hypothesis of noncirculative virus transmission. In: *Aphids as Virus Vectors*. Harris, K.F., ed. Academic Press, New York, 166-208.

HARRIS, K.F., BATH, J.E., 1972. The fate of pea enation mosaic virus in its aphid vector *Acyrthosiphon pisum*. *Virology*, 50, 778-790.

HARRIS, K.F., BATH, J.E., THOTTAPILLY, G., HOOPER, G.R., 1975. Fate of pea enation virus in PEMV-injected pea aphids. *Virology*, 65, 148-162.

HARRISON, B.D., 1981. Plant virus ecology: ingredients, interactions and environmental influences. *Ann. Appl. Biol.*, 99, 195-209.

HARRISON, B.D., ROBINSON, D.J., 1988. Molecular variations in vector-borne plant viruses: epidemiological significance. *Phil. Trans. R. Soc. Lond.*, B321, 447-462.

HASELOFF, J., GOELET, P., ZIMMERN, D., ALQUIST, P., DASGUPTA, R., KAESBERG, P., 1984. Striking similarities in amino acid sequence among nonstructural proteins encoded by RNA viruses that have dissimilar genomic organization. *Proc. Natl. Acad. Sci. USA*, 81, 4358-4362.

HAYES, R.J., BUCK, K.W., 1990. Complete replication of an eukaryotic virus RNA *in vitro* by a purified RNA-dependent RNA polymerase. *Cell*, 63, 363-368.

HAYES, R.J., TOUSCH, D., JACQUEMOND, M., PEREIRA, V.C., BUCK, K.W., TEPFER, M., 1992. Complete replication of a satellite RNA *in vitro* by a purified RNA-dependent RNA polymerase. *J. Gen. Virol.*, 73, 1597-1600.

HÉBRARD, E., FROISSART, R., LOUIS, C., BLANC, S., 1999. Les modes de transmission des virus phytopathogènes par vecteurs. *Virologie*, 3, 35-48.

HEINLEIN, M., EPEL, B.L., 2004. Macromolecular transport and signaling through plasmodesmata. *Int. Rev. Cytol.*, 235, 93-164.

HEINLEIN, M., EPEL, B.L., PADGETT, H.S., BEACHY, R.N. 1995. Interaction of tobamovirus movement proteins with the plant cytoskeleton. *Science*, 270, 1983-1985.

HEMMER, O., DUNOYER, P., RICHARDS, K., FRITSCH, C., 2003. Mapping of viral RNA sequences required for assembly of peanut clump virus particle. *J. Gen. Virol.*, 84, 2585-2594.

HEMMER, O., ONCINO, C., FRITSCH, C. 1993. Efficient replication of the *in vitro* transcripts from cloned cDNA of tomato black ring virus satellite RNA requires the 48K satellite RNA-encoded protein. *Virology*, 194, 800-806.

HENSON, J.M., FRENCH, R. 1993. The polymerase chain reaction and plant disease diagnosis. *Annu. Rev. Phytopathol.*, 31, 81-109.

HERRERA-ESTRELLA, A., CHEN, Z.M., MONTAGU, M., WANG, K., 1988. Vir D proteins of *Agrobacterium tumefaciens* are required for the formation of a covalent DNA-protein complex at the 5′ terminus of T-strand molecules. *EMBO J.*, 7, 4055-4062.

HEWITT, W.B., RASKI, D.J., GOHEEN, A.C., 1958. Nematode vector of soil borne fan-leaf virus of grapevines. *Phytopathology*, 48, 586-895.

HIEBERT, E., PURCIFULL, D.E., CHRISTIE, R.G., 1984. Purification and immunological analyses of plant viral inclusion bodies. In: *Methods in Virology*, Vol. VIII. Maramorosch, K., Koprowski, H., eds. Academic Press, London, 225-280.

HIGGINS, V.J., LU, H., XING, T., GELLI, A., BLUMWALD, E., 1998. The gene-for-gene concept and beyond: interactions and signals. *Can. J. Plant Pathol.*, 20, 150-157.

HILF, M.E., DAWSON, W.O., 1993. The tobamovirus capsid functions as a host specific determinant of long distance movement. *Virology*, 193, 106-114.

HILLS, G.J., PLASKITT, K.A., YOUNG, N.D., DUNIGAN, D.D., WATTS, J.W., WILSON, T.M.A., ZAITLIN, M., 1987. Immunogold localization of the intracellular sites of structural and non-structural tobacco mosaic virus proteins. *Virology*, 161, 488-496.

HIMBER, C., DUNOYER, P., MOISSIARD, G., RITZENTHALER, C., VOINNET, O., 2003. Transitivity-dependent and -independent cell-to-cell movement of RNA silencing. *EMBO J.*, 17, 4523-4533.

HÖFER, P., BEDFORD, I.D., MARKHAM, P.G., JESKE, H., FRISCHMUTH, T., 1997. Coat protein gene replacement results in whitefly transmission of an insect nontransmissible geminivirus isolate. *Virology*, 236, 288-295.

HOLLAND, J., DOMINGO, E., 1998. Origin and evolution of viruses. *Virus Genes*, 16, 13-21.

HOLMES, F.O., 1938. Inheritance of resistance to tobacco mosaic disease in tobacco. *Phytopathology*, 28, 553-561.

HOLMES, F.O., 1956. Elimination of aspermy virus from the Nightingale chrysanthemum. *Phytopathology*, 46, 599-600.

HOLT, C.A., BEACHY, R.N., 1991. *In vivo* complementation of infectious transcripts from mutant tobacco mosaic virus cDNA in transgenic plants. *Virology*, 181, 109-117.

HOLT, J., CHANCELLOR, T.C.B., 1997. A model of plant virus disease epidemics in asynchronously-planted cropping systems. *Plant Pathol.*, 46, 490-501.

HONG, Y., HUNT, A., 1996. RNA polymerase activity catalysed by a potyvirus-encoded RNA polymerase. *Virology*, 226, 146-151.

HONG, Y., SAUNDERA, K., HARTLEY, M.R., STANLEY, J., 1996. Resistance to geminivirus infection by virus-induced expression of dianthine in transgenic plants. *Virology*, 220, 119-127.

HORMUZDI, S.G., BISARO, D.M., 1993. Genetic analysis of beet curly top virus: evidence for three virion sense genes involved in movement and regulation of single- and double-stranded DNA levels. *Virology*, 193, 900-909.

HOURANI, H., ABOU-JAWDAH, Y., 2003. Immunodiagnosis of cucurbit yellow stunting disorder virus using polyclonal antibodies developed against recombinant coat protein. *J. Pl. Pathol.*, 85, 197-204.

HU, Y., FELDSTEIN, P.A., BOTTINO, P.J., OWENS, R.A., 1996. Role of the variable domain in modulating potato spindle tuber viroid replication. *Virology*, 219, 45-56.

HUET, H., GAL-ON, A., MEIR, E., LECOQ, H., RACCAH, B. 1994. Mutations in the helper component protease gene of zucchini yellow mosaic virus affect its ability to mediate aphid transmissibility. *J. Gen. Virol.*, 75, 1407-1414.

HUET, H., MAHENDRA, S., SIVAMANI, E., ONG, C.A., DE KOTCHKO, A., BEACHY, R.N., FAUQUET, C., 1999. Near immunity to rice tungro spherical virus achieved in rice by a replicase-mediated resistance strategy. *Phytopathology*, 89, 1022-1027.

HUGLIN, P., GUILLOT, R., VALAT, C., VUITTENEZ, A., 1980. Evaluation génétique et sanitaire du matériel clonal de la vigne. *Bull. Office Intl., Vigne Vin*, 597, 857-882.

HULL, R., 1996. Molecular biology of rice tungro viruses. *Annu. Rev. Phytopathol.*, 34, 275-297.

HULL, R., 2002. *Matthews' Plant Virology*, 4th ed. Academic Press, London, 1001 pp.

HUNT, M.D., RYALS, J.A., 1996. Systemic acquired resistance signal transduction. *Crit. Rev. Plant Sci.*, 15, 583-606.

HYAKAYAWA, T., ZHU, Y., ITHO, K., KIMURA, Y., IZAWA, T., SHIMAMOTO, K., TORIYAMA, S., 1992. Genetically engineered rice resistant to rice stripe virus, an insect-transmitted virus. *Proc. Natl. Acad. Sci. USA*, 89, 9865-9869.

INOUE-NAGATA, A.K., KOMELINK, R., NAGATA, R., KITAJIMA, E.W., GOLDBACH, R., PETERS, D., 1997. Temperature and host effect on the generation of tomato spotted wilt virus defective interfering RNAs. *Phytopathology*, 87, 1168-1173.

ISHIKAWA, M., DIEZ, J., RESTREPO-HARTWIG, M., AHLQUIST, P., 1997b. Yeast mutations in multiple complementation groups inhibit brome mosaic virus RNA replication and transcription and perturb regulated expression of viral polymerase-like genes. *Proc. Natl. Acad. Sci. USA*, 94, 13810-13815.

ISHIKAWA, M., JANDA, M., KROL., M., AHLQUIST, P., 1997a. *In vivo*, DNA expression of functional brome mosaic virus RNA replicons in *Saccharomyces cerevisiae*. *J. Virol.*, 71, 7781-7790.

ISHIKAWA, M., MESHI, T., OHNO, T., OKADA, Y., 1991. Specific cessation of minus-strand RNA accumulation at an early stage of tobacco mosaic virus infection. *J. Virol.*, 65, 861-868.

ISHIWATARI, Y., FUJIWARA, T., MCFARLAND, K.C., NEMOTO, K., HAYASHI, H., CHINO, M., LUCAS, W.J., 1998. Rice phloem thioredoxin has the capacity to mediate its own cell-to-cell transport through plasmodesmata. *Planta*, 205, 12-22.

ISKRA-CARUANA, M.L., LHEUREUX, F., TEYCHENEY, P.Y., 2003. Les pararétrovirus endogènes (EPRV), voie nouvelle de transmission des virus de plantes. *Virologie*, 7, 255-265.

IVANOVSKI, D., 1892. Üeber die Mosaikkrankheit des Tabakspflanze. *St. Petersb. Acad. Imp. Sci. Bul.*, 35, 67-70. (English translation : Concerning the mosaic disease of tobacco plant, in Johnson J., ed., Phytopathological Classics N°7, *American Phytopathological Society*, Saint Paul, Minnesota, 1942, 27-30).

JACKSON, A.O., ZAITLIN, M., SIEGEL, A., FRANCKI, R.I.B., 1972. Replication of tobacco mosaic virus. *Virology*, 48, 655-665.

JACOB, F., 1981. *Le jeu des possibles*. Fayard, Paris.

JACOBI, V., BACHAND, G.D., HAMELIN, R.C., CASTELLO, J.D., 1998. Development of a multiplex immunocapture RT-PCR assay for detection and differentiation of tomato and tobacco mosaic tobamoviruses. *J. Virol. Methods*, 74, 167-178.

JACQUEMOND, M., AMSELEM, J., TEPFER, M., 1988. A gene coding for a monomeric form of cucumber mosaic virus satellite RNA confers tolerance to CMV. *Mol. Plant-Microbe Interact.*, 1, 311-316.

JACQUEMOND, M., TEPFER, M., 1998. Satellite RNA-mediated resistance to plant viruses: are the ecological risks well assessed? In: *Plant Virus Disease Control*. Hadidi, A., Khetarpal, R.K., Koganezawa, H., eds. APS Press, St. Paul, Minnesota, 94-120.

JACQUET, C., DELÉCOLLE, B., RACCAH, B., LECOQ, H., DUNEZ, J., RAVELONANDRO, M., 1998. Use of plum pox virus coat protein genes developed to limit heteroencapsidation-associated risks in transgenic plants. *J. Gen. Virol.*, 79, 1509-1517.

JACQUOT, E., DAUTEL, S., LEH, V., GELDREICH, A., YOT, P., KELLER, M., 1997. Les pararétrovirus de plantes. *Virologie*, 1, 111-120.

JAKOWITSCH, J., METTE, M.F., VAN DER WINDEN, J., MATZKE, M.A., MATZKE, A.J., 1999. Integrated pararetroviral sequence define a unique class of dispersed repetitive DNA in plants. *Proc. Natl. Acad. Sci. USA*, 96, 13241-13246.

JOHANSEN, E., EDWARDS, M.C., HAMPTON, R.O., 1994. Seed transmission of viruses: current perspectives. *Annu. Rev. Phytopathol.*, 32, 363-386.

JOHANSEN, I.E., DOUGHERTY, W.G., KELLER, K.E., WANG, D., HAMPTON, R.O., 1996. Multiple viral determinants affect seed transmission of pea seed borne mosaic virus in *Pisum sativum*. *J. Gen. Virol.*, 77, 3149-3154.

JOHANSEN, I.E., LUND, O.S., HJULSAGER, C.K., LAURSEN, J., 2001. Recessive resistance in *Pisum sativum* and potyvirus pathotype resolved in a gene-for-cistron correspondence between host and virus. *J. Virol.*, 75, 6609-6614.

JOHNSON, J.E., 1996. Functional implications of protein-protein interaction in icosahedral viruses. *Proc. Natl. Acad. Sci. USA*, 93, 27-33.

JOHNSON, J.E., SPEIR, J.A., 1997. Quasi-equivalent viruses: a paradigm for protein assemblies. *J. Mol. Biol.*, 269, 665-675.

JOLLY, C.A., MAYO, M.A., 1994. Changes in the amino acid sequence of the coat protein readthrough domain of potato leaf roll luteovirus affect the formation of an epitope and aphid transmission. *Virology*, 201, 182-185.

JONARD, G., RICHARDS, K.E., GUILLEY, H., HIRTH, L., 1977. Sequence from the assembly nucleation region of tobacco mosaic virus RNA. *Cell*, 11, 483-493.

JONES, A.T., 1987. Control of virus infection in crop plants through vector resistance: a review of achievements, prospects and problems. *Ann. Appl. Biol.*, 11, 745-772.

JONES, A.T., 1998. Control of virus infection in crops through breeding plants for vector resistance. In: *Plant Virus Disease Control*. Hadidi, A., Khetarpal, R.K., Koganezawa, H., eds. APS Press, St. Paul, Minnesota, 41-55.

JONES, R.A., 2004. Using epidemiological information to develop effective integrated virus diseases management strategies. *Virus Res.*, 100, 5-30.

JORDAN, R., 1992. Potyviruses, monoclonal antibodies and antigenic properties. *Arch. Virol. Suppl.*, 5, 81-95.

JORGENSEN, R.A., ATKINSON, R.G., FORSTER, R.L., LUCAS, W.J., 1998. An RNA-based information superhighway in plants. *Science*, 279, 1486-1487.

JUPIN, I., DE KOUCHKOVSKY, F., JOUANNEAU, F., GRONENBORN, B., 1994. Movement of tomato yellow leaf curl geminivirus (TYLCV): involvement of the protein encoded by ORF C4. *Virology*, 204, 82-90.

JUPIN, I., HERICOURT, F., GRONENBORN, B., 1995. DNA replication specificity of TYLCV geminivirus is mediated by the amino-terminal 116 amino acids of the Rep protein. *Febs Lett.*, 362, 116-120.

KADARÉ, G., HAENNI, A.L., 1997. Les hélicases virales, une nouvelle famille d'enzymes. *Virologie*, 2, 459-469.

KAHN, T.W., LAPIDOT, M., HEINLEIN, M., REICHEL, C., COOPER, B., GAFNY, R., BEACHY, R.N., 1998. Domains of the TMV movement protein involved in subcellular localization. *Plant J.*, 15, 15-25.

KAMER, G., ARGOS, P., 1984. Primary structural comparison of RNA-dependent polymerases from plant, animal and bacterial viruses. *Nucl. Acids Res.*, 12, 7269-7282.

KANIEVSKY, W., LAWSON, C., 1998. Coat protein and replicase-mediated resistance to plant viruses. In: *Plant Virus Disease Control*. Hadidi, A., Khetarpal, R.K., Koganezawa, H., eds. APS Press, St. Paul, Minnesota, 65-78.

KAO, C.C., AHLQUIST, P. 1992. Identification of the domains required for direct interaction of the helicase-like and polymerase-like RNA replication proteins of brome mosaic virus. *J. Virol.*, 66, 7293-7302.

KAO, C.C., SUN, J.H., 1996. Initiation of minus-strand RNA synthesis by the brome mosaic virus RNA-dependent RNA polymerase: use of oligonucleotides primers. *J. Virol.*, 70, 6826-6830.

KAPLAN, I.B., SHINTAKU, M.H., LI, Q., ZHANG, L., MARSH, L.E., PALUKAITIS, P., 1995. Complementation of virus movement in transgenic tobacco plants expressing the cucumber mosaic virus 3a gene. *Virology*, 209, 188-199.

KAPLAN, I.B., ZHANG, L., PALUKAITIS, P., 1998. Characterization of cucumber mosaic virus. V. Cell-to-cell movement requires capsid protein but not virions. *Virology*, 246, 221-231.

KARASAWA, A., OKADA, I., AKASHI, K., CHIDA, Y., HASE, S., NAKASAWA-NASU, Y., ITO, A., EHARA, Y., 1999. One amino acid change in cucumber mosaic virus RNA polymerase determines virulent/avirulent phenotypes on cowpea. *Phytopathology*, 89, 1186-1192.

KASSANIS, B., 1949. Potato tubers freed from leafroll virus by heat. *Nature*, 164, 881.

KASSANIS, B., GOVIER, D.A., 1971. New evidence on the mechanism of transmission of potato virus C and potato aucuba mosaic viruses. *J. Gen. Virol.*, 10, 99-101.

KASSANIS, B., RUSSELL, G.E., WHITE, R.F., 1978. Seed and pollen transmission of beet cryptic virus in sugar beet plants. *Phytopathol. Z.*, 91, 76-79.

KASSCHAU, K.D., CARRINGTON, J.C., 1998. A counterdefensive strategy of plant viruses: suppression of posttranscriptional gene silencing. *Cell*, 95, 461-470.

KASSCHAU, K.D., CARRINGTON, J.C., 2001. Long-distance movement and replication maintenance functions correlate with silencing suppression activity of potyviral HC-Pro. *Virology*, 285, 71-81.

KASSCHAU, K.D., CRONIN, S., CARRINGTON, J.C., 1997. Genome amplification and long distance movement functions associated with the central domain of tobacco etch potyvirus helper component proteinase. *Virology*, 228, 251-262.

KASTEEL, D.T.J., PERBAL, C.M., BOYER, J.C., WELLINK, J., GOLDBACH, R.W., MAULE, A.J., VAN LENT, J.W.M., 1996. The movement proteins of cowpea mosaic virus and cauliflower mosaic virus induce tubular structures in plant and insect cells. *J. Gen. Virol.*, 77, 2587-2864.

KASTEEL, D.T.J., WELLINK, J., GOLDBACH, R.W., VAN LENT, J.W.M., 1997. Isolation and characterization of tubular structures of cowpea mosaic virus. *J.Gen. Virol.*, 78, 3167-3170.

KAUFFMANN, S., LEGRAND, M., GEOFFROY, P., FRITIG, B., 1987. Biological function of "pathogenesis-related" proteins: four PR proteins of tobacco have 1,3-β-glucanase activity. *EMBO J.*, 6, 3209-3212.

KAUSCHE, G., PFANKUCHE, E., RUSKA, H., 1939. Die sichtbarmachung von planzlichen virus im ubermicrokroskop. *Naturwissenschaften* 27, 292-299.

KAY, R., CHAN, A., DALY, M., MCPHERSON, J., 1987. Duplication of CaMV-35S promoter sequences creates a strong enhancer for plant genes. *Science*, 236, 1299-1301.

KEESE, P., SYMONS, R.H., 1985. Domains in viroids: evidence intermolecular RNA rearrangements and their contribution to viroid evolution. *Proc. Natl. Acad. Sci. USA*, 82, 4582-4586.

KELLER, K.E., JOHANSEN, I.E., MARTIN, R.R., HAMPTON, R.O., 1998. Potyvirus genome-linked protein (VPg) determines pea seed-borne mosaic virus pathotype-specific virulence in *Pisum sativum*. *Mol. Plant-Microbe Interact.*, 11, 124-130.

KEMPERS, R., PRIOR, D.A.M., VAN BEL, A.J.E., OPARKA, K.J., 1993. Plasmodesmata between sieve element and companion cell of extrafascicular stem phloem of *Cucurbita maxima* permit passage of 3 kDa fluorescent probes. *Plant J.*, 4, 567-575.

KENDALL, C., IONESCU-MATIU, I., DREESMAN, G.R., 1983. Utilization of the biotin-avidin system to amplify the sensitivity of the enzyme-linked immunosorbent assay (ELISA). *J. Immunol. Meth.*, 56, 329-339.

KENNEDY, J.S., DAY, M.F., EASTOP, V.F., 1962. *A Conspectus of Aphids as Vectors of Plant Viruses.* Commonwealth Institute of Entomology, London, 114 pp.

KHEYR-POUR, A., BANANEJ, K., DAFALLA, G.A., CACIAGLI, P., NORIS, E., AHOONMANESH, A., LECOQ, H., GRONENBORN, B., 2000. Watermelon chlorotic stunt virus from the Sudan and Iran: sequence comparison and identification of a whitefly-transmission determinant. *Phytopathology,* 90, 629-635.

KIM, M., KAO, C., 2001. Factors regulating template switch *in vitro* by viral RNA-dependent RNA polymerase: implications for RNA-RNA recombination. *Proc. Natl. Acad. Sci. USA,* 98, 4972-4977.

KIMURA, M., 1989. The neutral theory of molecular evolution and the world view of the neutralists. *Genome,* 31, 24-31.

KIRKEGAARD, K., BALTIMORE, D., 1986. The mechanisms of RNA recombination in poliovirus. *Cell,* 47, 433-443.

KISS-LASZLO, Z., BLANC, S., HOHN, T., 1995. Splicing of cauliflower mosaic 35S RNA is essential for viral infectivity. *EMBO J.,* 14, 3552-3562.

KJEMTRUP, S., SAMPSON, K.S., PEELE, C.G., NGUYEN, L.V., CONKLING, M.A., THOMPSON, W.F., ROBERTSON, D., 1998. Gene silencing from plant DNA carried by a Geminivirus. *Plant J.,* 14, 91-100.

KLEIN, P.G., KLEIN, R.R., RODRIGUEZ-CEREZO, E., HUNT, A.G., SHAW, J.G., 1994. Mutational analysis of the tobacco vein mottling virus genome. *Virology,* 204, 759-769.

KLESSIG, D.F., DURNER, J., SHAH, J., YANG, Y., 1998. Salicylic acid-mediated signal transduction in plant disease resistance. In: *Phytochemical Signals and Plant-Microbe Interactions.* Romeo et al., eds. Plenum Press, New York, 119-137.

KNORR, D.A., DAWSON, W.O., 1988. A point mutation in the tobacco mosaic virus capsid protein gene induces hypersensitivity in *Nicotiana sylvestris. Proc. Natl. Acad. Sci. USA,* 85, 170-174.

KOENIG, R., 1981. Indirect ELISA methods for the broad specificity detection of plant viruses. *J. Gen. Virol.,* 55, 53-62.

KOENIG, R., 1986. Plant viruses in rivers and lakes. *Adv. Virus Res.,* 31, 321-333.

KOENIG, R., AN, D., LESMANN, D.E., BURGERMEISTER, W., 1988. Isolation of carnation ringspot virus from a canal near a sewage plant. *J. Phytopathol.,* 121, 346-356.

KOEV, G., MILLER, W.A., 2000. A positive strand RNA synthesis with three very different RNA promoters. *J. Virol.,* 74, 5988-5996.

KOEV, G., MOHAN, B.R., MILLER, W.A., 1999. Primary and secondary structural elements required for synthesis of barley yellow dwarf virus subgenomic RNA1. *J. Gen. Virol.,* 78, 2876-2885.

KOFALVI, S.A., MARCOS, J.F., CANIZARES, M.C., PALLAS, V., CANDRESSE, T., 1997. Hop stunt viroid (HSVd) sequence variants from *Prunus* species: evidence for recombination between HSVd isolates. *J. Gen. Virol.,* 78, 3177-3186.

KOHLSTAEDT, L.A., WANG, J., FRIEDMAN, J.M., RICE, P.A., STEITZ, T.A., 1992. Crystal structure at 3'5Å resolution of HIV-1 reverse transcriptase complexed with an inhibitor. *Science,* 256, 1783-1790.

KÖHM, B.A., GOULDEN, M.G., GILBERT, J.E., KAVANAGH, T.A., BAULCOMBE, D.C., 1993. A potato virus X resistance gene mediates an induced nonspecific resistance in protoplasts. *Plant Cell,* 5, 913-920.

KOLLAR, A., DALMAY, T., BURGYAN, J., 1993. Defective interfering RNA-mediated resistance against Cymbidium ringspot tombusvirus in transgenic plants. *Virology*, 193, 313-318.

KONATE, G., FRITIG, B., 1984. An efficient microinoculation procedure to study plant virus multiplication at predetermined individual infection sites on the leaves. *Phytopathol. Z.*, 109, 131-138.

KOONIN, E.V., 1991a. The phylogeny of RNA-dependent RNA polymerases of positive-strand RNA viruses. *J. Gen. Virol.*, 72, 2197-2207.

KOONIN, E.V., 1991b. Similarities in RNA helicases. *Nature*, 352, 290.

KOONIN, E.V., DOLJA, V.V., 1993. Evolution and taxonomy of positive strand RNA viruses: implications of comparative analysis of aminoacid sequences. *Crit. Rev. Biochem. Mol. Biol.*, 28, 375-430.

KOURILSKY, P., VINEY, G. 2000. *Le principe de précaution*. Odile Jacob, La Documentation Francaise, pp. 99-115.

KRCZAL, G., ALBOUY, J., DAMY, I., KUSIAK, C., DEOGRATIAS, J.M., MOREAU, J.P., BERKELMAN, B., WOHANKA, W., 1994. Transmission of pelargonium flower break virus (PFBV) in irrigation systems and by thrips. *Plant Dis.*, 72, 163-166.

KUDO, H., UYEDA, I., SHIKATA, E., 1991. Viruses in the Reoviridae family have the same conserved terminal sequences. *J. Virol.*, 72, 2857-2866.

KUMAR, A., 1998. The evolution of plant retroviruses: moving to green pastures. *Research News*, 3, 371-374

KUNKEL, L.O., 1935. Heat treatment for the cure of yellows and rosette in peach. *Phytopathology*, 25, 24.

KURATH, G., ROBAGLIA, C., 1995. Genetic variation and evolution of satellite viruses and satellite RNAs. In: *Molecular Basis of Virus Evolution*. Gibbs, A.J., Calisher, C.H., Garcia-Arenal, F., eds. Cambridge University Press, Cambridge, UK, 385-403.

KURSTAK, E., ed., 1981. *Handbook of Plant Virus Infections and Comparative Diagnosis*, 23. Hollings M., Brunt, A.A.: *Potyviruses*. Elsevier, New York.

KYLE, M.M., DICKSON, M.H., 1988. Linkage of hypersensitivity to five potyviruses with the *B* locus for seed color in *Phaseolus vulgaris*. *J. Hered.*, 79, 308-311.

LABONNE, G., FAUVEL, C., LECLANT, F., QUIOT, J.B., 1982. Description d'un piège à succion: son emploi dans la recherche des aphides vecteurs de virus transmis sur le mode non persistant. *Agronomie*, 2, 773-776.

LABONNE, G., QUIOT, J.B., 1988. L'épidémiologie des virus transmis sur le mode non-persistant. *Ann. Assoc. Natl. de Prot. des Plantes*, 1, 245-272.

LAHSER, F.C., MARSH, L.E., HALL, T.C., 1993. Contribution of the brome mosaic virus RNA3 3'-untranslated region to replication and translation. *J. Virol.*, 67, 3295-3303.

LAI, M.M.C., 1992. RNA combination in animal and plant viruses. *Microbiol. Rev.*, 56, 61-79.

LAI, M.M.C., 1998. Cellular factors in the transcription and replication of viral RNA genomes: a parallel to DNA-dependent RNA transcription. *Virology*, 244, 1-12.

LAIN, S., MARTIN, M.T., RIECHMANN, J.L., GARCIA, J.A., 1991. Novel catalytic activity associated with positive strand RNA virus infection: nucleic acid stimulated ATPase activity of the plum pox potyvirus helicaselike protein. *J. Gen. Virol.*, 65, 1-6.

LANDSTEINER, K., POPPER, E., 1909. Übertragung der Poliomyelitis acuta auf Affen. *Zeitschr. Immunitätsforsch. Orig.*, 2, 377-390.

Laporte, C., Vetter, G., Loudes, A.M., Robinson, D.G., Hillmer, S., Stussi-Garaud, C., Ritzenthaler, C., 2003. Involvement of the secretory pathway and the cytoskeleton in intracellular targeting and tubule assembly of Grapevine fanleaf virus movement protein in tobacco BY-2 cells. *Plant Cell*, 15(9), 2058-2075.

Laterrot, H., 1989. Intérêt et utilisation des espèces sauvages pour la création variétale. *PHM-Rev. Horticole*, 295, 13-17.

Lauber, E., Bleykasten-Grosshans, C., Erhardt, M., Bouzoubaa, S., Jonard, G., Richards, K.E., Guilley, H., 1998. Cell-to-cell movement of beet necrotic yellow vein virus: I. Heterologous complementation experiments provide evidence for specific interactions among the triple gene block proteins. *Mol. Plant-Microbe Interact.*, 11, 618-625.

Laufs, J., Traut, W., Heyraud, F., Matzeit, W., Rogers, S.G., Schell, J., Gronenborg, B., 1995. *In vitro* cleavage and joining at the viral origin of replication by the replication initiator protein of tomato yellow leaf curl virus. *Proc. Natl. Acad. Sci. USA*, 92, 3879-3883.

Lawson, R.H., Hearon, S., 1971. The association of pinwheel inclusions with plasmodesmata. *Virology*, 44, 454-456.

Leather, V., Tangay, R., Kobayashi, M., Gallie, M., 1993. A phylogenetically conserved sequence within 3′ untranslated RNA pseudoknot regulates translation. *Mol. Cell. Biol.*, 13, 5331-5334.

Lebeurrier, G., Hirth, L., 1966. Effect of elevated temperatures on the development of two strains of tobacco mosaic virus. *Virology*, 29, 385-395.

Lebeurrier, G., Nicolaieff, A., Richards, K.E., 1977. Inside-out model for self assembly of tobacco mosaic virus. *Proc. Natl. Acad. Sci. USA*, 74, 149-153.

Leclant, F., 1982. Les effets nuisibles des pucerons sur les cultures. In: *Les Pucerons des Cultures*. ACTA, Paris, 35-56.

Lecoq, H., 1992. Les virus des cultures de melon et de courgette de plein champ. *PHM-Rev. Horticole*, 324, 15-25.

Lecoq, H., 1996. La dissémination des maladies à virus des plantes. *PHM-Rev. Horticole*, 365, 13-20.

Lecoq, H., 1998. Control of plant virus diseases by cross protection. In: *Plant Virus Disease Control*. Hadidi, A., Khetarpal, R.K., Koganezawa, H., eds. APS Press, St. Paul, Minnesota, 33-40.

Lecoq, H., Cohen, S., Pitrat, M., Labonne, G., 1979. Resistance to cucumber mosaic virus transmission by aphids in *Cucumis melo*. *Phytopathology*, 69, 1223-1225.

Lecoq, H., Lemaire, J.M., Wipf-Scheibel, C., 1991. Control of zucchini yellow mosaic virus in squash by cross protection. *Plant Dis.*, 75, 208-211.

Lecoq, H., Moury, M., Desbiez, C., Palloix, A., Pitrat, M., 2004. Durable virus resistance in plants through conventional approaches: a challenge. *Vir. Res.*, 100, 31-39.

Lecoq, H., Pitrat, M., 1982. Field experiments on the integrated control of aphid-borne viruses in musk-melon. In: *Plant Virus Epidemiology*. Tresh, J.M., Plumb, R.T., eds. Blackwell, Oxford, 169-176.

Lecoq, H., Pitrat, M., 1984. Strains of zucchini yellow mosaic virus in melon (*Cucumis melo* L.). *Phytopathol. Z.*, 111, 165-173.

Lecoq, H., Pitrat, M., 1989. Effects of resistance on the epidemiology of virus disease of cucurbits. In: *Cucurbitaceae 89: Evaluation and Enhancement of Cucurbit Germplasm*. Thomas, C., ed. Charleston, 40-48.

LECOQ, H., POCHARD, E., PITRAT, M., LATERROT, H., MARCHOUX, G., 1982. Identification et exploitation de résistances aux virus chez les plantes maraîchères. *Cryptogam. Mycol.*, 3, 333-345.

LECOQ, H., RAVELONANDRO, M., WIPF-SCHEIBEL, C., MONSION, M., RACCAH, M., DUNEZ, J., 1993., Aphid transmission of a non-aphid-transmissible strain of zucchini yellow mosaic potyvirus from transgenic plants expressing the capsid protein of plum pox potyvirus. *Mol. Plant-Microbe Interact.*, 6, 403-406.

LEE, W.M., ISHIKAWA, M., AHLQUIST, P., 2001. Mutation of host $\Delta 9$ fatty acid desaturase inhibits brome mosaic virus RNA replication between template recognition and RNA synthesis. *J. Virol.*, 75, 2097-2106.

LEGNANI, R., GEBRE SELASSIE, K., NONO WONDIM, R., GOGNALONS, P., MORETTI, A., LATERROT, H., MARCHOUX, G., 1995. Evaluation and inheritance of the *Lycopersicum hirsutum* resistance against potato virus Y. *Euphytica*, 86, 219-226.

LEGRAND, M., KAUFFMANN, S., GEOFFROY, P., FRITIG, B., 1987. Biological function of pathogenesis-related proteins: four tobacco pathogenesis-related proteins are chitinases. *Proc. Natl. Acad. Sci. USA*, 84, 6750-6754.

LEH, V., JACQUOT, E., GELDREICH, A., HERMANN, T., LECLERC, D., CERUTTI, M., YOT, P., KELLER, M., BLANC, S., 1999. Aphid transmission of cauliflower mosaic virus requires the viral PIII protein. *EMBO J.*, 18, 7077-7085.

LEH, V., YOT, P., KELLER, M., 2000. The cauliflower mosaic virus translational activator interacts with the 60 S ribosomal subunit protein L18 of *Arabidopsis thaliana*. *Virology*, 266, 1-7.

LEIGH BROWN, A., 1994. Methods of evolutionary analysis of viral sequences. In: *The Evolutionary Biology of Viruses*. Morse, S., ed. Raven Press, New York, 75-86.

LEISER, R.M., ZIEGLER-GRAFF, V., REUTENAUER, A., HERRBACH, E., LEMAIRE, O., GUILLEY, H., RICHARDS, K., JONARD, G., 1992. Agroinfection as an alternative to insects for infecting plants with beet western yellows virus. *Proc. Natl. Acad. Sci. USA*, 89, 9136-9140.

LELLIS, A.D., KASSCHAU, K.D., WHITHAM, S.A., CARRINGTON, J.C., 2002. Loss-of-susceptibility mutants of *Arabidopsis thaliana* reveal an essential role for eIF(iso)4E during potyvirus infection. *Curr Biol.*, 12, 1046-1050.

LÉONARD, S., PLANTE, D., WITTMANN, S., DAIGNEAULT, N., FORTIN, M., LALIBERTÉ, J.F., 2000. Complex formation between potyvirus VPg and translation eukaryotic initiation factor 4E correlates with virus infectivity. *J. Virol.*, 74, 7730-7737.

LÉONARD, S., VIEL, C., BEAUCHEMIN, C., DAIGNEAULT, N., FORTIN, M.G., LALIBERTÉ, J.F., 2004. Interaction of VPg-Pro of turnip mosaic virus with the translation initiation factor 4E and the poly(A)-binding protein in planta. *J. Gen. Virol.*, 85, 1055-1063.

LEPPIKK, E.E., 1970. Gene centers of plants as sources of disease resistance. *Annu. Rev. Phytopathol.*, 8, 323-344.

LEVINE, A., TENHAKEN, R., DIXON, R., LAMB, C., 1994. H_2O_2 from the oxidative burst orchestrates the plant hypersensitive disease response. *Cell*, 79, 583-593.

LEVIS, C., ASTIER-MANIFACIER, S., 1993. The 5′-untranslated region of PVY RNA, even located in an internal position, enables initiation of translation. *Virus Genes*, 7, 367-379.

LEWANDOWSKI, D.J., DAWSON, W.O., 2000. Functions of the 126 and 183 kDa proteins of tobacco mosaic virus. *Virology*, 271, 90-98.

LI, Q., PALUKAITIS, P., 1996. Comparison of the nucleic acid and NTP-binding properties of the movement protein of cucumber mosaic cucumovirus and tobacco mosaic tobamovirus. *Virology*, 216, 71-79.

LI, X.H., VALDEZ, P., OLVERA, R., CARRINGTON, J.C., 1997. Functions of the tobacco etch virus RNA polymerase (Nib): subcellular transport and protein-protein interaction with VPg-proteinase (NIa). *J. Virol.*, 71, 1598-1607.

LIMASSET, P., CORNUET, P., 1949. Recherche du virus de la mosaique du tabac dans les méristèmes des plantes infectées. *C.R. Acad. Sci. Paris.*, 228, 1971-1972.

LINDBO, J.A., DOUGHERTY, W.G., 1992. Untranslatable transcripts of the tobacco etch virus protein gene sequence can interfere with tobacco etch replication in transgenic plants and protoplasts. *Virology*, 189, 725-733.

LINDBO, J.A., SILVA-ROSALES, L., PROEBSTING, W.M., DOUGHERTY, W.G., 1993. Induction of a highly specific antiviral state in transgenic plants: implications for regulation of gene expression and virus resistance. *Plant Cell*, 5, 1749-1759.

LIU, H., BOULTON, M.I., THOMAS, C.L., PRIOR, D.A.M., OPARKA, K.J., DAVIES, J.W., 1999a. Maize streak virus coat protein is karyophyllic and facilitates nuclear transport of viral DNA. *Mol. Plant-Microbe Interact.*, 12, 894-900.

LIU, L., SAUNDERS, K., THOMAS, C.L., DAVIES, J.W., STANLEY, J., 1999b. Bean yellow dwarf virus RepA, but not Rep, binds to maize retinoblastoma protein, and the virus tolerates mutations in the consensus binding motif. *Virology*, 256, 270-279.

LOCKHART, B., MENKE, J., DAHAL, G., OLSZEWSKY, N., 2000. Characterization and genomic analysis of tobacco vein clearing virus, a plant pararetrovirus that is transmitted vertically and related to sequences integrated in the host genome. *J. Gen. Virol.*, 81, 1579-1585.

LOEFFLER, F., FROSCH, P., 1898. Berichte der Komission zur erforschung der Maul und klauenseuche bei dem Institut fur Infectios-Krankheiten in Berlin. *Ztlb. Bakt. Parasitkde.* Abt 1.

LOMONOSSOFF, G.P., 1995. Pathogen-derived resistance to plant viruses. *Annu. Rev. Phytopathol.*, 33, 323-343.

LONGSTAFF, M., BRIGNETI, G., BOCCARD, F., CHAPMAN, S.N., BAULCOMBE, D.C., 1993. Extreme resistance to potato virus X infection in plants expressing a modified component of the putative viral replicase. *EMBO J.*, 12, 379-386..

LOUGH, T.J., NETZLER, N.E., EMERSON, S.J., SUTHERLAND, P., CARR, F., BECK, D.L., LUCAS, W.J., FORSTER, R.L.S., 2000. Cell-to-cell movement of Potexviruses: evidence for a ribonucleoprotein complex involving the coat protein and first triple gene block protein. *Mol. Plant-Microbe Interact.*, 13, 962-974.

LOUGH, T.J., SHASH, K., XONOCOSTLE-CAZARES, B., HOFSTRA, K.R., BECK, D.L., BALMORI, E., FORSTER, R.L.S., LUCAS, W.J., 1998. Molecular dissection of the mechanism by which Potexvirus Triple Gene Block proteins mediate cell-to-cell transport of infectious RNA. *Mol. Plant-Microbe Interact.*, 11, 801-814.

LU, B., STUBBS, G., CULVER, J.N., 1998. Coat protein interactions involved in tobacco mosaic tobamovirus cross-protection. *Virology*, 248, 188-198.

LU, B., TARAPOREWALA, F.Z., STUBBS, G., CULVER, J.N., 1998. Intersubunit interactions allowing a carboxylate mutant coat protein to inhibit Tobamovirus disassembly. *Virology*, 244, 12-19.

LU, R., FOLIMONOV, A., SHINTAKU, M., LI, W.X., FALK, B.W., DAWSON, W.O., DING, S.W., 2004. Three distinct suppressors of RNA silencing encoded by a 20-kb viral RNA genome. *Proc. Natl. Acad. Sci. USA*, 101, 15742-15747.

LUCAS, W.J., BOUCHE PILLON, S., JACKSON, D.P., NGUYEN, L., BAKER, L., DING, B., HAKE, S., 1995. Selective trafficking of KNOTTED-1 homeodomain protein and its RNA through plasmodesmata. *Science*, 270, 1980-1983.

LUCAS, W.J., LEE, J.Y., 2004. Plasmodesmata as a supracellular control network in plants. *Nat. Rev. Mol. Cell Biol.*, 5(9), 712-726.

LUSTIG, A., LEVINE, A.J., 1992. One hundred years of virology. *J. Virol.*, 66, 4629-4631.

LWOFF, A., 1957. The concept of virus. *J. Gen. Microbiol.*, 17, 239-253.

MACFARLANE, S.A., 1997. Natural recombination among plant virus genomes: evidence from tobraviruses. *Sem. Virol.*, 8, 25-31.

MADDEN, L.V., CAMPBELL, C.L., 1986. Descriptions of virus epidemics in time and space. In: *Plant Virus Epidemics: Monitoring, Modelling and Predicting Outbreaks*. McLean, G.D., Garrett, R.G., Ruesink, W.G., eds. Academic Press, New York, 273-293.

MAIA, I.G., BERNARDI, F., 1996. Nucleic acid-binding properties of a bacterially expressed potato virus Y helper component-proteinase. *J. Gen. Virol.*, 77, 869-877.

MAIA, I.G., HAENNI, A.L., BERNARDI, F., 1996. Potyviral HC-Pro: a multifunctional protein. *J. Gen. Virol.*, 77, 1335-1341.

MAKKOUK, K.M., HSU, H.T., KUMARI, S.G., 1993. Detection of three plant viruses by dot-blot and tissue-blot immunoassay using chemiluminescent and chromogenic substrates. *J. Phytopathol.*, 139, 97-102.

MAKKOUK, K.M., KUMARI, S.G., 2002. Low cost paper can be used in tissue-blot immunoassay for detection of cereal and legume viruses. *Phytopathol. Medit.*, 41, 275-278.

MALLORY, A.C., REINHART, B.J., BARTEL, D., VANCE, V.B., BOWMAN, L.H., 2002. A viral suppressor of RNA silencing differentially regulates the accumulation of short interfering RNAs and micro-RNAs in tobacco. *Proc. Natl. Acad. Sci. USA*, 99, 15228-15233.

MALYSHENKO, S.I., KONDAKOVA, O.A., NAZAROVA, J.V., KAPLAN, I.B., TALIANSKY, M.E., ATABEKOV, J.G., 1993. Reduction of tobacco mosaic virus accumulation in transgenic plants producing non-functional viral transport proteins. *J. Gen. Virol.*, 74, 1149-1156.

MARATHE, R., ANANDALAKSHMI, R., SMITH, T.H., PRUSS, G.J., VANCE, V.B., 2000. RNA viruses as inducers, suppressors and targets of post-transcriptional gene silencing. *Plant Mol. Biol.*, 43, 295-306.

MARIE-JEANNE TORDO, V., CHACHULSKA, A.M., FAKHFAKH, H., LE ROMANCER, M., ROBAGLIA, C., ASTIER-MANIFACIER, S., 1995. Sequence polymorphism in the 5'NTR and in the P1 coding region of potato virus Y genomic RNA. *J. Gen. Virol.*, 76, 939-949.

MARSH, L.E., HUNTLEY, C.C., POGUE, G.P., CONNELL, J.P., HALL, T.C., 1991. Regulation of (+):(–) strand asymmetry in replication of Brome mosaic virus RNA. *Virology*, 182, 76-83.

MARTELLI, G.P., WALTER, B., 1998. Virus certification of Grapevines. In: *Plant Virus Disease Control*. Hadidi, A., Khetarpal, R.K., Koganezawa, H., eds. APS Press, St. Paul, Minnesota, 261-276.

MARTIENSSEN, R.A., 2003. Maintenance of heterochromatin by RNA interference of tandem repeats. *Nat. Genet.*, 35, 213-214.

MARTIN, B., COLLAR, J.L., TJALLINGII, W.F., FERERES, A., 1997. Intracellular ingestion and salivation by aphids may cause the acquisition and transmission of non-persistently transmitted plant viruses. *J. Gen. Virol.*, 78, 2701-2705.

MARTIN, C., GALLET, M., 1966. Nouvelles observations sur le phénomène d'hypersensibilité aux virus chez les végétaux. *C.R. Acad. Sci. Paris* (série D), 263, 1316-1318.

MARTIN, M.T., CERVERA, M.T., GARCIA, J.A., 1995. Properties of the active plum pox potyvirus RNA polymerase complex in defined glycerol gradient fractions. *Virus Res.*, 37, 127-137.

MARTIN, R.R., 1998. Advanced diagnostic tools as an aid to controlling plant virus diseases. In: *Plant Virus Disease Control*. Hadidi, A., Khetarpal, R.K., Koganezawa, H., eds. APS Press, St. Paul, Minnesota, 381-391.

MARTINEZ-SORIANO, J.P., GALINDO-ALONSO, J., MAROON, C.J., YUCEL, I., SMITH, D.R., DIENER, T.O., 1996. Mexican papita viroid: putative ancestor of crop viroids. *Proc. Natl. Acad. Sci. USA*, 93, 9397-9401.

MASMOUDI, K., DUBY, C., SUHAS, M., GUO, J.Q., GUYOT, L., OLIVIER, V., TAYLOR, J., MAURY, Y., 1994. Quality control of pea seed for pea seed borne mosaic virus. *Seed Sci. Technol.*, 11, 491-503.

MASMOUDI, K., SUHAS, M., KHETARPAL, R.K., MAURY, Y., 1994. Specific serological detection of the transmissible virus in pea seed infected by Pea seed-borne mosaic virus. *Phytopathology*, 84, 756-760.

MASUTA, C., NISHIMURA, M., MORISHITA, H., HATAYA, T., 1999. A single amino acid change in viral genome-associated protein of potato virus Y correlates with resistance breaking in "Virgin A mutant" tobacco. *Phytopathology*, 89, 118-123.

MATTHEWS, R.E.F., 1991. *Plant Virology*, 3d ed., Academic Press, San Diego, 835 pp.

MAULE, A.J., WANG, D., 1996. Seed transmission of plant viruses: a lesson in biological complexity. *Trends Microbiol.*, 4, 153-158.

MAURY, Y., BOSSENNEC, J.M., BOUDAZIN, G., HAMPTON, R.O., PIETERSEN, G., MAGUIRE, J.D., 1987. Factors influencing ELISA evaluation of transmission of pea seed borne mosaic virus in infected pea seed: seed group size and seed decortication. *Agronomie*, 7, 225-230.

MAURY, Y., DUBY, C., BOSSENNEC, J.M., BOUDAZIN, G., 1985. Group analysis using ELISA: determination of the level of transmission of *Soybean mosaic virus* in soybean seed. *Agronomie*, 5, 405-415.

MAURY, Y., DUBY, C., KHETARPAL, R.K., 1998. Seed certification for viruses: In: *Plant Virus Disease Control*. Hadidi, A., Khetarpal, R.K., Koganezawa, H., eds. APS Press, St. Paul, Minnesota, 237-248.

MAYER, A., 1886. Üeber die Mosaikkrankheit des Tabaks. *Die Landwirt. Versuchs-Stationen*, 32, 451-467. (English translation : Concerning the mosaic disease of tobacco, in Johnson J., ed., Phytopathological Classics N°7, *American Phytopathological Society*, Saint Paul, Minnesota, 1942, 11-24).

MAYO, M.A., JOLLY, C.A., 1991. The 5′ terminal sequence of potato leafroll virus RNA: evidence of recombination between virus and host RNA. *J. Gen. Virol.*, 72, 2591-2595.

MAYO, M., RYABOV, E., FRASER, G., TALIANSKY, M., 2000. Mechanical transmission of *potato leafroll virus*. *J. Gen. Virol.*, 81, 2791-2795.

MAYR, E. 1994. Driving forces in evolution. In: *The Evolutionary Biology of Viruses*. Raven Press, New York, 29-48.

MAZZOLINI, L., DABOS, P., CONSTANTIN, S., YOT, P., 1989. Further evidence that viroplasms are the site of cauliflower mosaic virus genome replication by reverse transcription during viral infection. *J. Gen. Virol.*, 70, 34-39.

MCCAFFERTY, J., GRIFFITHS, A.D., WINTER, G., CHISWELL, D.J., 1990. Phage antibodies-filamentous phage displaying antibodies variable domains. *Nature*, 348, 552-554.

MCLEAN, B.P., ZUPAN, J., ZAMBRYSKI, P.C., 1995. Tobacco mosaic virus movement protein associates with the cytoskeleton in tobacco cells. *Plant Cell*, 7, 2101-2114.

MCLEAN, M.A., CAMPBELL, R.N., HAMILTON, R.I., ROCHON, D.M., 1994. Involvement of the cucumber necrosis virus coat protein in the specificity of fungus transmission by *Olpidium bornovanus*. *Virology*, 204, 840.

MEISTER, G., TUSCHL, T., 2004. Mechanisms of gene silencing by double-stranded RNA. *Nature*, 431, 343-349.

MELCHER, U. 2000. The '30K' superfamily of viral movement proteins. *J. Gen. Virol.*, 81, 257-266.

MELLO, C.C., CONTE, D. JR., 2004. Revealing the world of RNA interference. *Nature*, 431, 338-342.

MERITS, A., RAJAMAKI, M., LINDHOLM, P., RUNEBERG-ROOS, P., KEKARAINEN, T., PUUSTINEN, P., MAKELAINEN, K., VALKONEN, J., SAARMA, M., 2002. Proteolytic processing of potyviral proteins and polyprotein processing intermediate in insects and plant cells. *J. Gen. Virol.*, 83, 1211-1221.

MERLET, J., LE HINRAT, Y., ELLISÈCHE, D., CROUAU, G., LANGLADE, P., 1996. Production du plant. In: *La Pomme de Terre*. Rousselle, P., Robert, Y., Crosnier, J.C., eds. INRA Paris, 415-448.

MESHI, T., MOTOYOSHI, F., ADACHI, A., WATANABE, Y., TAKAMATSU, N., OKADA, Y., 1988. Two concomitant base substitutions in the putative replicase gene of tobacco mosaic virus confer the ability to overcome the effects of a tomato resistance gene, *Tm-1*. *EMBO J.*, 7, 1575-1581.

MESHI, T., MOTOYOSHI, T., MAEDA, T., YOSHIWOKA, S., WATANABE, H., OKADA, Y., 1989. Mutations in the tobacco mosaic virus 30-kDa protein gene overcome *Tm-2* resistance in tomato. *Plant Cell*, 1, 515-522.

MESHI, T., WATANABE, Y., SAITO, T., SUGIMOTO, A., MAEDA, T., OKADA, Y., 1987. Function of the 30 kDa protein of tobacco mosaic virus: involvement in cell-to-cell movement and dispensability for replication. *EMBO J.*, 6, 2557-2563.

METTE, M., KANNO, T., AUFSATZ, W., JAKOWITSCH, J., VAN DER WINDEN, J., MATZE, M., MATZE, A., 2002. Endogenous viral sequences and their potential contribution to heritable virus resistance in plants. *EMBO J.*, 21, 461-469.

MEULEWAETER, F., DANTHINNE, X., MONTAGU, M.V., CORNELISSEN, M., 1998. 5′ and 3′ sequences of satellite tobacco necrosis virus RNA promoting translation in tobacco. *Plant J.*, 14, 169-176.

MEZITT, L.A., LUCAS, W.J., 1996. Plasmodesmal cell to cell transport of proteins and nucleic acids. *Plant Mol. Biol.*, 32, 251-273.

MILLER, E.D., KIM, K.H., HEMENWAY, C., 1999. Restoration of a stem-loop structure required for Potato virus X RNA accumulation indicates selection for a mismatch and a GNRA tetraloop. *Virology*, 260, 342-353.

MILLER, W.A., BUJARSKI, J.J., DREHER, T.W., HALL, T.C., 1986. Minus-strand initiation by brome mosaic virus replicase within the 3′-tRNA-like structure of native and modified RNA templates. *J. Mol. Biol.*, 187, 537-546.

MILLER, W.A., DREHER, T.W., HALL, T.C., 1985. Synthesis of brome mosaic virus subgenomic RNA *in vitro* by internal initiation on (-) sense genomic RNA. *Nature*, 313, 68-70.

MILNE, R.G., 1988. Quantitative use of the electron microscope decoration technique for plant virus diagnostics. *Acta Hort.*, 234, 321-329.

MILNE, R.G., LUISONI, E., 1975. Rapid high resolution immune electron microscopy of plant viruses. *Virology*, 68, 270-274.

MINK, G.I., 1993. Pollen and seed transmitted viruses and viroids. *Annu. Rev. Phytopathol.*, 31, 375-402.

MINK, G.I., WAMPLE, R., HOWELL, W.E., 1997. Heat treatment of perennial plants to eliminate phytoplasmas, viruses and viroids while maintaining plant survival. In: *Plant Virus*

Disease Control. Hadidi, A., Khetarpal, R.K., Koganezawa, H., eds. APS Press, St. Paul, Minnesota, 332-345.

MITRA, A., HIGGINS, D.W., LANGERBERG, W.C., NIE, H., SENGUPTA, D.N., SILVERMAN, R., 1996. A mammalian 2-5A system functions as an antiviral pathway in transgenic plants. *Proc. Natl. Acad. Sci. USA*, 93, 6780-6785.

MITTLER, R., SHULAEV, V., LAM, E., 1995. Coordinated activation of programmed cell death and defence mechanisms in transgenic tobacco plants expressing a bacterial proton pump. *Plant Cell*, 7, 29-42.

MOORE, C.J., SUTHERLAND, P.W., FORSTER, R.L., GARDNER, R.C., MACDIARMID, R.M., 2001. Dark green islands in plant virus infection are the result of posttranscriptional gene silencing. *Mol. Plant Microbe Interact.*, 14, 939-946.

MORASCO, B., SHARMA, N., PARILLA, J., FLANEGAN, J., 2003. Poliovirus cre(2C)-dependent synthesis of VpgpUpU is required for positive- but not negative-strand RNA synthesis. *J. Virol.*, 77, 5136-5144.

MORCH, M.D., JOSHI, R.L., DENIAL, T.M., HAENNI, A.L., 1988. Overlapping reading frame revealed by complete nucleotide sequencing of turnip yellow mosaic virus genomic RNA. *Nucl. Acids Res.*, 16, 6157-6173.

MOREAU-MHIRI, C., MOREL, J.B., AUDEON, C., FERAULT, M., GRANDBASTIEN, M.A., LUCAS, H., 1996. Regulation of expression of the tobacco Tnt1 retrotransposon in heterologous species following pathogen-related stresses. *Plant J.*, 9, 409-419.

MOREL, G., MARTIN, C., 1952. Guérison de dahlias atteints d'une maladie à virus. *C.R. Acad. Sci. Paris*, 235, 1324-1325.

MOREL, G., MARTIN, C., 1955. Guérison de pommes de terre atteintes de maladies à virus. *C.R. Acad. Sci. Fr.*, 41, 472-475.

MORIN, S., GHANIM, M., ZEIDAN, M., CZOSNEK, H., VERBEEK, M., VAN DEN HEUVEL, J.F., 1999. A GroEL homologue from endosymbiotic bacteria of the whitefly *Bemisia tabaci* is implicated in the circulative transmission of tomato yellow leaf curl virus. *Virology*, 256, 75-84.

MORITZ, G., KUMM, S., MOUND, L., 2004. Tospovirus transmission depends on thrips ontogeny. *Vir. Res.*, 100, 143-149.

MORSE, S.S., 1994a. Towards an evolutionary biology of viruses. In: *The Evolutionary Biology of Viruses*. Morse, ed., Raven Press, New York, 1-28.

MORSE, S.S., 1994b. The viruses of the future? Emerging viruses and evolution. In: *The Evolutionary Biology of Viruses*. Morse, ed., Raven Press, New York, 325-336.

MOSSOP, D.W., FRANCKI, R.I.B., 1977. Association of RNA3 with aphid transmission of cucumber mosaic virus. *Virology*, 81, 177-181.

MOUCHÈS, C., BOVÉ, C., BOVÉ, J.M., 1984. Turnip yellow mosaic virus RNA replicase: partial purification of the enzyme from the solubilized enzyme-template complex. *Virology*, 58, 409-423.

MOUCHÈS, C., CANDRESSE, T., BOVÉ, J.M., 1984. Turnip yellow mosaic virus RNA-replicase contains host and virus-encoded subunits. *Virology*, 134, 78-91.

MOURRAIN, P., BECLIN, C., ELMAYAN, T., FEUERBACH, F., GODON, C., MOREL, J.B., JOUETTE, D., LACOMBE, A.M., NIKIC, S., PICAULT, N., REMOUE, K., SANIAL, M., VO, T.A., VAUCHERET, H., 2000. Arabidopsis SGS2 and SGS3 genes are required for posttranscriptional gene silencing and natural virus resistance. *Cell*, 101, 533-542.

MOURY, B., MOREL, C., JOHANSEN, I., GUILBAUD, L., SOUCHE, S., AYME, V., CARANTA, C., PALLOIX, A., JACQUEMOND, M., 2004. Mutations in Potato virus Y genome-linked protein determine virulence towards recessive genes in *Capsicum annuum* and *Lycopersicon esculentum*. *Mol. Plant Microbe Interact.*, 17, 322-329.

MOURY, B., PALLOIX, A., GEBRE-SELASSIE, K., MARCHOUX, G., 1998. L'émergence des tospovirus. *Virologie*, 2, 357-367.

MUANGSAN, N., BECLIN, C., VAUCHERET, H., ROBERTSON, D., 2004. Geminivirus VIGS of endogenous genes requires SGS2/SDE1 and SGS3 and defines a new branch in the genetic pathway for silencing in plants. *Plant J.*, 38, 1004-1014.

MULLIN, R.H., SMITH, S.H., FRAZIER, R.W., SHLEGEL, D.E., MCCALL, S.R., 1974. Meristem tip culture freed strawberries of mild yellow edge, pallidosis and mottle disease. *Phytopathology*, 64, 1425-1429.

MURASHIGE, T., SKOOG, F., 1962. A revised medium for rapid growth and bioassays with tobacco tissue culture. *Physiol. Plant.*, 15, 473-497.

MURILLO, I., CAVALLARIN, L., SAN SEGUNDO, B., 1997. The maize pathogenesis-related PRms protein localizes to plasmodesmata in maize radicles. *Plant Cell*, 9, 145-156.

MURPHY, J.F., KYLE, M.M., 1995. Restricted systemic spread of pepper mottle virus in Capsicum annuum "Avelar" *Phytopathology*, 85, 561-566.

MUTTERER, J.D., STUSSI-GARAUD, C., MICHLER, P., RICHARDS, K.E., JONARD, G., ZIEGLER-GRAFF, V., 1999. Role of the beet western yellows virus readthrough protein in viral movement in *Nicotiana clevelandii*. *J. Gen. Virol.*, 80, 2771-2778.

NAGANO, H., OKUNO, T., MISE, K., FURUSAWA, I., 1997. Deletion of the C-terminal 33 amino acids of cucumber mosaic virus movement protein enables a chimeric brome mosaic virus to move from cell to cell. *J. Virol.*, 71, 2270-2276.

NAGATA, T., INOUE-NAGATA, A.K., SMID, H.M., GOLDBACH, R., PETERS, D., 1999. Tissue tropism related to vector competence of *Frankliniella occidentalis* for tomato spotted wilt tospovirus. *J. Gen. Virol.*, 80, 507-515.

NAGY, P.D., BUJARSKI, J.J., 1995. Efficient system of homologous RNA recombination in brome mosaic virus: Sequence and structure requirement and accuracy of crossovers. *J. Virol.*, 69, 131-140.

NAGY, P.D., BUJARSKI, J.J., 1997. Engineering of homologous recombination hotspots with AU sequence in brome mosaic virus. *J. Virol.*, 71, 3799-3810.

NAGY, P.D., SIMON, A.E., 1997. New insights into the mechanisms of RNA recombination. *Virology*, 235, 1-9.

NAKAJIMA, K., SENA, G., NAWY, T., BENFEY, P.N., 2001. Intercellular movement of the putative transcription factor SHR in root patterning. *Nature*, 413, 307-311.

NAKASHIMA, N., NODA, H., 1995. Nonpathogenic Nilaparvata lugens reovirus is transmitted to the brown planthopper through rice plant. *Virology*, 207, 303-307.

NAMBA, K., PATTANAYEK, R., STUBBS, G., 1989. Visualization of protein-nucleic acid interaction in a virus. Refined structure of intact tobacco mosaic virus at 2.9 Ä resolution by X-ray fiber diffraction. *J. Mol. Biol.*, 208, 307-325.

NAVARRO, J.A., DAROS, J.A., FLORES, R., 1999. Complexes containing both polarity strands of avocado sunblotch viroid: identification in chloroplasts and characterization. *Virology*, 253, 77-85.

NAVARRO, L., LLACER, G., CAMBRA, M., ARREQUI, J.M., JUAREZ, J., 1982. Shoot tip grafting *in vitro* for elimination of viruses in peach plants (*Prunus persica* Batsch). *Acta Hort.*, 130, 185-192.

NAVARRO, L., ROISTACHER, C., MURASHIGE, T., 1975. Improvement of shoot tip grafting *in vitro* for virus-free *Citrus. J. Am. Soc. Hort. Sci.*, 100, 471-479.

NDOWORA, T., DAHAL, G., LAFLEUR, D., HARPER, G., HULL, R., OLSZEWSKI, N.E., LOCKHART, B., 1999. Evidence that badnavirus infection in *Musa* can originate from integrated pararetroviral sequences. *Virology*, 255, 214-220.

NELSON, M.R., FELIX-GASTELUM, R., ORUM, T.V., STOWELL, L.J., MYERS, D.E., 1994. Geographic information systems and geostatistics in the design and validation of regional plant virus management programs. *Phytopathology*, 84, 898-905.

NG, M., TAN, S., SEE, E., OOI, E., LING, A., 2003. Proliferative growth of SARS coronavirus in Vero E6 cells. *J. Gen. Virol.*, 84, 3291-3303.

NGUYEN, M., RAMIREZ, B.C., GOLDBACH, R., HAENNI, A.L., 1997. Characterization of the *in vitro* activity of the RNA-dependent RNA polymerase associated with the ribonucleoproteins of the rice hoja blanca tenuivirus. *J. Virol.*, 71, 2621-2627.

NICAISE, V., GERMAN-RETANA, S., SANJUAN, R., DUBRANA, M.P., MAZIER, M., MAISONNEUVE, B., CANDRESSE, T., CARANTA, C., LEGALL, O., 2003. The eukaryotic translation initiation factor 4E controls lettuce susceptibility to the Potyvirus Lettuce mosaic virus. *Plant Physiol.*, 132, 1272-1282.

NIDERMAN, T., GENETET, I., BRUYÈRE, T., GEES, R., STINTZI, A., LEGRAND, M., FRITIG, B., MÖSINGER, E., 1995. Pathogenesis-related PR1 proteins are antifungal; isolation and characterization of three 14-kilodalton proteins of tomato and of a basic PR-1 of tobacco with inhibitory activity against *Phytophthora infestans. Plant Physiol.*, 108, 17-27.

NIEPEL, M., GALLIE, D.R., 1999. Identification and characterization of the functional elements within the tobacco etch virus 5′ leader required for cap-independent translation. *J. Virol.*, 73, 9080-9088.

NIKOLAEVA, O.V., KARASEV, A.V., GUMPF, D.J., LEE, R.F., GARNSEY, S.M., 1995. Production of polyclonal antibodies to the coat protein of *Citrus tristeza virus* expressed in *Escherichia coli*: application for immunodiagnosis. *Phytopathology*, 85, 691-694.

NIRENBERG, M., MATTHAEI, H., 1961. The dependence of cell free protein synthesis in *E. coli* upon naturally occurring or synthetic polyribonucleotides. *Proc. Natl. Acad. Sci. USA*, 47, 158-1602.

NISHIGUCHI, M., MOTOYOSHI, F., OSHIMA, N., 1978. Behaviour of a temperature sensitive strain of tobacco mosaic virus in tomato leaves and protoplasts. *J. Gen. Virol.*, 39, 53-61.

NORIS, E., VAIRA, A.M., CACIAGLI, P., MASENGA, V., GRONENBORN, B., ACCOTTO, G.P., 1998. Amino-acids in the capsid protein of tomato yellow leaf curl virus that are crucial for systemic infection, particle formation, and insect transmission. *J. Virol.*, 72, 10050-10057.

NOUEIRY, A., DIEZ, J., FALK, S., CHEN, J., AHLQUIST, P., 2003. Yeast Lsm1p-7p/Pat 1p deadenylation-dependent mRNA decapping factors are required for brome mosaic virus genomic RNA translation. *Mol. Cell. Biol.*, 23, 4094-4106.

NOUEIRY, A.O., AHLQUIST, P., 2003. Brome mosaic virus RNA replication: revealing the role of the host in RNA replication. *Ann. Rev. Phytopathol.*, 41, 77-98.

NOUEIRY, A.O., LUCAS, W.J., GILBERTSON, R.L., 1994. Two proteins of a plant DNA virus coordinate nuclear and plasmodesmatal transport. *Cell*, 76, 925-932.

NUTTER, F.W., 1997. Quantifying the temporal dynamics of plant virus epidemics: a review. *Crop Prot.*, 7, 603-618.

NYLAND, G., GOHEEN, A.C., 1969. Heat therapy of virus diseases of perennial plants. *Annu. Rev. Phytopathol.*, 7, 331-354.

O'Reilly, E., Wang, Z., French, R., Kao, C.C., 1998. Interactions between the structural domains of the RNA replication proteins of plant-infecting RNA viruses. *J. Virol.*, 72, 7160-7169.

O'Reilly, E.K., Kao, C.C., 1998. Analysis of RNA-dependent RNA polymerase structure and function as guided by known polymerase structures and computer predictions of secondary structure. *Virology*, 252, 287-303.

Offei, S.K., Coffin, R.S., Coutts, R.H., 1995. The tobacco necrosis virus p7 protein is a nucleic-acid binding protein. *J. Gen. Virol.*, 76, 1493-1496.

Ohira, K., Namba, S., Rozanov, M., Kusumi, T., Tsuchizaki, T., 1995. Complete sequence of an infectious full-length clone of citrus tatter leaf capillovirus: comparative sequence analysis of capillovirus genomes. *J. Gen. Virol.*, 76, 2305-2309.

Ohno, T., Takamatsu, N., Meshi, T., Okada, Y., Nishigushi, M., Kiho, Y., 1983. Single amino acid substitution in 30 K protein of TMV defective in virus transport function. *Virology*, 131, 255-258.

Ohta, T., Kreitman, M., 1996. The neutralist-selectionist debate. *BioEssays*, 18, 673-683.

Ohta, T., Kreitman, M., The neutralist-selectionist debate. *BioEssays*, 18, 673-683.

Olson, A.J., Bricogne, G., Harrison, S.C., 1983. Structure of the tomato bushy stunt virus. IV: the virus particle at 2.9 Ä resolution. *J. Mol. Biol.*, 171, 61-93.

Olsthoorn, R.C., Mertens, S., Brederode, F.T., Bol., J.F., 1999. A conformational switch at the 3' end of a plant virus RNA regulates viral replication. *EMBO J.*, 18, 4856-4864.

Omura, T., Yan, J., 1999. Role of the outer capsid proteins in transmission of phytoreovirus by insect vectors. *Adv. Virus Res.*, 54, 15-43.

Oparka, K.J., 2004. Getting the message across: how do plant cells exchange macromolecular complexes? *Trends Plant Sci.*, 9, 33-41.

Oparka, K.J., Boevink, P., Santa Cruz, S., 1996. Studying the movement of plant viruses using green fluorescent protein. *Trends Plant Sci.*, 1, 412-418.

Oparka, K.J., Cruz, S.S., 2000. The great escape: phloem transport and unloading of macromolecules. *Annu. Rev. Plant Physiol. Plant Mol. Biol.*, 51, 323-347.

Oparka, K.J., Duckett, C.M., Prior, D.A.M., Fisher, D.B., 1994. Real time imaging of phloem unloading in the root tip of *Arabidopsis. Plant J.*, 6, 759-766.

Oparka, K.J., Prior, D.A.M., Santa Cruz, S., Padgett, H.S., Beachy, R.N., 1997. Gating of epidermal plasmodesmata is restricted to the leading edge of expanding infection sites of tobacco mosaic virus (TMV). *Plant J.*, 12, 781-789.

Oparka, K.J., Roberts, A.G., Boevink, P., Santa Cruz, S., Roberts, I., Pradel, K.S., Imlau, A., Kotlizky, G., Sauer, N., Epel, B., 1999. Simple, but not branched, plasmodesmata allow the non-specific trafficking of proteins in developing tobacco leaves. *Cell*, 97, 743-754.

Orozco, B.M., Hanley-Bowdoin, L., 1996. A DNA structure is required for Geminivirus replication origin function. *J. Virol.*, 70, 148-158.

Oshima, K., Taniyama, T., Yamanaka, T., Ishikawa, M., Naito, S., 1998. Isolation of a mutant of *Arabidopsis thaliana* carrying two simultaneous mutations affecting tobacco mosaic virus multiplication within a single cell. *Virology*, 243, 472-481.

Osman, T.A.M., Buck, K.W., 1997. The tobacco mosaic virus RNA polymerase complex contains a plant protein related to the RNA binding subunit of yeast eIF3. *J. Virol.*, 71, 6075-6082.

Owens, R.A., Steger, G., Hu, Y., Fels, A., Hammond, R.W., Riesner, D., 1996. RNA structural features responsible for potato spindle tuber viroid pathogenicity. *Virology*, 222, 144-158.

PACOT-HIRIARD, C., CANDRESSE, T., LE GALL, O., DUNEZ, J. 1997. Les fonctions multiples des protéines de capsides des virus de plante à RNA simple brin positif. *Virologie*, 1, 375-382.

PADGETT, H.S., WATANABE, Y., BEACHY, R.N., 1997. Identification of the TMV replicase sequence that activates the N-gene mediated hypersensitive response. *Mol. Plant-Microbe Interact.*, 10, 709-715.

PADIDAM, M., SAWYER, S., FAUQUET, C.M., 1999. Possible emergence of new *Geminiviruses* by frequent recombination. *Virology*, 265, 218-225.

PALAUQUI, J.C., ELMAYAN, T., POLLIEN, J.M., VAUCHERET, H., 1997. Systemic acquired silencing: transgene-specific post-transcriptional silencing is transmitted by grafting from silenced stocks to non-silenced scions. *EMBO J.*, 16, 4738-4745. Erratum in: *EMBO J.*, 17, 21.

PALAUQUI, J.C., VAUCHERET, H., 1998. Transgenes are dispensable for the RNA degradation step of cosuppression. *Proc. Natl. Acad. Sci. USA*, 95, 9675-9680.

PALLAS, J.A., PALVA, N.L., LAMB, C., DIXON, R.A., 1996. Tobacco plants epigenetically suppressed in phenylalanine ammonia-lyase expression do not develop systemic acquired resistance in response to infection by tobacco mosaic virus. *Plant J.*, 10, 281-293.

PALLOIX, A., DAUBÈZE, A.M., LEFEBVRE, V., CARANTA, C., MOURY, B., PFLIEGER, S., GEBRE-SELASSIE, K., MARCHOUX, G., 1997. Construction de systèmes de résistance aux maladies adaptés aux conditions de culture chez le piment. *C.R. Acad. Agric. Fr.*, 83, 87-98.

PALUKAITIS, P., ZAITLIN, M. 1997. Replicase-mediated resistance to plant virus disease. *Adv. Virus Res.*, 48, 349-377.

PANG, S.Z., JAN, F.J., GONSALVES, D., 1997. Nontarget DNA sequences reduce the transgene length necessary for RNA-mediated resistance in transgenic plants. *Proc. Natl. Acad. Sci. USA*, 94, 8261-8266.

PAUL, A.V., VAN BOOM, J.H., FILIPPOV, D., WIMMER, E., 1998. Protein-primed RNA synthesis by purified poliovirus RNA polymerase. *Nature*, 393, 280-284.

PELHAM, H.B.,1978. Leaky UAG codon in TMV RNA. *Nature*, 272, 469-471.

PELHAM, J., FLETCHER, J.T., HAWKINS, J.H., 1970. The establishment of a new strain of tobacco mosaic virus resulting from the use of resistant varieties of tomato. *Ann. Appl. Biol.*, 65, 293-297.

PENG, Y.H., KADOURY, D., GAL-ON, A., HUET, H., WANG, Y., RACCAH, B., 1998. Mutations in the HC-Pro gene of zucchini yellow mosaic potyvirus: effects on aphid transmission and binding to purified virions. *J. Gen. Virol.*, 79, 897-904.

PENNAZIO, S., ROGGERO, P. 1998. Systemic acquired resistance against plant virus infections: a reality? *J. Plant Pathol.*, 80, 179-186.

PERBAL, M.C., THOMAS, C.L., MAULE, A.J., 1993. Cauliflower mosaic virus gene I product (P1) forms tubular structures which extend from the surface of infected protoplasts. *Virology*, 195, 281-285..

PERRING, T.M., FARRAR, C.A., MAYBERRY, K., BLUA, M.J., 1992. Research reveals pattern of virus spread. *Calif. Agric.*, 46, 35-40.

PERRY, K.L., ZHANG, L., PALUKAITIS, P., 1998. Amino acid changes in the coat protein of cucumber mosaic virus differentially affect the transmission by the aphids *Myzus persicae* and *Aphis gossypii*. *Virology*, 242, 204-210.

PETERS, S.A., MESNARD, J.M., KOOTER, I.M., VERVER, J., WELLINK, J., VAN KAMEN, A., 1995. The cowpea mosaic virus RNA 1-encoded 112 kDa protein may function as a VPg precursor *in vivo*. *J. Gen. Virol.*, 76, 1807-1813.

Peters, S.A., Verver, J., Nollen, E.A., van Lent, J.W., Wellink, J., van Kamen, A., 1994. The NTP-binding motif in cowpea mosaic virus B polyprotein is essential for viral replication. *J. Gen. Virol.*, 75, 3267-3176.

Peterson-Burch, B., Wright, D., Laten, H., Voytas, D., 2000. Retroviruses in plants? *Trends in Genetics*, 16, 151-152.

Petty, I.T.D., Edwards, M.C., Jackson, A.O., 1990. Systemic movement of an RNA plant virus determined by a point substitution in a 5' leader sequence. *Proc. Natl. Acad. Sci. USA*, 87, 8894-8897.

Petty, I.T.D., French, R., Jones, R.W., Jackson, A.O., 1990. Identification of barley stripe mosaic virus proteins involved in viral replication and movement. *EMBO J.*, 9, 3453-3457.

Pierangeli, A., Bucci, M., Forzan, M., Pagnotti, P., Equestre, M., Bercoff, R.P., 1999. Primer alignment-and-extension: a novel mechanism of viral RNA recombination responsible for the rescue of inactivated poliovirus cDNA clones. *J. Gen. Virol.*, 80, 1889-1897.

Pieterse, C.M.J., van Loon, L.C., 1999. Salicylic acid-independent plant defence pathways. *Trends Plant Sci.*, 4, 52-58.

Pinck, M., Yot, P., Chapeville, F., Duranton, H., 1970. Enzymatic binding of valine to the 3′ end of TYMV-RNA. *Nature*, 226, 954-956.

Pinto, Y.M., Kok, R.A., Baulcombe, D.C., 1999. Resistance to rice yellow mottle virus (RYMV) in cultivated African rice varieties containing RYMV transgenes. *Nature Biotechnol.*, 17, 702-707.

Pirone, T.P., Blanc, S., 1996. Helper-dependent vector transmission of plant viruses. *Annu. Rev. Phytopathol.*, 34, 227-247.

Pirone, T.P., Megahed, E., 1966. Aphid transmissibility of some purified viruses and viral RNA's. *Virology*, 30, 631-637.

Pirone, T.P., Thornburry, D.W., 1988. Quantity of virus required for aphid transmission of a potyvirus. *Phytopathology*, 78, 104-107.

Pochard, E., 1977. Méthodes pour l'étude de la résistance au virus de la mosaïque du concombre chez le piment. In: *Capsicum 77*. C.R. IIId Congres Eucarpia. Pochard, E. ed. INRA, Avignon, 93-104.

Pogue, G.P., Hall, T.C., 1992. The requirement for a 5′ stem-loop structure in brome mosaic virus replication supports a new model for viral positive-strand RNA initiation. *J. Virol.*, 66, 674-684.

Pogue, G.P., Marsh, L.E., Connell, J.P., Hall, T.C., 1992. Requirement for ICR-like sequences in the replication of brome mosaic virus genomic RNA. *Virology*, 188, 742-753.

Polston, J.E., Anderson, P.K., 1997. The emergence of whitefly-transmitted geminiviruses on tomato in the Western Hemisphere. *Plant Dis.*, 81, 1358-1369.

Ponz, F., Bruening, G., 1986. Mechanisms of resistance to plant viruses. *Annu. Rev. Phytopathol.*, 24, 355-381.

Ponz, F., Glascock, C.B., Bruening, G., 1988. An inhibitor of polyprotein processing with the characteristics of a natural resistance factor. *Mol. Plant Microbe Interact.*, 1, 25-31.

Powell, Abel P., Nelson, R.S., Barun De, Hoffman, N., Rogers, S.G., Fraley, R.T., Beachy, R.N., 1986. Delay of disease development in transgenic plants that express the tobacco mosaic coat protein gene. *Science*, 232, 738-743.

Powell, C.C., Schlegel., D.E., 1970. Factors influencing seed transmission of squash mosaic virus in Cantaloupe. *Phytopathology*, 60, 1466-1469.

PREISS, T., HENTZE, M.W., 1998. Dual function of the messenger RNA cap structure in polyA-tail-promoted translation in yeast. *Nature*, 392, 516-519.

PRINGLE, C.R., 1999. Virus taxonomy 1999: the universal system of virus taxonomy, updated to include the new proposals ratified by the International Committee on Taxonomy of Viruses during 1998. *Arch. Virol.*, 144, 422-429.

PRINGLE, C.R., 1999a. Virus taxonomy at the XIth International Congress of Virology. *Arch. Virol.*, 144, 2065-2070.

PROD'HOMME, D., JAKUBIEC, A., TOURNIER, V., DRUGEON, G., JUPIN, I., 2003. Targeting of the *Turnip yellow mosaic virus* 66K replication protein to the chloroplast envelope is mediated by the 140K protein. *J. Virol.*, 77, 9124-9135.

PROVVIDENTI, R., 1990. Inheritance of resistance to pea mosaic virus in *Pisum sativum*. *J. Hered.*, 81, 143-145.

PROVVIDENTI, R., ALCONERO, R., 1988. Inheritance of resistance to a lentil strain of pea seed-borne mosaic virus in *Pisum sativum*. *J. Hered.*, 79, 45-47.

PRÜFER, D., TACKE, E., SCHMITZ, J., KUUL, B., KAUFMAN, A., RHODE, W., 1992. Ribosomal frameshifting in plants: a novel signal directs the -1 frameshift in the synthesis of the putative viral replicase of potato leaf roll luteovirus. *EMBO J.*, 11, 1111-1117.

PRUSS, G., GE, X., SHI, X.M., CARRINGTON, J.C., VANCE, V.B., 1997. Plant viral synergism: the potyviral genome encodes a broad range pathogenicity enhancer that transactivates replication of heterologous viruses. *Plant Cell*, 9, 859-868.

PURCIFULL, D.E., HIEBERT, E., 1979. Serological distinction of *Watermelon mosaic virus* isolates. *Phytopathology*, 69, 112-116.

PUUSTINEN, P., MAKINEN, K., 2004. Uridylylation of the potyvirus VPg by viral replicase NIb correlates with the nucleotide binding capacity of VPg. *J. Biol. Chem.*, 279, 38103-38110.

QU, F., MORRIS, T.J., 1997. Encapsidation of turnip crinkle virus is defined by a specific packaging signal and RNA size. *J. Virol.*, 71, 1428-1435.

QU, F., MORRIS, T.J., 2000. Cap-independent translational enhancement of turnip crinkle virus genomic and subgenomic RNAs. *J. Virol.*, 74(3), 1085-1093.

QUADT, R., ISHIKAWA, M., JANDA, M., AHLQUIST, P., 1995. Formation of brome mosaic virus RNA-dependant RNA polymerase in yeast requires coexpression of viral proteins and viral RNA. *Proc. Natl. Acad. Sci. USA*, 92, 4892-4896.

QUADT, R., KAO, C.C., BROWNING, K.S., HERSHBERGER, R.P., AHLQUIST, P., 1993. Characterization of a host protein associated with brome mosaic virus RNA-dependant RNA polymerase. *Proc. Natl. Acad. Sci. USA*, 90, 1498-1502.

QUIOT, J.B., LABONNE, G., MARROU, J., 1982. Controlling seed and insect-borne viruses. In: *Pathogens, Vectors and Plant Diseases*. Maramorosch, K., ed. Academic Press, New York, 95-122.

QUIOT, J.B., LABONNE, G., QUIOT-DOUINE, L., 1983. The comparative ecology of cucumber mosaic virus in Mediterranean and tropical regions. In: *Plant Virus Epidemiology*. Plumb, R.T., Thresh, J.M., eds. Blackwell, Oxford, 177-183.

QUIOT, J.B., VERBRUGGHE, M., LABONNE, G., LECLANT, F., MARROU, J., 1979. Ecologie et épidémiologie du virus de la mosaïque du concombre dans le Sud-Est de la France. IV. Influence de brise-vent sur la répartition des contaminations virales dans une culture protégée. *Ann. Phytopathol.*, 11, 307-324..

RACCAH, B., 1986. Nonpersistent viruses: epidemiology and control. *Adv. Virus Res.*, 31, 387-429.

RAMIREZ, B.C., HAENNI, A.L., 1994. Molecular biology of Tenuiviruses, a remarkable group of plant viruses. *J. Gen. Virol.*, 75, 467-475..

RATCLIFF, F., HARRISON, B.D., BAULCOMBE, D.C., 1997. A similarity between viral defense and gene silencing in plants. *Science*, 276, 1558-1560.

RATCLIFF, F.G., MACFARLANE, S.A., BAULCOMBE, D.C., 1999. Gene silencing without DNA: RNA-mediated cross-protection between viruses. *Plant Cell*, 11, 1207-1215.

RATHJEN, J.P., KARAGEORGOS, L.E., HABILI, N., WATERHOUSE, P.M., SYMONS, R.H., 1994. Soybean dwarf luteovirus contains the third variant type in the luteovirus group. *Virology*, 198, 671-679.

REDDY, D.V.R., 1998. Control measures for the economically important peanut viruses. In: *Plant Virus Disease Control*. Hadidi, A., Khetarpal, R.K., Koganezawa, H., eds. APS Press, St. Paul, Minnesota, 541-546.

REED, W., CARROL, J., AGRAMONTE, A., LAZEAR, W., 1901. The etiology of yellow fever: a preliminary note. *Philadelphia Med.*, 6, 790-796.

REGISTER, J.C., BEACHY, R.N., 1988. Resistance to TMV in transgenic plants results from interference with an early event in infection. *Virology*, 166, 524-532.

REGLES INTERNATIONALES POUR LES ESSAIS DE SEMENCES., 1999. *Seed Sci. Technol.*, 27, Supplement Rules, 340 pp.

REINBOLD, C., HERRBACH, E., BRAULT, V., 2003. Posterior midgut and hindgut are both sites of acquisition of Cucurbit aphid-borne yellows virus in *Myzus persicae* and *Aphis gossypii*. *J. Gen. Virol.*, 84, 3473-3484.

RESTREPO-HARTWIG, M., AHLQUIST, P., 1999. Brome mosaic virus RNA replication proteins 1a and 2a colocalize and 1a independently localizes on the yeast endoplasmic reticulum. *J. Virol.*, 73, 10303-10309.

RESTREPO-HARTWIG, M.A., AHLQUIST, P., 1996. Brome mosaic virus helicase and polymerase-like proteins colocalize on the endoplasmic reticulum at sites of RNA synthesis. *J. Virol.*, 70, 8908-8916.

RESTREPO-HARTWIG, M.A., CARRINGTON, J.C., 1994. The tobacco etch potyvirus 6-kilodalton protein is membrane-associated and involved in viral replication. *J. Virol.*, 68, 2388-2397.

REVERS, F., LEGALL, O., CANDRESSE, T., MAULE, A.J., 1999. New advances in understanding the molecular biology of plant/potyvirus interactions. *Mol. Plant Microbe Interact.*, 12, 367-376.

REYMOND, P., FARMER, E.E., 1998. Jasmonate and salicylate as global signals for defense gene expression. *Curr. Opin. Plant Biol.*, 1, 404-411.

RICHERT-PÖGGELER, K.R., SHEPHERD, R.J., 1997. Petunia vein-clearing virus: a pararetrovirus with the core sequence for an integrase function. *Virology*, 236, 137-146.

RICHMOND, K.E., CHENAULT, K., SHERWOOD, J.L., GERMAN, T.L. 1998. Characterization of the nucleic acid binding properties of tomato spotted wilt virus nucleocapsid protein. *Virology*, 248, 6-11.

RIECHMAN, J.L., LAIN, S., GARCIA, J.A., 1992. Highlights and prospects of potyviral molecular biology. *J. Gen. Virol.*, 73, 1-16.

ROBAGLIA, C., DURAND-TARDIF, M., TRONCHET, M., BOUDAZIN, G., ASTIER-MANIFACIER, S., CASSE-DELBARD, F., 1989. Nucleotide sequence of potato virus Y (N strain) genomic RNA. *J. Gen. Virol.*, 70, 935-947.

ROBERTS, A.G., SANTA CRUZ, S., ROBERTS, I.M., PRIOR, D.A.M., TURGEON, R., OPARKA, K.J., 1997. Phloem unloading in sink leaves of *Nicotiana benthamiana*: comparison of a fluorescent solute with a fluorescent virus. *Plant Cell*, 9, 1381-1396.

ROBERTS, D.A., CHRISTIE, R.G., ARCHER, M.C., 1970. Infection of apical initials in tobacco shoot meristems by tobacco ringspot virus. *Virology*, 42, 217-220.

ROBERTS, I.M., WANG, D., FINDLAY, K., MAULE, A.J., 1998. Ultrastructural and temporal observations of the potyvirus cylindrical inclusions (CIs) shows that the CI protein acts transiently in aiding virus movement. *Virology*, 245, 173-181.

ROBERTSON, D., 2004. VIGS vectors for gene silencing: many targets, many tools. *Annu. Rev. Plant Biol.*, 55, 495-519.

ROCHON, D., KAKANI, K., ROBBINS, M., READE, R., 2004. Molecular aspects of plant virus transmission by Olpidium and Plasmidiophorid vectors. *Annu. Rev. Phytopathol.*, 42, 211-241.

ROCHOW, W.F., 1970. Barley yellow dwarf virus: phenotypic mixing and vector specificity. *Science*, 167, 875-878.

RODRIGUEZ-ALVARADO, G., ROOSSINK, M.J., 1997. Structural analysis of a necrogenic strain of cucumber mosaic virus satellite RNA *in planta*. *Virology*, 236, 155-166.

RODRIGUEZ-CEREZO, E., ELENA, S.F., MOYA, A., GARCIA-ARENAL, F., 1991. High genetic stability in natural populations of the plant RNA virus tobacco mild green mosaic virus. *J. Mol. Evol.*, 32, 328-332.

RODRIGUEZ-CEREZO, E., FINDLAY, K., SHAW, J.G., LOMONOSSOFF, G.P., QIU, S.G., LINSTEAD, P., SHANKS, M., RISCO, C., 1997. The coat and cylindrical inclusion proteins of a potyvirus are associated with connections between plant cells. *Virology*, 236, 296-306.

ROJAS, M.R., ZERBINI, F.M., ALLISON, R.F., GILBERTSON, R.L., LUCAS, W.L., 1997. Capsid protein and helper component-proteinase function as potyvirus cell-to-cell movement proteins. *Virology*, 237, 283-295.

ROOSSINCK, M., 1997. Mechanism of plant virus evolution. *Annu. Rev. Phytopathol.*, 35, 191-209.

ROOSSINK, M.J., KAPLAN, I., PALUKAITIS, P., 1997. Support of a cucumber mosaic virus satellite RNA maps to a single amino acid proximal to the helicase domain of the helper virus. *J. Virol.*, 71, 608-612.

ROOSSINK, M.J., SLEAT, D., PALUKAITIS, P., 1992. Satellite RNAs of plant viruses: structures and biological effects. *Microbiol. Rev.*, 56, 265-279.

ROSS, A.F., 1961a. Localized acquired resistance to plant virus infection in hypersensitive hosts. *Virology*, 14, 329-339.

ROSS, A.F., 1941. The concentration of alfalfa mosaic virus in tobacco plants at different periods of time after inoculation. *Phytopathology*, 31, 410-421.

ROSS, A.F., 1961b. Systemic acquired resistance induced by localized virus infection in plants. *Virology*, 14, 340-358

ROSSMANN, M.G., JOHNSON, J.E., 1989. Icosahedral RNA virus structure. *Annu. Rev. Biochem.*, 58, 533-573.

ROTH, B.M., PRUSS, G.J., VANCE, V.B., 2004. Plant viral suppressors of RNA silencing. *Virus Res.*, 102, 97-108.

RUBIO, T., BORJA, M., SCHOLTHOF, H.B., JACKSON, A.O., 1999. Recombination with host transgenes and effects on virus evolution: an overview and opinion. *Mol. Plant-Microbe Interact.*, 12, 87-92.

RUDENKO, G.N., ONO, A., WALBOT, V., 2003. Initiation of silencing of maize MuDR/Mu transposable elements. *Plant J.*, 33, 1013-1025.

RUESINK, W.G., IRWIN, M.E., 1986. Soybean mosaic virus epidemiology: a model and some implications. In: *Plant Virus Epidemics—Monitoring, Modelling and Predicting Outbreaks.* McLean, G.D., Garrett, R.G., Ruesink, W.G., eds. Academic Press, New York, 295-313.

RUFFEL, S., DUSSAULT, M.H., PALLOIX, A., MOURY, B., BENDAHMANE, A., ROBAGLIA, C., CARANTA, C., 2002. A natural recessive resistance gene against potato virus Y in pepper corresponds to the eukaryotic initiation factor 4E (eIF4E). *Plant J.*, 32, 1067-1075.

RUFFEL, S., GALLOIS, J.L., MOURY, B., ROBAGLIA, C., PALLOIX, A., CARANTA, C., 2006. Simultaneous mutations in translation initiation factors eIF4E and eIF(iso)4E are required to prevent pepper veinal mottle virus infection of pepper. *J. Gen. Virol.*, 8, 2089-2098.

RUIZ-MEDRANO, R., XOCONOSTLE-CAZARES, B., LUCAS, W.J., 1999. Phloem long-distance transport of CmNACP mRNA: implications for supracellular regulation in plants. *Development*, 126, 4405-4419.

RYALS, J.A., NEUENSCHWANDER, U.H., WILLITS, M.G., MOLINA, A., STEINER, H.Y., HUNT, M.D., 1996. Systemic acquired resistance. *Plant Cell*, 8, 1809-1819.

SACHS, A.B., SARNOW, P., HENTZE, M.X., 1997. Starting at the beginning, middle, and end: translation in Eucaryotes. *Cell*, 89, 831-838.

SACRISTAN, S., MALPICA, J., FRAILE, A., GARCIA-ARENAL, F., 2003. Estimation of population bottlenecks during systemic movement of *Tobacco mosaic virus* in tobacco plants. *J. Virol.*, 77, 9906-9911.

SAENZ, P., SALVADOR, B., SIMON-MATEO, C., KASSCHAU, K.D., CARRINGTON, J.C., GARCIA, J.A., 2002. Host-specific involvement of the HC protein in the long-distance movement of potyviruses. *J. Virol.*, 76, 1922-1931.

SAIKI, R.K., SCHARF, S., FALOONA, F., MULLIS, K.B., HORN, G.T., HERLICH, H.A., 1985. Enzyme amplification of B-globin genomic sequences and restriction site analysis for diagnosis of sickle cell anaemia. *Science*, 230, 1350-1354.

SAMUEL, G., 1934. The movement of tobacco mosaic virus within the plant. *Ann. Appl. Abiol.*, 21, 90-111.

SANFORD, J.C., JOHNSTON, S.A., 1985. The concept of parasite-derived resistance: deriving resistance genes from the parasite's own genome. *J. Theor. Biol.*, 113, 395-405.

SANGARÉ, A., DENG, D., FAUQUET, C.M., BEACHY, R.N., 1999. Resistance to *African cassava mosaic virus* conferred by a mutant of the putative NTP-binding domain of the Rep gene in *Nicotiana benthamiana. Mol. Biol. Rep.*, 5, 95-102.

SANGER, M., PASSMORE, B., FALK, B.W., BRUENING, G., DING, B., LUCAS, W.J., 1994. Symptom severity of beet western yellows virus strain ST9 is conferred by the ST9-associated RNA and is not associated with virus release from the phloem. *Virology*, 200, 48-55.

SANO, T., CANDRESSE, T., HAMMOND, R.W., DIENER, T.O., OWENS, R.A., 1992. Identification of structural domains regulating viroid pathogenicity. *Proc. Natl. Acad. Sci. USA*, 89, 10104-10108.

SANO, Y., VAN DER VLUGT, R., DE HAAN, P., TAKAHASHI, A., KAWAGAMI, M., GOLDBACH, R., KOJIMA, M., 1992. On the variability of the 3′ terminal sequence of the turnip mosaic virus genome. *Arch. Virol.*, 126, 231-238.

SANTA CRUZ, S., ROBERTS, A.G., PRIOR, D.A.M., CHAPMAN, S., OPARKA, K.J., 1998. Cell to cell and phloem mediated transport of potato virus X: the role of virions. *Plant Cell*, 10, 495-510.

Sato, M., Nakahara, K., Yoshii, M., Ishikawa, M., Uyeda, I., 2005. Selective involvement of members of the eukaryotic initiation factor 4E family in the infection of *Arabidopsis thaliana* by potyviruses. *FEBS Lett.*, 579, 1167-1171.

Saunders, K., Stanley, J., 1999. A nanovirus-like component associated with yellow vein disease of *Ageratum conyzoides*: evidence for interfamilial recombination between plant DNA viruses. *Virology*, 264, 142-152.

Schaad, M.C., Anderberg, R.J., Carrington, J.C., 2000. Strain-specific interaction of the tobacco etch virus NIa protein with the translation initiation factor eIF4E in the yeast two-hybrid system. *Virology*, 273, 300-306.

Schaad, M.C., Haldeman-Cahill, R., Cronin, S., Carrington, J.C., 1996. Analysis of the VPg-proteinase (NIa) encoded by tobacco etch potyvirus: effects of mutations on subcellular transport, proteolytic processing, and genome amplification. *J. Virol.*, 70, 7039-7048.

Schaad, M.C., Jensen, P., Carrington, J.C., 1997a. Formation of plant RNA virus replication complexes on membranes: role of an endoplasmic reticulum-targeted viral protein. *EMBO J.*, 13, 4049-4059.

Schaad, M.C., Lellis, A.D., Carrington, J.C., 1997b. VPg of Tobacco Etch Potyvirus is a host genotype-specific determinant for long distance movement. *J. Virol.*, 71, 8624-8631.

Schärer-Hernandez, N., Hohn, T., 1998. Nonlinear ribosome migration on cauliflower mosaic virus 35 S RNA in transgenic tobacco plants. *Virology*, 242, 403-413.

Schiebel, W., Hass, B., Marinkovic, S., Klanner, A., Sanger, H., 1993. RNA-directed RNA polymerase from tomato leaves. *J. Biol. Chem.*, 268, 11851-11867.

Schiebel, W., Pelissier, T., Riedel, L., Thalmeir, S., Schiebel, R., Kempe, D., Lottspeich, F., Sanger, H.L., Wassenegger, M., 1998. Isolation of an RNA-directed RNA polymerase specific c-DNA clone from tomato. *Plant Cell*, 10, 2087-2101.

Schindler, I.M., Mühlbach, H.P., 1992. Involvement of nuclear DNA-dependent RNA polymerases in potato spindle tuber viroid replication: a reevaluation. *Plant Sci.*, 84, 221-229.

Schmitz, J., Stussi-Garaud, C., Tacke, E., Prüfer, D., Rohde, W., Rohfritsch, O., 1997. *In situ* localization of the putative movement protein (pr17) from potato leafroll luteovirus (PLRV) in infected and transgenic potato plants. *Virology*, 235, 311-322.

Scholthof, H.B., Scholthof, K.B.G., Kikkert, M., Jackson, A.O., 1995. Tomato bushy stunt spread is regulated by two nested genes that function in cell to cell movement and host dependent systemic invasion. *Virology*, 213, 425-438.

Schwartz, M., Chen, J., Janda, M., Sullivan, M., den Boon, J., Ahlquist, P., 2002. A positive-strand RNA virus replication complex parallels form and function of retrovirus capsid. *Mol. Cell* 9, 505-514.

Scully, B.T., Federer, W.T., 1993. Application of genetic theory in breeding for multiple viral resistance. In: *Resistance to Viral Diseases of Vegetables*. Kyle, M.M., ed. Timber Press, Portland, 167-195.

Shaw, J.G., Plaskitt, K.A., Wilson, T.M.A., 1986. Evidence that tobacco mosaic virus particles disassemble cotranslationally *in vivo*. *Virology*, 148, 326-336.

Shepardson, S., Esau, K., McCrum, R., 1980. Ultrastucture of potato leaf phloem infected with potato leafroll virus. *Virology*, 105, 379-392.

Sherwood, J.L., 1987. Mechanisms of cross-protection between plant virus strains. In: *Plant Resistance to Viruses*. Evered, D., Harnett, S., eds. John Wiley and Sons, Chichester, New York, 136-150.

SHINKAI, A., 1962. Studies on insect transmission of rice virus diseases. *Jap. Natl. Inst. Agric. Sci. Bull.*, C14: 1-112.

SHIRAKO, Y., WILSON, T.M.A., 1993. Complete nucleotide sequence and organization of the bipartite RNA genome of soil-borne wheat mosaic virus. *Virology*, 195, 16-32.

SHUKLA, D.D., TRIBBICK, G., MASON, T.J., HEWISH, D.R., GEYSEN, H.M., WARD, C.W., 1989. Localization of virus-specific and group-specific epitopes of plant potyviruses by systematic immunochemical analysis of overlapping peptide fragments. *Proc. Natl. Acad. Sci. USA*, 86, 8192-8196.

SHUKLA, D.D., WARD, C.W., 1989. Structure of potyvirus coat protein and its application to the taxonomy of potyvirus group. *Adv. Virus Res.*, 36, 273-314.

SIEGEL, A., HARI, V., KOLACZ, K., 1978. The effect of tobacco mosaic virus infection on host and virus specific protein synthesis in protoplasts. *Virology*, 85, 494-503.

SIEGEL, R., ADKINS, S., KAO, C.C., 1997. Sequence specific recognition of a subgenomic promoter by a viral RNA polymerase. *Proc. Natl. Acad. Sci. USA*, 94, 11238-11243.

SIMON, A.E., NAGY, P.D., 1996. RNA recombination in turnip crinkle virus. *Semin. Virol.*, 7, 373-379.

SIMON, A.E., NAGY, P.D., CARPENTER, C.D., 1997. Studies on RNA recombination *in vivo* and *in vitro*. In: *Virus-Resistant Transgenic Plants: Potential Ecological Impact*. Tepfer, M., Balazs, E., eds. Springer-INRA, Berlin, 33-38.

SIMON-BUELA, L., GARCIA-ARENAL, F., 1999. Virus particles of cucumber green mottle mosaic tobamovirus move systemically in the phloem of infected cucumber plants. *Mol. Plant-Microbe Interact.*, 12, 112-18.

SIMON-BUELA, L., GUO, H.S., GARCIA, J.A., 1997. Cap-independent leaky scanning as the mechanism of translation initiation of a plant viral genomic RNA. *J. Gen. Virol.*, 78, 2691-2699.

SINGH, R.P., DHAR, A.K., 1998. Detection and management of plant viroids. In: *Plant Virus Disease Control*. Hadidi, A., Khetarpal, R.K., Koganezawa, H., eds. APS Press, St. Paul, Minnesota, 428-447.

SINGH, R.P., NIE, X., SINGH, M., 1999. Tomato chlorotic dwarf viroid: an evolutionary link in the origin of pospoviroids. *J. Gen. Virol.*, 80, 2823-2328.

SJÖLUND, R.D. 1997. The phloem sieve element: a river flows through it. *Plant Cell*, 9, 1137-1146.

SKAF, J.S., RUCKER, D.G., DEMLER, S.A., WOBUS, C.E., DE ZOETEN, G.A., 1997. The coat protein is dispensable for the establishment of systemic infections by Pea enation mosaic enamovirus. *Mol. Plant-Microbe Interact.*, 10, 929-932.

SKUZECSKI, J.M., NICHOLS, M.L., GESTELAND, R.F., ATKINS, J.F., 1991. The signal for a leaky UAG stop codon in several plant viruses includes the two downstream codons. *J. Mol. Biol.*, 15, 65-79.

SMITH, O.P., HARRIS, K.F., 1990. Potato leaf roll virus 3′ genome organization: sequence of the coat protein gene and identification of viral subgenomic RNA. *Phytopathology*, 80, 609-614.

SOLOVYEV, A.G., SAVENKOV, E.I., GRDZELISHVILI, V.Z., KALININA, N.O., MOROZOV, S. YU., SCHIEMANN, J., ATABEKOV, J.G., 1999. Movement of hordeivirus hybrids with exchanges in the triple gene block. *Virology*, 253, 278-287.

SOLOVYEV, A.G., ZELENINA, D.A., SAVENKOV, E.I., GRDZELISHVILI, V.Z., MOROZOV, S.YU., LESEMANN, D.E., MAISS, E., CASPER, R., ATABEKOV, J.G., 1996. Movement of a barley stripe mosaic virus chimera with a tobacco mosaic virus movement protein. *Virology*, 217, 435-441.

STANLEY, J., FRISCHMUTH, T., ELLWOOD, S., 1990. Defective viral DNA ameliorates symptoms of geminivirus infection in transgenic plants. *Proc. Natl. Acad. Sci. USA*, 87, 6291-6295.

STANLEY, J., LATHAM, J.R., PINNER, M.S., BEDFORD, I., MARKHAM, P.G., 1992. Mutational analysis of the monopartite geminivirus beet curly top virus. *Virology*, 191, 396-405.

STANLEY, W.M., 1935. Isolation of a crystalline protein possessing the properties of tobacco mosaic virus. *Science*, 81, 644-645.

STEITZ, T.A., 1998. A mechanism for all polymerases. *Nature*, 391, 231-232.

STENGER, D.C., HEIN, G.L., GILDOW, F.E., HORKEN, K.M., FRENCH, R., 2005. Plant virus HC-Pro is a determinant of eriophyid mite transmission. *J. Virol.*, 79, 9054-9061.

STONE, O.M., 1968. The elimination of four viruses from carnation and sweet William by meristem-tip culture. *Ann. Appl. Biol.*, 62, 119-122.

STUPINA, V., SIMON, A.E., 1997. Analysis *in vivo* of turnip crinkle virus satellite RNA C variants with mutations in the 3′-terminal minus strand promoter. *Virology*, 238, 470-477.

STUSSI-GARAUD, C., MUTTERER, J., DELECOLLE, B., ROHFRITSCH, O., VANTARD, M., 1998. Routes intra- et intercellulaires utilisées par les virus de plantes pour leur mouvement de cellule à cellule. *Virologie*, 2, 269-283.

SULZINSKI, M.A., ZAITLIN, M., 1982. Tobacco mosaic virus replication in resistant and susceptible plants: in some resistant species, virus is confined to a small number of initially infected cells. *Virology*, 121, 12-19.

SUN, J.H., KAO, C.C., 1997a. Characterisation of RNA products associated with or aborted by viral RNA-dependent RNA polymerase. *Virology*, 236, 348-353.

SUN, J.H., KAO, C.C., 1997b. RNA synthesis by the brome mosaic virus RNA-dependent RNA polymerase transition from initiation to elongation. *Virology*, 233, 63-73.

SUTULA, C.L., GILLETT, J.M., MORRISSEY, S.M., RAMSDELL, D.C., 1986. Interpreting ELISA data and establishing the positive-negative threshold. *Plant Dis.*, 70, 722-726.

SUZUKI, M., KUWATA, S., MASUTA, C., TAKANAMI, Y., 1995. Point mutations in the coat protein of cucumber mosaic virus affect symptom expression and virion accumulation in tobacco. *J. Gen. Virol.*, 76, 1791-1799.

SYLVESTER, E.S., 1956. Beet mosaic and beet yellows virus transmission by the green peach aphid. *J. Econ. Entomol.*, 49, 789-800.

SYLVESTER, E.S., RICHARDSON, J., 1969. Additional evidence of multiplication of the sowthistle yellow vein virus in an aphid vector-serial passage. *Virology*, 37, 26-31.

SYMONS, R.H., 1997. Plant pathogenic RNAs and RNA catalysis. *Nucl. Acids Res.*, 25, 2683-2689.

TABLER, M., TSAGRIS, M., HAMMOND, J., 1998. Antisense RNA and ribozyme-mediated resistance to plant viruses. In: *Plant Virus Disease Control*. Hadidi, A., Khetarpal, R.K., Koganezawa, H., eds. APS Press, St. Paul, Minnesota, 79-93.

TACKE, E., PRÜFER, D., SCHMITZ, J., ROHDE, W., 1991. The potato leafroll luteovirus 17K protein is a single-stranded nucleic acid binding protein. *J. Gen. Virol.*, 72, 2035-2038.

TAGU, D., 1999. *Principes des Techniques de Biologie Moléculaire*. INRA, Paris, 131 pp.

TAKATSUJI, H., HIROCHIKA, H., FUKUSHI, T., IKEDA, J., 2003. Expression of Cauliflower mosaic reverse transcriptase in yeast. *Nature*, 319, 240-243.

TAKEBE, I., OTSUKI, Y., 1969. Infection of tobacco mesophyll protoplasts by tobacco mosaic virus. *Proc. Natl. Acad. Sci. USA*, 64, 843-848.

TALIANSKI, M.E., GARCIA-ARENAL, F., 1995. Role of cucumovirus capsid protein in long distance movement within the infected plant. *J. Virol.*, 69, 916-922.

TAMADA, T., SCHMITT, T., SAITO, C., GUILLEY, H., RICHARDS, K., JONARD, G., 1996. High resolution analysis of the readthrough domain of beet necrotic yellow vein virus readthrough protein: a KTER motif is important for efficient transmission of the virus by the fungus *Polymyxa betae*. *J. Gen. Virol.*, 77, 1359-1367.

TANNER, N.K., 1998. Ribozymes: caractéristiques et applications. *Virologie*, 2, 127-137.

TARAPOREWALA, Z.F., CULVER, J.N., 1997. Structural and functional conservation of the Tobamovirus coat protein elicitor active site. *Mol. Plant-Microbe Interact.*, 10, 597-604.

TARUN, S.Z., SACHS, A.B., 1996. Association of the yeast poly(A) tail binding protein with translation initiation factor eIF-4G. *EMBO J.*, 15, 7168-7177.

TAYLOR, C.E., BROWN, D.J.F., 1997. *Nematode Vectors of Plant Viruses*. Cab International, Wallingford, UK, 286 pp.

TAYLOR, C.E., ROBERTSON, W.M., 1974. Electron microscopy evidence for the association of tobacco severe etch virus with the maxillae in *Myzus persicae* (Sulz.). *Phytopathol. Z.*, 80, 257-266.

TEAKLE, D.S., 1962. Transmission of tobacco necrosis virus by a fungus, *Olpidium brassicae*. *Virology*, 18, 224-231.

TEMIN, H.M., 1980. Origin of retroviruses from cellular genetic elements. *Cell*, 21, 599-600.

TEPFER, M., 2002. Risk assessment of virus-resistant transgenic plants. *Annu. Rev. Phytopathol.*, 467-491.

THOMPSON, J.R., GARCIA-ARENAL, F., 1998. The bundle sheath-phloem interface of *Cucumis sativus* is a boundary to systemic infection by tomato aspermy virus. *Mol. Plant-Microbe Interact.*, 11, 109-114.

THORNBURY, D.W., HELLMAN, G.M., RHOADS, R.E., PIRONE, T.P., 1985. Purification and characterization of potyvirus helper component. *Virology*, 144, 260-267.

THRESH, J.M., 1978. The epidemiology of plant virus diseases. In: *Plant Disease*. Scott, P.R., Bainbridge, A., eds. Blackwell, Oxford, 79-91.

THRESH, J.M., 1988. Eradication as a virus disease control measure. In: *Control of Plant Diseases, Costs and Benefits*. Clifford, B.C., Lester, E., eds. Blackwell Scientific Publiscations, Oxford, 155-194.

TIAN, T., RUBIO, L., YEH, H.H., CRAWFORD, B., FALK, B.W., 1999. Lettuce infectious yellows virus: *in vitro* acquisition analysis using partially purified virions and the whitefly *Bemisia tabaci*. *J. Gen. Virol.*, 80, 1111-1117.

TIMCHENKO, T., DE KOUCHKOVSKY, F., KATUL, L., DAVID, C., VETTEN, H.J., GRONENBORN, G., 1999. Single rep protein initiates replication of multiple genome components of faba bean necrotic yellows virus, a single-stranded DNA virus of plants. *J. Virol.*, 73, 10173-10182.

TOMENIUS, K., CLAPHAM, D., MESHI, T., 1987. Localisation by immunogold cytochemistry of the virus-coded 30 K protein in plasmodesmata of leaves infected with tobacco mosaic virus. *Virology*, 160, 363-371.

TOMITA, Y., MIZUNO, T., DIEZ, J., NAITO, S., AHLQUIST, P., ISHIKAWA, M., 2003. Mutation of host *dnaJ* homologs inhibits brome mosaic virus negative-strand RNA synthesis. *J. Virol.* 77, 2990-2997.

TOMLINSON, J.A., 1988. Chemical control of *Spongospora* and *Olpidium* in hydroponic systems and soil. In: *Viruses with Fungal Vectors*. Cooper, J.I., Asher, M.J., eds. AAB, Wellesbourne, UK, 293-303.

TOUSSAINT, A., KUMMERT, J., MAROQUIN, C., LEBRUN, A., ROGGEMANS, J., 1993. Use of Virazole to eradicate odontoglossum ringspot virus from *in vitro* cultures of Cymbidium Sw. *Plant Cell, Tissue Organ Cult.*, 32, 303-309.

TSAGRIS, M., TABLER, M., MÜHLBACH, H.P., SANGER, H.L., 1987. Linear oligomeric potato spindle tuber viroid (PSTV) RNAs are accurately processed *in vitro* to the monomeric circular viroid proper when incubated with a nuclear extract from healthy potato cells. *EMBO J.*, 6, 2173-2183.

TSAI, M.S., HSU, Y.H., LIN, N.S., 1999. Bamboo mosaic potexvirus satellite-encoded P20 protein preferentially binds to satBaMV RNA. *J. Virol.*, 73, 3032-3039.

TSUCHIZAKI, T., HIBINO, H., 1971. Seed transmission of viruses in cowpea and azuki bean plants. IV. Relations between seed transmission and virus distribution in apical meristem of flower bud. *Ann. Phytopathol. Soc. Jap.*, 37, 17-21.

TSUJIMOTO, Y., NUMAGA, T., OHSHIMA, K., YANO, M., OHSAWA, R., GOTO, D., NAITO, S., ISHIKAWA, M., 2003. *Arabidopsis TOBAMOVIRUS MULTIPLICATION (TOM)2* locus encodes a transmembrane protein that interacts with TOM 1. *EMBO J.*, 22, 335-343.

TURNAGE, M.A., MUANGSAN, N., PEELE, C.G., ROBERTSON, D., 2002. Geminivirus-based vectors for gene silencing in Arabidopsis. *Plant J.*, 30, 107-114.

UCKO, O., COHEN, S., BEN JOSEPH, R., 1998. Prevention of virus epidemics by a crop-free period in the Arava region of Israel. *Phytoparasitica*, 26, 313-321.

ULLMAN, D.E., CHO, J.J., MAU, R.F.L., WESCOT, D.M., CUSTER, D., 1993. Tospovirus replication in insect vector cells: immunocytochemical evidence that the nonstructural protein encoded by the S RNA of tomato spotted wilt tospovirus is present in thrips vector cells. *Phytopathology*, 83, 456-463.

UYEMOTO, J.K., GROGAN, R.G., 1977. Southern bean mosaic virus: evidence for seed transmission in bean embryos. *Phytopathology*, 67, 1190-1196.

VALLE, R.P.C., DRUGEON, G., DEVIGNES-MORCH, M.D., LEGOCKI, A.B., HAENNI, A.L., 1992. Codon context effect in virus translational readthrough. A study *in vitro* of the determinants of TMV and Mo-MuLV amber suppression. *FEBS Lett.*, 306, 133-139.

VALLEAU, W.D., 1941. Experimental production of symptoms in so-called recovered ring-spot tobacco plants and its bearing on acquired immunity. *Phytopathology*, 31, 522-533.

VALLEAU, W.D., 1952. The evolution of susceptibility to tobacco mosaic in *Nicotiana* and the origin of *Tobacco mosaic virus*. *Phytopathology*, 42, 40-42.

VAN DEN BOOGAART, T., LOMONOSSOFF, G.P., DAVIES, J.W., 1998. Can we explain RNA-mediated virus resistance by homology-dependent gene silencing? *Mol. Plant-Microbe Interact.*, 11, 717-723.

VAN DEN HEUVEL, J.F., BRUYÈRE, A., HOGENHOUT, S.A., ZIEGLER-GRAFF, V., BRAULT, V., VERBEEK, M., VAN DER WILK, F., RICHARDS, K., 1997. The N-terminal region of the luteovirus readthrough domain determines virus binding to *Buchnera* GroEL and is essential for virus persistence in the aphid. *J. Virol.*, 71, 7258-7265.

VAN DEN HEUVEL, J.F., VERBEEK, M., VAN DER WILK, F., 1994. Endosymbiotic bacteria associated with circulative transmission of potato leafroll virus by *Myzus persicae*. *J. Gen. Virol.*, 75, 2559-2565.

VAN DER BIEZEN, E.A., JONES, J.D.G., 1998a. Plant disease-resistance proteins and the gene-for-gene concept. *Trends Biol. Sci.*, 23, 454-456.

VAN DER BIEZEN, E.A., JONES, J.D.G., 1998b. The NB-ARC domain: a novel signaling motif shared by plant resistance gene products and regulators of cell death in animals. *Curr. Biol.*, 8, R226-227.

VAN LENT, J., STORMS, M., VAN DER MEER, F., WELLINK, J., GOLDBACH, R., 1991. Tubular structures involved in movement of cowpea mosaic virus are also formed in infected cowpea protoplasts. *J. Gen. Virol.*, 72, 2615-2623.

VAN LENT, J.W.M., WELLINK, J., GOLDBACH, R., 1990. Evidence for the involvement of the 58K and 48K proteins in the intercellular movement of cowpea mosaic virus. *J. Gen. Virol.*, 71, 219-223.

VAN LOON, L.C., 1997. Induced resistance in plants and the role of pathogenesis-related proteins. *Eur. J. Plant Pathol.*, 103, 753-765.

VAN REGENMORTEL, M.H.V., 1986. Tobacco mosaic virus: antigenic structure. In: *The Plant Viruses*, Vol. 2. Van Regenmortel, M.H.V., Fraenkel-Conrat, H., eds. Plenum Press, New York and London, 79-104.

VAN REGENMORTEL, M.H.V., 1989. Applying the species concept to plant viruses. *Arch Virol.*, 104, 1-17.

VAN REGENMORTEL, M.H.V., BISHOP, D.H.L., FAUQUET, C.M., MAYO, M.A., MANILOFF, J., CALISHER, C.H., 1997. Guidelines to the demarcation of virus species. *Arch. Virol.*, 142, 1507-1518.

VAN REGENMORTEL, M.H.V., FAUQUET, C.M., 2000. Progrès en taxonomie virale. *Virologie*, 4, 29-37.

VAN REGENMORTEL, M.H.V., FAUQUET, C.M., BISHOP, D.H.L., CARSTENS, E., ESTES, M.K., LEMON, S., MANILOFF, J., MAYO, M.A., MCGEOCH, D.J., PRINGLE, C.R., WICKNER, R., 2000. Virus taxonomy. Classification and nomenclature of viruses. *VIIth Report of the International Committee of Taxonomy of Viruses*. Academic Press, San Diego, 1160 pp.

VANDERPLANK, J.E., 1960. Analysis of epidemics. In: *Plant Pathology: an Advanced Treatise*. Horsfal, J.G., Dimond, A.E., eds. Academic Press, New York, 229-289.

VANDERPLANK, J.E., 1963. *Plant Diseases: Epidemics and Control*. Academic Press, New York, 344 pp.

VANDERPLANK, J.E., 1968. *Disease Resistance in Plants*. Academic Press, New York, 206 pp.

VANDERPLANK, J.E., 1975. *Principles of Plant Infection*. Academic Press, New York, 216 pp.

VARRELMANN, M., PALKOVICS, L., MAISS, E., 2000. Transgenic or plant expression vector-mediated recombination of *Plum pox virus*. *J. Virol*, 74, 7462-7469.

VAUCHERET, H., BECLIN, C., FAGARD, M., 2001. Post-transcriptional gene silencing in plants. *J. Cell Sci.*, 114, 3083-3091.

VERMA, H.N., BARANWAL, V.K., SRIVASTAVA, S., 1997. Antiviral substances of plant origin. In: *Plant Virus Disease Control*. Hadidi, A., Khetarpal, R.K., Koganezawa, H., eds. APS Press, St. Paul, Minnesota, 154-162.

VERNOOIJ, B., FRIEDRICH, L., MORSE, A., REIST, R., KOLDITZ-JAWHAR, R., WARD, E., UKNES, S., KESSMANN, H., RYALS, J., 1994. Salicylic acid is not the translocated signal responsible for inducing systemic acquired resistance but is required in signal transduction. *Plant Cell*, 6, 959-965.

VISVADER, J.E., SYMONS, R.H., 1986. Replication of *in vitro* constructed mutants. Location of the pathogenicity-modulating domain of citrus exocortis viroid. *EMBO J.*, 5, 2051-2055.

VLOT, A., LAROS, S., BOL, J., 2003. Coordinate replication of Alfalfa mosaic virus RNA1 and 2 involves *cis*- and *trans*-acting functions of the encoded helicase-like and polymerase-like domains. *J. Virol.*, 77, 10790-10798

VOINNET, O., PINTO, Y.M., BAULCOMBE, D.C., 1999. Suppression of gene silencing: a general strategy used by diverse DNA and RNA viruses of plants. *Proc. Natl. Acad. Sci. USA*, 96, 14147-1452.

VOS, P., VERVER, J., JEAGLE, M., WELLINK, J., VAN KAMEN, A., 1988. Two viral proteins involved in the proteolytic processing of the cowpea mosaic virus polyproteins. *Nucl. Acids Res.*, 16, 1967-1985.

VOYTAS, D.F., CUMMINGS, M.P., KONIECZNY, A., AUSUBEL, F.M., RODERMEL, S.R., 1992. Copia-like retrotransposons are ubiquitous among plants. *Proc. Natl. Acad. Sci. USA*, 89, 7124-7128.

VUITTENEZ, A., DALMASSO, M.M., 1978. Problèmes de replantation de la vigne et de désinfection du sol dans les pays tempérés. *Bull. Off. Int. Vigne Vin*, 567, 337-351.

WAGNER, J.D., JACKSON, A.O., 1997. Characterization of components and activity of Sonchus yellow net rhabdovirus polymerase. *J. Gen. Virol.*, 71, 2371-2382.

WAIGMANN, E., LUCAS, W.J., CITOVSKY, V., ZAMBRYSKI, P., 1994. Direct functional assay for tobacco mosaic virus cell to cell movement protein and identification of a domain involved in increasing plasmodesmal permeability. *Proc. Natl. Acad. Sci. USA*, 91, 1433-1437.

WAIGMANN, E., ZAMBRYSKI, P., 1995. Tobacco mosaic virus movement protein-mediated protein transport between trichome cells. *Plant Cell*, 7, 2069-2079.

WALTER, B., 1998. Virus et viroses de la vigne: diagnostic et methodes de lutte. *Virologie*, 2, 435-444.

WALTER, B., BASS, P., LEGIN, R., COLLAS, A., VESSELLE, G., 1990. Amélioration du dépistage des maladies de type viral de la vigne. Indexage à l'aide de la méthode de la greffe-bouture herbacée. *Prog. Agric. Vitic.*, 107, 367-370.

WANG, D., MACFARLANE, S.A., MAULE, A.J., 1997. Viral determinants of pea early browning virus seed transmission in pea. *Virology*, 234, 112-117.

WANG, D., MAULE, A.J., 1992. Early embryo invasion as a determinant in pea of the seed transmission of pea seed borne mosaic virus. *J. Gen. Virol.*, 73, 1615-1620.

WANG, D., MAULE, A.J., 1995. Inhibition and host genes expression associated with plant virus replication. *Science*, 267, 229-231.

WANG, J., CARPENTER, C.D., SIMON, A.E., 1999. Minimal sequence and structural requirements of a subgenomic RNA promoter for turnip crinkle virus. *Virology*, 253, 327-336.

WANG, J.Y., CHAY, C., GILDOW, F.E., GRAY, S.M., 1995. Readthrough protein associated with virions of barley yellow dwarf luteovirus and its potential role in regulating the efficiency of aphid transmission. *Virology*, 206, 954-962.

WANG, M.B., BIAN, X.Y., WU, L.M., LIU, L.X., SMITH, N.A., ISENEGGER, D., WU, R.M., MASUTA, C., VANCE, V.B., WATSON, J.M., REZAIAN, A., DENNIS, E.S., WATERHOUSE, P.M., 2004. The role of RNA silencing in the pathogenicity and evolution of viroids and viral satellites. *Proc. Natl. Acad. Sci. USA*, 101, 3275-3280.

WANG, R.Y., AMMAR, E.D., THORNBURY, D.W, LOPEZ-MOYA, J.J., PIRONE, T.P., 1996. Loss of potyvirus transmissibility and helper-component activity correlate with non-retention of virions in aphid stylets. *J. Gen. Virol.*, 76, 861-867.

WANG, S., BROWNING, K.S., MILLER, W.A., 1997. A viral sequence in the 3′ untranslated region mimics a 5′ cap in facilitating translation of uncapped mRNA. *EMBO J.*, 16, 4107-4116.

WANG, S., GUO, L., ALLEN, E., MILLER, W.A., 1999. A potential mechanism for selective control of Cap-independent translation by a viral RNA sequence in *cis* and in *trans*. *RNA*, 5, 728-738.

WANG, W., TRONCHET, M., LARROQUE, N., DORION, N., ALBOUY, J., 1988. Production of virus-free Dahlia: meristem-culture and virus detection through cDNA probes and ELISA. *Acta Hort.*, 234, 421-428.

WANG, X., ULLAH, Z., GRUMET, R., 2000. Interaction between zucchini yellow mosaic potyvirus RNA-dependent polymerase and host Poly-(A) binding protein. *Virology*, 275, 433-443.

WARD, C.D., STOCKES, M.A.M., FLANEGA, J.B., 1988. Direct measurement of the poliovirus RNA polymerase error frequency *in vitro*. *J. Virol.*, 62, 558-562.

WARD, C.W., 1993. Progress towards a higher taxonomy of viruses. *Res. Virol.*, 144, 419-453.

WARD, E., FOSTER, S.J., FRAAIJE, B.A., MCCARTNEY, H.A., 2004. Plant pathogen diagnostics: immunological and nucleic acid-based approaches. *Ann. Appl. Biol.*, 145, 1-16.

WARTIG, L., KHEYR-POUR, E., NORIS, E., DE KOUTCHOVSKY, F., JOUANNEAU, B., GRONENBORN, B., JUPIN, I., 1997. Genetic analysis of the monopartite tomato yellow leaf curl geminivirus: roles of V1, V2, and C2 ORFs in viral pathogenesis. *Virology*, 228, 132-140.

WATANABE, T., HONDA, A., IWATA, A., UEDA, S., HIBI, T., ISHIHAMA, A., 1999. Isolation from tobacco mosaic virus-infected tobacco of a solubilized template specific RNA-dependent RNA polymerase containing a 126k/123k protein heterodimer. *J. Virol.*, 73, 2633-2640.

WATANABE, Y., EMORI, Y., OOSHIKA, I., MESHI, T., OHNO, T., OKADA, Y., 1984. Synthesis of TMV specific RNAs and proteins at the early stage of infection in tobacco protoplasts: transient expression of the 30K protein and its mRNA. *Virology*, 133, 18-24.

WATANABE, Y., OGAWA, T., TAKAHASHI, H., ISHIDA, I., TAKEUCHI, Y., YAMAMOTO, M., OKADA, Y., 1995. Resistance against multiple plant viruses in plants mediated by a double-stranded RNA-specific ribonuclease. *FEBS Lett.*, 372, 165-168.

WATSON, J.D., 1954. The structure of tobacco mosaic virus. I. X-ray evidence of a helical arrangement of sub-units around the longitudinal axis. *Biochim. Biophys. Acta*, 13, 10-19.

WATSON, M.A., ROBERTS, F.M., 1939. A comparative study of the transmission of *Hyoscyamus* virus 3, potato virus Y and cucumber virus 1 by the vectors *Myzus persicae, M. circumflexus* and *Macrosiphum gei*. *Proc. R. Soc. Lond.*, ser. B, 127, 543-573.

WEBER, H., PFITZNER, J.P., 1998. *Tm2*^{2} resistance in tomato requires recognition of the carboxy terminus of the movement protein of tomato mosaic virus. *Mol. Plant-Microbe Interact.*, 6, 498-503.

WEBER, H., SCHULTZE, S., PFITZNER, A.J.P., 1993. Two amino acid substitutions in the tomato mosaic virus 30-kilodalton movement protein counter the ability to overcome *Tm2*^{2} resistance gene in tomato. *J. Virol.*, 67, 6432-6438.

WEBSTER, R.G., BEAN, X.J., GORMAN, O.T., KAWAOKA, Y., 1992. Evolution and ecology of influenza A viruses. *Microbiol. Rev.*, 56, 152-178.

WEILAND, J.J., DREHER, T.W., 1993. Cis-preferential replication of the turnip yellow mosaic virus RNA genome. *Proc. Natl. Acad. Sci. USA*, 90, 6095-6099.

WEILAND, J.J., EDWARDS, M.C., 1996. A single nucleotide substitution in the aa gene confers oat pathogenicity to barley stripe mosaic virus strain ND18. *Mol. Plant-Microbe Interact.*, 9, 62-67.

WELLS, S.E., HILLNER, P.E., VALE, R.D., SACHS, A.B., 1998. Circularization of mRNA by eukaryotic translation initiation factors. *Mol. Cell*, 2, 135-140.

WEN, F., LISTER, R.M., 1991. Heterologous encapsidation in mixed infections among four isolates of barley yellow dwarf virus. *J. Gen. Virol.*, 72, 217-2223.

WETZEL, T., CANDRESSE, T., MACQUAIRE, G., RAVELONANDRO, M., DUNEZ, J., 1992. A highly sensitive immunocapture polymerase chain reaction method for *Plum pox virus* detection. *J. Virol. Methods*, 39, 27-37.

WHITE, P.S., MORALES, F.J., ROOSINCK, M.J., 1995. Interspecific reassortment in the evolution of a cucumovirus. *Virology*, 207, 334-337.

WHITHAM, S., DINESH-KUMAR, S.P., CHOI, D., HEHL, R., CORR, C., BAKER, B., 1994. The product of the tobacco mosaic resistance gene N: similarity to Toll and the interleukin-1 receptor. *Cell*, 78, 1101-1105.

WHITHAM, S., MCCORMICK, S., BAKER, B., 1996. The N-gene of tobacco confers resistance to tobacco mosaic virus in transgenic tomato. *Proc. Natl. Acad. Sci. USA*, 93, 8776-8781.

WHITHAM, S.A., ANDERBERG, R.J., CHISHOLM, S.T., CARRINGTON., J.C., 2000. *Arabidopsis* RTM2 gene is necessary for specific restriction of tobacco etch virus and encodes an unusual small heat shock-like protein. *Plant Cell*, 12, 569-582.

WHITHAM, S.A., YAMAMOTO, M.L., CARRINGTON, J.C., 1999. Selectable viruses and altered susceptibility mutants in *Arabidopsis thaliana*. *Proc. Natl. Acad. Sci. USA*, 96, 772-777.

WIECZOREK, A., SANFAÇON, H., 1993. Characterization and subcellular location of tomato ringspot nepovirus putative movement protein. *Virology*, 194, 734-743.

WIERZCHOSLAWSKI, R., DZIANNOT, A., KUNIMALAYAN, S., BUJARSKI, J., 2003. A transcriptionally active subgenomic promoter supports homologous crossovers in a plus-strand RNA virus. *J. Virol.*, 77, 6769-6776..

WILSON, T.M.A., 1984. Co-translational disassembly of tobacco mosaic virus *in vitro*. *Virology*, 137, 255-265.

WILSON, T.M.A., 1993. Strategies to protect crop plants against viruses: pathogen-derived resistance blossoms. *Proc. Natl. Acad. Sci. USA*, 90, 3134-3141.

WINTERMANTEL, W.M., SCHOELZ, J.E., 1996. Isolation of recombinant viruses between cauliflower mosaic virus and a viral gene in transgenic plants under conditions of moderate selection pressure. *Virology*, 223, 156-164.

WISLER, G.C., DUFFUS, J.E., LIU, H.Y., LI, R.H., 1998. Ecology and epidemiology of whitefly-transmitted closteroviruses. *Plant Dis.*, 82, 270-280.

WITTMANN, S., CHATEL, H., FORTIN, M.G., LALIBERTÉ, J.F., 1997. Interaction of the viral genome-linked protein of turnip mosaic potyvirus with the translational eukaryotic initiation factor (iso) 4E of *Arabidopsis thaliana* using the yeast two-hybrid system. *Virology*, 234, 89-92.

WOROBEY, M., HOLMES, E.C., 1999. Evolutionary aspects of recombination in RNA viruses. *J. Gen. Virol.*, 80, 2535-2543.

WU, B., WHITE, K.A., 1999. A primary determinant of cap-independent translation is located in the 3′-proximal region of the tomato bushy stunt virus genome. *J. Virol.*, 73, 8982-8988.

WU, X., SHAW, J., 1996. Bidirectional uncoating of the genomic RNA of a helical virus. *Proc. Natl. Acad. Sci. USA*, 93, 2981-2984.

WU, X., SHAW, J.G., 1997. Evidence that a viral replicase protein is involved in the disassembly of tobacco mosaic virus particles *in vivo*. *Virology*, 239, 426-434.

XIE, Q., SUAREZ-LOPEZ, P., GUTTIERREZ, C., 1995. Identification and analysis of a retinoblasma binding motif in the replication protein of plant DNA viruses: requirement for efficient viral DNA replication. *EMBO J.*, 14, 4073-4082.

XIN, H.W., DING, S.W., 2003. Identification and molecular characterization of a naturally occurring RNA virus mutant defective in the initiation of host recovery. *Virology*, 317, 253-262.

XIONG, Y., EICKBUSH, T.H., 1990. Origin and evolution of retroelements based on their reverse transcriptase sequence. *EMBO J.*, 9, 3353-3362.

XONOCOSTLE-CAZARES, B., XIANG, Y., RUIZ-MEDRANO, R., WANG, H.L., MONZER, J., YOO, B.C., MCFARLAND, K.C., FRANCESCHI, V.R., LUCAS, W.J., 1999. Plant paralog to viral movement protein that potentiates transport of mRNA into the phloem. *Science*, 283, 94-98.

Yang, A.F., Hamilton, R.I., 1974. The mechanism of seed transmission of tobacco ringspot virus in soybean. *Virology*, 62, 26-37.

Yang, S.J., Carter, S.A., Cole, A.B., Cheng, N.H., Nelson, R.S., 2004. A natural variant of a host RNA-dependent RNA polymerase is associated with increased susceptibility to viruses by *Nicotiana benthamiana*. *Proc. Natl. Acad. Sci. USA*, 101, 6297-6302.

Yang, Y., Klessig, D.F., 1996. Isolation and characterization of a tobacco mosaic virus-inducible *myb* oncogene homolog from tobacco. *Proc. Natl. Acad. Sci. USA*, 93, 14972-14977.

Yepes, L.M., Fuchs, M., Slightom, J.L., Gonsalves, D., 1996. Sense and antisense coat protein gene constructs confer high level of resistance to tomato ringspot nepovirus in transgenic *Nicotiana* species. *Phytopathology*, 86, 417-424.

Yoo, B.C., Kragler, F., Varkonyi-Gasic, E., Haywood, V., Archer-Evans, S., Lee, Y.M., Lough, T.J., Lucas, W.J., 2004. A systemic small RNA signaling system in plants. *Plant Cell*, 16: 1979-2000.

Yoshii, M., Nishikiori, M., Tomita, K., Yoshioka, N., Kozuka, R., Naito, S., Ishikawa, M., 2004. The Arabidopsis cucumovirus multiplication 1 and 2 loci encode translation initiation factors 4E and 4G. *J. Virol.*, 78, 6102-6111.

Yot, P., Pinck, M., Haenni, A.L., Duranton, H., Chapeville, F., 1970. Valine-specific rRNA-like structure in turnip yellow mosaic virus RNA. *Proc. Natl. Acad. Sci. USA*, 67, 1345-1352.

Yusibov, V., Loesch-Fries, L.S., 1995. High-affinity RNA-binding domains of alfalfa mosaic virus coat protein are not required for coat protein-mediated resistance. *Proc. Natl. Acad. Sci. USA*, 92, 1-5.

Zaccomer, B., Cellier, F., Boyer, J.C., Haenni, A.L., Tepfer, M., 1993. Transgenic plants that express genes including the 3′ non translated region of the turnip yellow mosaic virus (TYMV) genome are partially protected against TYMV infection. *Gene*, 136, 87-94.

Zaccomer, B., Haenni, A.L., Macaya, G., 1995. The remarkable variety of plant virus genomes. *J. Gen. Virol.*, 76, 231-247.

Zadoks, J.C. Schein, R.D., 1979. *Epidemiology and Plant Disease Management*. Oxford University Press, Oxford, 427 pp.

Zaenen, I., Van Larebeke, N., Teuchy, H., Van Montagu, M., Schell, J., 1974. Supercoiled circular DNA in crown-gall inducing *Agrobacterium* strains. *J. Mol. Biol.*, 86, 109-127.

Zambryski, P., Tempe, J., Schell, J., 1989. Transfer and function of T-DNA genes from *Agrobacterium* Ti and Ri plasmids in plants. *Cell*, 56, 193-201.

Zamore, P.D., 2001. RNA interference: listening to the sound of silence. *Nat. Struct. Biol.*, 8, 746-750.

Zanotto, P., Gibbs, M.J., Gould, E.A., Holmes, E.C., 1996. A reevaluation of the higher taxonomy of viruses based on RNA polymerases. *J. Virology*, 70, 6083-6096.

Zhao, X., Fox, J.M., Olson, N.H., Baker, T.S., Young, M.J., 1995. *In vitro* assembly of cowpea chlorotic mottle virus from coat protein expressed in *Escherichia coli* and *in vitro* transcribed viral cDNA. *Virology*, 207, 486-494.

Zheng, H., Li, Y., Yu, Z., Li, W., Chen, M., Ming, X., Casper, R., Chen, Z., 1997. Recovery of transgenic rice plants expressing the rice dwarf virus outer coat protein gene. *Theor. Appl. Gen.*, 94, 522-527.

Zhou, X., Liu, Y., Calvert, L., Munoz, C., Otim-Nape, C., Robinson, D.J., Harrison, B.D., 1997. Evidence that DNA-A of a geminivirus associated with severe cassava mosaic disease in Uganda has arisen by interspecific recombination. *J. Gen. Virol.*, 78, 2101-2111.

ZHOU, X., LIU, Y., ROBINSON, D.J., HARRISON, B.D., 1998. Four DNA-A variants among Pakistani isolates of cotton leaf curl virus and their affinities to DNA-A of geminivirus isolates from okra. *J. Gen. Virol.*, 79, 915-923.

ZIEGLER-GRAFF, V., BRAULT, V., MUTTERER, J.D., SIMONIS, M.T., HERRBACH, E., GUILLEY, H., RICHARDS, K.E., JONARD, G., 1996. The coat protein of Beet Western Yellows luteovirus is essential for systemic infection but the viral gene products P29 and P19 are dispensable for systemic infection and aphid transmission. *Mol. Plant-Microbe Interact.*, 9, 501-510.

ZITTER, T.A., SIMONS, J.N., 1980. Management of viruses by alterations of vector efficiency and by cultural practices. *Annu. Rev. Phytopathol.*, 18, 289-310.

Photo Credits

Unless otherwise stated, the photographs illustrating this book were kindly provided by our colleagues from INRA:

J. Albouy: color plate I.1; color plate II.7; color plate VII.2b and 2c.
D. Blancard: color plate III.1; fig. 8.12 (left).
J.M. Bossenec: fig. 1.1(10); color plate I.6; color plate VI.2.
G. Boudazin: fig. 4.11; color plate I.6; color plate VII.1c.
D. Bourdin: fig. 1.1(1).
R. Coutin: fig. 8.7(a).
B. Delécolle: fig. 1.1(2, 3, 6, 7, 9 and 12b); fig. 9.2; color plate VI.1c.
G. Della Giustina: fig. 8.7 (b).
G. Demangeat: fig. 8.10 (left); color plate II.1.
J.M. Déogratias: color plate VIII.2.
S. Dinant: color plate VIII.3 and 4.
M. Fouchard: color plate III.4; color plate IV.1c, 1d, 1e and 2; color plate V.
M. Fuchs: color plate VIII.5 and 6.
G. Genestier: fig. 8.8.
B. Gronenborn and M. Lesemann (CNRS, Gif-sur-Yvette): fig. 1.1(5).
C. Kusiak: fig. 9.23
C. Lamarque: color plate VI.2.
H. Lecoq: fig. 1.1 (11), 10.6, 10.7; color plate I.4 and 5; color plate II.2 and 3; color plate III.2, 3 and 5; color plate VI.1a and 1b; color plate VIII.7.
H. Lot: fig. 1.1 (8).
Y. Maury: color plate III.6; color plate IV.2; color plate VII.2a.
J.P. Moreau: fig. 8.3 (b), fig. 8.9.
M. Pitrat: fig. 11.1.
C. Putz: fig. 8.12.
Y. Robert: fig. 8.3 (a).
J. Rougier: color plate I.3; color plate II.6; color plate VIII.1.
B. Walter: color plate VII.1a.

W.C. Wang: fig. 9.21.
J. Weber: color plate II.4.
C. Wipf-Scheibel: fig. 1.1 (4); fig. 9.18; fig. 12.4.
U. Wyss: fig. 8.10 (right).

Cover illustrations:

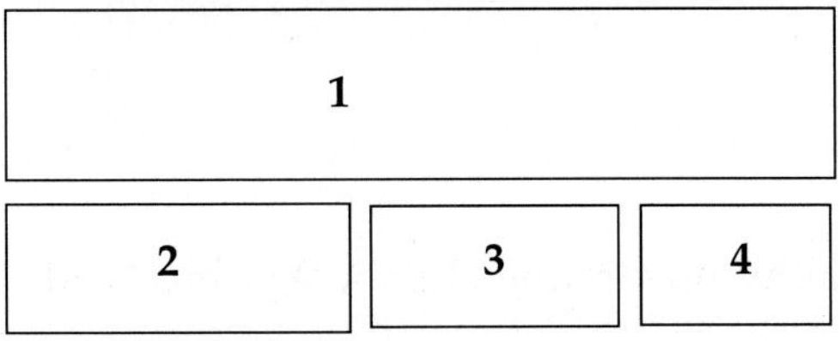

1. Court-noué of grapevine caused by *Grapevine fanleaf virus* (Photo G. Demangeat)
2. Mosaic on tobacco caused by *Tobacco mosaic virus* (Photo H. Lecoq)
3. Discoloration of tomato fruits caused by *Tomato mosaic virus* (Photo D. Blancard)
4. *Trialeurodes vaporariorum*, a whitefly vector of criniviruses (adults and larvae) (Photo R. Coutin)

Index

C

Q

R

S

U

V

W

WSMV (*Wheat streak mosaic virus, Tritimovirus*) 179, 187, 261, 341, 377
WSSMV (*Wheat spindle streak mosaic virus, Bymovirus*) Plate V.5
WTV (*Wound tumor virus, Phytoreovirus*) 341, 391

X

X-ray diffraction **12**, 13, 16, 26

Y

yellowing, yellows 208, **209**, 211, 335
YMV (*Yam mosaic virus, Potyvirus*) 322

Z

ZYMV (*Zucchini yellow mosaic virus, Potyvirus*) **15**, 66, **195**, 210, 218, 219, **236**, 260, **268-271**, 274, 288, **289**, 292, 299, 308, **309**, 312, 376, Plates III.3, VI.1, VIII.5